Soil Biology

Volume 44

Series Editor
Ajit Varma, Amity Institute of Microbial Technology,
Amity University Uttar Pradesh, Noida, UP, India

More information about this series at
http://www.springer.com/series/5138

Irena Sherameti • Ajit Varma
Editors

Heavy Metal Contamination of Soils

Monitoring and Remediation

Springer

Editors
Irena Sherameti
Institut für Allgemeine Botanik und Pflanzenphysiologie
Universität Jena
Jena
Germany

Ajit Varma
Amity Institute of Microbial Technology
Amity University Uttar Pradesh
Noida
Uttar Pradesh
India

ISSN 1613-3382 ISSN 2196-4831 (electronic)
Soil Biology
ISBN 978-3-319-14525-9 ISBN 978-3-319-14526-6 (eBook)
DOI 10.1007/978-3-319-14526-6

Library of Congress Control Number: 2015937019

Springer Cham Heidelberg New York Dordrecht London

Printed on acid-free paper

Springer International Publishing AG Switzerland is part of Springer Science+Business Media (www.springer.com)

Foreword

When the esteemed editors, Irena Sherameti and Ajit Varma, asked me to contribute the foreword for this book, I felt very honored and agreed spontaneously. The initial elation gave way very soon a certain disillusionment. Even temporary panic arose after a first look on the list with the competent authorship, the subject being "heavy metals and soil" from the perspective of a researcher who for the last several years was mainly concerned with microbial life in must and wine.

What gradually calmed me is the idea that all the knowledge of our time is available on the Internet, which can be easily absorbed, mixed, and presented as our own new insights. After some reflection, however, I became aware that at least metals, the definition of "heavy metal" is not as easy as it sounds, have met me here and there over the course of my scientific career. Therefore, I decided to forego "Dikipedia and Woogle," resorting as far as possible to own experiences.

Is there any relation between heavy metals and winemaking? Yes!

The metal content in wines originates from primary and secondary sources. The primary source is, in this case, the natural content, which comes from the vineyard soil going up through the vine roots into the grape, and finally into the wine. The secondary sources are impurities of geogenic and anthropogenic origins. These include fertilizers, pesticides, grape harvest, wine treatment, and wine storage. For example, chemical formulations of copper and zinc are included in fungicides. In former times, problems developed from inputs of iron by rusty nails in wood barrels and lead by motorcar traffic near vineyards. Nowadays, higher amounts of aluminum may enter the wine *via* metal tanks without internal lining and more previously by wine bottles and caps.

In mature grapes, the following average elemental concentrations (mg l^{-1}) were measured by our colleagues at the Institute of Nuclear Chemistry of the University of Mainz: potassium (1,600), calcium (80), magnesium (60), sodium (4.0), manganese (0.9), aluminum (0.6), copper (0.5), zinc (0.4), and rubidium (0.3).

In berry ripening and the alcohol fermentation by the wine yeast *Saccharomyces cerevisiae*, these metals play an important role, especially zinc as a cofactor or structural component in nearly 300 proteins. It is a constituent of the alcohol dehydrogenase and a deficiency can lead to stuck grape must fermentation. For the malolactic fermentation (also called secondary fermentation), namely the conversion of malate to lactate through the wine bacterium *Oenococcus oeni*, another metal is necessary: manganese as part of the malolactic enzyme and as a growth factor.

Copper (Cu) is the only "real" heavy metal, which is regularly detectable in must and wine. It may serve as an example for the ambivalent properties and the problems related with the use of heavy metals illuminated in this book. Copper is an essential trace element in living systems, where it serves as a cofactor in enzymes that function in energy generation, oxygen transport, detoxification, melanization, blood clotting, signal transduction, and many other processes. The physiological oxidation states of copper are Cu^{1+} and Cu^{2+}, whereas Cu^{3+} is not a biologically relevant species because of the high redox potential of the Cu^{3+}/Cu^{2+} couple. The copper in the active sites of redox proteins (e.g., tyrosinase, laccase, cytochrome-*c*-oxidase, superoxide dismutase) has been assigned to three main classes.

Our peers at the Institute for Molecular Biophysics of the University of Mainz found that copper in the blue blood of insects (hemocyanins) could mediate oxygen transport as well as induce tyrosinase activity. The evolution of the aerobic organic world might not be the same without the oxygen activation by inorganic copper. Similarly, iron is involved in various redox reactions: as cofactor of enzymes (in Fe-S clusters) and in the hemoglobin. Furthermore, according to established theories of Wächtershäuser and others, the origin of life was driven by catalytic activities at the surface of iron-sulfur minerals (pyrite).

The World Health Organization (WHO) has calculated a daily demand of about 0.08 mg Cu kg^{-1} body weight for infants and children and of 0.03 mg Cu kg^{-1} body weight for adults.

For drinking water, the guidelines of the WHO recommend an upper concentration of 2.0 mg l^{-1}; in the United States, the permissible limit is 1.0 mg l^{-1}. Agriculture has become one of the most important fields of copper contamination in soil and plant material, where it is used as one of the earliest fungicides. The false mildew, a plant disease caused by the fungus *Plasmopara viticola*, could be successfully inhibited with the so-called Bordeaux mixture, consisting of burnt lime (CaO) and a copper sulfate solution. Soils of wine-growing areas, which have been treated over several decades with these compounds, are enriched with high copper concentrations, which sometimes exceed considerably the permitted limits of the European community regulations (140 mg kg^{-1}).

Thus, it is not surprising that copper is among the most abundant heavy metals in wine, originating from the soil, plant protection products, and winemaking equipment. Moreover, addition of copper sulfate (up to 10 mg l^{-1}) is a common oenological practice to remove sulfurous off-flavor compounds by precipitation and subsequent filtration. In a recent study, the copper content determined in 72 wines averaged 0.18 mg l^{-1} with a maximum of 0.55 mg l^{-1}. The copper concentration limit recommended by the Organization International de la Vigne et du Vin (OIV) in wines is equal to 1.0 mg l^{-1}. National regulations allow the presence of 2.0 mg Cu l^{-1} in German wines and drinking water.

Elevated Cu concentrations, especially in combination with other heavy metals found in wine, e.g., iron, manganese, zinc, nickel, lead, and vanadium, bear some unaccountable health risks for the consumers. As typical for heavy metals, copper inhibits enzyme activities when present at higher concentrations. Wine-relevant microorganisms show a strain-dependent growth inhibition at different Cu concentrations, e.g., *Oenococcus oeni* (5–10 mg l^{-1}), *Saccharomyces cerevisiae* (32–75 mg l^{-1}), *Lactobacillus fermentum* (75–300 mg l^{-1}), or *Lactobacillus mesenteroides* (55–150 mg l^{-1}). High amounts (>20 mg l^{-1}) can even have an impact on grape must fermentations.

Furthermore, Cu concentrations >0.5 mg l^{-1} exert various undesirable effects in wine like a metallic taste and haze formation. The browning of wine, primarily due to the enzymatic and chemical oxidation of the phenolic compounds, represents one of the most feared processes, which may arise during winemaking. Wine grapes possess a very high phenolic content compared to many other fruits and crops. The phenolic compounds significantly affect the color, odor, and taste of the wine. In addition to the visual and gustatory properties, they serve as free radical scavengers and have antioxidant effects. Some polyphenols have been associated with anticarcinogenic, anti-inflammatory, antibacterial, and antihepatotoxic properties. Therefore, they have a high impact on the nutrient quality of wine. The catalytic effect of inorganic copper and utmost the release of laccase, a multicopper enzyme, into must from grapes infected with the phytopathogenic fungus *Botrytis cinerea* cause a significant reduction of the content of beneficial phenolic compounds.

If the copper content in wine exceeds the permissible limits, procedures must be applied for metal removal. A frequently used method is the "blue fining" with potassium hexacyanoferrate (II) which forms an insoluble complex with copper ions. The remaining cyanides react with the existing iron to insoluble compounds and precipitate. In case of an iron deficiency, an excess of highly toxic cyanides remains in the wine, making it commercially unsalable. In addition, the toxic waste raises environmental problems. Although the use of bentonites and artificial ion-exchange resins has been demonstrated to be effective in reducing the metal contents in wines, these procedures may cause off-taste and color loss of the product.

Alternative methods of metal removal use biomass (bacteria, yeasts, fungi, and algae) as sorbents. A relatively slow active process by which metal ions are transported inside the cells characterizes the bioaccumulation. Biosorption is defined as a passive, nonmetabolic process involving complexation, chelation, ion-exchange, adsorption, and microprecipitation. Bacterial and yeast cells are such able to accumulate, in either active or passive ways, a large variety of toxic heavy metals like cadmium, lead, and even the radionuclide uranium. Recently we found different strains of lactobacilli useful to remove copper from wine samples. The negatively charged proteinaceous cell-wall envelope (S-layer) of these Gram-positive bacteria serves as the initial binding matrix.

These tools are useful for aquatic systems, but for soil environments, improved phytoremediation strategies have to be developed. All the more so because heavy metals by no means exist inert in nature, but come *via* the food chain, conjugated with organic molecules. In this manner, new element species can arise with dramatically increased toxicity. Members of our working group, for example, could demonstrate the formation of high toxic methyl mercury from inorganic mercury by intestinal bacteria in the gut of earthworms. Furthermore, increasing anthropogenic production of metallic nanoparticles leads to compounds with new properties. An almost uncontrolled application and dissemination of metallic materials, for example, titanium oxide in sunscreens, already leads to unpredictable consequences for human health and environment. The responsible use and careful scientific monitoring in the development of such novel metallic materials is the need of the hour. Only in this way we can profit from their beneficial properties without harming nature and ourselves.

Mainz, Germany

Harald Claus
Institute of Microbiology and Wine Research,
Johannes Gutenberg-University Mainz

Preface

Heavy metals occur naturally as constituents of the Earth's crust (Singh et al. 2011) and are introduced to ecosystems *via* natural processes (Mahmood et al. 2012). Since they are nonbiodegradable, they can accumulate in the environment (Ali et al. 2013) and cause contamination. Nowadays, contamination of soils with heavy metals has become a worldwide problem and a serious threat to the environment because of increasing anthropogenic activities like mining, uses of fertilizers, wastes, sewage sludge, pesticides, wastewater irrigation, coal combustion residues, atmospheric deposition, etc. (Wuana and Okieimen 2011). In a critical review, Su et al. (2014) conclude that heavy metal contamination refers to the excessive deposition of toxic heavy metals in the soil caused by human activities.

This new volume will focus on the achievements of the last years on sources of heavy metals in soils and monitoring strategies, adaption strategies of plants and bacteria in response to heavy metals, approaches for the remediation of contaminated soils, and genetic engineering as a tool to cleaning up contaminated soils. The volume is organized in 5 parts and contains 23 chapters.

Part I: Sources of Heavy Metals in Soils

Chapters 1–4 will deliver general information on heavy metals in the post-catastrophic and agricultural soils, contamination and its impact on soil biological quality in urban agriculture, multiple-phase evaluation of copper geochemistry, sources of heavy metals, and factors influencing their contents in soils.

Chapter 1, by *Vesna Stankov Jovanović*, *Violeta Mitić*, *Snežana Nikolić Mandić*, *Marija Ilić*, and *Strahinja Simonović*, focuses on heavy metals in the post-catastrophic soils and discusses the distribution pattern of heavy metals in soils after catastrophic events and extreme phenomena, which can be induced by human activities or natural disasters. The authors conclude that metals can be easily transferred from one area to another by water and air. This transfer is more dynamic and more intense in case of catastrophic events like flooding, hurricanes, tsunamis, volcanoes, and fires. Chapter 2, written by *Zeshan Ali*, *Alvina G. Kazi*, *Riffat N. Malik*, *Mehreen Naz*, *Tayyaba Khan*, *Asim Hayat*, and *Abdul M. Kazi*, focuses on heavy metal buildup in agricultural soils of Pakistan, sources, ecological consequences, and possible remediation strategies. The authors conclude that the majority of the agricultural soils in Pakistan have higher metal concentrations than the standard/limits accepted by international agencies, which is leading to food chain contamination. Through genetic engineering, the abilities of plants, fungi, and bacteria can be enhanced to attain the desired bioremediation results in polluted environments either in situ or ex situ. Good agricultural practices and reduced chemical inputs can slow down further heavy metal buildup in soils. Chapter 3 (Multiple-Phase Evaluation of Copper Geochemistry), written by *Jean B. Diatta*, *Ewa Chudzińska*, *Leszek Drobek*, *Aleksandra Wojcicka-Półtorak*, *Bernd Markert*, and *Simone Wünschmann,* analyzes geochemical processes of mineral elements at geological and soil levels and the challenges that geochemists are facing for decades to quantify both natural and anthropogenic soil Cu concentrations. This chapter is enriched with a case study which outlines in detail multiple steps involved in the evaluation of copper geochemical processes. Chapter 4 (Potentially Toxic Element Contamination and Its Impact on Soil Biological Quality in Urban Agriculture: A Critical Review), written by *Avanthi Deshani Igalavithana*, *Sabry M. Shaheen*, *Jae Nam Park*, *Sang Soo Lee*, and *Yong Sik Ok*, focuses on the urban ecosystem as a complex composite of both natural and anthropogenic factors. The local climate, geology, and geographical characteristics are the main natural factors that influence the urban environment. The increased population density, intense industrialization, and excessive exploitation of natural resources characterize a megacity as a unique system, but they have consequential deteriorating impacts on the overall structure and function of the ecosystem.

Part II: Monitoring Heavy Metals in Soils

Soil enzymes, eggs as useful biomonitors of soil contamination, the adsorption equilibrium of Pb(II) in soil in the presence of humic substances, the availability and effect of arsenic (As) on microbial diversity, as well as heavy metals in Mediterranean grasslands on abandoned mines are discussed in Chaps. 5–9.

Chapter 5 (The Adsorption Equilibrium of Pb and Cd on Several Soils in Various pH Solutions or in the Presence of Dissolved Humic Substances), by *Takeshi Kobayashi*, *Yuyu Liu*, *Yukari Takahashi*, *Takashi Kameya*, and *Yuko Yamada,* discusses the requirement of adsorption isotherms for understanding the mobility and fate of heavy metals in soils and how various environmental factors influence the adsorption and elution of hazardous metals from contaminated soil. The authors conclude that preparing a database of Q_{max} (the maximum ion-exchange capacity of the soil) and selectivity coefficients for hydrogen ions and hydroxide ions of various soils will be helpful to analyze the permeability of contaminated groundwater and to understand the contamination states. Chapter 6, by *Silvia Rita Stazi*, *Rosita Marabottini*, *Ruxandra Papp*, and *Maria Cristina Moscatelli*, focuses on As in soil, its availability, and interactions with soil microorganisms. Soil microorganisms may be involved in As pollution issues in an active and passive way at the same time. Soil microbial biomass is an important indicator of soil quality and can be used as a tool to assess the impact of heavy metal pollution on soil. The authors conclude that many aspects are still to be elucidated in order to assess the complex and diverse interactions of soil microorganisms and As. Eggs as a biomonitor of heavy metals in soil is the focus of Chap. 7, written by *Muhammad Zaffar Hashmi*, *Naeem Akhtar Abbasi*, *Xianjin Tang*, and *Riffat Naseem Malik*. Eggs as biological indicators to monitor environmental pollution have gained considerable interest by scientific communities a few decades ago. These authors conclude that eggs are sensitive indicators of geographic and temporal trends of heavy metals in soil and that before using them as a biomonitoring tool the characteristic and history of local soil especially the turnover rate of sediments are needed. In Chap. 8—Enzyme Activities in Soils Contaminated with Heavy Metals in Varying Degrees—*Krystyna Ciarkowska* addresses the issue how the activity of soil enzymes responds to the pollution of soils caused by heavy metals. The role of soil enzymes, the natural factors which influence enzyme activity of soils, the mechanisms connected with the impact of heavy metals on the enzyme activity, and the use of soil enzyme activity as an indicator of changes in soil properties are discussed. The author concludes that soil enzyme activity reflects the degree and size of heavy metal pollution in soil, is a good indicator of both short- and long-term heavy metal contamination of the environment, and allows estimating the effects of reclamation of degraded ecosystems. Heavy metals in native Mediterranean grassland species growing at abandoned mine sites, ecotoxicological assessment, and phytoremediation of polluted soils, by *Mª J. Gutiérrez-Ginés*, *J. Pastor*, and *A. J. Hernández*, are the focus of Chap. 9. The heavy metal and trace element pollution of soils were examined at eight abandoned mine sites in central Spain. The Mediterranean grassland communities are able to grow in soils polluted with metals and other trace elements in the central Spain. The species analyzed were tolerant to the stress caused by soil pollution in the areas where they grew.

Part III: Adaptation Strategies of Plants and Bacteria in Response to Heavy Metals

Adaptation strategies of plants and bacteria in response to heavy metals and their impacts on environmentally sound combustion of ligneous plants grown in heavy metal-contaminated soil, adaptation mechanisms of *Pinus sylvestris* in industrial areas, the impact of soil Cd contamination on microbial dynamics in the rhizosphere of *Medicago sativa*, symbioses of plants with Rhizobia and mycorrhizal fungi in heavy metal-contaminated tropical soils, and phylogenetic diversity and physiological profiles of soil microbial communities impacted by heavy metal pollution, mycorrhiza, and heavy metals in soil treated with sewage sludge are discussed in Chaps. 10–13.

Chapter 10 (Heavy Metal Uptake and Transport in Plants), written by *Tayyaba Komal*, *Midhat Mustafa*, *Zeshan Ali*, and *Alvina G. Kazi,* analyzes the role of different plant metal transporters. The authors focus on restriction of metal ion uptake from soil and metal ion bindings. Copper transporter family; ZIP family; MATE protein transporters; HMA, ABC, and NRAMP family of transporters; oligopeptide transporters; and cation-diffusion facilitator family of transporters are also discussed. Chapter 11 (Adaptation Mechanisms of *Pinus sylvestris* L. in Industrial Areas), written by *Ewa Chudzińska*, *Aleksandra Wojnicka-Półtorak*, *Wiesław Prus-Głowacki*, *Konrad Celinski*, *Jean B. Diatta*, and *Leszek Drobek,* emphasizes *Pinus sylvestris* as one of the most frequently used bioindicators in the European forests. Although the tolerance mechanisms of forest trees are still unclear, the increased use of molecular techniques should have a significant impact on our understanding of heavy metal tolerance and be helpful in the use of plants to remove contaminants from the environment by dendroremediation. This method offers a cost-effective alternative for large areas of unused lands that have been contaminated by heavy metals. Chapter 12, written by *Fatima Maria de Souza Moreira*, *Paulo Ademar Avelar Ferreira*, *Laíze Aparecida Ferreira Vilela*, and *Marco Aurélio Carbone Carneiro*, focuses on symbioses of plants with rhizobia and mycorrhizal fungi in heavy metal-contaminated tropical soils which till now have been poorly studied. Their results demonstrate the beneficial effects of mycorrhizal fungi and rhizobia for promoting the establishment and survival of plants in soils contaminated with heavy metals. However, the specific mechanisms involved in the benefits of these organisms under these conditions are not yet understood. The authors conclude that for the application in remediation programs for contaminated soils, field and molecular studies should be conducted to enable the understanding of the complex dynamics of mycorrhizal fungi and their effects on plants in contaminated areas. Functional diversity of microorganisms in heavy metal polluted soils is discussed in Chap. 13 by *Anelia Kenarova* and *Silvena Boteva*. The authors summarize the results from studies designed to assess the impacts of heavy metal pollution on the physiology of soil microorganisms. Microbial functional responses to metal stress were observed as reduced catabolic activity and functional diversity, preferential community shifts from one carbon substrate use to another, and/or increased pollution-induced community tolerance. The authors conclude that the commercially available assays based on substrate use patterns of microbial communities were fast, reliable, and easy to use.

Part IV: Approaches for the Remediation of Contaminated Soils

Different applications and strategies to cleaning up soils from heavy metal contamination are presented in Chaps. 14–21.

Chapter 14, written by *Helga Kovacs*, *Katalin Szemmelveisz*, and *Arpad B. Palotas,* focuses on environmentally sound combustion of ligneous plants grown in heavy metal-contaminated soil. They discuss incineration as a viable option for safe disposal, which allows the utilization of the contaminated biomass waste as secondary fuel source. To develop a sound combustion technology for the disposal of heavy metal-contaminated biomass, it is necessary to assess the environmental impacts of process-related by-products and have reliable measurements. The authors conclude that bioremediation is one of the most eco-friendly ways of removing heavy metals from contaminated soils. In Chap. 15, the role of earthworms on phytoremediation of heavy metal polluted soils is discussed. The authors, *My Dung Jusselme*, *Edouard Miambi*, *Thierry Lebeau*, and *Corinne Rouland-Lefevre*, analyze the positive effects of earthworms on plant production and their effects on heavy metal solubility and availability. The interactions between heavy metals and earthworms depend on the earthworm species, the metal, and the physical and chemical properties of the soil. The activation by earthworms of microorganisms producing compounds which target heavy metals reduces the toxic effect of these metals and increases the heavy metal availability for the plants. Chapter 16 (Phytochemical Removal of Heavy Metal-Contaminated Soils), by *Ali A. Abdel-Salam*, *Haythum M. Salem*, *Mohamed A. Abdel-Salam,* and *Mahmoud F. Seleiman*, focuses on two strategies used for remediation: phytoremediation and chemical remediation. In the last years, particular attention is paid to phytoremediation since it is not costly and also because the biomass produced from cultivation of contaminated lands can be used for bioenergy purposes. Sludge amendment and tree vegetation could improve the quality of mine soils is the focus of Chap. 17, written by *Verónica Asensio*, *Alfonso Rodríguez-Vila*, and *Emma F. Covelo.* To increase the quality of chromium- and copper-polluted soils in Northwest Spain, some particular areas have been vegetated with eucalyptus or pines, amended with sludges, or received both treatments. The soil samples were analyzed for several physicochemical characteristics, and a soil quality index was determined. As a conclusion, the authors recommend the regular addition of sewage sludges and paper mill residues to degraded sites because their favorable effect on soils decreases over time. Phytoextraction of heavy metals: the potential efficiency of conifers is the focus of Chap. 18 written by *Gaëlle Saladin.* The author recommends uses of conifers, particularly for polluted sites localized in cold areas or even in dry zones. Conifers with a fast growth rate could be interesting and useful for soil remediation under certain conditions. The selection of hybrid conifer species can be of interest since they are often more vigorous than their parents and thus could grow better in contaminated soils. Chapter 19, written by *Nathalie Diagne*, *Mariama Ngom*, *Pape Ibrahima Djighaly*, *Daouda Ngom*, *Babou Ndour*, *Maimouna Cissokho*,

Mathieu Ndigue Faye, *Alioune Sarr*, *Mame Ourèye SY*, *Laurent Laplaze*, and *Antony Champion*, focuses on remediation of heavy metal-contaminated soils and enhancement of their fertility with actinorhizal plants. The use of actinorhizal plants for remediation of heavy metal-contaminated soils, the impact of mycorrhizal fungi and/or nitrogen fixing bacteria on plants adaptation in heavy metal-contaminated sites, and the positive effect of actinorhizal plants on the fertility of heavy metal-contaminated soil are discussed as well. Chapter 20 (Chelating Agents and Heavy Metal Phytoextraction), written by *Ali Akbar Safari Sinegani*, *Iman Tahmasbian*, and *Mahboobe Safari Sinegani*, focuses on the categories of chelating agents and their role in increasing metal bioavailability and soil productivity. This is achieved through enhancing soil with available nutrients which results in a more efficient phytoremediation. The authors conclude that application of chelating agents along with electrical fields raises phytoremediation efficiency. *Domen Lestan*, author of Chap. 21, focuses on remediation of toxic metal-contaminated soil using EDTA soil washing. This method where chelant and process waters are recycled in a closed loop with no wastewater generated aims at sustainable remediation and reclamation of contaminated soils. At the same time, soil washing with EDTA deteriorates some physical soil properties and deprives the soil of important micronutrients. However, the revitalization measures have the potential to restore the quality of remediated soil as a plant substrate.

Part V: Genetic Engineering: A Tool to Cleaning Up Contaminated Soils

The last two chapters of this volume focus on the role of transgenic plants and plant growth-promoting rhizobacteria as a tool for soil remediation.

Chapter 22, written by *Umesh B. Jagtap* and *Vishwas A. Bapat*, focuses on the genetic engineering of plants and their uses for heavy metal removal from the soil. Transgenic plants expressing various genes from different sources are developed to increase heavy metal tolerance. Potential target genes, plant transformation methods, model systems for transgenic studies, optimization of transgene expressions in transgenic plants, along with risk assessment and mitigation strategies are discussed. Characterization of PGPR: a perspective of conventional versus recent techniques is the topic of Chap. 23, written by *Ruchi Agrawal*, *Alok Satlewal*, and *Ajit Varma*. Soil bioremediation by using plant growth-promoting rhizobacteria (PGPR) is a promising cost-effective and eco-friendly technology. This review deals with various methods for PGPR detection and characterization with a focus on comparative evaluation of conventional *versus* recent techniques of PGPR detection/characterization.

Heavy Metal Contamination of Soils: *Monitoring and Remediation* promises to be useful for researchers, students, and other academicians.

We would like to thank the contributing authors for their stimulating cooperation to prepare this volume. It was a pleasure to work with them. We wish to thank Dr. Jutta Lindenborn at Springer Heidelberg for her generous assistance and patience in finalizing the volume. Finally, we give specific thanks to our families for their nice support.

Jena
Noida

Irena Sherameti
Ajit Varma

References

Su C, Jiang LQ, Zhang WJ (2014) A review on heavy metal contamination in the soil worldwide: situation, impact and remediation techniques. Environ Skep Crit 3(2):24–38

Mahmood Q, Rashid A, Ahmad SS, Azim MR, Bilal M (2012) Current status of toxic metals addition to environment and its consequences. In: Anjum NA et al (eds) The plant family Brassicaceae: contribution towards phytoremediation, vol 21, Environmental pollution. Dordrecht: Springer

Ali H, Khan E, Sajad MA (2013) Phytoremediation of heavy metals - concepts and applications. Chemosphere 91(7):869–881

Singh R, Gautam N, Mishra A, Gupta R (2011) Heavy metals and living systems: an overview. Ind J Pharmacol 43(3):246–253

Wuana RA, Okieimen FE (2011) Heavy metals in contaminated soils: a review of sources, chemistry, risks and best available strategies for remediation. ISRN Ecol. http://dx.doi.org/10.5402/2011/402647

Contents

Part I Sources of Heavy Metals in Soils

1 **Heavy Metals in the Post-catastrophic Soils** 3
Vesna Stankov Jovanović, Violeta Mitić, Snežana Nikolić Mandić, Marija Ilić, and Strahinja Simonović

2 **Heavy Metal Built-Up in Agricultural Soils of Pakistan: Sources, Ecological Consequences, and Possible Remediation Measures** 23
Zeshan Ali, Alvina G. Kazi, Riffat N. Malik, Mehreen Naz, Tayyaba Khan, Asim Hayat, and Abdul M. Kazi

3 **Multiple-Phase Evaluation of Copper Geochemistry** 43
Jean B. Diatta, Ewa Chudzińska, Leszek Drobek, Aleksandra Wojcicka-Półtorak, Bernd Markert, and Simone Wünschmann

4 **Potentially Toxic Element Contamination and Its Impact on Soil Biological Quality in Urban Agriculture: A Critical Review** 81
Avanthi Deshani Igalavithana, Sabry M. Shaheen, Jae Nam Park, Sang Soo Lee, and Yong Sik Ok

Part II Monitoring Heavy Metals in Soils

5 **The Adsorption Equilibrium of Pb and Cd on Several Soils in Various pH Solutions or in the Presence of Dissolved Humic Substances** 105
Takeshi Kobayashi, Yuyu Liu, Yukari Takahashi, Takashi Kameya, and Yuko Yamada

6 **Arsenic in Soil: Availability and Interactions with Soil Microorganisms** 113
Silvia Rita Stazi, Rosita Marabottini, Ruxandra Papp, and Maria Cristina Moscatelli

7 **Egg as a Biomonitor of Heavy Metals in Soil** 127
Muhammad Zaffar Hashmi, Naeem Akhtar Abbasi, Xianjin Tang, and Riffat Naseem Malik

8 **Enzyme Activities in Soils Contaminated with Heavy Metals in Varying Degrees** 145
Krystyna Ciarkowska

9 **Heavy Metals in Native Mediterranean Grassland Species Growing at Abandoned Mine Sites: Ecotoxicological Assessment and Phytoremediation of Polluted Soils** 159
M[a] J. Gutiérrez-Ginés, J. Pastor, and A.J. Hernández

Part III Adaptation Strategies of Plants and Bacteria in Response to Heavy Metals

10 **Heavy Metal Uptake and Transport in Plants** 181
Tayyaba Komal, Midhat Mustafa, Zeshan Ali, and Alvina G. Kazi

11 **Adaptation Mechanisms of *Pinus sylvestris* L. in Industrial Areas** 195
Ewa Chudzińska, Aleksandra Wojnicka-Półtorak, Wiesław Prus-Głowacki, Konrad Celiński, Jean B. Diatta, and Leszek Drobek

12 **Symbioses of Plants with Rhizobia and Mycorrhizal Fungi in Heavy Metal-Contaminated Tropical Soils** 215
Fatima Maria de Souza Moreira, Paulo Ademar Avelar Ferreira, Laíze Aparecida Ferreira Vilela, and Marco Aurélio Carbone Carneiro

13 **Functional Diversity of Microorganisms in Heavy Metal-Polluted Soils** 245
Anelia Kenarova and Silvena Boteva

Part IV Approaches for the Remediation of Contaminated Soils

14 **Environmentally Sound Combustion of Ligneous Plants Grown in Heavy Metal-Contaminated Soil** 261
Helga Kovacs, Katalin Szemmelveisz, and Arpad B. Palotas

15 **Role of Earthworms on Phytoremediation of Heavy Metal-Polluted Soils** 279
My Dung Jusselme, Edouard Miambi, Thierry Lebeau, and Corinne Rouland-Lefevre

16 **Phytochemical Removal of Heavy Metal-Contaminated Soils** 299
Ali A. Abdel-Salam, Haythum M. Salem, Mohamed A. Abdel-Salam, and Mahmoud F. Seleiman

17 Sludge Amendment and Tree Vegetation Could Improve the Quality of Mine Soils 311
Verónica Asensio, Alfonso Rodríguez-Vila, and Emma F. Covelo

18 Phytoextraction of Heavy Metals: The Potential Efficiency of Conifers 333
Gaëlle Saladin

19 Remediation of Heavy Metal-Contaminated Soils and Enhancement of Their Fertility with Actinorhizal Plants 355
Nathalie Diagne, Mariama Ngom, Pape Ibrahima Djighaly, Daouda Ngom, Babou Ndour, Maimouna Cissokho, Mathieu Ndigue Faye, Alioune Sarr, Mame Ourèye SY, Laurent Laplaze, and Antony Champion

20 Chelating Agents and Heavy Metal Phytoextraction 367
Ali Akbar Safari Sinegani, Iman Tahmasbian, and Mahboobe Safari Sinegani

21 Remediation of Toxic Metal-Contaminated Soil Using EDTA Soil Washing 395
Domen Lestan

Part V Genetic Engineering: A Tool to Cleaning Up Contaminated Soils

22 Genetic Engineering of Plants for Heavy Metal Removal from Soil 433
Umesh B. Jagtap and Vishwas A. Bapat

23 Characterization of Plant Growth-Promoting Rhizobacteria (PGPR): A Perspective of Conventional Versus Recent Techniques 471
Ruchi Agrawal, Alok Satlewal, and Ajit Varma

Index 487

Part I
Sources of Heavy Metals in Soils

Chapter 1
Heavy Metals in the Post-catastrophic Soils

Vesna Stankov Jovanović, Violeta Mitić, Snežana Nikolić Mandić, Marija Ilić, and Strahinja Simonović

1.1 Effect of Accidents Caused by Metal Mining on Heavy Metal Content in Soil

Under normal circumstances and in the geologically short periods, metal mines do not disturb significantly the content of heavy metals into the surrounding soil. However, if for some reason any incident occurs that is caused by meteorological conditions (collapsing tailings dam due to the enormous precipitation and outpouring of nearby rivers in the excavation or tailings) or human error (fault on plants for waste water treatment and mine raw sewage spill), it leads to a significant increase in the content of heavy metals in the surrounding soil and river sediments.

The content of heavy metals around the mine varies depending on atmospheric conditions and geochemical characteristics of the tailings and the surrounding area. Metals/metalloids can be transported into the surrounding soil due atmospheric conditions or erosion of the tailings. The ways of pollution can be air and water (waste water discharged into the river recipients or through groundwater by direct transport). Also, the soil can become contaminated with heavy metals when water from the river, into which mine wastewater is discharged, is used for irrigation. Major threat to the environment is represented by abandoned mines and unsecured tailings.

V.S. Jovanović (✉) • V. Mitić • M. Ilić
Faculty of Science and Mathematics, Department of Chemistry, University of Niš, Visegradska 33, 18 000 Niš, Serbia
e-mail: sjvesna@pmf.ni.ac.rs

S.N. Mandić
Faculty of Chemistry, Universty of Belgrade, Studentski trg14-16, 11 000 Belgrade, Serbia

S. Simonović
Faculty of Science and Mathematics, Department of Chemistry, University of Priština (Kosovska Mitrovica), Ive Lole Ribara 29, 38220 Kosovska Mitrovica, Serbia

I. Sherameti, A. Varma (eds.), *Heavy Metal Contamination of Soils*, Soil Biology 44, DOI 10.1007/978-3-319-14526-6_1

In the case of the active mine Panasqueira in Portugal, where wolframite, cassiterite, and chalcopyrite were exploited, as a by-product of the ore treatment process, the waste products rich in arsenic are produced, which are stored in open reservoirs. Determination of heavy metals in rhizosphere soils collected in S. Francisco de Assis village, downstream of the Barroca Grande tailings deposit, showed a high content of As 223.6, Ba 192.8, Cd 1.3, Co 10.7, Cr 37.8, Cu 163.7, Mn 531.3, Mo 0.6, Ni 24.9, Pb 58.6, Sb 0.6, and Zn 323.2 (average concentrations in ppm), which are in the case of As, Cu, and Mn several times higher than those prescribed by Ontario Soils Quality Guidelines (Candeias et al. 2014).

Nearby the two mines, the northern part of Jequetepeque River Basin, Peru, was investigated for the contents of As 20.55, Cd 3.17, Cu 68.93, Cr 5.50, Hg 0.34, Ni 7.90, Pb 12.33, Sb 1.10, and Zn 70.53 ppm. A comparative study of metal content in the sediments showed that the highest concentrations are present in the points that are closest to the mine in both cases and are substantially higher than in unpolluted areas (Yacoub et al. 2012).

In the vicinity of Dabaoshan Mine (China), research was conducted in order to determine the degree of contamination by heavy metals using the sequential indicator simulation, in order to distinguish between the spatial patterns derived from data on land; adjustment of multiple linear regression models for the intake of heavy metals by crops; and interpretation of land use with the images obtained by remote sensing and integrating spatial pattern, models of input, and land use in the dose–response model for the risk of heavy metals to human health. Areas with higher concentrations of heavy metals in soil are mainly in areas of Dabaoshan Mine in the river basin of Hengshi, Tielong, and Chuandu. Average concentrations of Cu, Zn, Cd, and Pb in the soil from the studied area are above natural background levels in soils. Of the various land uses in the studied areas, agricultural and residential land uses have the greatest health risk, because food intake is the main route of exposure to heavy metals. Spatial patterns of heavy metal concentrations and soil pH indicate that areas with the highest risk for people's health are directly matched with the areas with the highest concentrations of heavy metals but coincide with areas of low pH soils. Pollution with high concentrations of heavy metals is a potential source of risk, but the combination of high concentrations of heavy metals, low pH, and agricultural or residential land use is necessary for the existence of real threats to human health. The spatial distribution risk coefficient indicates that the Cd is the most important pollutant that contributes to human health risks (Zhao et al. 2012).

The central region of Minas Gerais (Brazil) is rich in important mineral reserves; Fe, Mn, Cu, Sb, As, Au, Al, and U are intensively exploited, emitting tons of waste into the air, water, sediment, and soil. A significant amount of these metals and toxic elements penetrate into the soil, groundwater, rivers, and lakes, affecting even distant farm regions by contaminated water and sediments due to the frequent flooding of the river Das Velhas. The contents of Al, As, Au, Ba, Br, Ca, Cl, Co, Cr, Cs, Cu, Fe, Hg, K, La, Mg, Mn, Na, Nd, Rb, Sb, Sc, Sm, Th, and Zn were determined in samples of water, sediment, animal feed, and fish in unpolluted areas, mining areas, and agricultural lands surrounding the river Das Velhas. Farming areas of the Das Velhas river basin, though hundreds of miles away from the mine, at the time of floods, showed the high level of contamination, mainly in the sediments (As 650, Co 15, Cr 260, Cu 60, and Mn 1,030 ppm) (Veado et al. 2006).

In the area which is located in Agra and Guadiamar river valleys and associated with tailings impoundment of mine Aznalcóllar, the samples of soils were analyzed after spill from the pyrite's mine in 1998 (after removing tailings), 1999 (after cleaning very contaminated areas), and 2004 (after plowing the upper soil layers 20–25 cm). It was found that the removal of mineral waste has led to mixed forms of metals distribution, where heavily polluted soils interchanged with less polluted soils. In relation to that in 1998, average pollutant concentrations in the upper soil layers were 918 ppm for Zn, 2.7 ppm for Cd, 156 ppm for As, and 440 ppm for Pb and decreased from 19 to 30 % in 1999 amounting to 675 ppm, 2.1 ppm, 126 ppm, and 305 ppm, respectively, for Zn, Cd, As, and Pb, while they fell in 2004 from 24 to 30 % to 487 ppm, 1.6 ppm, 89 ppm, and 224 ppm, respectively, for Zn, Cd, As, and Pb. Remedial actions did not significantly change the concentration of pollutants in the upper 50 cm of land, but subsequent measures of cleansing in 1999 considerably reduced the concentration of pollutants in heavily polluted areas. Considering the high concentration of pollutants in the fields (especially As), plowing the upper 20–25 cm, despite the decrease in the mean concentration of pollutants in the upper layers, did not significantly reduce the percentage of soils that exceed a concentration of 40 ppm As and nearly doubled the percentage of soils that exceed this concentration at a depth of 10–30 cm. How remediation measures are mainly affecting the upper layers of soil, the concentrations of Zn, Cd, As, and Pb in the deeper layers have hardly changed over the years. Very moving elements Zn and Cd, by their movement through the soil accumulate in the depth, lead to a reduction in the concentration of contaminants in the surface layer (10 cm) and lower percentage of soils that excess the reference concentrations of these elements (900 ppm Zn and 2 ppm Cd), while do not increase the percentage of soils that exceed the concentration at a depth of 10–50 cm (Simóna et al. 2008).

Samples of soil and corresponding vegetation, which originated from Monchegorsk, Northwest Russia, and Naruska, North Finland, were analyzed on heavy metal content (Ag, As, Co, Cs, Fe, Ir, Ni, Rb, Sb, and Sc), since they are located in the vicinity of a Cu–Ni smelter in which emitted species were mostly accumulated in the topsoil. The metal concentrations in soil and vegetation samples waned with distance from the smelter, except Cs and Rb, whose concentrations increase (Haugland et al. 2002).

The total concentrations of heavy metals in the soil in the drainage system of the mine (comprised of the river Atrai, Little Jamuna, Karatoa, Bandai) and surrounding agricultural fields located in Barapukuria (Bangladesh) coal basin and neighboring agricultural fields were investigated. For the irrigation of agricultural fields, the water from the main drainage system of the mine is used, despite that it contains high levels of potentially toxic elements. Average concentrations of Ti, Mn, Zn, Pb, As, Fe, Rb, Sr, Nb, and Zr exceeded the world average values, and in some cases, Mn, Zn, As, and Pb exceeded the limit of toxicity for the given metals. The coefficients of enrichment (EF) of 0.05–1.5 show that metal comes from materials of the earth's crust or natural processes, while EF values greater than 1.5 indicate anthropogenic sources. The soils showed a significant enrichment in Ti, Mn, Zn, Pb, As, Fe, Sr, and Nb, as consequences of mining activities (Zhang and Liu 2002).

The influence and extent of heavy metal contamination of rice fields affected by metal mining activities were examined, involving the soil of rice fields at different depths, rice plants, and irrigation water along six traverse lines in the vicinity of the mine and the surrounding checkpoint. The analysis of 25 elements including Cd, Cu, Pb, and Zn was performed. The concentrations of Cd, Cu, Pb, and Zn in the soil (0.9–1.7 ppm, 16–31 ppm, 25–90 ppm, 188–465 ppm), plants (0.1–0.5, 1.2–5.8, 0.1–0.6, and 19.0–37.0), and water for irrigation (1.0–50.0, 1.0–53.0, 1.0–67.0, 1.0–49,100) in the immediate vicinity of the mine were relatively high due to the leakage of metals from mining waste dumps. Increased concentrations of the metals in all parts of the rice plants are the result of increased metal content in the soil surface (Jung and Thornton 1997).

Watering with water from rivers in which wastewater of the mine is released can significantly increase the amount of heavy metals in soils. That situation is in the area around Dabaoshan Mine in Guangdong province in China, and therefore soil contamination (land with rice fields and garden soil) with heavy metals (Cu, Zn, Pb, and Cd) was investigated, which is located in the immediate vicinity of the mine Dabaoshan, irrigated with water from the river Hengshi and Chuandu, in which waste from the mine was discharged. Concentrations of Cu, Zn, Pb, and Cd in soils with rice fields (502 ppm, 498 ppm, 278 ppm, 3.92 ppm) and garden soil (271 ppm, 349 ppm, 190 ppm, 3.13 ppm) exceed the maximum permitted value of Chinese agricultural fields (50 ppm, 200 ppm, 250 ppm, 0.3 ppm). In addition, content of heavy metals in rice and vegetables from the gardens was investigated. It was found that rice can accumulate unexpectedly high concentrations of Pb and Cd (1.44 and 0.82 ppm), which exceed the maximum allowable limit for grains in China (0.2 ppm) (Zhuang et al. 2009).

In the soils of Les Malines mining district, 40 km from the city of Montpellier (France), the contents of Zn, Pb, and Cd were determined. Sampling was conducted at three sites of Petra Alba mine, which is located at the beginning of Crenze valley upstream from the village of Saint-Laurent-le-Minier; Les Avenieres, 1 km downstream from Saint-Laurent-le-Minier and 3.5 km from the Ganges; and Les Malines Mine, 2.5 km upstream of Saint-Laurent-le-Minier and 6 km from the Ganges. The highest concentration of heavy metals was found in the tailings pond from Les Avenieres (Zn 161,000 ppm, Pb 92,700 ppm, Cd 1,382 ppm) (Escarré et al. 2011).

Tailings from abandoned mines due to the effects of water and/or wind erosion can be an important source of heavy metal pollution, years after the end of economic exploitation of the mine. Results of the analysis of abandoned mine Fedj Lahdoum in Tunisia show that the total concentrations of metals in the accumulation were 10,460 ppm of zinc, 2,100 ppm of lead, and 62.08 ppm of cadmium, while by geochemical analysis of soil obtained, contents of Pb, Zn, and Cd were 3,646 ppm, 3,236 ppm, and 17 ppm, respectively. These high contents registered in the land surpassed both Tunisian and Canadian standards (Babbou-Abdelmalek et al. 2011).

In the soil in the vicinity of the abandoned iron mine of Daduk in Korea, increased contents of Cd, Cu, Pb, and Zn, with average values of 8.57 ppm, 481 ppm, 4,450 ppm, and 753 ppm, respectively, were recorded (Lee et al. 2001).

In the waste material of the abandoned chalcopyrite mine of San Finx in the Galicia (NW Spain), Fe with a total concentration of 4,315–31,578 ppm is the most abundant, then Cu from 273 to 5,241 ppm, Mn from 294 to 2,105 ppm, Zn from 73 to 894 ppm, and Cr from 0.01 to 30 ppm. The concentration of Fe in the bioavailable form ranged from 40 to 1,550 ppm, while Zn is present at least in this form of 2–100 ppm. Copper was the most abundant heavy metal in the exchange complex and in the aqueous extracts, followed by Zn, Mn, and Fe. Cu in exchangeable form ranged between 17.7 and 1,866 ppm, whereas the maximum concentrations of exchangeable Zn, Mn, and Fe did not exceed 140 ppm. The concentration of heavy metals in plants that grow on the studied area was determined, and significant increase in the content of heavy metals in some tested plants was recorded, which contributes to the natural detoxification of contaminated waste materials (Alvarez et al. 2003).

In the areas near the mine and tailings, flood can lead in a large extent to more prominent mobilization of heavy metals and their transfer to agricultural land, enabling in that way their further spreading into the plants, animals, and finally humans.

In Gyongyosoroszi in Hungary, analysis of the content of heavy metals in vegetable gardens to evaluate the effects of floods was performed, since they are located near the abandoned lead and zinc mine. There was a significant increase in the concentration of heavy metals in the flooded vegetable gardens in relation to non-flooded. The greatest differences were registered in the content of cadmium and lead, and concentrations in the flooded vegetable gardens were 0.33–13.6 ppm for cadmium and 29.2–694 ppm for lead. These concentrations of cadmium and lead were 19 and 17 times higher than the non-flooded vegetable garden. However, such a high concentration of heavy metals in the soil did not represent an unacceptable risk to human health because the plants grown in the soil absorb quantities of heavy metals that are not posing a threat. This indicates that the risk of soil contamination is not necessarily an unacceptable risk to human health, and influence of heavy metals greatly depends on human health on the plant barriers (Sipter et al. 2008).

Los Frailes mines in Aznalcóllar are polymetallic deposits of the Iberian pyrite belt. They are generally consisting of pyrite with a small part of sphalerite, chalcopyrite, galenite, arsenopyrite, and some sulfate salts. After the collapse of the embankment dam at the tailings impoundment (1998), which consists mainly of pyrite and other minor parts of sulfide tailings, pond spilled approximately 4,286 ha of alluvial soils in the valleys of the rivers Agrio and Guadiamar. The area is flooded with about 60,00,000 m^3 of barren land that is mainly composed of acidic water, full of traces of metals, metal sulfides, and materials used in the flotation process. Soil analysis of a few selected areas along the river valley Guadiamar after toxic flood and removing deposited sludge showed serious contamination with metals in the surface soil layer. Overall mean concentrations of nine elements

(As, Au, Bi, Cd, Cu, Pb, Sb, Tl, and Zn) of 23 elements analyzed were higher in soil-covered mud (2,878, 0.55, 61.8, 25.1, 1,552, 7,888, 669, 51.6, and 7,096 ppm) than in soils that have not been affected by the floods. The overall mean of these elements in the soil were higher than the upper limit of normal soils worldwide (0.1–40, 0.01–0.02, 0.1–13, 0.01–2, 2–250, 2–300, 0.2–10, 0.1–0.8, 1–900). In all areas that are analyzed, serious heavy metal contamination in the surface layer (0–20 cm) is observed for the most soils (Cabrera et al. 1999).

After a spill of acidic water from the Los Frailes mine in Aznalcóllar (Spain) containing trace elements, metal sulfides, and materials used in the process of refining/flotation, soil analysis before and after remediation (removal of sludge treatment and afforestation) showed a serious contamination of the surface layers by microelements (mostly As, Cd, Cu, Pb, Tl, and Zn). The aim of this study was to determine the influence of tailings spill and remedial measures on the contents of Hg in the soil. The soil was sampled at several locations along the valley of the river Guadiamar. The amount of Hg immediately after the spill (0.512 ppm) was eight times higher than the background (0.061 ppm) on the surface (0–5 cm) and about 3–4 times higher in the deeper layers (0–20 cm, 0172 ppm, and 0–50 cm, 0253 ppm). After the applied measures of rehabilitation (2002), mean values of Hg (0.404 ppm) and other elements (As, Cd, Cu, Pb, and Zn) were still higher than background values and sometimes greater than the value before rehabilitation. This is a consequence of the rest of the sludge on the surface or sludge that is buried in the depths during the remedial measures. The total concentration of trace elements was very changeable, indicating a very irregular distribution of these elements on the surface along the Guadiamar river basin. The highest values of less mobile elements were observed within 5 km from the tailings dam (up to 176 ppm As, 2.36 ppm Hg, and 1,556 ppm Pb) (Cabrera et al. 2008).

The Dawu river that flows through the Dexing copper mine (DCM) and the Jishui river that runs near the Yinshan mine (YLM) (lead and zinc mine) and several smelters, which are located in the basin of the Lean river, flooded a large area. A survey was conducted in order to determine the content of heavy metals (Cd, Ni, Cu, Pb, and Zn) and total sulfur in surficial sediments and neighboring flooded fields, in order to understand the consequences of flooding on heavy metals in soil, sediment quality evolution, and transfer of metals in sediments. Moderate concentrations of metals in tailings from the DCM were Cu (5,957 ppm), Pb (67.7 ppm), Zn (55.9 ppm), Ni (27.9 ppm), and Cd (0.11 ppm). In tailings from YLM, most abundant metals are Zn (1,836 ppm) and Pb (1,604 ppm). The sediments of the Lean river contained mean concentrations of Cu, Pb, Zn, Ni, and Cd of 271.9 ppm, 241.8 ppm, 119.4 ppm, 18.7 ppm, and 0.31 ppm, respectively. In the soil, mean concentrations of Cu, Pb, Zn, Ni, and Cd were 190.2 ppm, 45.5 ppm, 64.4 ppm, 15.4 ppm, 0.08 ppm respectively. At the most upstream and downstream locations, the concentration of metals in sediments is similar to the content of metals in the soil. In places near the Dexing copper mine, flooding brought in the floodplain system clay poor in copper, while in places that are located along the Yinshan lead–zinc mine, suspended solids contained high concentrations of iron and magnesium oxides that absorb large amount of dissolved Cd, Pb, and Zn and are deposited on the floodplain during floods. In spite of the increased production of the Dexing copper mine, there is a significant reduction in the concentration of Cu in the sediment compared to that of 10 years ago. The location in the vicinity of the Yinshan lead–zinc mine contained increased amounts of Pb and Zn in the sediments (Xiao et al. 2011).

The impact of accidental spills and accompanying floods of pyrite fluids and sludge from the flotation facility in Aznalcóllar mine (Spain) on the soil was examined, and the degree of contamination of soil at various depths was studied also, at various depths and over time. The analysis was carried out 15 days and 3 months after the flood. Total metal content was analyzed after aqua regia extraction from six different locations at various distances from the source of spills along the rivers Agrio and Guadiamar. Land affected by flooding is polluted with Zn, Cu, Pb, As, and Cd in different degrees and at various depths. The soil samples that were tested 15 days after a flood showed that level of contamination varies and have elevated concentrations values in comparison to the unpolluted soil—Zn (73.8–986.2 ppm at 300 m from the river Guadiamar, 49.7–4,045 ppm near the dam, 820.4 ppm nearby agricultural fields), Cu (21.3–193.9 ppm at 300 m from the river Guadiamar, 29.4–626.2 ppm near the dam, 154.5 ppm surrounding agricultural fields), Pb (42.3–835.1 ppm at 300 m from the river Guadiamar, 10.2–2,320 ppm near the dam, 104.7 ppm surrounding agricultural fields), As (27.2–452.8 ppm at 300 m from the river Guadiamar, 14.2–1,266 ppm near the dam, 39.1 ppm surrounding agricultural fields), and Cd (0.25–3.71 ppm at 300 m from the river Guadiamar, 0.18–16.35 ppm near the dam, 2.88 ppm surrounding agricultural fields). In soil samples after 3 months of flooding, it can be seen that the degree of contamination is the same or even higher than in the first case (Sierra et al. 2000).

The area around the mine Aznalcóllar was affected by accidental spill of acid water and sludge from pyrite flotation plant. Potentially toxic elements (PTE) were extracted with aqua regia and analyzed by ICP to determine the level of residual contamination at selected spots along the Agrio and Guadiamar river. Analysis was performed 1 year after the removal of the sludge. The analysis showed that the soil is still contaminated with several PTE in various degrees and at different depths. The degree of contamination of soil with Zn, Cu, Pb, and As is high at a depth of up to 30 cm (1,452–1,543 ppm, 222.2–310.3 ppm, 815.9–1,721 ppm, 488.2–720.1 ppm respectively). In deeper layers of soil (up to 300 cm), the degree of contamination is moderate (Sierra et al. 2003).

The zinc content in the bottom sediment mud and the suspended particulate matter along the river Guadalquivir, main tributary of the river Guadiamar, was also determined. The analysis was carried out 1 month after the spill and was recorded a high zinc content, with the mean value of 780.71 ppm in bottom sediment and 996.1 ppm in suspended particulate matter (Palanques et al. 1999).

Heavy rainfall has led to severe erosion and malfunction of the tailings impoundment of Abaroa antimony mines in Bolivia, when about 5,500 m^3 of contaminated waste dump in Rio Chilco–Rio Tupiza drainage system was released. Of primary importance are contaminated floodplain soils that are downstream from the Rio Tupiza, which have been found to contain mean concentrations of metals Pb, Zn, and Sb that exceed Canadian (200 ppm, 400 ppm, 20 ppm), German (500 ppm, 300 ppm, –), and Dutch (530 ppm, 720 ppm, –) regulations for agricultural land use. The sampling spot located 50 m from the river, at different depths, contained the concentration of Pb, Zn, and Sb ranging from 20.7 to 326.2 ppm, 52.6 to 750.7, and 6.3 to 111.3 ppm, respectively (Villarroel et al. 2006).

The contamination of soil with heavy metals (Cd, Cu, Pb, and Zn) in the vicinity of the gold and silver mine in Imcheon, Korea, was investigated. After the closing of the mine in 1978, a large amount of waste, including tailings rested as they were. This material has been spread by surface erosion, wind, and effluents into the deeper soil layers, which are used as rice fields and domestic gardens. Increased concentrations of metals (Cd, Cu, Pb, and Zn) were found in the soil (extracted by 0.1 N HCl), which is located in the center of the tailing (1.35 ppm, 26.4 ppm, 70.3 ppm, and 410 ppm, respectively), still remaining below the allowable limit of concentrations for metals in soil in Korea (30 ppm for Cd, 500 ppm for Cu, and 1,000 ppm for Pb). High concentrations of heavy metals were found in the soil (aqua regia extraction) in the center of the tailing (9.4 ppm, 229 ppm, 6,160 ppm, and 1,640 ppm for Cd, Cu, Pb, and Zn, respectively). As a result of the spreading of metals from this soil, samples from locations around the tailing have a higher metal content (0.8–2.6, 9.8–57.2, 100–225, and 104–465 ppm for Cd, Cu, Pb, and Zn, respectively) compared to the control land (0.8 ppm, 33.4 ppm, 51 ppm, and 108 ppm for Cd, Cu, Pb, and Zn respectively) (Jung 2001).

During January and March 2000, a dam collapse occurred in two tailings in Maramures County, northwestern Romania, which released about 2,00,000 m^3 of contaminated water and about 40,000 t of waste material to the tributaries of the Tisza River, which is a large tributary of the Danube. High concentrations of cyanide and metals, due to these accidents, contaminated and caused poisoning of fish, not only in Romania but also in Hungary, Serbia, and Bulgaria. The research program was conducted in northwestern Romania, to determine the metal content in the rivers affected by the leakage of water and solid tailings caused by the dam collapse, and the values obtained were compared with samples originated from mining and industrial regions, which were not affected by this disaster. In July 2000, from the region affected by these accidents, 65 samples of surface water and river sediments and 45 sediment samples from agricultural areas on the metal content were analyzed. Pb, Zn, Cu, and Cd pollution was drastically reduced downstream from the current active mines and tailings dumps. Concentrations of the heavy metals in river water and sediments on the borders of Hungary and Ukraine are mostly below the legal standards of the European Union, although the contents of Zn, Cu, and Cd in river sediments are approaching or exceeding these limits on Romanian territory. Therefore, there is a necessity for the long-term monitoring of heavy metals in this area, in order to facilitate content, fate, and environmental impacts of heavy metals, released by these accidents (Macklina et al. 2003).

In Chenzhou lead and zinc mine in Hunan province in southern China, the tailings dam collapsed and tailings waste speeded to the agricultural fields in the coastal area of the river Dong. In some places immediately after the accident, emergency cleanup was conducted. Seventeen years later, analysis of heavy metals was carried out in some types of cereals, pulses, and vegetable crops as well as land on which they grow, at four sites in the area of mining, land that is still covered by the flood tailings deposits, and land that was covered by flood tailings coat and later cleaned (which served as control). This study was conducted to comprehend the long-term consequences of spill of large amounts of heavy metals in agricultural crops and the potential risks to human health. The results showed that the physicochemical properties of the agricultural land changed depending upon the application of different processing techniques from various farmers. The effect of leaching and extraction of heavy metals from the some soils were very weak. Some lands were still heavily contaminated with As, Cd, Zn, Pb, and Cu. The concentration of heavy metals was lower in soils, where the cleaning procedure was carried out, although the maximum allowed concentrations of heavy metals by Chinese standards are still very high, especially for As and Cd (followed by Zn, Pb, and Cu), with mean concentrations of 709 and 7.6 ppm, exceeding prescribed amount 24 times for As and 13 times for Cd. Generally, the heavy metal content in all tested plants exceeds the standards for RDA (recommended dietary allowance), which indicates that the crops grown in the vicinity of the mine in Chenzhou, affected by the spill of tailings, are not safe for human health, and it is necessary to apply additional measures for cleaning of As, Cd, Pb, Zn, and Cu to the land (Liu et al. 2005).

1.2 Effect of Floods on Heavy Metal Content in Soil

Due to frequent flooding in the river Niger Delta in Nigeria Ubeji region, the growing contamination of arable soils with heavy metals is registered, especially for mercury, whose concentration in parts of studied area reaches a value up to 14,200 ppm, and iron, whose concentration approximates to 4,300 ppm. Increased content of heavy metals originated from petrochemical companies, from where the floods distribute heavy metals through arable land (Achudume 2007).

The total amount and available forms of metals (Cd, Cr, Cu, Ni, Pb, and Zn) in the fluvial sediments were determined in surface soil samples (0–30 cm), which were collected before (in 2009) and after a flood of the Odra river in western Poland (in 2010). The soil samples after the flood were changed in regard to physicochemical properties, as well as the content of the tested metals. In the samples from 2009, total concentrations of the following metals were determined as follows: Zn (1,270 ppm), Pb (340 ppm), Cu (243 ppm), Ni (96.8 ppm), Cr (83.5 ppm), and Cd (20.2 ppm). After flooding in 2010, these concentrations are increased (with the exception of Ni and Cu), and their amounts were 1,544 ppm of Zn, 404 ppm of Pb, 234 pm for Cu, 80.2 ppm for Ni, 133 ppm of Cr, and 86.7 ppm of Cd, showing a statistically insignificant increase in the total content of Cd, Cr, Cu, and Pb. However, the total concentrations of Cu, Cd, Cr, Pb, and Zn in sediments before and after the flooding exceed the allowable values in Poland. The content of available metal is strongly correlated with the total metal content, and the only significant increase in the concentration was recorded for Cr in available fraction after the flood (Ibragimow et al. 2013).

Apart from the significance of determining the total amount of metals present in the soil, it is also important to determine the forms in which they exist, because their availability depends on the way they are bound, and it can help to better understand the impact on the environment and human health. A study conducted by Khaokaew et al. aimed to determine not only the total amount of metals present in the soil that is periodically flooded (the four periods of flooding were 1, 7, 30, and 150 days) but also the speciation of Cd. Metal content in the analyzed soil was 142 for Cd, 3,050 for Zn, 18,328 for Fe, 34,360 for Ca, 25 for Cu, 878 for Mn, and 5,793 for Mg. For most of the samples, it was found that Cd is less bound to Zn than to Ca and thus is more likely to find Cd and Ca present together in the same mineral phase than Zn and Cd, which are found together along the sediment in the vicinity of zinc mine in Mae Sot district, Tak province, Thailand (Khaokaew et al. 2011).

The prediction of the distribution of heavy metals on the border of solid–liquid for flooded lowland soils is very important. The mechanic geochemical modeling was compared to a statistical approach. In order to characterize heavy metal pollution of land protected by dikes in the Netherlands, 194 soil samples from 133 sites distributed in the Dutch part of the Rhine and Meuse river systems were analyzed. Total amounts of As, Cd, Cr, Cu, Ni, Pb, and Zn in soil samples were determined in soil fraction extracted with 2.5 mM $CaCl_2$. The high correlation in the content of heavy metals and organic matter content was found, which was almost identical for both systems (Schröder et al. 2005).

1.3 Effect of the Hurricanes on Heavy Metal Content in Soil

Hurricanes and tropical storms are formed over warm tropical seas and may be defined as a storm system, with low pressure in the center and stormy weather around it, which causes extremely strong winds and torrential rains. Cyclone denotes a characteristic of the system when the air turns around the center. In different parts of the world, the same phenomenon has different names (hurricane, over the Atlantic; typhoons, over the Pacific Ocean; or cyclone, over the Indian Ocean). Hurricanes besides strong winds and heavy rainfalls cause also high waves, storm, and tornado, threatening primarily coastal areas. The US government appoints each tropical storm over the Atlantic.

In 2005 Catherine hurricane struck New Orleans, inflicting great damage in southeastern Louisiana, where there are 21 oil refineries that have been damaged or completely damaged and it came to spill more than 70,00,000 gallons of oil. The content of heavy metals in the soil after the hurricane in the Greater New Orleans Region was determined. Study results showed significant contamination with As and V whose maximum concentrations were 49.07 ppm and 92.06 ppm respectively. The highest concentration of vanadium was found near the oil refinery, which can be explained by its use in refining industry (Su et al. 2008).

The leaching of pollutants (As, Cd, Cr, Cu, Hg, Pb, and V) from the soil under simulated acid rain conditions was determined in sediments and associated soils in New Orleans and the Louisiana Peninsula. The maximum amounts of leachable metals observed for As and Pb were 293 pg/L and 72 pg/L, respectively. Thus, these levels could potentially pose a health threat, if significant exposure occurred (Adams et al. 2007).

In January 2006, from 75 sites in the New Orleans (LA, USA), area soil and sediment samples were collected and analyzed on metals. The study design was intended to provide a spatial pattern of metal concentrations within the city, following hurricanes Katrina and Rita. Throughout the city, concentrations of Pb and As exceeded the criteria unequivocally. Nineteen percent of all analyzed samples exceeded the soil screening criteria for Pb, while 97 % exceeded the criteria for As. Fifty-seven percent of samples coincided with a previous sampling event in October 2005. Metal concentrations were evaluated for temporal comparisons, and it was found that As concentrations are significantly different over time, but Pb concentrations are not (Abel et al. 2007).

It is expected for urban areas to have elevated concentrations of heavy metals, and New Orleans is no exception. Generally, samples of street mud and suspended sediment and samples from the 17th Street Canal area had higher heavy metals concentrations, in comparison to samples in other media or from other different locations. As a control, mid-lake reference site (MID) served. Some of the samples had elevated concentrations of Ag, Cu, Pb, and Zn, which all of them (except Ag) are elevated in street dust and have vehicular sources. Elevated Ag might originate from wastewater contamination. The bottom-sediment samples, collected near the 17th Street contained elevated levels of trace elements (Cd, Cu, Pb, Zn, and Hg). Levels of analyzed metals varied among the sampling sites in a short time, indicating sediment redistribution by Hurricane Rita. In the case of samples farthest from the coast and most of the lake sediment samples, relatively low heavy metal concentrations were registered. Earlier studies of Lake Pontchartrain revealed similar distributions of heavy metals in the bottom sediments. In the early 1980s decrease in Pb content, within 5–10 km from New Orleans, was recorded. Elevated Cu and Pb values were recorded in bottom sediments from the 17th Street Canal in September 1996 but not in August 1997, characterizing the contamination as temporary. In the same time, elevated Ba, Cu, and Zn off the mouth of the Industrial Canal were also reported. In surface and deeper samples, following Hurricane Rita, most elemental concentrations decreased, indicating that a large amount of sediment deposited by Rita obviously did not come from contaminated sources (Van Metre et al. 2006). After the tsunami in 2004 that hit most of the countries around the Indian Ocean, the analysis of the content of heavy metals in sediments occurring due to the impact of the tsunami in the coastal areas of Thailand was carried out. The most noticeable difference in the concentrations of bioavailable heavy metals between sediments that were affected by the tsunami and the reference sample was recorded for Cd, Cu, Zn, and Pb, especially for Zn and Pb whose maximum concentrations are several times higher in sediments (49.1 ppm and 46.3 ppm for Zn and Pb) in comparison to the reference sample (2.7 ppm for Zn and 1.1 ppm for Pb) (Szczuciński et al. 2005).

And along the southeast coast of India, the content of heavy metals in sediments brought on by the tsunami in 2004 was determined, where an increase in the concentration of some heavy metals was recorded. High concentrations of heavy metals can be explained by the relocation of pollutants of anthropogenic origin. The data also suggested that increased pollution by heavy metals was also controlled by the sediment source (Srinivasalu et al. 2008).

Tsunami does not necessarily cause contamination by heavy metals in soils of stricken regions. The analysis of sediments on Upolu, Samoa, after a South Pacific tsunami as well as sediments in Sendai Plain, Japan, after the Tohoku–Oki tsunami in 2011 revealed that there are no significant changes in the concentrations of heavy metals. This can be explained by the lack of sources of pollution in areas affected by the tsunamis, as well as with the nature of the source from which sediments came (Chagué-Goff et al. 2011).

Large areas of farmland in the Sendai Plain, Japan, were inundated by the 11 March 2011 Tohoku–Oki tsunami and covered by a discontinuous 0.2–30 cm thick sediment layer consisting of sand and/or mud. Two months after the tsunami, numerous rice paddy fields and depressions remained ponded with brackish or saline water. Field analysis in May, August, and October 2011 was performed, with an aim to estimate the environmental impact of the tsunami. Tsunami sediments, underlying soil, and soil beyond the tsunami inundation limit were collected at 43 sites along and near a transect extending over 5 km inland and analyzed for grain size, organic content, water leachable ions, acid leachable metals, and exchangeable metalloids. Anion and cation concentrations in water leachable fraction were elevated in sandy and muddy tsunami deposits and soils, particularly in areas where seawater had stagnated for a longer period of time after the tsunami. Vertical variations were also recorded, with higher concentrations often measured in the surface samples. A similar trend could be observed for some of the metalloids (As) and metals (Zn, Cu, and Ni), although, in general, maximum concentrations of metals and metalloids were not much higher than in soils not inundated by the tsunami and were within background levels for uncontaminated Japanese soils (Chagué-Goff et al. 2012).

1.4 Effect of Volcano Eruptions on Heavy Metal Content in Soil

The soils in Kanagawa prefecture, which had previously been affected by the eruption of Mt. Fuji, was analyzed in order to determine heavy metal content. Mean values of heavy metals in soil were Ni 36.3, Cu 128, Zn 119, Cd 0.59, and Pb 19.8 ppm of dry soil (Okamoto et al. 1997).

The effect of clouds of gas emissions, due to activities of Masaya volcano (Nicaragua) on metal deposition in surrounding soils, was investigated. Preliminary studies showed that rapid deposition of metals occurs in the soils, which are located close to the source, while the metals concentration decreases with distancing from the emission source. Cr and As follow this trend, with maximum concentrations of 20.71 ppm and 7.61 ppm, respectively, near the volcano crater. The concentrations of Mn, Co, Ni, Cu, and Zn (959.30, 21.57, 13.44, 152.85, and 72.73 ppm) were registered at some distance from the crater, indicating that these metals are transferred further. It was also found that the concentrations of Cr, Co, Al, Mn, and Ni increase with increasing depth, while the concentration of Zn decreases. Metal concentrations in control soils were lower than metal concentrations in soils in the vicinity of the volcano, which confirms that volcano smoke contributes to increasing concentrations of metals in the soil (Hinrichs et al. 2011).

Evidence from volcanic plumes and fumarolic sublimates indicates that certain elements are fractionated into the vapor phase during eruption and shallow degassing of magmas. Especially important are high concentrations of metals (As, Sb, Hg, Bi, Cd, Cu, In, Ag, Au, Re, MO, Sn, W, and Pb) in vapor associated with some intermediate-to-silicic systems. Understanding the timing of release and natural fluxes of these elements in volcanic systems is important, both because of their potentially negative impact on human health and because of the significant economic importance of some of these metals. The possible release and redistribution of volatile metals (particularly Pb) during the eruption process and during subsequent post-emplacement degassing, crystallization, and cooling of the Bandelier Tuff (Stimac et al. 1996) were studied.

1.5 Effect of Wildfire on Heavy Metal Content in Soil

Fires are a frequent cause of changes in environmental (soil, air, waters) characteristics and vegetation area coverage and are classified as prescribed (controlled) fires and wildfires. The consequences of fires are mostly estimated by intensity and duration (Certini 2005).

The direct effects of fire on soil characteristics appear to be observable several decades after the accident, and it is very important to monitor changes of soil characteristics, especially concerning heavy metal content (Belanger et al. 2004).

There is limited number of studies dealing with influence of fire on micronutrients fate such as Fe, Mn, Cu, Zn, B, and Mo. In the after-fire area, on the habitat of *P. pinaster*, there was a significant increase in Mn content in the fraction of total and easily reducible fractions, while in the soil exchangeable fraction, Mn content remained the same. It is assumed that the Fe, Cu, and Zn also follow this pattern (Gonzalez et al. 1994).

Fire on the Vidlic Mountain in eastern Serbia, which occurred in summer 2007, lasted for 10 days, and it was caused by human factor. The study was focused on four metals, i.e., Cd, Cu, Pb, and Zn. Intake of metals generally depends both on soil and plants characteristics. It was shown that total heavy metal concentrations in soils after fire were increased for all analyzed metals, except for Cd, which can be explained by its presence in the soil, rather than consequence of fire. For both areas, not affected and affected by wildfire, metal content did not exceed average Earth values for each metal, indicating that both of them are still unpolluted by heavy metals (Stankov Jovanovic et al. 2011).

After the fire that completely destroyed the Brazilian research base in Antarctica, the concentration of potentially toxic elements (Cd, Cr, Cu, Mn, Ni, Pb, V, and Zn) in soil samples that were collected before 2008 and after 2012 fire was examined. The total of 34 samples after the fire at locations within the Comandante Ferraz's ruins around the research station was analyzed. For Ni it was found that its concentration did not differ significantly between the three groups of samples: samples collected before the fire (8 ppm), samples collected after the fire and outside the station (7 ppm), and samples collected after the fire, inside the station (15 ppm). The highest concentrations within the ruins of research base were observed for Cu, Pb, and Zn (34,000, 13,700 and 42,200 ppm), which are 85, 46, and 42 times higher than the values for residential areas proposed by the Brazilian National Environment Council (Guerra et al. 2013).

The municipal landfill of Tagarades which is located about 20 km from Thessaloniki has already suffered from several small fires, but in 2006, large-scale fire occurred, involving about 50,000 t of waste, and more than a week, a thick cloud of smoke covered the area of about 10 km^2 around the landfill. The effect of atmospheric pollution on the concentration of metals in the surrounding soils was investigated. Soil samples were collected at a depth of 0–5 cm from the fields in the vicinity of the landfill and contained similar concentrations of metals and reference soils and soils from other locations, not affected by fire. Surprisingly, the soil of the landfill was particularly enriched by Zn, Cu, Sn, Sb, and Pb (Chrysikou et al. 2008).

To determine the concentrations of heavy metals in the soils from four locations after the wildfire in forest (1986 and 1992) and meadows (2006 and 2008) in Lower Silesia (Poland), soil samples were collected from two sites in the woods and two in the meadows, at different depths. It was found that there is an increase of some heavy metals, mainly in upper layers. Maximal concentrations of Zn, Cu, Pb, Ni, and Cr in the soil of the forest areas were 2.77 g m^{-2}, 1.06 g m^{-2}, 4.82 g m^{-2}, 1.40 g m^{-2}, and 0.40 g m^{-2} and of the meadows area 5.58 g m^{-2}, 3.70 g m^{-2}, 5.34 g m^{-2}, 2.65 g m^{-2}, and 2.24 g m^{-2}, respectively (Bogacz et al. 2011).

The concentrations of Hg in soil after the fire were determined, in samples collected from several locations that have been exposed to fire in CA (USA). Total Hg concentrations in soil samples from the Malibu Creek watershed ranged from 1 to 37 ppb Hg, with a mean concentration of 17 ppb of Hg; in soil samples from the Arroyo Seco watershed (2006 Pines Fire, ASB1-4) ranged from 14 to 349 ppb of Hg, with a mean concentration of 134 ppb of Hg; and in soil samples from the Piru Creek watershed (2006 Day Fire, PCB1-6) ranged from 2 to 44 ppb of Hg, with a mean concentration of 18.5 ppb of Hg (Burke et al. 2010).

Tsagan–Daban crest (Tarbagatayskiy rayon, Republic of Buryatia) in Siberia (Russian federation) was subject to constant fire, and soddy-podbur soil samples were analyzed on heavy metals. It was observed that the fires of intermediate and high intensities result in a considerable increase in the content of chemical elements in the horizon of the siero humic soils which consequently leads to the increase in Zn, Co, Cd, and Pb, while the amount of Cu and Ni is decreased. In the pine forest belt, the content of Pb and Cd in the litter ash is slightly increased. One year after the fire, the total content of Mn, Zn, Cu, Ni, Cr, Pb, and Cd was increased 1.2–2-fold in the upper soil layer, and the total content of Cr and Cd was decreased. Five years after the fire, the content of Mn, Zn, and Cu is still relatively high in the 0–1 cm-deep soil layer (Sosorova et al. 2013).

1.6 Conclusion

Metals can be relatively easily transferred from one area to another by means of water and air, depending on the form in which they are bound in the soil. These processes were studied a lot and allow creation of distribution matrices and transfer models, taking into account known environmental conditions. Processes of metal transfer are becoming more dynamic and more intense in case of catastrophic events, where they can largely deviate from the previously created models and predictions.

The most prominent effect on metal content in soil has, as expected, vicinity of metal mines. This influence may be observed over the possibility of spreading metals from the tailings, where the metal content is uneconomic but is significantly higher than the average presence of metals in the Earth's crust. Unsecured tailings, tailings dam collapse accidents, spills, and floods by water located nearby, as well as the production process either due to outdated and obsolete equipment, or due to human error, cause the mobilization of large amounts of metals and their transfer to broad areas. In general, the content of heavy metals in soil is inversely related to the distance of the source (the mine or tailings), but the factors that can further affect and possibly change this rule, e.g., the nature of the metal, a form in which it is bound in soil as well as the nature of the soil receiver, should also be taken into account.

In flooding areas, where excess of heavy metals is not present (i.e., which are not near the mines, smelters, or industries that produce waste, are burdened with heavy metals) changes in distribution of heavy metals in soil also may be expected, although to a lesser extent than in case of metal mines and smelters, regardless of the fact that floods as a phenomenon are potentially much more represented in almost all regions of the Earth. Also, in the coastal river areas, agricultural lands are generally found, and through the crops, heavy metals can enter the food chain, posing a serious threat to the health of humans and domestic animals. On the other hand, applying measures of flood management, these accidents becoming controllable, and their influence is in most cases reduced to acceptable limits.

Atmospheric phenomena, in particular in the tropic ocean coastal areas, may also cause disturbances in the distribution of heavy metals in soil. Hurricanes and tsunamis are natural disasters that usually can lead to an increase of heavy metals in soil, and the degree of the impact depends on the distance to which the ocean flood wave penetrates into the land, the composition of the local soil, and the existence of potential natural and anthropogenic sources of heavy metals in the endangered area.

The influence of volcanoes on the content of heavy metals in soil is relatively little studied, since there are a small number of active volcanoes in the world. In contrast to the above-discussed disasters, in which the transport of heavy metals predominantly took place through water, in the case of the volcano, the main medium for transport of heavy metals is air, thus enabling very rapid spread of polluting matter and covering much larger surfaces.

Fires, whether caused by natural or human factor, change the distribution of heavy metals in soil, and their effects can be recorded for a long time after the accident. Dominant role in the content of heavy metals after the fire has their representation in the affected area prior to the accidents. In some cases fires may cause a reduction of heavy metals in soil affected by fire, because they are converted during the fire in a water-soluble form, which is transmitted via water flows further to the environment.

All the disasters provoke modified distribution of heavy metals in soil. Whether it will significantly affect the environment and human health depends on their primary presence in a particular area. It is essential knowing their contents, the form in which they are bound in soil and sediments, as well as their nature, and the surrounding soil characteristics as potential receptors in order to be able to control and keep within limits that represent the acceptable risks to the environment and human population.

It has been shown that the measures of remediation of land contaminated with heavy metals as a result of any of the aforementioned catastrophes provide encouraging results in the reduction of heavy metal pollution, although there is an imperative for their further study and improvement in order to protect the land from contamination in the most effective way and preserved as a resource for the future.

References

Abel MT, Presley SM, Rainwater TR et al (2007) Spatial and temporal evaluation of metal concentrations in soils and sediments from New Orleans, Louisiana, USA, following hurricanes Katrina and Rita. Environ Toxicol Chem 26:2108–2114

Achudume AC (2007) Assessment of farmland sediments after flooding in Ubeji Land in Niger Delta of Nigeria. Environ Monit Assess 135:335–338

Adams C, Witt EC, Wang J et al (2007) Chemical quality of depositional sediments and associated soils in New Orleans and the Louisiana Peninsula following Hurricane Katrina. Environ Sci Technol 41:3437–3443

Alvarez E, Fernández Marcos ML, Vaamonde C et al (2003) Heavy metals in the dump of an abandoned mine in Galicia (NW Spain) and in the spontaneously occurring vegetation. Sci Total Environ 313:185–197

Babbou-Abdelmalek C, Sebei A, Chaabani F (2011) Incurred environmental risks and potential contamination sources in an abandoned mine site. Afr J Environ Sci Technol 5:894–915
Belanger N, Cote B, Fyles JW et al (2004) Forest regrowth as the controlling factor of soil nutrient availability 75 years after fire in a deciduous forest of Southern Quebec. Plant Soil 262:363–372
Bhuiyana MAH, Parvez L, Islam MA et al (2010) Heavy metal pollution of coal mine-affected agricultural soils in the northern part of Bangladesh. J Hazard Mater 173:384–392
Bogacz A, Woźniczka P, Łabaz B (2011) Concentration and pools of heavy metals in organic soils in post-fire areas used as forest and meadows. J Elem 16:515–524
Burke MP, Hogue TS, Ferreira M et al (2010) Effect of wildfire on soil mercury concentrations in southern California watersheds. Water Air Soil Pollut 212:369–385
Cabrera F, Clemente L, Díaz Barrientos E et al (1999) Heavy metal pollution of soils affected by the Guadiamar toxic flood. Sci Total Environ 242:117–129
Cabrera F, Ariza J, Madejón P et al (2008) Mercury and other trace elements in soils affected by the mine tailing spill in Aznalcóllar (SW Spain). Sci Total Environ 390:311–322
Candeias C, Melo R, Freire Ávila P et al (2014) Heavy metal pollution in mine–soil–plant system in S. Francisco de Assis – Panasqueira mine (Portugal), João Paulo Teixeira. Appl Geochem 44:12–26
Certini G (2005) Effects of fire on properties of forest soils: a review. Oecologia 143:1–10
Chagué-Goff C, Schneider J-L, Goff JR et al (2011) Expanding the proxy toolkit to help identify past events -lessons from the 2004 Indian Ocean Tsunami and the 2009 South Pacific Tsunami. Earth Sci Rev 107:107–122
Chagué-Goff C, Niedzielski P, Wong HKY et al (2012) Environmental impact assessment of the 2011 Tohoku-oki tsunami on the Sendai Plain. Sediment Geol 282:175–187
Chrysikou L, Gemenetzis P, Kouras A et al (2008) Distribution of persistent organic pollutants, polycyclic aromatic hydrocarbons and trace elements in soil and vegetation following a large scale landfill fire in northern Greece. Environ Int 34:210–225
Escarré J, Lefèbvre C, Raboyeau S et al (2011) Heavy metal concentration survey in soils and plants of the Les Malines Mining District (Southern France): implications for soil restoration. Water Air Soil Poll 216:485–504
Gonzalez MJ, Ramos L, Hernandez LM (1994) Distribution of trace metals in sediments and their relationship with their accumulation in earthworms. Int J Environ Anal Chem 57:135–150
Guerra MBB, Neto EL, Prianti MTA et al (2013) Post-fire study of the Brazilian Scientific Antarctic Station: toxic element contamination and potential mobility on the surrounding environment. Microchem J 110:21–27
Haugland T, Steinnes E, Frontasyeva MV (2002) Trace metals in soil and plants subjected to strong chemical pollution. Water Air Soil Poll 137:343–353
Hinrichs M, Rymer H, Gillman M et al (2011) Characterisation and distribution of heavy metals at Masaya volcano, Nicaragua. Paper presented at AGU Fall Meeting abstracts Dec 2011
Ibragimow A, Walna B, Siepak M (2013) Effects of flooding on the contamination of floodplain sediments with available fractions of trace metals (Western Poland). Pol J Environ Stud 22:131–140
Jung MC (2001) Heavy metal contamination of soils and waters in and around the Imcheon Au–Ag mine, Korea. Appl Geochem 16:1369–1375
Jung MC, Thornton I (1997) Environmental contamination and seasonal variation of metals in soils, plants and waters in the paddy fields around a Pb-Zn mine in Korea. Sci Total Environ 198:105–121
Khaokaew S, Chaney RL, Landrot G et al (2011) Speciation and release kinetics of cadmium in an alkaline paddy soil under various flooding periods and draining conditions. Environ Sci Technol 45:4249–4255
Lee CG, Chon H-T, Jung MC (2001) Heavy metal contamination in the vicinity of the Daduk Au–Ag–Pb–Zn mine in Korea. Appl Geochem 16:1377–1386

Levin BW (2004) Tsunamis: causes, consequences, prediction and response. Natural disasters. Encyclopedia of life support systems. 1. UNESCO
Liu H, Probsta A, Liao B (2005) Metal contamination of soils and crops affected by the Chenzhou lead/zinc mine spill (Hunan, China). Sci Total Environ 339:153–166
Macklina MG, Brewera PA, Balteanub D et al (2003) The long term fate and environmental significance of contaminant metals released by the January and March 2000 mining tailings dam failures in Maramures¸ County, upper Tisa Basin, Romania. Appl Geochem 18:241–257
Okamoto T, Kamiyama K, Wachi K (1997) Background levels of heavy metals in Kanagawa prefecture farm soils impacted by volcanic ash from Mt. Fuji. In: Ando T, Fujita K, Mae T, Matsumoto H, Mori S, Sekyia J (eds) Plant nutrition for sustainable food production and environment Developments in plant and soil sciences, vol 78. Springer, Dordrecht, pp 553–554
Palanques A, Puig P, Guillen J et al (1999) Zinc contamination in the bottom and suspended sediments of the Guadalquivir estuary after the Aznalcollar spill (south-western Spain). Control of hydrodynamic processes. Sci Total Environ 242:211–220
Schröder TJ, Hiemstra T, Vink JPM et al (2005) Modeling of the solid-solution partitioning of heavy metals and arsenic in embanked flood plain soils of the rivers Rhine and Meuse. Environ Sci Technol 39:7176–7184
Sierra J, Montserrat G, Martí E et al (2000) Contamination levels in the soils affected by the flood from Aznalcóllar (Spain). Soil Sediment Contam 9:311–329
Sosorova SB, Merkusheva MG, Ubugunov LL (2013) Pyrogenic changes in microelement content in soils and plants from the pine forests of Western Transbaikal. Contem Probl Ecol 6:499–508
Stankov Jovanovic VP, Ilic MD, Markovic MS (2011) Wild fire impact on copper, zinc, lead and cadmium distribution in soil and relation with abundance in selected plants of Lamiaceae family from Vidlic Mountain (Serbia). Chemosphere 84:1584–1591
Stimac J, Hickmott D, Abel R et al (1996) Redistribution of Pb and other volatile trace metals during eruption, devitrification, and vapor-phase crystallization of the Bandelier Tuff, New Mexico. J Volcanol Geoth Res 73:245–266
Su T, Shu S, Shi H et al (2008) Distribution of toxic trace elements in soil/sediment in post-Katrina New Orleans and the Louisiana Delta. Environ Pollut 156:944–950
Sierra J, Montserrat G, Martí E et al (2003) Contamination levels remaining in Aznalcóllar spill-affected soils (Spain) following pyritic sludge removal. Soil Sediment Contam 12:523–539
Simóna M, García I, Martín F et al (2008) Remediation measures and displacement of pollutants in soils affected by the spill of a pyrite mine. Sci Total Environ 407:23–39
Sipter E, Rózsa E, Gruiz K et al (2008) Site-specific risk assessment in contaminated vegetable gardens. Chemosphere 71:1301–1307
Srinivasalu S, Thangadurai N, Jonathan MP et al (2008) Evaluation of trace-metal enrichments from the 26 December 2004 tsunami sediments along the southeast coast of India. Environ Geol 53:1711–1721
Szczuciński W, Niedzielski P, Rachlewicz G et al (2005) Contamination of tsunami sediments in a coastal zone inundated by the 26 December 2004 tsunami in Thailand. Environ Geol 49:321–331
Van Metre PC, Horowitz AJ, Mahler BJ et al (2006) Effects of Hurricanes Katrina and Rita on the chemistry of bottom sediments in Lake Pontchartrain, Louisiana, USA. Environ Sci Technol 40:6894–6902
Veado MARV, Arantes IA, Oliveira AH et al (2006) Metal pollution in the environment of Minas Gerais State-Brazil. Environ Monit Assess 117:157–172
Villarroel LF, Miller JR, Lechler PJ et al (2006) Lead, zinc, and antimony contamination of the Rio Chilco-Rio Tupiza drainage system, Southern Bolivia. Environ Geol 51:283–299
Xiao HY, Zhou WB, Wu DS et al (2011) Heavy metal contamination in sediments and floodplain topsoils of the lean river Catchment, China. Soil Sediment Contam 20:810–823
Yacoub C, Pérez-Foguet A, Miralles N (2012) Trace metal content of sediments close to mine sites in the Andean Region. Sci World J 2012:732519. doi:10.1100/2012/732519

Zhang J, Liu CL (2002) Riverine composition and estuarine geochemistry of particulate metals in China-weathering features, anthropogenic impact and chemical fluxes. Estuar Coast Shelf Sci 54:1051–1070
Zhao H, Xia B, Fan C et al (2012) Human health risk from soil heavy metal contamination under different land uses near Dabaoshan Mine, Southern China. Sci Total Environ 417–418:45–54
Zhuang P, Zou B, Li NY et al (2009) Heavy metal contamination in soils and food crops around Dabaoshan mine in Guangdong, China: implication for human health. Environ Geochem Health 31:707–715

Chapter 2
Heavy Metal Built-Up in Agricultural Soils of Pakistan: Sources, Ecological Consequences, and Possible Remediation Measures

Zeshan Ali, Alvina G. Kazi, Riffat N. Malik, Mehreen Naz, Tayyaba Khan, Asim Hayat, and Abdul M. Kazi

2.1 Introduction

Soil is the principal constituent of earth ecosystem that comprises of a complicated mixture of organic matter, liquids, minerals, gases, and a diversity of organisms that sustain life (Huang et al. 1998). The composition and proportion of these constituents affect the physical, chemical, and biological properties of soil which in turn affect its agricultural suitability. Soil interfaces with lithosphere, hydrosphere, atmosphere, and biosphere playing important role in nutrient and organic wastes recycling, inhabiting microflora that aid decomposition processes, provide medium for plant growth and water storage. Soils continue to modify over time as a product of climate change, weathering, anthropogenic, and other biotic activities. It acts as sink for all chemicals generated from anthropogenic and natural activities (Jonathan

Z. Ali
Ecotoxicology Research Institute, National Agricultural Research Centre, Park Road, Islamabad PO 45500, Pakistan

A.G. Kazi • M. Naz • T. Khan
Atta-ur-Rahman School of Applied Biosciences, National University of Sciences & Technology (NUST), Islamabad, Pakistan

R.N. Malik (✉)
Environmental Biology and Ecotoxicology Laboratory, Department of Environmental Sciences, Faculty of Biological Sciences, Quaid-i-Azam University, Islamabad PO 45320, Pakistan
e-mail: r_n_malik2000@yahoo.co.uk; eco4nd@yahoo.com

A. Hayat
Soil Fertility Lab, Land Resources Research Institute, National Agricultural Research Center (NARC), Park Road, Islamabad, Pakistan

A.M. Kazi
Wheat Wide Crosses, National Agricultural Research Centre, Park Road, Islamabad PO 45500, Pakistan

I. Sherameti, A. Varma (eds.), *Heavy Metal Contamination of Soils*, Soil Biology 44, DOI 10.1007/978-3-319-14526-6_2

et al. 2004). Retention time of different substances in soil ecosystems is longer than in hydrosphere and atmosphere because contaminants accumulate quickly in soils and deplete at a slow rate.

Agricultural soils in this regard are greatly prone to anthropogenic substances which are used to enhance agricultural productivity. Wastewater irrigation, fossil fuel combustion, vehicle emission, mining/smelting activities, atmospheric deposition from municipal and industrial sectors, and application of fertilizers, pesticides, and sewage sludge have resulted in metal contamination of agricultural soils (Guvencc et al. 2003; Ali et al. 2014). Increased rate of heavy metal addition in soils has also accelerated corresponding metal biogeochemical cycles. Along with metals, many other organic substances, i.e., dieldrin, aldrin, lindane, pentachlorobenzene, heptachlor, polychlorinated dibenzo-*p*-dioxins, endrin, mirex, hexachlorobenzene, endosulfans, dichlorodiphenyltrichloroethane (DDT), polychlorinated biphenyls (PCBs), toxaphene, chlordane, polychlorinated dibenzofurans, and chlordecone, are found in agricultural soils generated from a diverse set of municipal and industrial activities. Soil contamination/pollution by anthropogenic activities is an established phenomenon with reports which date back to 100 BC (Eney and Petzold 1987).

Heavy metals have relatively high density and are poisonous at extremely low concentrations. Elevated heavy metal concentrations in agricultural soils are particularly important due to their persistence, toxicity, long half-lives, and bioaccumulation potential (Ali et al. 2014). They are a natural constituent of the earth's crust, and many are necessary for normal metabolic functioning in plants, animals, and humans. In nature, their concentrations rarely exceed toxic levels unless intervened by any anthropogenic or natural activity. Their elevated concentration in surface soils and plants constitutes a general and recognized indication of environmental pollution (Panek 2000). Numerous studies to date have been conducted on the metal levels in soils and therein growing plants establishing environmental health concerns, i.e., food chain contamination.

Pakistan is an agricultural country with two thirds of the population depending on the agriculture sector for their livelihood. An increase in agricultural production is directly related with the regional soil fertility and quality. For this reason, agricultural soils of Pakistan are facing an ever-increasing pressure not only to ensure sustainable food supplies to rapidly expanding population that is over 160 million now but also to support the livelihood of two thirds of the population. Besides salinity, fertility, and erosion problems in Pakistan soils, heavy metal accretion is an emerging issue with least concern from the legislatory bodies and policy makers. Wastewater irrigation, sewage sludge application, vehicle emissions, atmospheric deposition, and fertilizer/pesticide application are the major factors of heavy metal built-up in the agricultural soils of Pakistan (Ali et al. 2014). Agricultural soils near urban and peri-urban areas are more affected due to the complex anthropogenic activities. Recent studies in the agricultural soils of Sialkot, Gujranwala, Lahore, Kasur, Peshawar, Karachi, etc., have described heavy metal concentrations exceeding threshold levels (Midrar-ul-Haq et al. 2003; Tariq et al. 2005; Tariq et al. 2006; Malik et al. 2010a; Mushtaq et al. 2010; Ali et al. 2014). Heavy metals are known to affect crop quality/production, threatening human and livestock health through plant produce consumption. Such ecological risks associated with heavy metal contamination of agricultural soils are grave and urge remediation measures. Globally, food safety remains a major concern after food security. Plant metal uptake/contamination depends on the species type, metal loads in soils, bioavailability, and soil characteristics, i.e., pH, electrical conductivity, salinity, organic matter, texture, cation exchange capacity, sodium absorption ratio, and redox conditions. Atmospheric deposition can also be vital in metal deposition on plant surfaces via atmospheric dust thus contributing to crop contamination.

Soil reclamation strategies in Pakistan are often ill practiced. Therefore, metal enrichment in agricultural soils continues indefinitely. Different lab-based studies have been continuously reported in literature for the metal removal from contaminated soils but unfortunately are not replicated in polluted areas. Compared to the high-tech metal removal technologies, more focus has been on the bioremediation-related techniques owing to their cost-effectiveness and environment-friendly nature. Chemometric and geo-statistical approaches can be of momentous help in evaluating national metal loads and the identification of sources and hotspots that need to be reclaimed on top priority basis.

2.2 Common Heavy Metal Sources in Agricultural Soils of Pakistan

Heavy metal sources in soils can be categorized into point metal sources and nonpoint metal sources. Point metal sources indicate localized and discrete contamination source for heavy metals, whereas nonpoint metal sources epitomize diffuse processes covering large areas. Examples for point metal sources in soil include mining/smelting and industrial and/or municipal activities; however, nonpoint metal sources include adverse agricultural practices, fossil fuel burning, and atmospheric deposition. In point metal sources, usually excess heavy metal concentration is present in the immediate soils, whereas fairly less metal levels are found in soils having nonpoint metal sources due to dilution effects. Some of the major sources of heavy metal in Pakistan soils are described hereunder.

2.2.1 Pesticide/Fertilizer Application

Pakistan is an agricultural country with a rapidly expanding population. To meet the growing population demands, excessive pesticides and fertilizers are utilized to enhance agricultural productivity. In irrigated areas, these chemicals (pesticides and fertilizers) are more excessively employed to boost agricultural production as compared to rain-fed areas. Pesticides and other agrochemicals coupled with fertilizers are a significant source of heavy metals in the agricultural soils of Pakistan. Pakistan has witnessed a fourfold increase in pesticides from the last two decades similar to fertilizers (WB-PCWRAS 2005). According to some recent estimates, around 56,00,000 t of fertilizer and 70,000 t of pesticides are used each year in Pakistan for agricultural purposes (Martin et al. 2006). These chemicals (pesticides and fertilizers) promise an increase in agricultural productivity at the cost of substantial environmental damage by the heavy metals present in them. Heavy metals are not only toxic for the plant growth and metabolism but, on accumulation in the edible parts, are a serious threat for human and livestock consumption. Agricultural soils contain different heavy metals based on the fertilizer and pesticide application history. Metal-based pesticides, i.e., arsenic acid, mercury, and arsenates of chromium, lead, and copper, have been used in the agroecosystems as defoliants, herbicides, and fungicides. Copper-based pesticides are used in the molluscicides, algicides, and crustacean's control. Their excess application increases respective metal levels in the agricultural soils. Metal-based pesticides have serious human and environmental health repercussions. According to UN reports, around 5 lac Pakistani individuals suffered from agrochemical (pesticides) poisoning in Pakistan of which approximately 10,000 ultimately died (Mustafa et al. 2007).

Commonly employed fertilizers in Pakistan are nitrogen (N), phosphorus (P), and potassium (K), but along with them, boron (B)-, zinc (Zn)-, and sulfur (S)-based fertilizers are also used nowadays. Most abundantly employed fertilizers are nitrogenous followed by phosphate and potash. Small farmers who cannot afford commercial fertilizers due to high cost apply industrial wastes, sewage sludge, and animal manure in their agricultural fields to ensure good nutrient supply to growing crops. Fertilizers generated from the industrial and municipal wastes often contain appreciable amounts of heavy metals. Excess application of fertilizers in the agricultural soils not only results in heavy metal built-up in soils but also leads to other environmental problems, i.e., nutrient leaching from the agricultural lands to the underground aquifers and nearby water bodies causing eutrophication and environmental health problems. It is estimated that approximately 50 % of nutrients are lost from the fertilizer application sites.

Heavy metals commonly present in the fertilizers include chromium, cadmium, zinc, lead, arsenic, and cobalt. Certain heavy metals have been reported to generate from specific fertilizers, i.e., arsenic from NPK, iron, and liming fertilizers; cadmium from NPK, phosphate, liming, and zinc fertilizers; zinc from NPK fertilizers; nickel and mercury from liming materials; and lead from NPK, zinc, iron, and liming fertilizers. Cadmium levels in phosphate fertilizers usually range between 0.1 and 200 mg/kg, and addition of these fertilizers certainly increases Cd in the agricultural soils (Wagner 1993). Similarly, incorporation of other fertilizers in the agricultural soils for an indefinite period continues to load heavy metals in treated soils. These metals can easily biomagnify via crops leading to human health hazards. The accumulation of inorganic heavy metals in the soils is a persistent environmental problem, and a legislation that governs metal levels in the inorganic fertilizers is lacking in Pakistan.

2.2.2 Wastewater Irrigation

The agriculture sector is the largest consumer of available fresh water supplies (96 %) in Pakistan followed by domestic and industrial sectors with 2 % each, respectively (Pakistan Economic survey 2009–2010). Increased water usage in the municipal and industrial sectors has deprived agriculture sector of its basic water share. Such declining water levels have attracted farming communities in Pakistan to use wastewater in irrigation purposes. Wastewater irrigation provides a good alternative to fresh water whose supplies are inconsistent due to various reasons, i.e., water scarcity, faulty canal system, and conveyance losses. Excess nutrients (N, P, and organic matter), continuous supplies, and ready availability are further advantages of using wastewater for agricultural production (Ali et al. 2013).

Wastewater is a complicated mixture of inorganic/organic substances and microbial pathogens. Organic substances in the municipal wastewater are usually not persistent and easily degraded by the microbial diversity. However, synthetic organic constituents in the industrial and agricultural effluents are persistent and are not easily biodegradable. Inorganic constituents (heavy metals) in municipal and industrial wastewaters are frequently present in large quantities with higher persistence and toxicity. Irrigation either by municipal or industrial wastewater is a potential heavy metal source in the agricultural soil of Pakistan (Ali et al. 2014). Various researchers have documented higher metal loads in the agricultural soils in response to wastewater irrigation practices. Mushtaq et al. (2010) investigated heavy metal contents in the Rawalpindi soils irrigated with municipal and industrial wastewaters. They reported total metal levels of Cd, Ni, and Fe which exceeded threshold concentrations in investigated arid soils. Midrar-ul-Haq et al. (2003) reported metal contamination of peri-urban agricultural lands of Karachi by nearby industrial activities of Korangi Industrial Area (KIA). Composite samples from two depths, 0–20 and 21–40 cm, were analyzed for Cd, Cr, Pb, and Ni. Levels of Cr, Cd, and Pb were considerably higher ($p > 0.0001$) in the surface soil layer (0–20 cm) as compared to the subsurface layer (21–40 cm).

Mahmood-ul-Hassan et al. (2012) analyzed Cd, Cu, Cr, Pb, and Ni in surface (0–15 cm) and subsurface (15–30 cm) agricultural soils of Hyderabad, Mirpur Khas, Gujranwala, and Sialkot irrigated with municipal effluents. Investigated metals were reported in higher concentrations in the surface layer (0–15 cm) which is consistent with the results of Haq et al. (2003). Cadmium and Ni in both surface and subsurface soil samples of Hyderabad, Mirpur Khas, Gujranwala, and Sialkot were many times higher than the permissible standards of FAO/WHO (2001). Copper was recorded as potential metal pollutant in the agricultural soils of Gujranwala and Sialkot; however, its concentration remained below FAO/WHO standards in the surface and subsurface soil samples of Hyderabad and Mirpur Khas. Mean Pb concentration in 11 % agricultural soil samples of Hyderabad, 44 % of Sialkot, 79 % of Gujranwala, and 100 % of Mirpur Khas exceeded recommended permissible limit for Pb. Except subsurface soil samples of Hyderabad, Cr levels were significantly higher than the FAO/WHO standards in the studied four agricultural locations. Mohammadzai et al. (2001) reported elevated Ni and Cr in Amangarh and Nagoman areas of Khyber Pakhtunkhwa (KPK), respectively, in response to municipal and industrial (ghee and tannery industrial) effluents. Qadir et al. (2000) reported increased Cd levels (0.29 mg/kg) in the agricultural soils of Faisalabad irrigated with municipal effluents. Cadmium concentration decreased with the increase in soil depth having adverse impacts on soil biota and functions. Peri-urban agriculture largely in Pakistan depends on the wastewater (of municipal and industrial origin) irrigation which is responsible for increasing metal loads in the peri-urban soils.

2.2.3 Atmospheric Deposition

Atmospheric deposition from the industrial hubs, urban centers, and dense traffic areas is a common and diffuse source of heavy metals in surrounding soils. Global annual estimates of soil heavy metal accretion through atmospheric deposition are presented in Table 2.1 (Nriagu and Pacyna 1988). Airborne sources of heavy metals from the aforementioned anthropogenic activities are in two forms, i.e., fugitive (dust) and stack/duct emissions (gases, air, or vapor streams) (Simonson 1995). Stack/duct emissions usually transport heavy metals to distant areas, whereas fugitive emissions distribute metals to a considerably smaller area (Wuana and Okieimen 2011).

Also metal loads carried by the fugitive emissions are much less than the stack/dust emissions. Heavy metals in both emission forms after covering a distance from

Table 2.1 Global estimates of soil heavy metal built-up through atmospheric fallouts (adopted from Nriagu and Pacyna 1988)

S. no	Heavy metals	10^6 kg/year
1	Arsenic	8.4–18.0
2	Cadmium	2.2–8.4
3	Chromium	5.1–38
4	Copper	14–36
5	Mercury	0.63–4.3
6	Manganese	7.4–46
7	Molybdenum	0.55–4.0
8	Nickel	11–37
9	Lead	202–263
10	Antimony	1.0–3.9
11	Selenium	1.3–2.6
12	Vanadium	3.2–21
13	Zinc	49–135

the originating source are deposited on land or water body. High metal loads, i.e., Zn, Pb, Cu, and Cd, are reported in the nearby agricultural areas of intense traffic hubs (Al-Khashman et al. 2004). In Pakistan, agricultural lands near to large cities and industrial hubs are suffering from an ever-increasing metal load from atmospheric deposition processes. Ali et al. (2014) reported atmospheric input to be an important contributor of heavy metals in the surface agricultural soils of Sialkot besides intentional and/or unintentional usage of tannery wastewaters for irrigation. Lead and Ni were shown to be deposited directly from the atmosphere generated from the tannery industries. Rind et al. (2013) described Zn, Fe, and Cu contamination of agricultural lands through atmospheric deposition from thermal power station, Jamshoro, Sindh. Wind direction, wind speed, precipitation, and related climatic factors strongly influenced the rate and intensity of atmospheric deposition. Fossil fuel burning is an important contributor of heavy metals in Pakistan atmosphere. Prevailing power shortage at industrial and domestic levels has led to increase reliance on the fossil fuels to fulfill the energy demands in both sectors. This scenario has further worsened the air quality with respect to metal enrichment followed by its deposition on soils.

2.3 Ecological Risks Associated with Increasing Metal Loads in Agricultural Soils

Increased metal built-up in agricultural soils has led to serious ecological consequences in Pakistan, i.e., phytotoxicity, risks to soil dwelling organisms, food chain contamination, and public health problems. Food safety is currently the uprising concern which is at stake by various anthropogenic engagements.

2.3.1 Food Chain Contamination and Human Health Concerns

To date, various researchers have documented phytotoxicity, food chain contamination, and mounting human health concerns from different agroecological zones of Pakistan. Mahmood and Malik (2014) described metal contamination of vegetables and associated human health risk from different regions of Lahore. Considerable higher levels of Zn, Mn, Cd, Cu, Pb, Ni, Cu, and Cr were found in the leafy vegetables (i.e., *Spinacia oleracea* and *B. campestris*) as compared to tuber/bulb vegetables. Exceeding metal levels in vegetables placed public health at significant risk on consumption. Another study from the peri-urban areas of the same district showed 100 % contamination of spinach, tomato, cauliflower, and carrot with Cd, Ni, Pb, Mn, Zn, Fe, and Cu higher than the FAO/WHO critical limits (Khan et al. 2013a). These researchers also confirmed higher accumulation of metals in leaves than fruits. Saif et al. (2005) reported metal contamination of spinach (*Spinacia oleracea*) grown in soils contaminated with industrial activities from Korangi Industrial Area (KIA), Karachi.

Ahmed et al. (2012) investigated metal levels in the vegetables from wastewater-irrigated farms and local market from district Rawalpindi. Higher metal levels (Cd, Cr, Cu, Fe, Mn, Ni, Pb, and Zn) were reported in bitter melon, cucumber, bell pepper, eggplant, and tomato grown in soils irrigated with wastewater. Elevated metal levels deteriorated plant produce quality and safety in contaminated soils. Similar authors reported transportation, marketing, and air pollution to be the other main causes of vegetable contamination with toxic metals (Al Jassir et al. 2005). In a comprehensive study conducted in four cities (Hyderabad, Mirpur Khas, Gujranwala, and Sialkot), Mahmood-ul-Hassan et al. (2012) reported metal contamination of vegetables, i.e., cucumber, turnip, bottle gourd, bitter gourd, cauliflower, brinjal, chillies, coriander, gourd, okra, lettuce, spinach, and sponge gourd. Among the five metals (Cr, Pb, Cd, Cu, and Ni) analyzed in the vegetables, only Ni and Cu concentrations were within the FAO/WHO permissible limits (FAO/WHO 2001). Cadmium and Pb levels were beyond permissible limits (0.2 mg/kg and 0.3 mg/kg dry weight, respectively) in nearly all vegetables from the four locations. Chromium concentration in more than 50 % vegetable samples from Gujranwala/Sialkot and 70 % from Hyderabad/Mirpur Khas were beyond permissible limit of 2.3 mg/kg dry weight. Continued human and animal exposure to such metal-contaminated vegetables from Gujranwala, Sialkot, Hyderabad, and Mirpur Khas can lead to serious health implications. Similarly, Hussain et al. (2006) reported higher Cd (0.2–1.0), Cr (0–106), and Pb (3.3–34.6) levels in mg/kg dry weight in vegetables from metal-contaminated agricultural soils of Faisalabad.

Nergus et al. (2005) investigated human health risk from heavy metals in fodder, vegetables, and fruits when grown in contaminated soils in Karachi. Metals including Cr, Hg, Pb Cd, and Ni were found much higher than the permissible limits in the studied contaminated vegetables. Qadir et al. (2000) documented Cd in all the studied vegetables, i.e., spinach (*Spinacia oleracea*), eggplant (*Solanum melongena*), okra (*Abelmoschus esculentus*), pumpkin (*Cucurbita pepo*), onion (*Allium cepa*), fenugreek (*Trigonella foenum-graecum*), cauliflower (*Brassica oleracea*), and bitter gourd (*Momordica charantia*). Their results revealed twofold Cd accumulation in leafy vegetables as compared to others. Cadmium ranged between 0.17–0.24 and 0.07–0.18 in mg/kg fresh weight in leafy vegetables and fruits, respectively. Sattar et al. (2013) described wastewater irrigation to be the most significant contributor in the heavy metal contamination of agricultural soils and vegetables, i.e., brinjal, okra, spinach, and cauliflower. Metal residues including Cr, As, Cd, Hg, and Pb were found in these vegetables exceeding permissible limits defined by the FAO/WHO. These researchers indicated that an appreciable amount of heavy metals is also deposited on the vegetable surfaces; careful washing can reduce surface-deposited heavy metal residues ultimately decreasing human health risks.

Abbas et al. (2010) investigated toxic metals (As, Hg, Pb, and Cd) in 21 vegetables categorized in four groups, i.e., leafy, cucurbits, fruity, and root/tubers collected from the different districts of Sindh province. Their results revealed higher Cd, Pb, and Hg accumulation in the leafy vegetables as compared to the remaining groups. Consequently, human health risk associated with the consumption of leafy vegetables is more pronounced. Higher accumulation of metals in leafy vegetables is attributable to their high transpiration and translocation efficiencies. Jan et al. (2010) described comparative human health risk from the consumption of food crops (vegetables, fruits, and cereals) grown in contaminated agricultural soil of Peshawar and Lower Dir, Khyber Pakhtunkhwa. Among the seven heavy metals (Zn, Ni, Cr, Cu, Cd, Mn, and Pb) analyzed, only Zn exceeded its permissible limit (100 mg/kg dry weight) in *Portulaca oleracea*, *Solanum tuberosum*, *Pisum sativum*, *Daucus carota*, *Coriandrum sativum*, *Hibiscus esculentus*, *Brassica rapa*, *Mentha viridis*, *and Spinacia oleracea*. Rehman et al. (2013) reported Pb contamination of vegetables from industrially wastewater-fed regions of Faisalabad. Elevated Pb levels were beyond the FAO/WHO limits for human consumption, and Cr levels were near to toxic concentrations.

Husaini et al. (2010) reported rice (Se 1.28, Pb 3.84 μg/g), tomato (As 0.44, Co 0.5 μg/g), and bitter gourd (Mn 45.3, Sb 0.1 μg/g) contamination with respective metals grown in industrial areas of Gujranwala. Found levels were higher than the allowable limits for human consumption. Khan et al. (2013b) reported high human health risk associated with the consumption of *Allium sativum*, *Solanum melongena*, *Mentha spicata*, *Triticum aestivum*, *Capsicum annuum*, *Allium cepa*, *Coriandrum sativum*, *Raphanus sativus*, *Lycopersicum esculentum*, *Daucus carota*, *Spinacia oleracea*, and *Abelmoschus esculentus* due to elevated Cd and Pb levels in the edible portions of these vegetables from Gujranwala and Sialkot districts.

All these reported studies seriously indicate human health risks from the consumption of heavy metal-contaminated vegetables, cereals, and fruits. This alarming metal pollution is resulting in a variety of public health concerns especially toxicological conditions in Pakistan. Excess Cd intake through foods results in cancer, renal failure, diarrhea, nausea, and incurable vomiting (Mudgal et al. 2010). Arsenic excess dosage results in hypertension, cardiovascular problems, hepatic dysfunction, neural damage, and dermal disorders (Yoshida et al. 2004). Lead intoxication can lead to anemia, cancer, reproductive problems in males, mental impairment, and hormonal imbalance (Tandon et al. 2001). Mercury is a lot more toxic as compared to As, Cd, and Pb as reported from frequent casualties around the world owing to Hg exposure. Its toxicological penalties and physiological disorders include cardiovascular problems, pink disease, neurotoxicity, nephritic syndrome, gastritis, constipation, abdominal pain, vomiting, and renal toxicity (Mudgal et al. 2010). Chromium is also a potential neurotoxic, teratogenic, carcinogenic, and mutagenic element which suppresses immune response on uptake and produces such deleterious effects (Kumar et al. 2008; Monteiro-Neto et al. 2010). Similarly, other heavy metals produce abnormal biochemical responses in humans on exposure through contaminated food.

2.3.2 Fodder Contamination and Livestock Safety

Heavy metal toxicity has not only affected humans through food chain contamination but has also placed livestock health at considerable risk through the fodder by virtue of contaminated soils in Pakistan. Ahmad et al. (2014) analyzed Cr, Cd, and Pb in *Dactyloctenium aegyptium*, *Calotropis procera*, *Parthenium hysterophorus*, and *Eragrostis pilosa* in two seasons (summer and winter) to investigate their aptness for the livestock consumption. Significant variations in metal levels were found in both seasons with relatively higher metal levels in the summer season in the studied area soil and fodder samples. Researchers recommended periodic analyses of fodders in summer on which ruminants graze to prevent toxicological threats. Besides general animal health threats, there are studies which have indicated accumulation of excess metals in the animal body tissues (edible portions, i.e., muscles) and fluids (milk).

Farid and Baloch (2012) reported heavy metal levels in milk from two sites: one with fodder grown on wastewater and second irrigated with clean water. Fodder crops analyzed included sorghum (*Sorghum bicolor*), berseem (*Trifolium alexandrinum*), and maize (*Zea mays*) grown on both sites. Higher Cd (0.06), Cr (0.29), Pb (0.11), and Ni (0.11) in ruminant milk were recorded from the site with wastewater-grown fodder. Cadmium, Cr, Pb, and Ni were transported from wastewater to soil to fodder crops to ruminants ultimately accumulating in milk. Higher metal levels detected in the milk were declared unsuitable for humans if consumed for longer durations. Aslam et al. (2011) investigated heavy metals in the goat and cattle milk from wastewater drain sites in Faisalabad. Concentrations of Hg, As, Pb, and Cr were high in goat milk, whereas Ni and Cd were recorded high in cattle milk. Reported high metal levels render milk unfit for human consumption; also the high metal level coming from the fodder is toxic for animals.

2.4 Chemometric and GIS Approaches in Evaluating National Soil Metal Loads

Chemometric (multivariate) and geo-statistical approaches are popular nowadays for source apportionment of pollutants, tracing complicated patterns, mapping soil metal loads, finding spatial variability, etc. (White et al. 1997a; Romic and Romic 2003; McGrath et al. 2004). A large number of researchers from Pakistan have employed chemometric and geo-statistical approaches in evaluating national metal loads in soils (Tariq et al. 2005; Tariq et al. 2006; Jan et al. 2010; Ali and Malik 2011; Malik et al. 2010a; Ali et al. 2014). Ali and Malik (2011) employed kriging method (geo-statistics) for the identification of metal-contaminated hotspot areas in Islamabad. Identified heavy metals included Zn, Ni, and Pb originating from the two main anthropogenic sources in Islamabad soils, i.e., municipal and industrial waste disposal coupled with vehicular exhaust.

Malik et al. (2010a) employed both chemometric and GIS approaches for determining heavy metal source apportionment and spatial distribution patterns in soils of Sialkot having industrial and agricultural significance. Chemometric approaches, i.e., hierarchical agglomerative cluster analyses (HACA) and principal component analyses based on factor analyses (PCA/FA), identified the association of Cd, Co, and Pb with traffic and Zn, Cr, Cu, and Ni with industrial operations in the soils of Sialkot. Geospatial maps showed high metal loads near the industrial areas and dense traffic routes. Ali et al. (2014) traced the origin of heavy metals (Cr, Na, and Pb) in soils of peri-urban agricultural areas of Sialkot in tannery operations and vehicular emissions. Similarly, many other researchers from Pakistan have employed chemometric and GIS approaches to evaluate national metal loads to help in the identification of hotspots and provide basis to initiate regulatory procedures for metal abatement.

2.5 Mitigation Measures

Agricultural soil pollution has become critical in Pakistan. Soil reclamation strategies are not potentially implemented as in developed countries. Research on exploring prospective methods regarding soil remediation is in the process. Biological methods (bioremediation) hold a special place in soil reclamation strategies and are more extensively researched as compared to physical, chemical, or mechanical methods. Bioremediation technology is not only cost-effective and environment friendly, but it is also the best alternative to other conventional treatments including immobilization, volatilization, and incineration (Iram et al. 2009a, b). Recent soil reclamation research in Pakistan has featured fungal, plant, and bacterial usage in soil bioremediation.

2.5.1 Mycoremediation

Due to national industrialization, heavy metal contamination has become a chief environmental issue owing to the dangerous effects on human and environmental health. In industrialized countries like Pakistan, heavy metals are the main source of soil pollution. Air and water pollution by metals differs from soil pollution, because heavy metals remain in soil for a great time period as compared with water and air (Lasat 2002). In the recent years, the annual global emission of heavy metals has reached to 22,000 mt (metric ton) for cadmium, 939,000 mt for copper, 783,000 mt for lead, and 1,350,000 mt for zinc. Cadmium (Cd) is a major lethal metal pollutant of soils, damaging plant growth and yield production. Many phytoremediation techniques have been used for the reclamation of polluted soils in Pakistan. Phytoremediation is a cost-benefit method almost 1/4 of the other physical and chemical treatments used for soil reclamation. The key benefits of this process are good soil quality, as it is taken by solar energy, low cost, etc. (Singh et al. 2003).

Soluble and insoluble metal species were reported to be eliminated from solid waste by heterotrophic fungi including *Yarrowta*, *Penicillium*, *Aspergillus*, and *Mucor* (White et al. 1997b). Screening of heavy metal-resistant fungus from the effluent of Kohinoor Textile Mills showed that the three fungal genera that were found in alkaline pH were *Penicillium*, *Rhizopus*, and *Candida* (Faryal and Hameed 2005). Bioremediation potential of *A. niger* has been reported by Sayer et al. (1997) for carrying out the solubilization of insoluble metallic compounds ($CaCO_3$, Mn $(CO_3)_2$, PbS, and CuO_2) in solid media.

The comparison of the industrial and sewerage waste from Lahore, Pakistan, and mine waste from Wiesloch, Germany, showed that heavy metal concentration in mine waste of Germany was higher for Zn, As, and Pb concentration. The occurrence of fungal strains in Pakistan soil was reported to be higher than the soil of Germany. Different genera that were found in Pakistan soil which include *Acremonium* sp., *Alternaria* sp., *Aspergillus niger*, *Aspergillus* sp., *Aspergillus nidulans*, *Aureobasidium* sp., *Chaetomium* sp., *Coniothyrium* sp., *Curvularia* sp., *Fusarium* sp., *Humicola* sp., *Monilia* sp., *Monocillium* sp., and *Mortierella* sp. were isolated. *Aspergillus niger*, *Aspergillus flavus*, and *Aspergillus nidulans*. The radial growth of these fungi at different concentrations of heavy metals was also measured, and these fungi were reported to be resistant to $ZnCl_2$ at 25 ppm and $NiCl_2$ at 12 ppm Iram et al. (2009a, b).

The diversity of fungi was determined in heavy metal-contaminated soil of peri-urban agricultural areas of Lahore, Faisalabad, Kasur, Islamabad, Wah Cantt, and Multan (Pakistan). *Aspergillus niger* was among the most frequent species of *Aspergillus*. In reference to bioremediation of soil, these fungal strains will be subjected to tolerance and biosorption analysis according to Iram et al. (2011).

Faisalabad is the industrial city of Pakistan and Rawalpindi, and its different regions including Sihala, Gorakhpur, and Dahgal are polluted areas. Heavy metal (Zn, Pb, Cd, Ni, and Co) concentration was found more in Faisalabad as compared to the different regions of Rawalpindi. Fungal strains that were found in contaminated soils include *Aspergillus niger*, *Aspergillus* sp., *Fusarium* sp., and *Penicillium* sp. Experiments for tolerance studies indicate that *Aspergillus niger* was the most tolerant strains for heavy metals tested. The tolerance level of strains was determined in the presence of each metal by minimum inhibitory concentration (MIC) (Iram et al. 2009a, b). *Aspergillus niger* was also isolated from the contaminated soil of Multan. The tolerance potential for this filamentous fungi was determined for different salts of heavy metals Cr $(NO_3)_3$, Pb $(NO_3)_2$, and $ZnCl_2$. High level of resistance was reported for *A. niger* against chromium and lead. Therefore, *A. niger* is considered as a potential candidate for the removal of heavy metals from contaminated soil (Iram et al. 2012a, b).

The effect of biosorption potential of *Aspergillus niger* van. Tieghem for Cu(II) and Ni(II) was studied by Javaid et al. (2011). Pretreatment of fungal biomass with the different types of alkaline salts (NaOH, $NaHCO_3$, Na_2CO_3, NaCl, and $CaCl_2$) and acids (HCl and H_2SO_4) was carried out. Increase in adsorption efficiency was observed when biomass is treated with Na_2CO_3 and NaOH. Metal sequestering efficiency of adsorbent was reduced by 10–40 % when pretreated with $NaHCO_3$, detergent, NaCl, and $CaCl_2$. Drastic loss (80 %) in metal uptake efficiency was observed when treated with different acids. Pretreatment of Na_2CO_3 is suggested for the improved removal of heavy metal from *A. niger* (Javaid et al. 2011).

Comparison of three different basidiomycetous fungi (*Schizophyllum commune*, *Ganoderma lucidum*, and *Pleurotus ostreatus*) for removing metal ions Cu(II), Cr(VI), Ni(II), and Zn(II) from electroplating industrial effluents was investigated by Javaid and Bajwa (2008). Results for biosorption assay stated that the maximum efficacy was of *S. commune* followed by *G. lucidum* and *P. ostreatus*. Predilection order of metal ions was of Ni (II) > Cu (II) > Cr (VI) > Zn (II) (Javaid and Bajwa 2008).

2.5.2 Phytoremediation

The study of contaminated areas of Islamabad industrial area showed that metal concentrations of Pb, Zn, Cu, Co, Ni, and Cr in soils varied between 2.0 and 29.0, 61.9 and 172.6, 8.9 and 357.4, 7.3 and 24.7, 41.4 and 59.3, and 40.2 and 927.2 mg/kg. Plant species in contaminated areas were not hyperaccumulators, but still the accumulation for Cu in shoots was higher. The concentration of zinc was reported to be higher in grasses. *Parthenium hysterophorus* L. and *Amaranthus viridis* L. were suggested for the phytoextraction of Pb and Ni, whereas *Portulaca oleracea* L., *Brachiaria reptans* (L.), *Solanum nigrum* L., and *Xanthium strumarium* L. for the phytostabilization of soils contaminated with Pb and Cu (Malik et al. 2010b).

The cost-effective biosorption and phytoremediation potential of *Euphorbia prostrata* was carried out in a study by Husnain et al. Wastewater from sugar industry effluent was collected and analyzed for heavy metal concentrations (Cd, Cr, and Pb). In situ and ex situ experiments for the heavy metal reduction were carried out, and Euphorbia that is an indigenous plant was grown at different concentrations of (100, 50, and 0 %) wastewater. The accumulation of heavy metals over a period of 45 days was 20–55 %. Results of biosorption experiment showed that more than 70 % of heavy metals were adsorbed. Therefore, indigenous species of *Euphorbia* was suggested as a promising plant for biosorption and phytoremediation of heavy metals.

Yousaf et al. (2010) investigate the effects of industrial effluents on the growth of different soybean varieties. Effluents were rich in heavy metal concentrations. The growth of five varieties of *Glycine max*, viz., PSC-82, NARC-2, NARC-5, NARC-7, and William-82, was observed in heavy metal-contaminated effluents. Seedling length was found longer in paper and board effluent for PSC-62, NARC-2, and NARC-5 than in other effluents. Lower concentrations of effluents enhanced the growth of *Glycine max*. The growth of these varieties of soybean indicates its phytoremediation potential for contaminated soil.

Phytoremediation potential of five wild plant species from Jijal, Dubair, and Alpuri areas of Kohistan region, northern Pakistan, was investigated for macro and trace metal potential. The analysis of heavy metal concentrations in soil samples of these areas, through atomic spectrometer, showed significant ($p < 0.001$) contamination level. *Selaginella jacquemontii*, *Rumex hastatus*, and *Plectranthus rugosus* have multifold enrichment factor of Fe, Mn, Cr, Ni, and Co. The concentration of heavy metals in indigenous plant species is an indication of environmental hazards, but at the same time, it acts for the remediation of heavy metals from soil and thus lowers the toxicological concerns of soil. It was suggested on the basis of these studies that these wild plants are best suited for land reclamation (Muhammad et al. 2013).

Lemna is reported as a hyperaccumulator when the performance of bio-treatment ponds was studied for 1 year (Iram et al. 2012a, b). *Lemna* is planted around these ponds, and Iram et al. (2012a, b) reported that high accumulation of heavy metals was present in plant body as compared to wastewater. Therefore, it may act as best candidate for phytoremediation purpose.

Effective use of less-cost, quick-growing indigenous or native plants with superb biomass-producing plant species such as *Arundo donax* L. is greatly worth-able for the phytoremediation of metal-contaminated soils. It is grown all over Asia, Southern Europe, North Africa, and the Middle East for thousands of years with local names of "giant reed," "nurr," or "nurru." It is regarded as one of the highly biologically productive of all major communities. Earlier studies on giant reed showed that the stem height and diameter, number of nodes, fresh and dry weight of leaves, and whole photosynthesis were not affected, showing that plants are resistant to the high amount of Cd and Ni. Giant reed plants are very high-energy plants; they can grow in polluted soils to supply biomass for energy production fields (Kamran et al. 2014).

Lead is used as a key part in bullet manufacture along with arsenic and Ni in Pakistan. In firings sputter out in the type of refined and rough particulate and get putted on closed soil, that's why polluting it richly. Greatly, these metals deposit in the nearby specific area and the target of contamination of target area less rapidly with depth. In Pakistan, firing ranges are used mostly for daily practices of the armed forces (Tariq and Ashraf 2013).

In Pakistan, the amount of contamination of the soil of firing ranges has also been found to be higher with the increase in time period of firing history. There is a dreadful requirement to remove the lethal metals from these polluted areas to control the lethal effects generated due to leaching to closed agricultural soils and groundwater. Nowadays, phytoextraction has become a cheap technique to mitigate the metal-contaminated soils in Pakistan. The best plant species for phytoextraction are those having the power to gather and bear high amount of metals in tissue and exhibit a fast growth ratio. Metal deposit by the plants is ruled by their growth rate and ability to translocate metals to the aboveground tissue (Khan et al. 2013a, b). Various varieties of plant species have been exercised for their phytoextraction abilities for different metals, e.g., various *Brassica* spp., clover (*Trifolium pratense* L), panicgrass (*Panicum antidotale*), *Salix populus*, and *Nicotiana* spp. (Abdel-Sabour and Al-Salama 2007).

In Pakistan, crops like *Cucurbita pepo*, *Amaranthus* sp., *Raphanus sativus* var. *oleiformis*, and *Zea mays* have been greatly used as metal accumulators. *Brassica napus* and *R. sativus* grown on various polluted soils have shown to be good to mitigate a marginally contaminated soil. *Brassica juncea* (L.), Czern, proved to have the best ability to deposit lead in roots and then move it to the shoots from soils having sulfates and phosphates as fertilizers. In the same way, amid the several varieties of grasses, vetiver grass was found to greatly tolerate the lead-contaminated soils (Marchiol et al. 2004).

In Pakistan, phytoextraction period is the key cost factor for phytoextraction. The phytoextraction period of a particular heavy metal-contaminated soil is calculated by determining a linear relationship between the adsorbed heavy metal amount in the soil and the heavy metal amount in the plant shoots. In the majority of the examples, the best metal uptake by remediation plants is limited by less phytoavailability of the specific metals. Various chelants, e.g., EDTA, have been used to increase the bioavailability of the metals in soil as much as 100 times by forming soluble M-chelant complexes (Houston 2007).

The danger of metal leaching from the soil may be lessened by using appropriate chelates in the right quantity (Tariq and Ashraf 2013). The chelant usage at various phases of plant growth produces various results. EDTA usage before seed germination effectively lessens *Helianthus annuus* seedling emergence and dry weight. Soil available lead and Pb amount in plant biomass were found to increase with EDTA quantity, but the real amount of phytoextracted Pb decreased at high EDTA quantity because of high growth depression. The bioavailability and uptake of lead by coffee depends on the nature and kind of lead-polluted soil being mitigated. Weed was increased by changing the soil with the right chelates especially after the plants have reached high biomass (Miller et al. 2008).

2.5.3 Bacterial Bioremediation

Mercury-resistant bacteria were isolated from Lahore and around polluted area of Sheikhupura road. Six mercury-resistant bacteria (SHg-7, SHg-8, SHg-9, SHg-10, SHg-11, and SHg-12) were isolated and could also tolerate 450 $\mu g\ ml^{-1}$ of $HgCl_2$. This study showed that the ability of these strains for the detoxification of mercury in the medium by transforming into volatile form or by utilizing through any mechanism can help in the restoration/reclamation of the polluted environment. Further investigations revealed that Hg resistant in these bacteria is not plasmid encoded or it may not be transferred by plasmid (Afrasayab et al. 2002). Remediation of contaminated soils could also be enhanced by horizontal transfer of genes.

The effect of lead-resistant bacteria on growth and remediation potential of *Vigna mungo* L. (Hepper) was studied. These resistant bacteria were isolated from industrial effluents and could tolerate the concentration of lead up to 1,000 μg/ml. It was reported by Meryem and Yasmin (2013) that seed germination and seedling length were drastically reduced in the presence of heavy metal-contaminated industrial effluents. Lead-resistant bacteria that were isolated from the tannery effluent enhanced the growth of three selected varieties of *V. mungo* (NARC-Mash-2, NARC-Mash-3, and NARC-Mash-97) (Meryem and Yasmin 2013). Plant growth-promoting bacteria not only enhance the plant growth but also decrease the contaminant levels in soil by the uptake of these contaminants through different mechanisms.

Dichromate-resistant strain (ATCC 700729) was isolated from the effluent of tanneries of Kasur (Pakistan). The reduction potential of this strain showed that in the medium containing 20 mg ml^{-1} of dichromate, there was a reduction of more than 87 % of chromate within 72 h. Shakoori et al. (2000) also reported that strain (ATCC 700729) is feasible for the exploitation of the cleanup operations of dichromates in industrial effluents.

Copper-resistant bacterial strains were isolated from the wastewater of tanneries. These strains were highly tolerant to copper at high concentrations of 380 mg/L up to 400 mg/mL. Three resistant strains belong to genera *Salmonella*, *Streptococcus*, and *Vagococcus*. Atomic absorption spectrophotometer analysis showed the accumulation of Cu^{2+} with 30–45 % inside the cells (Shakoori and Muneer 2002).

2.6 Conclusion

The increased knowledge regarding the deleterious impacts of heavy metals has led to the basic legislation in Pakistan to combat mounting metal levels in different environmental compartments. Still, however, soil background metal levels (preindustrial) are not established in the agricultural, urban, industrial, or rural soils of Pakistan. Similarly, soil metal standards are not available at the federal and/or provincial level to maintain healthy soil ecosystems. Based on the recent studies from irrigated and rain-fed areas of Pakistan, it can be concluded that most of the agricultural soils have higher metal concentrations than the allowable standard/limits set by international agencies leading to food chain contamination. Public and livestock health is under serious toxicological threat where high metal levels are recorded both in the soils and the food crops. Such adverse soil ecological conditions urge reclamation procedures that befit Pakistan's agricultural environments. Bioremediation (i.e., mycoremediation, phytoremediation, and bacterial bioremediation) has been extensively studied by scientists in Pakistan for devising metal abatement strategies in the contaminated soils. Legislation at the federal and provincial level needs to be revised and strengthened to conserve this precious resource. Farmer's awareness, good agricultural practices, improved performance of the agricultural extension departments, and strict implementation of legislation regarding metal diminution can bring reduction in the soil metal pollution loads.

References

Abbas M, Parveen Z, Iqbal M, Riazuddin M, Iqbal S, Ahmed M, Bhutto R (2010) Monitoring of toxic metals (cadmium, lead, arsenic and mercury) in vegetables of Sindh, Pakistan. KUJSET 6 (2):60–65

Abdel-Sabour M, Al-Salama Y (2007) Zinc and cobalt phytoextraction by different plant species. Remediation 18(1):109–119

Afrasayab S, Yasmin A, Hasnain S (2002) Characterization of some indigenous mercury resistant bacteria from polluted environment. Pak J Biol Sci 5(7):792–797

Ahmad K, Shaheen M, Khan Z, Bashir H (2014) Heavy metals contamination of soil and fodder: a possible risk to livestock. Sci Tech Dev 32(2):140–148

Ahmed W, Ahmed A, Ahmad A, Randhawa M, Ahmad R, Khalid N (2012) Heavy metal contamination in vegetables grown in Rawalpindi, Pakistan. J Chem Soc Pak 34(4):914–919

Al Jassir M, Shaker A, Khaliq M (2005) Deposition of heavy metals on green leafy vegetables sold on roadsides of Riyadh City, Saudi Arabia. Bull Environ Contam Toxicol 75(5):1020–1027

Ali SM, Malik RN (2011) Spatial distribution of metals in top soils of Islamabad City, Pakistan. Environ Monit Assess 172(1–4):1–16

Ali Z, Ali B, Mohammad A, Ahmad M, Ahmad I, Napar AA, Kazi AG, Ali A, Shah SS, Mujeeb-Kazi A (2013) Combating water scarcity for global food security. In: Raza A (ed) Agricultural systems in the 21st century. Nova, Hauppauge, NY, pp 1–30

Ali Z, Malik RN, Shinwari ZK, Qadir A (2014) Enrichment, risk assessment, and statistical apportionment of heavy metals in tannery-affected areas. Int J Environ Sci Tech, 1–14. Available from: http://dx.doi.org/10.1007/s13762-013-0428-4.

Al-Khashman O et al (2004) Heavy metal distribution in dust, street dust and soils from the work place in Karak Industrial Estate, Jordan. Atmos Environ 38(39):6803–6812

Aslam B, Javed I, Khan F et al (2011) Uptake of heavy metal residues from sewerage sludge in the milk of goat and cattle during summer season. Pakistan Vet J 31(1):75–77

Eney A, Petzold D (1987) The problem of acid rain: an overview. Environmentalist 7(2):95–103

FAO/WHO (2001) Codex alimentarius commission. Food additives and contaminants. Joint FAO/WHO Food Standards Program 2001

Farid S, Baloch M (2012) Heavy metal ions in milk samples collected from animals feed with city effluent irrigated fodder. GJPS 2(2):010–017

Faryal R, Hameed A (2005) Isolation and characterization of various fungal strains from textile effluent for their use in bioremediation. Pak J Bot 37(4):1003–1008

Guvencc N, Alagha O, Tuncel G (2003) Investigation of soil multi-element composition in Antalya, Turkey. Environ Int 29(5):631–640

Houston M (2007) The role of mercury and cadmium heavy metals in vascular disease, hypertension, coronary heart disease, and myocardial infarction. Alt Ther Health Med 13(2):128–133

Huang P, Adriano D, Chlopecka A, Kaplan D (1998) Role of soil chemistry in soil remediation and ecosystem conservation. In: Huang PM (ed.), Soil chemistry and ecosystem health. Madison, WI: Soil Science Society of America. Available from: http://dx.doi.org/10.2136/sssaspecpub52.c13

Husaini S, Zaidi J, Akram M, Subhan K et al (2010) Trace elements analysis of crops and vegetables grown around industrial areas of Faisalabad and Gujranwala cities using INAA and AAS. Nucleus 47(3):239–246

Hussain S, Ghafoor A, Ahmad S, Murtaza G, Sabir M (2006) Irrigation of crops with raw sewage: hazard assessment of effluent, soil and vegetables. Pak J Agr Sci 43:3–4

Iram S, Ahmad I, Javed B, Yaqoob S, Akhtar K, Kazmi M, Rand Zaman B (2009a) Fungal tolerance to heavy metals. Pak J Bot 41(5):2583–2594

Iram S, Ahmad I, Stuben D et al (2009b) Analysis of mines and contaminated agricultural soil samples for fungal diversity and tolerance to heavy metals. Pak J Bot 41(2):885–895

Iram S, Ahmad I, Nasir K, Akhtar S (2011) Study of fungi from the contaminated soils of peri-urban agricultural areas. Pak J Bot 43(4):2225–2230

Iram S, Ahmad I, Riaz Y, Zahra A (2012a) Treatment of wastewater by *Lemna minor*. Pak J Bot 44 (2):553–557

Iram S, Arooj A, Parveen K et al (2012b) Tolerance potential of fungi isolated from polluted soil of Multan, Pakistan. Int J Biosci 2(10):27–34

Jan F, Ishaq M, Khan S, Ihsanullah I, Ahmad I, Shakirullah M (2010) A comparative study of human health risks via consumption of food crops grown on wastewater irrigated soil (Peshawar) and relatively clean water irrigated soil (lower Dir). J Hazard Mater 179(1):612–621

Javaid A, Bajwa R (2008) Biosorption of electroplating heavy metals by some basidiomycetes. Mycopathologia 6(1 and 2):1–6

Javaid A, Bajwa R, Manzoor T (2011) Biosorption of heavy metals by pretreated biomass of *Aspergillus niger*. Pak J Bot 43(1):419–425

Jonathan M, Ram-Mohan V, Srinivasalu S (2004) Geochemical variations of major and trace elements in recent sediments, off the Gulf of Mannar, the southeast coast of India. Environ Geol 45(4):466–480

Kamran M, Mufti R, Mubariz N, Syed JH, Bano A, Javed M, Munis M, Tan Z, Chaudhary H et al (2014) The potential of the flora from different regions of Pakistan in phytoremediation: a review. Environ Sci Pollut Res 21(2):801–812

Khan A, Javid S, Muhmood A, Mjeed T, Niaz A, Majeed A et al (2013a) Heavy metal status of soil and vegetables grown on peri-urban area of Lahore district. Soil Environ 32(1):49–54

Khan MU, Malik RN, Muhammad S (2013b) Human health risk from Heavy metal via food crops consumption with wastewater irrigation practices in Pakistan. Chemosphere 93 (10):2230–2238

Kumar V, Majumdar C, Roy P (2008) Effects of endocrine disrupting chemicals from leather industry effluents on male reproductive system. J Steroid Biochem Mol Biol 111(3):208–216

Lasat MM (2002) Phytoextraction of toxic metals: a review of biological mechanisms. J Environ Qual 31(1):109–120

Mahmood A, Malik RN (2014) Human health risk assessment of heavy metals via consumption of contaminated vegetables collected from different irrigation sources in Lahore, Pakistan. Arab J Chem 7(1):91–99

Mahmood-ul-Hassan M, Suthor V, Rafique E, Ahmad R, Yasin M (2012) Metal contamination of vegetables grown on soils irrigated with untreated municipal effluent. Bull Environ Contam Toxicol 88(2):204–209

Malik RN, Jadoon WA, Husain S (2010a) Metal contamination of surface soils of industrial city Sialkot, Pakistan: a multivariate and GIS approach. Environ Geochem Health 32(3):179–191

Malik RN, Husain S, Nazir I (2010b) Heavy metal contamination and accumulation in soil and wild plant species from industrial area of Islamabad, Pakistan. Pak J Bot 42(1):291–301

Marchiol L, Assolari S, Sacco P, Zerbi G (2004) Phytoextraction of heavy metals by canola (*Brassica napus*) and radish (*Raphanus sativus*) grown on multicontaminated soil. Environ Pollut 132(1):21–27

Martin P, Nishida J, Afzal J, Akbar S, Damania R, Hanrahan D (2006) Pakistan strategic country environmental assessment, South Asia Region, World Bank Vol-I: 36946-PK

McGrath D, Zhang C, Carton O (2004) Geostatistical analyses and hazard assessment on soil lead in Silvermines area, Ireland. Environ Pollut 127(2):239–248

Meryem SS, Yasmin A (2013) Effects of lead resistant bacteria on the early growth of *Vigna Mungo* L. (Hepper) under lead stress. Pak J Bot 45(S1):481–485

Midrar-ul-Haq, Puno HK, Khattak RA, Saif MS (2003) Agricultural land contamination by industrial activities. Int J Agr Biol 5(2): 150–153

Miller G, Begonia G, Begonia M, Ntoni J, Hundley O (2008) Assessment of the efficacy of chelate-assisted phytoextraction of lead by coffeeweed (Sesbania exaltata Raf.). Int J Environ Res Public Health 5(5):428–435

Monteiro-Neto M, Lazaro C, Tavares D, Cecchi A (2010) Frequency of chromosomal aberrations in peripheral lymphocytes of tannery workers in Brazil. Environ Toxicol Pharmacol 29(1):3–6

Mudgal V, Madaan N, Mudgal A, Singh R, Mishra S (2010) Effect of toxic metals on human health. Open Nutraceut J 3:94–99

Muhammad S, Shah M, Khan S, Saddique U, Gul N, Khan M, Malik R, Farooq M, Naz A (2013) Wild plant assessment for heavy metal phytoremediation potential along the mafic and ultramafic terrain in northern Pakistan. BioMed Res Int, Article ID 194765, 9 pages, Available from http://dx.doi.org/10.1155/2013/194765

Mohammadzai I, Khan N, Khan MJ, Khattak RA (2001) Heavy metals (Cu, Fe, Mn, Zn, Cd, Co, Ni, Mo, Cr and Pb) status in some selected soils of Peshawar and Nowshera districts. J Chem Soc Pak 23:23–30

Mushtaq N, Khan K et al (2010) Heavy metals contamination of soils in response to wastewater irrigation in Rawalpindi region. Pak J Agr Sci 47(3):215–224

Mustafa K, Shah M, Khan N, Khan R, Khan I (2007) Resource degradation and environmental concerns in Pakistan's agriculture. Sarhad J Agr 23(4): 1159

Nergus Y, Ahmed SI, Sharif SI (2005) Impact of contaminated vegetables, fruits and fodders on human health by Malir River farms Karachi. J Chem Soc Pak 6:561–574

Nriagu J, Pacyna J (1988) Quantitative assessment of worldwide contamination of air, water and soils by trace metals. Nature 333(6169):134–139

Panek E (2000) Metale sladowe w glebach i wybranych gatunkach roslin obszaru polskiej czesci Karpat. IGSMiE PAN, Krakow, Poland

Pakistan Country Water Resources Assistance Strategy Water Economy: Running Dry (2005) South Asia Region Agriculture and Rural Development Unit, South Asia Region, Report No: 34081-PK

Pakistan Economic Survey (2010) 2009–10; Finance Division, EA Wing, Government of Pakistan, Islamabad: Pakistan Press

Qadir M, Ghafoor A, Murtaza G, Murtaza G (2000) Cadmium concentration in vegetables grown on urban soils irrigated with untreated municipal sewage. Environ Dev Sustain 2(1):13–21

Rehman K, Ashraf S, Rashid U, Ibrahim M, Hina S, Iftikhar T, Ramzan S (2013) Comparison of proximate and heavy metal contents of vegetables grown with fresh and wastewater. Pak J Bot 45(2):391–400

Rind A, Mastoi G, Hullio A (2013) Impacts of Jamshoro thermal power station on soil of the surrounding area. Ind J Sci Res Technol 1(2):65–71

Romic M, Romic D (2003) Heavy metals distribution in agricultural topsoils in urban area. Environ Geol 43(7):795–805

Saif MS, Midrar-ul-haq, Memon KS (2005) Heavy metals contamination through industrial effluent to irrigation water and soil in Korangi Area of Karachi (Pakistan). Int J Agr Biol 7 (4): 646–648

Sattar M, Khan M, Khalil A, Amir R et al (2013) Mitigation of heavy metals in vegetables through washing with house hold chemicals. Int J Agr Sci Res 3(5):1–11

Sayer J, Kierans M, Gadd G (1997) Solubilization of some naturally occurring metal-bearing minerals, lime scale and lead phosphate by *Aspergillus niger*. FEMS Microbiol Lett 154 (1):29–35

Shakoori A, Muneer B (2002) Copper-resistant bacteria from industrial effluents and their role in remediation of heavy metals in wastewater. Folia Microbiol 47(1):43–50

Shakoori A, Makhdoom M, Haq R (2000) Hexavalent chromium reduction by a dichromate-resistant gram-positive bacterium isolated from effluents of tanneries. Appl Microbiol Biotechnol 53(3):348–351

Simonson R (1995) Airborne dust and its significance to soils. Geoderma 65(1–2):1–43

Singh O, Labana S, Pandey G, Budhiraja R, Jain R (2003) Phytoremediation: an overview of metallic ion decontamination from soil. Appl Microbiol Biotechnol 61(5–6):405–412

Tandon S, Chatterjee M, Bhargava A, Shukla V, Bihari V (2001) Lead poisoning in Indian silver refiners. Sci Total Environ 281(1):177–182

Tariq SR, Ashraf A (2013) Comparative evaluation of phytoremediation of metal contaminated soil of firing range by four different plant species. Arabian Journal of Chemistry. Available from: http://dx.doi.org/10.1016/j.arabjc.2013.09.024

Tariq SR, Shah MH, Shaheen N, Khalique A, Manzoor S, Jaffar M (2005) Multivariate analysis of selected metals in tannery effluents and related soil. J Hazard Mater 122(1):17–22

Tariq SR, Shah MH, Shaheen N, Khalique A, Manzoor S, Jaffar M (2006) Multivariate analysis of trace metal levels in tannery effluents in relation to soil and water: a case study from Peshawar, Pakistan. J Environ Manage 79(1):20–29

Wagner GJ (1993) Accumulation of cadmium in crop plants and its consequences to human health. Adv Agron 51:173–212

White J, Welch R, Norvell W (1997a) Soil zinc map of the USA using geostatistics and geographic information systems. Soil Sci Soc Am J 61(1):185–194

White C, Sayer J, Gadd G (1997b) Microbial solubilization and immobilization of toxic metals: key biogeochemical processes for treatment of contamination. FEMS Microbiol Rev 20 (3–4):503–516

Wuana R, Okieimen F (2011) Heavy metals in contaminated soils: a review of sources, chemistry, risks and best available strategies for remediation. ISRN Ecol (2011), Article ID 402647, 20 pages, Available from: http://dx.doi.org/10.5402/2011/402647

Yoshida T, Yamauchi H, Fan-Sun G (2004) Chronic health effects in people exposed to arsenic via the drinking water: dose-response relationships in review. Toxicol Appl Pharmacol 198 (3):243–252

Yousaf I, Ali S, Yasmin A (2010) Germination and early growth response of Glycine max varieties in textile and paper industry effluents. Pak J Bot 42(6):3857–3863

Chapter 3
Multiple-Phase Evaluation of Copper Geochemistry

Jean B. Diatta, Ewa Chudzińska, Leszek Drobek, Aleksandra Wojcicka-Półtorak, Bernd Markert, and Simone Wünschmann

3.1 Geological Aspects

3.1.1 Geogenesis

Considerations given to the Earth's geochemical structure might be referred to one of the most widely recognised hypotheses on its creation, i.e., initially the Earth was a melted mass having been maintained in its semifluid state by bombing planetoids. The chemistry of the Earth at the early forming stage was similar to an average compound of meteorites with the predominance of magnesium and iron silicates, nickel-bearing iron in the metallic state and numerous other elements. Referring to the commonly agreed threshold of 0.01 %, elements such as O, Si, Al, Fe, Ca, Na, Mg, K, C are considered as major since their concentrations are higher than this value. In the case of Ni, Zn, Cu, Co, Pb, Sn, As and Mo, their level is below 0.01% mm^{-1} and they are designated as rare earth metals or trace elements. Copper levels in the rocks of the Earth's crust are estimated to range from 5 to 100 mg kg^{-1},

J.B. Diatta (✉)
Department of Agricultural Chemistry and Environmental Biogeochemistry, Poznan University of Life Sciences, ul. Wojska Polskiego 71F, 60-625 Poznan, Poland
e-mail: Jeandiatta63@yahoo.com

E. Chudzińska • A. Wojcicka-Półtorak
Department of Genetics, Institute of Experimental Biology, Adam Mickiewicz University, ul. Umultowska 89, 61-614 Poznan, Poland

L. Drobek
Department of Environment Monitoring, Central Mining Institute, Plac Gwarków 1, 40-166 Katowice, Poland

B. Markert • S. Wünschmann
Environmental Institute of Scientific Networks (EISN-Institute), Fliederweg 17, 49733 Haren (Ems), Lower Saxony, Germany

I. Sherameti, A. Varma (eds.), *Heavy Metal Contamination of Soils*, Soil Biology 44, DOI 10.1007/978-3-319-14526-6_3

with the average of 55 mg kg^{-1} (Konstantynowicz 1979; Kabata-Pendias and Pendias 1999).

Copper as free metal is rarely encountered in the environment. Due to its chemical affinity for sulphur, copper is classified as the so-called chalcophylic elements and is mostly found in the form of sulphide minerals. In terms of usefulness, the greatest significance is attributed to chalcopyrite $CuFeS_2$, chalcocite Cu_2S and bornite Cu_5FeS_4. These minerals originate from magma processes whose quantitative predominance in nature is recognised, although the richest concentrations of copper are related to its secondary concentration at the weathering stage (Polański and Smulikowski 1969). Copper, being a weak basic oxide, forms also minerals of hydroxycarbonate nature with high utility significance such as azurite $Cu_3(CO_3)_2(OH)_2$ and malachite $Cu_2CO_3(OH)_2$. In this case they are mainly found in sedimentary rocks as products of hydrothermal transformations.

3.1.2 Geological CO_2 Sequestration Mediated Mineral Solubilisation

Processes of rock and minerals' weathering in conditions of underground injection of CO_2 to deep geological structures are considered to be new and of local character. In fact, these practices have occurred in different parts of our globe for over 40 years (Drobek et al. 2008). Carbon dioxide injection to deep geological structures including uncultivated saline aquifers initiates chemical processes in the pattern: CO_2-rock, CO_2-brine and brine (saturated CO_2)-rock, resulting in chemical changes of both underground waters and surrounding rocks. The effect of the diphasic system of CO_2-brine on geological milieu is based on bicarbonate balance according to the following reactions:

$$H_2CO_3 + Ca^{2+} \rightarrow CaCO_3 + 2H^+ \quad (3.1)$$

$$CaCO_3 + H_2O + CO_2 \leftrightarrow Ca(HCO_3)_2 \leftrightarrow Ca^{2+} + 2HCO_3^- \quad (3.2)$$

A targeted example of such process is the surface aquifer of copper deposits in the Legnicko-Glogowski Okręg Miedziowy (LGOM, Poland) with average amounts of chalcophylic elements Cu 0.08, Zn 0.50 and Pb 0.045 mg dm^{-3} (Wilk and Bocheńska 2003). For metals of similar concentrations in the places of carbon dioxide injection in underground waters, copper and other metals will precipitate in the form of virtually dissoluble or hardly soluble hydroxycarbonate salts which are subject to the process of secondary dissolving, due to hydrogeochemical balance (Drobek et al. 2008).

The effect of two-phase brine (saturated CO_2)-rock system on geological environment is a subsequent mechanism of chemical processes occurring during CO_2 deposition under the ground. Carbonate minerals will gradually dissolve causing an increase in permeability of the surrounding rocks; thus, in the environment of brine supersaturated with carbon dioxide, precipitation of metals including dissoluble and hardly soluble copper hydroxycarbonates will take place. In this case the bicarbonate balance in brine environment will be again a potential source of co-precipitation of the hydroxycarbonate form of copper, loamy minerals and silicate (Huijgen et al. 2006; Ibsen and Jacobsen 1996).

The reported cases of chemical processes accompanying CO_2 injection to deep geological structures point out that such anthropogenic interference into the environment may generate changes in the forms of elements' mineralogy, copper among others. Taking into account great dispersions of copper in the Earth's crust, these processes are rather of local character and their magnitude should not generate significant changes in the environment at present.

3.2 Solid-Soil Solution Interface: Processes and Evaluation

Soil in its whole entity is one of the more complex naturally formed medium. The heterogeneity of its constituents, their mutual interactions and last but not the least the direct or indirect influence of external factors make this complexity quite unbelievable to define. In all natural heterogeneous soil environments, the following generalised mechanisms and phenomena occur simultaneously: (a) dissolution, (b) adsorption/desorption, (c) complexation, (d) migration, (e) precipitation, (f) occlusion, (g) diffusion, (h) absorption by microbiota and (i) volatilisation. The developments in soil chemistry started with the pioneering studies of Thomas Way on Base Exchange published in 1850 as *On the Power of Soils to Absorb Manure* (Forrester and Giles 1971). Detailed informations on geochemical reactions are reported by Gapon (1933), Gaines and Thomas (1953), Sokołowska (1989), Sposito (1984), Sparks (1989), Mulder and Cresser (1994) and Barrow (1999).

Soil chemistry has traditionally focused on the chemical reactions in soils that affect plant growth and plant nutrition. However, as concerns increased about inorganic and organic contaminants in water and soil, the emphasis of soil chemistry is now on environmental soil chemistry. This holds to the definition suggested by McBride (1994) and Sparks (1995) as the study of chemical reactions between soils and environmentally important plant nutrients, radionuclides, heavy metals and organic chemicals.

Environmental concern in the 1970s led to the adaptation of simulative/predictive models for metals solid-phase/aqueous-phase interactions and their probable behaviour in natural environments. Transport of heavy metals and copper among others through soil layers has long presented great interest to both environmental and soils scientists (Murali and Aylmore 1983a, b; Ogwada and Sparks 1986; Diatta et al. 2000). The complexity of the soil matrix makes it difficult to selectively choose interactions, which mostly contribute to the adsorption of a specific metal (Diatta et al. 2012). This problem creates more difficulty in the process of formulating meaningful soil models for the prediction of metal transport.

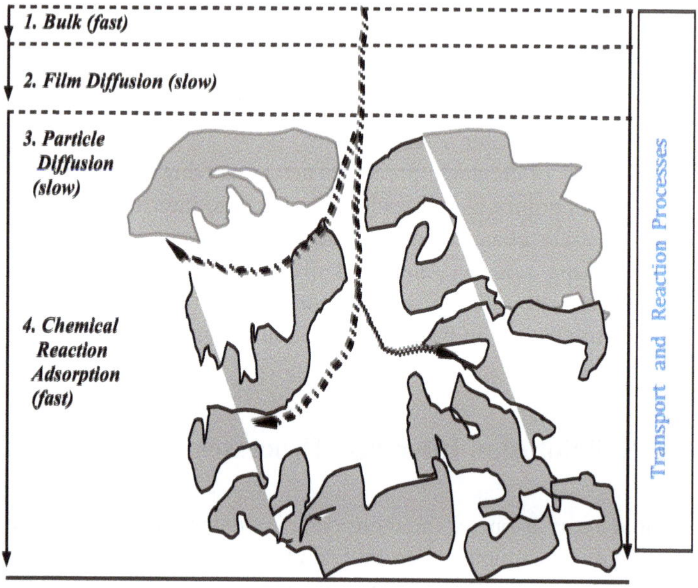

Fig. 3.1 Rate-determining steps in soil chemical reactions (Weber 1984 with permission from ASCE)

Most of soil chemical reactions are time dependent and may differ notably in rates. The frequently concerned mechanisms of ion movement through soil layer are diffusion-based processes, the rates of which are closely related to the homogeneity and heterogeneity of the systems (Lai and Mortland 1968; Sparks 2000). The following illustration (Weber 1984) shows rate-determining steps of soil chemical processes (Fig. 3.1):

There are three basic ways to determine rate constants (Skopp 1986; Sparks 1989, 1995). The first is the use of initial rates by plotting the concentration of a reactant or product over a short reaction time period during which the concentrations of the reactants change so little that the instantaneous rate is hardly affected. The second way is by directly applying integrated equations and graphing the data, and the last one is by using the nonlinear least square analysis.

In recent years, research on metal (copper among others) adsorption/desorption processes has been based mainly on the study of equilibrium conditions. The kinetic approach of these processes in soil systems is also reported (Bunzl et al. 1976; Elkhatib et al. 1992; Violante et al. 2008). The kinetics of metal adsorption (copper among others) at its first phase is found to be basically influenced by sorbent amounts, types and pH as well (Basta and Tabatabai 1992). The impact of organic matter on the kinetics of copper sorption by peat humic substances was extensively reported by Sapek (1976). The author reported that the time course of these processes is partly attributed to the occurrence of specific high- and low-energy adsorbing sites at the surface of soil organic colloids. It is of common view that the pool of high-energy sites is more developed than that expected for the low energy one. This acts as a limiting factor during adsorption process. At a longer time, more and more other low-energy sites become successively saturated (Tiller et al. 1984). On the basis of kinetic studies, it was formulated that kinetic adsorption is characterised by a two-phase process, a rapid exchange followed by a slower reaction of copper ions with organic sorbents (Sapek and Zebrowski 1976; Boehringer 1980; Aringhieri et al. 1985).

The reaction rate theory assumes that colliding molecules (e.g. reactions between solution ions and ions adsorbed in the exchange complex) must be in a high-energy state before a reaction can occur. This energy of activation is of van der Waals repulsive forces that take place as two ions approach each other (Sparks 1989). Then the energy of activation may be considered as a measure of the magnitude of the forces to be overcome during the process of adsorption/desorption. Thermodynamic approach can predict only the final state of a soil system from an initial nonequilibrium state and has been used to characterise exchange equilibria on clay and soil particle surfaces (El-Sayed et al. 1970). An alternative that should be examined is the use of kinetics to follow the course of adsorption/desorption reactions. This technique has the potential of providing both the mechanistic and thermodynamic informations. However, application of kinetic theory to determine thermodynamic parameters (e.g. for phosphorus and potassium) has received more attention (Evans and Jurinak 1976; Pavlatou and Polyzopoulos 1988; Ogwada and Sparks 1986; Liu and Huang 2000; Zhang et al. 2000) than for heavy metals (El-Sayed et al. 1970; Elkhatib et al. 1993).

A number of models have been elaborated for describing adsorption/desorption reactions of various elements in soils (Gapon 1933; Stawiński 1977; Mattigod and Sposito 1979; Kinniburgh 1986; Amacher et al. 1986; Sokołowska 1989; Dube et al. 2001; Diatta et al. 2003). The most well-known sorption models may be classified into two groups: (1) models which take account of electrostatic forces and bonding and (2) models which do not include these parameters. Models of the first group have some bug: they are complicated—it means that they require many parameters at the beginning of modelling—or they are simply physical equalisations of transport inside the soil profile without consideration of interaction between contaminants and soil matrix. To the second group belong the classical models, i.e. Freundlich and Langmuir isotherms with their suggestive extensions (Veith and Sposito 1977; Sposito 1982, 1984; Sparks 1995; Sarbak 2000). These isotherms are generally based on the amount of adsorbed element (quantity) plotted against its equilibrium concentration (intensity) and the resulting slopes and intercepts to be indicative of various soil sorption/desorption parameters (Fig. 3.2).

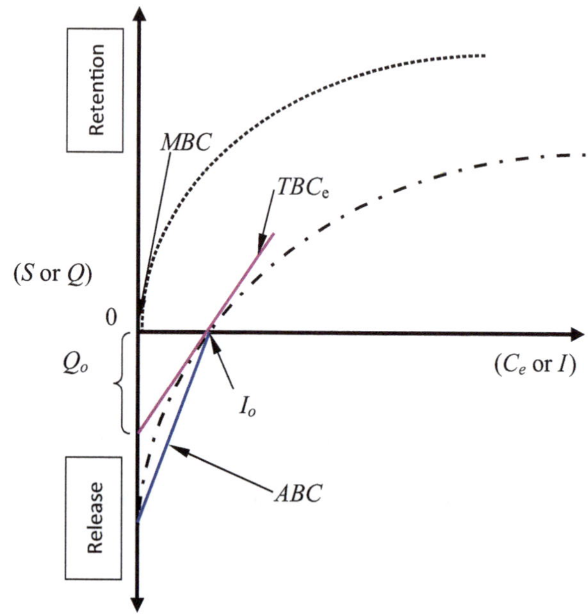

Fig. 3.2 Schematic representation of various sorption and buffering parameters that could be characterised during equilibration studies. S = amount retained (i.e. adsorbed) or released (desorbed). Q = quantity, Q_o = intercept on the Q axis of the TBC_e tangent, C_e = equilibrium concentration, I = intensity, I_o = no retention (i.e. adsorption) and no release (desorption), MBC = maximum buffering capacity ($C_e \rightarrow 0$), TBC_e = tangential buffering capacity, ABC = average buffering capacity

3.3 Copper Status in Contaminated and Uncontaminated Soils

Dynamic development of civilisation during the past century led to excessive accumulation of heavy metals, among others copper, in the environment, creating serious ecological problems. Gradual increase of soil contamination with copper and its relative mobility in the environment threatens with the risk of this metal introduction into trophic cycle (Kabata-Pendias 1978; Markert and Friese 2000; Markert et al. 2003; Nriagu et al. 2012). Several phosphate fertilisers and pests' products contain copper either as impurity or as active constituent (Alloway 1995; Robert 1996), but one of the major copper pollution sources is attributed to mining activity and dust emissions from metallurgical industries (Geiger et al 1993; Kabata-Pendias A. 1993; Grzebisz and Chudziński 1996; Diatta et al. 2000).

Heavy metals (copper among others) are ubiquitous and fundamental components of life on Earth and part of all food chain. Because their levels in soils naturally range between 0.1 and 100.0 mg kg^{-1} (Cox 1989; Baker 1990; Kabata-Pendias and Pendias 1992), they are also called "trace elements" (Adriano 1986; Duffus 2002). Some metals such as copper and zinc are essential to life, but in high concentrations, however, the same metals can exhibit toxicity effects. The natural concentration of copper in soil solution is found to be very low, being in the range of 1×10^{-8} to 60×10^{-8} moles dm^{-3}. Hodgson et al. (1966) reported that more than 98 % of soil solution copper was complexed with organic matter and more strongly bound to organic matter than are other micronutrient cations like Zn^{2+} and

Mn^{2+}. The affinity to organic matter regulates Cu mobility and availability in the soil. In a fractionation study of Cu in soils, McLaren and Crawford (1973a, b) have found that a high proportion of total copper was in occluded or lattice form. Irrespective of the forms in which copper may occur, its behaviour and shift from natural copper nontoxicity to anthropogenic copper toxicity are intrinsically related to metal specificity and man activities (Adriano 1986; Kabata-Pendias 1993; Chudziński 1995; Prasad 2008).

The behaviour of heavy metals and copper among others is governed mainly by two groups of adversely acting phenomena. The first group involves processes such as: sorption of copper by soil mineral and organic sorptive complex, precipitation of insoluble compounds and biogenic accumulation, which aim at reducing copper solubility and mobility (Adriano 1986; Brümmer et al. 1986; Logan 1990). The second group consists of processes that tend to increase copper mobility, i.e. desorption, solubility and mineralisation of organic compounds (Ram and Verloo 1985). Copper solubility and desorption in soil systems are generally strengthened by a decrease of pH (Lindsay 1979) which increases its activity manifold than for soil with neutral or weakly acid soils (Xian and Shokohifard 1989; Msaky and Calvet 1990; McBride 2001).

Copper retention by the organo-mineral soil complex has been a topic of several reports (McLaren and Crawford 1973a, b; Cavallaro and McBride 1978; Sapek and Sapek 1980; Logan et al. 1997; Diatta et al. 2000). Several works dealt with copper organic complexes of soil systems in which the organic part is quite dominating. Clay minerals are also potential binding agents for copper as they exhibit large specific surface area and enhanced cation exchange capacity. In general clay minerals are found to possess both a pH-dependent and permanent negative surface charge, which is among others a resultant of isomorphous substitution. The permanent charge is for the most part compensated by cations intercalated in the interstitial space (McBride 1994; Sparks 1995). Therefore, the metal cations bound in such a way can readily be exchanged by other cations, e.g. Ca^{2+}, Mg^{2+} and NH_4^+ (Van Bladel et al. 1993).

The mechanisms of copper adsorption are pH dependent and may additionally depend on the concentrations of copper in the soil solution. At relatively high pH and low copper concentrations, specific adsorption is often the predominant binding mechanism (McBride 1989; Diatta et al. 2004, 2012). Surface complex is thought to occur via interaction of a surface functional group with an ion or molecule present in the soil solution, which creates a stable molecular entity. If water molecule is present between the surface functional group and the bound ion or molecule, the surface complex is termed outer sphere (Sposito 1984). If there is not a water molecule present between the ion or molecule and the surface functional group to which it is bound, this is an inner sphere complex. It is important to note that inner sphere complexes can be monodentate (metal is bound to only one oxygen) and bidentate (metal is bound to two oxygens) as illustrated below (Fig. 3.3).

The prediction of the mobility of metals in soils may also be assessed by simulative natural diffusion studies (Helios-Rybicka and Jędrzejczyk 1995) or desorption. It is often observed that desorption is a more difficult process than

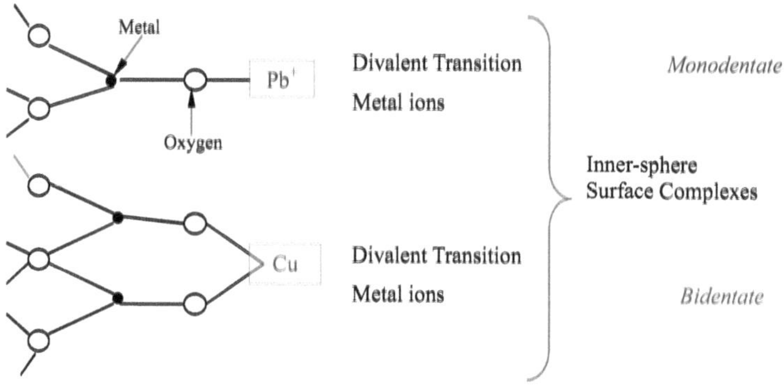

Fig. 3.3 Illustration of surface complexes' formation between inorganic ions and hydroxyl groups of an oxide surface (with permission from Elsevier for Sparks 1995: Environmental Soil Chemistry, p. 99–139)

adsorption and that the adsorbate is not entirely desorbed. So the reactions (adsorption process) appear to be irreversible. Ion exchange is in essence a reversible reaction in which cations are stoichiometrically exchanged between the electrolyte solution and the surface phase of the exchanger material. Exchange irreversibility is sometimes observed and has been attributed to colloidal aggregation and the formation of quasi-crystals (Van Bladel and Laudelout 1967), which could make exchange sites inaccessible. Such apparent irreversibility is commonly referred to as hysteresis that means adsorption and desorption isotherms corresponding to the forward and backward reaction would not coincide (Verburg and Baveye 1994). Hysteresis is often reported for soil-pesticide interactions (Barriuso et al. 1994; Zhu and Selim 2000) and also for metal reactions with soils or soil components (Ainstworth et al. 1994).

Informations about hysteretic behaviour of copper in soils are scarce. Schultz et al. (1987) reported about a substantial hysteresis in the desorption of Cu, Ni, Pb and Zn and Cr sorbed to hydrous ferric oxide. They found that the magnitude of the hysteresis was dependent on pH at which the adsorption process took place and the length of time of contact of sorbent-solute prior to initiation of the desorption process. However, Maqueda et al. (1998) have found that copper desorption isotherms on a montmorillonite showed a very high hysteresis, and in the same order a substantial sorption/desorption hysteresis was also observed by Wu et al. (1999), who reported that Cu has formed high-energy bonds with organic matter and layer silicate surfaces.

3.4 Site-Specific Copper Geochemistry: A Case Study

3.4.1 Site Characterisation and Soil Sampling Procedure

The study area is located in the Głogów proglacial stream valley (N 51°41′03″ and E 15°57′12″) situated between Sudetes (North) and the Małopolska Plateau (South) (Kondracki 1978). Soil samplings were performed within the years 2000–2001. The investigated area is represented by two major soil groups: Fluvisols and Luvisols (WRB 1998), as reported in Table 3.1. The Głogów copper smelter (GCS) was established in the 1970s, and the primary cause of high metal accumulation was the emission of dusts and gases by smelters. In the 1970s the Głogów copper smelter annually discharged into the atmosphere over 15,000 tons of dusts containing heavy metals, sulfuric acid and sulphur dioxide, and from early 1990s a rapid reduction of dust emissions occurred. Copper content of dust discharges amounted to 2,968, 204, 23, 9.7 tons for the years 1980, 1990, 2000 and 2012, respectively (www.kghm.pl/index. Access 20.07.2014).

These tremendous amounts of copper accumulated progressively in soils for many years, and their negative environmental impact is still lasting. The zone mostly contaminated by heavy metals (copper mainly) was taken out of agricultural utilisation and afforested (poplars and maples), establishing since 1987 the "sanitary belt". Earlier investigations carried out in this area showed a copper distribution pattern consistent with the wind rose, where southwest (SW) and west (W) winds—referring to Głogów copper smelter position—were dominating (Kabata-Pendias 1978; Grzebisz et al. 1997; Geochemical Atlas 1999) (Fig. 3.4).

It should be mentioned in the past all soils of the area under study were arable, some have currently changed management purposes. Since most of sampling sites were located in arable soils, the sampling depth of 0–20 cm was applied due to agricultural practices. The same sampling depth was considered also for sampling sites other than those under agricultural practices.

3.4.2 Physical and Chemical Soil Tests

Prior to basic analyses, soil samples were air-dried and crushed to pass through a 1 mm mesh sieve. Total copper (Cu_{Tot}) content was determined in soil samples (International Standard 1995). On the basis of total copper content (Cu_{Tot}), it was decided to elaborate a "consensual" share of soils into two operational groups: Group I represents the moderate and highly contaminated soils (from the sanitary belt) with total copper content varying from 76.1 to 334.0 mg kg^{-1}, except one soil sample with Cu_{Tot} of 1,048 mg kg^{-1}. To the group I belong soil Nos. 1–8 considered as contaminated. The group II consists of soils collected beyond the "Sanitary Belt", where Cu_{Tot} was below 30.0 mg kg^{-1}. The mentioned group involved soil Nos. 9–16 and is designated as uncontaminated (Table 3.2). Copper from soils was

Table 3.1 Sampling sites, positioning towards the Głogów copper smelter and soils' taxonomical description

Soil sample (No.)	Distance from the main emitter (km)	Position towards the main emitter	Sampling site[a]	Particle size (mm) Sand (1.0–0.05)	Silt (0.05–0.002) (g kg^{-1})	Clay (<0.002)	Soil texture[b]	WRB (1998)[c]	Locality
1	5.0	E	A	660	220	120	Sandy loam	Calci-Stagnic Luvisols	Rapocin
2	1.5	S	Fa	820	130	50	Loamy sand	Haplic Luvisols	Zukowice
3	3.5	NW	A	860	120	20	Loamy sand	Gleyic Fluvisols	Wróblin Głogowski
4	4.0	E	A	460	490	50	Sandy loam	Gleyic Fluvisols	Biechów
5	4.8	NE	Fa	420	420	160	Loam	Molli-Gleyic Fluvisols	Bogomice
6	4.0	NE	M	780	190	30	Loamy sand	Gleyic Fluvisols	Bogomice
7	3.5	NE	M	560	360	80	Sandy loam	Haplic Luvisols	Bogomice
8	0.5	NE	Fo	790	180	30	Loamy sand	Dystri-Gleyic Fluvisols	Sanitary belt
9	8.0	SW	A	820	160	20	Loamy sand	Gleyic Fluvisols	Domanice
10	7.5	W	M	750	160	90	Sandy loam	Stagnic Luvisols	Brzeg Głogowski
11	9.0	E	A	820	130	50	Loamy sand	Haplic Luvisols	Serby
12	7.0	E	A	750	210	40	Loamy sand	Gleyic Fluvisols	Grodziec Mały

13	7.0	W	A	790	160	50	Loamy sand	Calci-Stagnic Luvisols	Brzeg Głogowski
14	8.5	E	A	820	140	40	Loamy sand	Albic Luvisols	Serby
15	8.5	W	M	900	80	20	Sand	Areni-Gleyic Fluvisols	Kromolin
16	6.0	SE	A	770	140	90	Sandy loam	Stagni-Cutanic Luvisols	Brzostów

RED line for delimitating the group I (contaminated) from the and II (uncontaminated)
[a]*A* arable, *Fa* fallow, *Fo* forest, *M* meadow
[b]Soil taxonomy. USDA (1975)
[c]WRB-84, World Soil Resources Reports, FAO-ISSIC-ISRIC (1998)

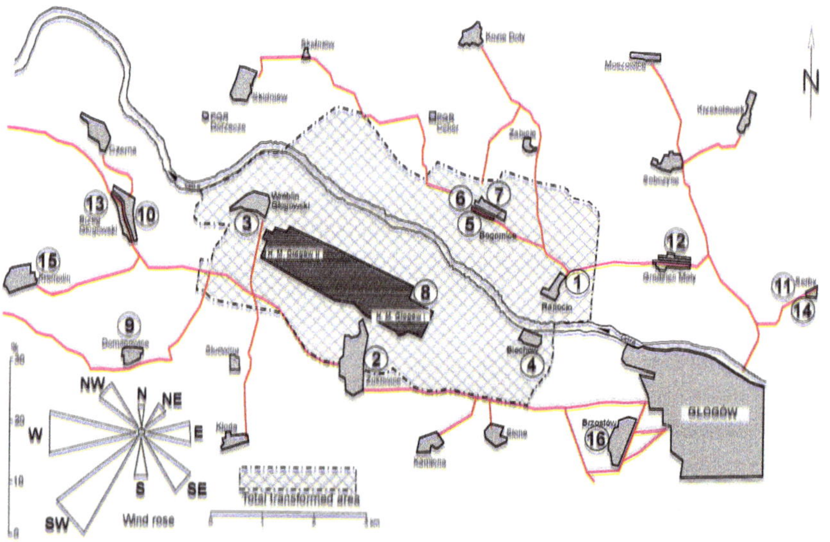

Fig. 3.4 Spatial location of sampling sites (*numbers in circles*) within the investigated area

additionally extracted by using DTPA solution (diethylenetriaminepentaacetic acid) (Lindsay and Norvell 1978; Kociałkowski et al. 1999). Copper in filtrates was determined by the FAAS method (Flame Atomic Absorption Spectrometry, Varian SpectrAA 250 plus).

Particle-size distribution was determined according to the method of Bouyoucos-Casagrande (Ryan et al. 2001). The texture classes were established according to USDA classification (Soil Taxonomy 1975) on the basis of the following separates: sand (1.0–0.05 mm), silt (0.05–0.002 mm) and clay (<0.002 mm). Organic carbon was determined by the dichromate wet oxidative method (Nelson and Sommers 1996) and soil pH potentiometrically, according to Polish Standard (1994). Soil samples were additionally analysed for carbonate content (Loeppert and Suarez 1996), whereas the cation exchange capacity (CEC) was assessed by the barium chloride method as suggested by Hendershot and Duquette (1986). The cation exchange capacity of the soils was obtained by summation of exchangeable alkaline cations and acidity according to Thomas (1982). The specific surface area (SSA) of the soils was determined according to the ethylene glycol monoethyl ether (EGME) procedure, whereas the external surface area was measured by the BET method (Gregg and Sing 1967; Carter et al. 1986; de Jong 1999).

Table 3.2 Some physical and chemical properties of investigated soils

Soil texture	pH ($CaCl_2$)	Total copper (Cu_{Tot})	Cu-DTPA	C_{org} ($g\ kg^{-1}$)	SSA_{EGME}[a] ($m^2\ g^{-1}$)	SSA_{BET}[b] ($m^2\ g^{-1}$)	SSA_{INT}[c] ($m^2\ g^{-1}$)	CEC[d] ($cmol_c\ kg^{-1}$)	$CaCO_3$ (%)
Sandy loam (1)	7.3	76.1	16.5	6.5	25.9	6.7	19.1	15.0	4.6
Loamy sand (2)	6.5	79.5	22.0	8.9	14.7	1.3	13.4	11.7	–[e]
Loamy sand (3)	6.4	84.9	27.7	6.8	15.5	3.4	12.1	13.4	–
Sandy loam (4)	6.8	117.8	40.3	9.9	37.4	4.3	33.2	13.2	–
Loam (5)	6.9	162.8	52.3	18.3	72.0	17.1	54.9	21.8	–
Loamy sand (6)	6.6	235.3	93.2	7.7	13.6	1.6	12.0	11.7	–
Sandy loam (7)	6.9	334.0	131.5	14.9	25.2	3.9	21.3	17.5	–
Loamy sand (8)	5.0	1048.0	472.2	5.6	9.4	2.1	7.4	9.8	–
Loamy sand (9)	7.0	22.3	6.8	8.9	15.7	1.5	14.2	11.3	–
Sandy loam (10)	7.3	22.6	5.0	9.4	26.9	4.9	22.0	14.5	0.6
Loamy sand (11)	6.2	23.1	5.2	5.8	9.4	2.6	6.9	9.6	–
Loamy sand (12)	7.6	23.4	3.5	12.0	37.4	10.1	27.3	17.5	2.2
Loamy sand (13)	7.4	25.6	7.7	8.8	16.9	2.5	14.4	12.9	0.9
Loamy sand (14)	5.1	26.0	5.9	12.2	17.5	1.4	16.0	11.6	–

(continued)

Table 3.2 (continued)

Soil texture	pH ($CaCl_2$)	Total copper (Cu_{Tot})	Cu-DTPA	C_{org} (g kg^{-1})	SSA_{EGME}[a] (m^2 g^{-1})	SSA_{BET}[b] (m^2 g^{-1})	SSA_{INT}[c] (m^2 g^{-1})	CEC[d] ($cmol_c$ kg^{-1})	$CaCO_3$ (%)
Sand (15)	6.4	28.2	9.0	8.1	8.4	1.1	7.3	10.1	–
Sandy loam (16)	6.8	28.5	10.9	9.0	26.2	4.0	22.2	12.2	–

Red line delimitates group I (contaminated) and group II (uncontaminated) soils
[a]Specific surface area: EGME—ethylene glycol monoethyl ether
[b]BET—Brunauer, Emmett and Teller N_2 adsorption
[c]INT—Internal surface area ([a]EGME – [b]BET)
[d]Cation exchange capacity according to $BaCl_2$ method (Hendershot and Duquette 1986)
[e]Not detected

3.4.3 Dynamic Phase Evaluation of Copper in Soils

3.4.3.1 Batch Tests

Copper concentrations of 0.2, 0.4, 0.6, 0.8, 1.2, 1.6, 2.0, 2.4, 2.8 and 3.2 mmol$_c$ dm^{-3} designated as initial solutions (C_i) were prepared by dissolving appropriate amounts of $Cu(NO_3)_2{\cdot}3H_2O$ p.a. in 0.010 mole $Ca(NO_3)_2{\cdot}4H_2O$ (p.a.) as a background electrolyte. The use of 0.010 mole $Ca(NO_3)_2{\cdot}4H_2O$ was an attempt to approximate the electrolyte environment of soils in the field (Jopony and Young 1994). Nitrate was chosen because of its poor ability to complex metallic cations (Sillén and Martell 1971). Solutions were added to homogenised soils at soil/solution ratios (*w*/*v*) of 1:25 into a series of polyethylene centrifuge tubes and shaken in a rotative shaker for 2 h. Suspensions were allowed to equilibrate for 22 h after which filtration followed. The concentrations of Cu of the filtrates at equilibrium concentration (C_e) were determined by FAAS method (Flame Atomic Absorption Spectrometry, Varian SpectrAA 250 plus). The solid-phase Cu was calculated as the difference between the initial Cu concentration (C_i) and that remaining in the solution after equilibration (i.e. C_e) accordingly:

$$S = (C_i - C_e)V/W \tag{3.3}$$

where

S—amount of Cu adsorbed (mmol$_c$ kg^{-1})
C_i—initial Cu concentration in solution before reaction (mmol$_c$ dm^{-3})
C_e—equilibrium Cu concentration in solution after reaction (mmol$_c$ dm^{-3})
V—volume of the solution (dm^3)
W—weight of soil sample (kg)

Langmuir and Freundlich adsorption constants were computed by using Sorption Softwares (IZOTERMY© 1993), especially elaborated for sorption studies.

3.4.3.2 Kinetics of Copper Adsorption and Desorption

Studies on the kinetics of adsorption were carried out using soil samples described earlier. Prior to these studies, soil samples were separated into two sets on the basis of copper equilibrium concentrations (C_e) at the highest initial copper concentration (C_i) as indicated in Table 3.3.

Appropriate triplicate 1.00 g of soil samples were placed in 100 cm^3 polyethylene centrifuge tubes and equilibrated with 50 cm^3 of 0.8, 1.6, 2.4 and 3.2 mmol$_c$ dm^{-3} as solution of $Cu(NO_3)_2{\cdot}3H_2O$ in 0.010 mole $Ca(NO_3)_2{\cdot}4H_2O$ used as a background electrolyte to reduce variability and simulate natural soil conditions (Table 3.4). The suspensions were vigorously shaken for 30 s at the beginning of the kinetic experiment to ensure a uniform mixture and allowed to settle for 96 h (whole kinetic adsorption time) at temperature conditions of 21.0 °C (294 K). Two (2) cm^3 aliquots for Cu determination (FAAS method) were collected

Table 3.3 Soil/solution ratios (kinetic studies) applied for appropriate soils on the basis of Cu equilibrium concentration (C_e) at the highest initial copper concentration (C_i)

Soil number	Ratio (soil/solution) (g cm^{-3})	Equilibrium concentration (C_e, mmol$_c$ dm^{-3}) at $C_i = 3.2$ mmol$_c$ dm^{-3}
1, 5, 7, 10, 12, 13	1/50	$C_e < 0.5$
2, 3, 4, 6, 8, 9, 11, 14, 15, 16	1/25	$C_e > 0.5$

Table 3.4 Amounts of copper adsorbed (%, mean value) at the end of 8 h of kinetic studies

	Initial copper concentration (C_i, mmol$_c$ dm^{-3})			
Soils	0.8	1.6	2.4	3.2
Contaminated	93	88	85	84
Soil no. 8	76	57	51	45
Uncontaminated	90	85	84	78
Soil no. 15	60	51	47	48

at the following time intervals: 0.5, 2, 8, 24, 48, 72 and 96 h. Amounts of Cu adsorbed during the adsorption kinetic phases were calculated by using Eq. (3.3).

The kinetics of desorption studies started immediately at the end of the kinetics of adsorption and were conducted on all samples used above. The supernatants at 96 h (end of kinetic adsorption study) were centrifuged at 4,000 rpm for 5 min, 5 cm^3 aliquots were collected, and the remaining decant was discarded. The polyethylene centrifuge tubes with soil samples were weighted and 50 cm^3 of 0.010 mole $Ca(NO_3)_2{\cdot}4H_2O$ Cu-free solution was placed after which the suspensions were shaken for 30 min. After this time the suspensions were centrifuged and 5 cm^3 aliquots were collected for Cu analysis. This step was repeated six more times using 0.010 mole $Ca(NO_3)_2{\cdot}4H_2O$ Cu-free solution. Amounts of copper desorbed at each desorption stage were calculated on the basis of the amounts of copper initially adsorbed (at $t = 96$ h).

3.4.3.3 Langmuir and Freundlich Models (Isotherms)

Solute adsorption mechanisms at solid surfaces are observed to bear several shapes basically divided into four main classes, namely, *S*, *L*, *H* and *C*, as widely outlined by Giles et al. (1960). With an *S*-type isotherm, the slope initially increases with adsorbate concentration, but eventually decreases and becomes zero as vacant adsorbent sites are occupied. The *L*-type (Langmuir) isotherm is characterised by a decreasing slope as concentration increases since vacant adsorption sites decrease as the adsorbent surface becomes covered. The *H*-type (high affinity) isotherm is indicative of strong adsorbate-adsorbent interactions such as inner sphere complexes. The *C*-type isotherms are indicative of a partitioning mechanism whereby adsorbate ions or molecules are linearly distributed or partitioned between the solid phase (adsorbent) and the solution phase without any specific bonding between the adsorbent and adsorbate.

Of these classes, the *L*-type isotherm is described mathematically by the Langmuir equation (Sposito 1984), as follows:

$$S = \frac{a_{\max} \cdot b \ C_e}{1 + b \ C_e} \tag{3.4}$$

where

S—amount of Cu adsorbed ($\text{mmol}_c \, \text{kg}^{-1}$)
C_e—equilibrium Cu concentration in solution after reaction ($\text{mmol}_c \, \text{dm}^{-3}$)
$a_{\max}$—adsorption maximum ($\text{mmol}_c \, \text{kg}^{-1}$)
b—bonding term relative to interaction energies ($\text{dm}^3 \, \text{mmol}_c^{-1}$)

In the linear form, Eq. (3.4) may be rewritten as follows:

$$\frac{C_e}{S} = \frac{1}{a_{\max} \cdot b} + \frac{C_e}{a_{\max}} \tag{3.5}$$

where

$a_{\max} \cdot b$ expresses the maximal buffering capacity (MBC, $\text{dm}^3 \, \text{kg}^{-1}$) exhibited by an adsorbent (soil) for a given adsorbate (Cu) when the equilibrium concentration is very low (tends to 0). The slope is equal to $\frac{1}{a_{\max}}$ and the intercept (when C_e tends closely to 0) is expressed by $\frac{1}{a_{\max} \cdot b}$. The value of the Langmuir term related to the interaction energies (b) may be calculated from the ratio of the slope and the intercept and is equal to $\frac{\text{slope}}{\text{intercept}}$ Diatta (2002). It is noteworthy that the calculation of the maximum sorption capacity $a_{\max}$ can involve errors of 50 % and more (Harter 1984) if the isotherm does not have correct Langmuir shape and only low concentration data (Schulthess and Dey 1996) is used for the calculation.

The Freundlich isotherm as an empirical equation is given by

$$S = K_F \ C_e^{\,m} \tag{3.6}$$

where

K_F—Freundlich partition constant ($\text{dm}^3 \, \text{kg}^{-1}$)
C_e—equilibrium Cu concentration in solution after reaction ($\text{mmol}_c \, \text{dm}^{-3}$)
m—sorption constant (shape intensity factor, dimensionless) expressing the slope of the adsorption isotherm. This positive-valued empirical constant is reported to lie between 0 and 1 (Elzinger et al. 1999; Diatta et al. 2003). By taking the log of Eq. (3.6), the linear form of the Freundlich equation is obtained as

$$\log S = \log K_F + m \log C_e \tag{3.7}$$

3.4.3.4 Sorption and Surface Charge Densities

Adsorption and desorption processes predominantly occur at surfaces of organic and inorganic colloids. Therefore, the expression of these processes in terms of "concentration" of substances (e.g. ions) per unit area may give more practical approach. Copper sorption densities at any point were calculated as reported by Schulte and Beese (1994) and Zehetner and Wenzel (2000):

$$\mathrm{SD_{SSA}} = \frac{SN_{\mathrm{A}}}{\mathrm{SSA}} \tag{3.8}$$

where

$\mathrm{SD_{SSA}}$—sorption density based on the specific surface area (ions m^{-2})
S—amount of Cu adsorbed ($\mathrm{mmol_c\,kg^{-1}}$)
N_{A}—Avogadro's number (6.023×10^{23} ions $\mathrm{mol^{-1}}$)
SSA—specific surface area ($\mathrm{m^2\,g^{-1}}$)

Sorption processes, especially non-specific adsorption through ion exchange, are influenced by the surface charge, of which *CEC* is a measure. The surface charge density (*SCD*) as pointed out by Poonia and Talibudeen (1977) and Laird et al. (1992) is expressed as follows:

$$\mathrm{SCD} = \frac{\mathrm{CEC}}{\mathrm{SSA}} \tag{3.9}$$

where

SCD—surface charge density ($\mathrm{mmol_c\,m^{-2}}$)
CEC—cation exchange capacity ($\mathrm{cmol_c\,kg^{-1}}$)

By dividing $\mathrm{SD_{SSA}}$ (Eq. 3.8) by the surface charge density (SCD), we obtain the charge-based sorption density ($\mathrm{SD_{CEC}}$):

$$\mathrm{SD_{CEC}} = \frac{SN_{\mathrm{A}}}{\mathrm{CEC}} \tag{3.10}$$

where
$\mathrm{SD_{CEC}}$—charge-based sorption density (ions $\mathrm{mol_c^{-1}}$)

3.4.3.5 Development of the Simplified Elovich Equation

The simplest form of Elovich equation (Allen and Scaife 1966) is generally expressed as

$$\frac{dS}{dt} = \alpha \exp^{(-\beta S)} \tag{3.11}$$

where S is the amount of Cu adsorbed at time t and α and β are constants during one experiment. The constant α can be regarded as initial rate since $\frac{dS}{dt} \to \alpha$ as $S \to 0$, i.e. a rapid adsorption not governed by the exponential law (Low 1960). The integrated form of the Eq. (3.11) gives

$$S = \frac{1}{\beta}\ln(\alpha\beta) + \frac{1}{\beta}\ln(t + t_o) \tag{3.12}$$

in which t_o is the integration constant for which sometimes the symbol k is given (to avoid confusion with the time variable). Assuming $t_o \approx 0$ indicates that no other processes besides Elovichian ones are occurring, that is, the boundary condition $S = 0$ at $t = 0$ applies (Allen and Scaife 1966; Sparks 1989). In this case the assumption $t_o \approx 0$ leads directly to the simplified equation:

$$S = \frac{1}{\beta}\ln(\alpha\beta) + \frac{1}{\beta}\ln t \tag{3.13}$$

Chien and Clayton (1980) reached the simplified Eq. (3.13) by obtaining the form:

$$S = \frac{1}{\beta}\ln(1 + \alpha\beta t) \tag{3.14}$$

from Eq. (3.12), by first assuming that the boundary condition $S = 0$ at $t = 0$ applies and then by making the assumption that $\alpha\beta t \gg 1$.

However, as pointed out by Allen and Scaife (1966), t_o can never be exactly zero without loss of the physical significance of Eq. (3.11). If this approximation is used, important information may be lost. Equation (3.13) was used to test the applicability of the simplified Elovich equation for the kinetics of Cu adsorption and desorption of the soils. Thus, a plot of S versus $\ln t$ should give a linear relationship with the slope of $\frac{1}{\beta}$ and intercept of $\frac{1}{\beta}\ln(\alpha\beta)$.

3.4.3.6 Kinetics of Copper Adsorption and Desorption and Hysteresis Approach

It was assumed that copper adsorption during the kinetic course should be monitored at some selected time intervals by sampling given equilibrium solution volumes for Cu determination. Therefore, it should be necessary to correct for this volume loss in the calculation of the amount of metal adsorbed at each subsequent sampling time (Amacher et al. 1986) as shown below:

$$\Delta S_{\text{kin}} = \left(C_{e\,\text{prev}} - C_{e\,\text{pres}}\right)\,\frac{(V - vn)}{W} \tag{3.15}$$

and

$$S_{\text{kin}} = S_{\text{prev}} + \Delta S_{\text{kin}} \tag{3.16}$$

where

ΔS_{kin}—change in amount of Cu adsorbed by the soil from one sampling time to the next ($\text{mmol}_c\,\text{kg}^{-1}$)
S_{kin}—amount of Cu adsorbed by the soil at each sampling time ($\text{mmol}_c\,\text{kg}^{-1}$)
$C_{e\,\text{prev}}$—concentration of Cu in solution at previous sampling time ($\text{mmol}_c\,\text{dm}^{-3}$)
$C_{e\,\text{pres}}$—concentration of Cu in solution at present sampling time ($\text{mmol}_c\,\text{dm}^{-3}$)
V—initial total solution volume (dm^3)
v—volume of solution removed at each sampling time for Cu determination (dm^3)
n—number of times sampled
S_{prev}—amount of Cu adsorbed by the soil at previous sampling time ($\text{mmol}_c\,\text{kg}^{-1}$)

Copper partition parameters during the kinetics of the adsorption phase were calculated at each sampling time on the basis of Freundlich equation:

$$S_{\text{kin}} = K_{\text{kin-ads}} C_e^{N_{\text{kin-ads}}} \tag{3.17}$$

where

$K_{\text{kin-ads}}$—Freundlich partition parameter for copper kinetics of adsorption ($\text{dm}^3\,\text{kg}^{-1}$)
$N_{\text{kin-ads}}$—empirical kinetics of adsorption parameter (dimensionless)

It was assumed that the starting point of the desorption isotherm is a point on the adsorption isotherm and that Freundlich adsorption and desorption coefficients are not independent as outlined by O'Connor et al. (1980) and Laird et al. (1992). Then:

$$S_{\text{kin-max}} = K_{\text{kin-ads}}\left(C_{e\text{-max}}\right)^{N_{\text{kin-ads}}} = K_{\text{kin-des}}\left(C_{e\text{-max}}\right)^{N_{\text{kin-des}}} \tag{3.18}$$

where

$S_{\text{kin-max}}$—amount of Cu adsorbed at the end of kinetic adsorption process (i.e. $t = 96$ h) ($\text{mmol}_c\,\text{kg}^{-1}$)
$C_{e\text{-max}}$—equilibrium copper concentration at the end of kinetic adsorption process (i.e. $t = 96$ h) ($\text{mmol}_c\,\text{dm}^{-3}$)

By solving Eq. (3.18) for $K_{\text{kin-des}}$, it comes

$$K_{\text{kin-des}} = K_{\text{kin-ads}}\, C_{e\text{-max}}^{\;N_{\text{kin-ads}}\left(1 - N_{\text{kin-des}}/N_{\text{kin-ads}}\right)} \tag{3.19}$$

where

$K_{kin\text{-}des}$—Freundlich partition constant for copper kinetics of desorption ($dm^3\ kg^{-1}$)
$N_{kin\text{-}des}$—empirical kinetics of desorption constant (dimensionless)

By taking into account the amount of copper adsorbed, Eq. (3.19) bears the form

$$K_{kin\text{-}des} = K_{kin\text{-}ads}{}^{N_{kin\text{-}des}/N_{kin\text{-}ads}}\, S_{max}{}^{(1-N_{kin\text{-}des}/N_{kin\text{-}ads})} \tag{3.20}$$

The ratio of Freundlich exponents ($N_{kin\text{-}des}/N_{kin\text{-}ads}$) gives an indication of adsorption/desorption hysteresis (O'Connor et al. 1980).

3.4.3.7 Activation Energies: Processes Evaluation

All adsorption and desorption kinetic studies were run in triplicate under isothermal conditions at 294 K. Activation energies were calculated based on the simplified Elovich equation using the following relationship as reported by Pavlatou and Polyzopoulos (1988) and Zhang et al. (2000):

$$E_{aa/ad} = \frac{S_{kin}RT}{E_{slope}} \tag{3.21}$$

where

$E_{aa/ad}$—activation energy for kinetics of Cu adsorption/desorption processes ($J\ mol^{-1}$)
S_{kin}—amount of Cu adsorbed or desorbed during the kinetic study ($mmol_c\ kg^{-1}$)
R—gas constant ($8.314\ J\ mol^{-1}\ K^{-1}$)
T—absolute temperature (273 K)
E_{slope}—slope of the simplified Elovich linear part of S_{kin} versus ln t plots for Cu adsorption and desorption processes ($mmol_c\ kg^{-1}$)

3.5 Evaluation and Quantification

The basic physical and chemical properties of investigated soils are reported in Tables 3.1 and 3.2. Contaminated soils (Nos. 1–8) were predominantly neutral to alkaline with pH ranging from 6.4 to 7.3 (except for soil Nos. 8 with pH = 5.0). Organic carbon (C_{org}) content fluctuated within the range 5.6–18.3 g kg^{-1}, whereas the specific surface area (SSA_{EGME}), external (SSA_{BET}) and internal (SSA_{INT}), varied from 9.4 to 72.0, 1.3 to 17.1 and 7.1 to 54.9 $m^2\ g^{-1}$, respectively. Soil particle distribution showed a general prevalence of the sand fraction varying from 420 to 860 g kg^{-1} over the silt and clay ones. Clay fractions were relatively low and were in the range 20–160 g kg^{-1} in which most of soils were characterised by clay content below 100 g kg^{-1}. The particle distribution and the relatively low organic carbon content promoted a cation exchange capacity (CEC), which varied from 9.8

to 21.8 $cmol_c\,kg^{-1}$ with quite all soils characterised by a CEC higher than 10.0 $cmol_c\,kg^{-1}$ (except for soils No. 8).

Most of uncontaminated soils (Nos. 9–16) were slightly alkaline but their pH values fluctuated from pH 5.1 to 7.6. The organic carbon content ranged within the interval 5.8–12.2 g kg^{-1}, where 75 % of soil samples did not exceed 10.0 g kg^{-1}. The clay fraction content of soils was below 100 g kg^{-1} (from 20 to 90 g kg^{-1}) with a relatively high content of the sand fraction varying from 750 to 900 g kg^{-1}. About 75 % of soils were characterised by CEC values higher than 10.0 $cmol_c\,kg^{-1}$ and relatively high specific surface area (SSA_{EGME}) with 75 % of soils possessing more than 15.0 $m^2\,g^{-1}$. The occurrence of carbonates in some soils could be attributed to frequent liming in order to reduce acidity.

The average copper content in uncontaminated soils around the world is estimated to vary from 20 to 30 mg kg^{-1}, whereas in Poland the mean level is about 12 mg kg^{-1} (Kabata-Pendias and Pendias 1992). On the other hand, there is great concern of copper contamination from metal processing plants, where a site pollution of 1,400–3,700 mg Cu kg^{-1} in Ontario was reported by Freeman and Hutchinson (1980). Local copper pollution in Poland was also pointed out by Roszyk and Szerszeń (1988) at the surroundings of Legnica copper smelter, with a copper content ranging from 25 to 9,800 mg kg^{-1}, and in the case of Głogów copper smelters, authors have reported a relatively lower level of 30–3,280 mg kg^{-1}. Copper chemistry in soils is ruled by several factors which may be classified into two main groups: (1) soil properties, in which pH plays one of the key roles (Cavallaro and McBride 1980; Harter and Lehmann 1983; Msaky and Calvet 1990), organic matter and clay content (Eriksson 1988) and redox potential (McBride 1989; Diatta 2008), and (2) copper content and its forms such as Cu^{2+}, $CuOH^+$, $Cu(OH)_2$, CuO, $CuCO_3$ and $Cu(OH)_2CO_3$ (Adriano 1986; McBride 1989).

The distribution of sorptive sites on soil colloids is generally not uniform (Sparks 1995; Wu et al. 1999; Diatta et al. 2000). For adsorption, cations are held either through electrostatic attraction, this giving rise to ion exchange with surrounding ions or by specific adsorption through surface complexation on organic and mineral substrates (Benjamin and Leckie 1981). The distribution of charges per square metre of the specific surface area (SSA) is reported in Table 3.5 for contaminated and uncontaminated soils.

As it could be observed, the higher the reactive surface, the apparently lower the number of free active sites, as pointed out especially for soils No. 4 and No. 5 (contaminated soils) with surface charge density (SCD_{SSA}) amounting to 3.5 and 3.0×10^{-3} $mmol_c\,m^{-2}$, characterised relatively by SSA of 37.4 and 72.0 $m^2\,g^{-1}$, respectively. A reverse case occurred for soils No. 3, No. 6 and No. 8, which developed apparently higher SCD_{SSA}, accordingly 8.7, 8.6 and 10.4×10^{-3} $mmol_c\,m^{-2}$. If one considers the $SCD_{SSA} = 5.0 \times 10^{-3}$ $mmol_c\,m^{-2}$ as targeted reference, it comes out that only two contaminated soils should be tentatively classified as "highly" charged and the remaining as "moderately" charged. Uncontaminated soils behave similarly as the contaminated ones, that is, the lowest the specific surface area, the apparently highest the surface charge density as shown for soils No. 11 and No. 15 with proper SCD_{SSA} values of 10.2 and 12.0×10^{-3} $mmol_c\,m^{-2}$. Referring to the targeted $SCD_{SSA} = 5.0 \times 10^{-3}$ $mmol_c\,m^{-2}$

Table 3.5 Specific surface area (SSA)-based surface charge density (SCD_{SSA}) of investigated soils

Soil texture	SCD_{SSA} (10^{-3} mmol$_c$ m^{-2})
Contaminated	
Sandy loam (1)	5.8
Loamy sand (2)	7.9
Loamy sand (3)	8.7
Sandy loam (4)	3.5
Loam (5)	3.0
Loamy sand (6)	8.6
Sandy loam (7)	6.9
Loamy sand (8)	10.4
Uncontaminated	
Loamy sand (9)	7.2
Sandy loam (10)	5.4
Loamy sand (11)	10.2
Loamy sand (12)	4.7
Loamy sand (13)	7.6
Loamy sand (14)	6.6
Sand (15)	12.0
Sandy loam (16)	4.7

as earlier reported, only two of eight soils, basically, i.e. No. 12 and No. 16 (and soil No. 10 inclusive), should be considered as "highly" charged in opposite to the "moderately" ones.

Plots of these sorption densities (SD) versus copper equilibrium concentration (C_e) resulted in at least one linear regression for most of the soils (Fig. 3.5a, b). Such plots are indicative of the occurrence of at least one pool of adsorption sites, the first highly energetic and the consecutive with lower adsorption energy.

The use of surface-based sorption density isotherms for heavy metal ecotoxicology assessment should be based on the relationship of metal ion densities at the soil colloid surface versus equilibrium concentration (C_e), as suggested by Schulte and Beese (1994) for Cd and Zehetner and Wenzel (2000) for Cu and Ni. In the current work, plots of SD_{SSA} versus C_e for the studied soils gave the best visualisation of copper adsorption process. A general analysis of plots revealed that the lower the SD_{SSA} and C_e, the highest the potential reactive sites for copper, and inversely. A simulative computation based on regressions of soils No. 1, No. 5 and No. 14, for instance, has shown that these soils attained a $C_e = 2.0$ μmol$_c$ dm^{-3} at an average SD_{SSA} of 1.2×10^{17} ions m^{-2} (soil No. 1), whereas for soils No. 5 and No. 14, it was 0.4×10^{17} and 2.3×10^{17} ions m^{-2}, respectively.

Practically a marked increase in the solution concentration following copper additions (for soils No. 1 and No. 5) is not to be expected as long as a targeted ion density (e.g. 2.3×10^{17} ions m^{-2} for soil No. 14) limit has not been reached. These approaches emphasise on the possibility of copper equilibrium monitoring based on sorption densities' application.

Copper adsorption and its equilibrium concentrations, both equilibria are pH dependent, pointed out that soils exhibited in fact more charges than evaluated on

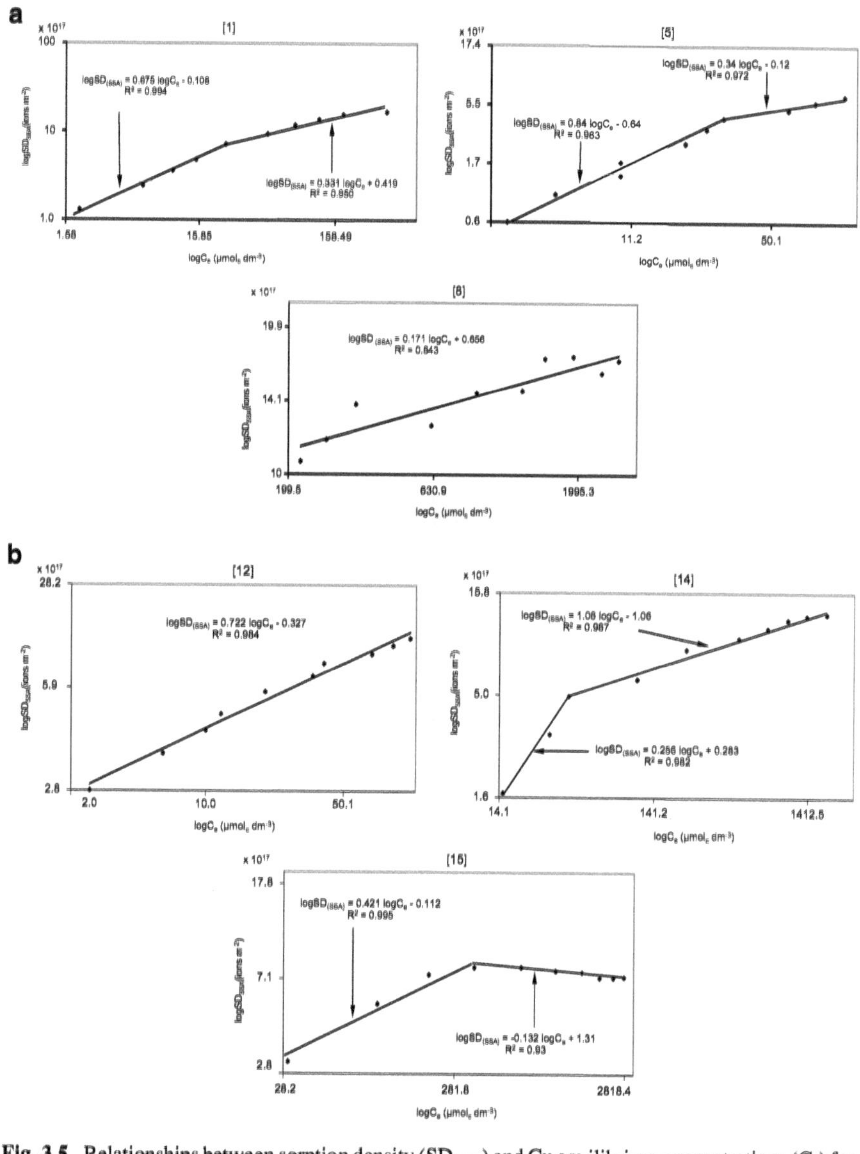

Fig. 3.5 Relationships between sorption density (SD_{SSA}) and Cu equilibrium concentrations (C_e) for selected (**a**) contaminated soils (Nos. 1, 5 and 8) and (**b**) uncontaminated soils (Nos. 12, 14 and 15)

the basis of SCD_{SSA} (Table 3.5). Therefore the complex reaction of copper ions with deprotonated surface OH and COOH groups as bidentate and monodentate ligands could be expectable as pointed out by Schindler et al. (1976) and Davis and Leckie (1978). The adsorption of copper at energetically high sorptive sites can be formulated:

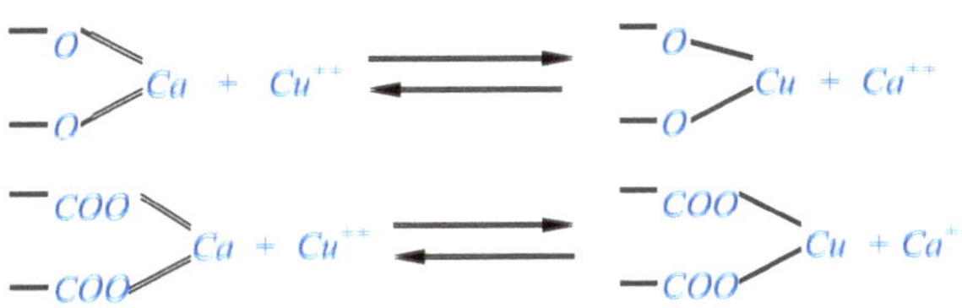

The saturation of the energetically high sorptive sites runs with the emergence of energetically low sorptive sites, and the latter ones may be presumably attributed to isomorphous substitution charge (Forbes et al. 1976; Bowden et al. 1977) as indicated below:

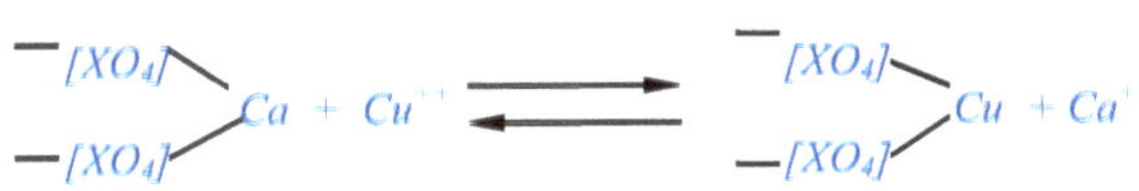

$[XO_4]$—exchange site due to isomorphous substitution charge, X to represent Si, Al

Such approach is in agreement with the reports of Msaky and Calvet (1990), who stated that copper adsorption systematically decreased as the surface coverage increased. So it may be deduced that copper ions were primarily attracted by high energetically charged sites, what coincided to the first intensive adsorption process leading to a marked removal of copper ions from the equilibrium solution. These sorption mechanisms showed some limitation of the formerly applied Langmuir and Freundlich isotherms for soil systems (Diatta et al. 2004, 2012).

Langmuir constants, i.e. a_{max} (adsorption maximum) and b (constant related to bonding energies) spread widely, mostly in the case of b values (Table 3.6). For contaminated soils, it was noted that the lowest b value was obtained for the soil No. 8 and accounted for 6.7 $dm^3\ mmol_c^{-1}$, whereas the highest one, amounting for 40.2 $dm^3\ mmol_c^{-1}$, was attributed to soil No. 7. It is also important to note that both a_{max} and b constants directly influence the whole soils' maximal buffering capacities (MBC) towards Cu. Soil No. 5 with the highest a_{max} (111.7 $mmol_c\ kg^{-1}$) was characterised by an MBC of 2,412 $dm^3\ kg^{-1}$ with a respective $b = 21.6$ $dm^3\ mmol_c^{-1}$, whereas soil No. 7 at $a_{max} = 81.5$ $mmol_c\ kg^{-1}$ has developed the highest maximal buffering capacity of 3,276 $dm^3\ kg^{-1}$ (for $b = 40.2$ $dm^3\ mmol_c^{-1}$). Soil No. 8 deserves special attention, since all sorption constants are strikingly low as a result of combined low pH, organic carbon content (C_{org}) and specific surface area on one hand and high copper content (1,048.0 $mg\ kg^{-1}$), on the other hand. Uncontaminated soils were also characterised by a great variation of both Langmuir constants, which for adsorption maximum—a_{max}—ranged from 14.0 to 115.1 $mmol_c\ kg^{-1}$ and for the bonding energy term, b, from 11.4 to 26.0 $dm^3\ mmol_c^{-1}$ and exceptionally to 236.6 $dm^3\ mmol_c^{-1}$, obtained for soil No. 15 with a resulting high buffering capacity (MBC = 2,454 $dm^3\ kg^{-1}$). Assuming that small amounts of copper should be strongly fixed by soil colloids, therefore, the notably low a_{max} simultaneously with high b explains the case encountered for this soil. The involved mechanisms

Table 3.6 Langmuir one-site and Freundlich adsorption parameters calculated for the contaminated and uncontaminated soils

Statistical description	Langmuir one-site a_{max} (mmol$_c$ kg^{-1})	b (dm^3 mmol$_c$$^{-1}$)	MBC[a] (dm^3 kg^{-1})	R^2	Freundlich K_F (dm^3 kg^{-1})	m[b] (dimensionless)	R^2
Contaminated							
Min	27.5	6.7	185.0	0.972	7.4	0.35	0.836
Max	111.7	40.2	3276.0	0.998	704.9	0.87	0.981
Mean	60.9	22.0	1492.0	0.993	166.2	0.51	0.915
SD	26.9	10.8	1087.7	0.009	230.9	0.18	0.046
Uncontaminated							
Min	10.4	11.4	408.0	0.975	10.3	0.17	0.825
Max	115.1	236.6	2454.0	0.998	569.4	0.84	0.966
Mean	54.2	47.4	1405.1	0.992	131.2	0.44	0.913
SD	33.6	76.6	764.8	0.009	184.7	0.21	0.052

SD standard deviation
[a]$a_{max}\cdot b$ = MBC (Maximal Buffering Capacity)
[b]Adsorption constant

resulted basically from the heterogeneous character of soil sorptive surfaces (Sokołowska 1989; Sparks 1995; Diatta et al. 2003).

The Freundlich constant m expresses the intensity of isotherms slope (Table 3.6). Most of m values were $\ll 1$, that means isotherm shapes were less steepy (occurrence of curvature), typically for soils in which most of high energetic sorptive sites have been just occupied by copper ions, the remaining ones being characterised by low adsorption energies with increasing surface coverage. Such cases were frequently reported also for Langmuir isotherms (Wu et al. 1999). Since slope intensity clearly indicated differences in the capacities of soils for copper adsorption, it appeared necessary to find out any relationships which may be helpful in determining Cu adsorption maximum as equated below:

$$a_{max} = 137.3m - 9.4, \quad R^2 = 0.82, \text{ for contaminated soils}$$

$$a_{max} = 159.7\,m - 15.7, \quad R^2 = 0.98, \text{ for uncontaminated soils}$$

where a_{max} is the Langmuir adsorption maximum constant and m is the Freundlich slope intensity.

Table 3.7 Stepwise variable selection for simplified Elovich rate constant (β) versus selected soil properties for contaminated and uncontaminated soils

Simplified Elovich rate constant (β) for	Contaminated		Uncontaminated	
	Stepwise variable selection			
Adsorption at [a]$C_i = 2.4$ mmol$_c$ dm^{-3}	$\beta = 0.01$ CEC + 0.01 Ca$_{ex}$ − 0.012 C_{org} + 0.20 pH − 0.8	[b]$R^2 = 59$	$\beta = 0.04$ Ca$_{ex}$ − 0.10 C_{org} − 0.09 pH − 0.005 Clay + 1.9	$R^2 = 95$
Desorption at $C_i = 2.4$ mmol$_c$ dm^{-3}	$\beta = 0.02$ Ca$_{ex}$ − 0.01 CEC − 0.04 C_{org} + 0.002 Clay + 0.6	$R^2 = 91$	$\beta = 0.06$ CEC − 0.06 Ca$_{ex}$ − 0.11 pH − 0.004 Clay + 0.9	$R^2 = 78$

[a]Initial copper concentration
[b]Coefficients of determination, R^2 (%)

3.6 Kinetic Reaction (Adsorption/Desorption) Rate Constants

Kinetics is the study of the time-dependent processes, which in heterogeneous systems such as soil components occur simultaneously and consecutively over a wide time scale (Amacher 1991). The time-interval set for kinetics of copper adsorption in this study ranged from 0.5 up to 96 h. The 8 h time used for the estimation showed that amounts of adsorbed copper varied from 84 to 93 % for contaminated soils (with 57 % for soil No. 8) and from 78 to 90 % for uncontaminated soils (with soil No. 15 about 52 %). Experimental data fitted the simplified Elovich equation that proves its utility for empirical prediction. According to Sparks (1989), the chemical significance of these constants has not been clearly resolved β constants, but they can be used for comparing adsorption rates in different soils (Chien and Clayton 1980; Elkhatib et al. 1992; Dang et al. 1994; Taylor et al. 1995).

The constants β calculated from the slopes $\frac{1}{\beta}$ of the Elovich equation were also a function of both soil textural group and initial copper concentration. The values of the slopes $\frac{1}{\beta}$ increased gradually when increasing copper concentrations in the solution with a simultaneous decrease of β values. When comparing contaminated and uncontaminated soils, it was found that the decrease in β between $C_i = 0.8$ and 1.6 mmol$_c$ dm^{-3} averaged 69 % in comparison with the relatively small decrease (31 %) observed between $C_i = 2.4$ and 3.2 mmol$_c$ dm^{-3}. The "sensitivity" of the slope and hence constants β relied on the degree of surface coverage by copper ions. The relatively high coefficients of determination (R^2) resulting from multiple variable interactions (Table 3.7) on the constant β pointed out the fact that copper chemistry of contaminated as well as uncontaminated soils should be similarly described by simplified Elovich equation, which is generally considered to characterise a number of different processes including bulk diffusion.

Table 3.8 Freundlich kinetic adsorption and desorption (at $C_i = 3.2$ mmol$_c$ Cu dm^{-3}) parameters at adsorption time $t = 96$ h and hysteresis indices based on the ratio of Freundlich exponents[a] for adsorption and desorption isotherms for contaminated and uncontaminated soils

Statistical description	Adsorption $K_{kin\text{-}ads}$	$N_{kin\text{-}ads}$	Desorption $K_{kin\text{-}des}$	$N_{kin\text{-}des}$	Hysteretic index ($N_{kin\text{-}des}/N_{kin\text{-}ads}$)
Contaminated					
Min	8.0	0.20	6.8	0.37	1.21
Max	716.7	0.33	566.6	1.02	5.04
Mean	160.5	0.25	135.4	0.87	3.65
SD	235.2	0.05	184.0	0.22	1.33
Uncontaminated					
Min	12.8	0.21	11.2	0.60	1.78
Max	623.8	0.61	500.7	1.60	4.66
Mean	137.7	0.32	111.5	0.96	3.30
SD	203.0	0.14	162.3	0.31	1.13

SD standard deviation
[a]For $K_{kin\text{-}ads}$; $K_{kin\text{-}des}$; $N_{kin\text{-}ads}$; $N_{kin\text{-}des}$ parameters, refer to Eqs. (3.17, 3.19 and 3.20)

3.7 Hysteresis of Copper in Investigated Soils

Ion exchange reactions in soils are by nature reversible, but this reversibility may appear to be apparently irreversible (Van Bladel and Laudelout 1967), or quite irreversible, that means occurrence of hysteresis (Fripiat et al. 1965; Verburg and Baveye 1994; Zhu and Selim 2000). The total amount of copper desorbed at the end of the desorption process was relatively high and averaged 22 and 26 % at $C_i = 2.4$ and 3.2 mmol$_c$ dm^{-3}, respectively, for contaminated soils, except for soil No. 1, where copper was less desorbed (i.e. 6.0 and 10 % at the respective C_i levels). Interestingly higher desorption rates, i.e. 28 and 35 %, were obtained even for uncontaminated soils at $C_i = 2.4$ and 3.2 mmol$_c$ dm^{-3}, respectively, whereas soil No. 12 desorbed accordingly low copper in the order 4.0 and 8 %. On the other hand Wu et al. (1999) have found that 35 % of formerly adsorbed copper was desorbed from a fine clay fraction, while 27 and 25 % of copper were desorbed from medium and coarse clay, respectively. The relatively high levels of copper desorbed in the current study did not confirm the occurrence of any irreversibility (hysteresis) as earlier reported by Ainstworth et al. (1994). The presence of carbonates in the case of soils No. 1 and No. 12 resulted in a limited copper desorbability (i.e. limited irreversibility).

According to O'Connor et al. (1980) and Barriuso et al. (1994), if there is no hysteresis the ratio $N_{kin\text{-}des}/N_{kin\text{-}ads} = 1$ and $K_{kin\text{-}ads} > K_{kin\text{-}des}$ and in case of negative hysteresis, i.e. enhanced desorbability, the ratio $N_{kin\text{-}des}/N_{kin\text{-}ads} > 1$ and $K_{kin\text{-}ads} > K_{kin\text{-}des}$, as also observed in this study by the fact that the increase of $N_{kin\text{-}des}/N_{kin\text{-}ads}$ induced an increase of $K_{kin\text{-}ads}$ parameter (Table 3.8). Such conditions could explain the relatively high copper desorbability (exception for soils No. 1 and No. 12). As surface charge density (SCD$_{SSA}$) increases, the $N_{kin\text{-}des}/N_{kin\text{-}ads}$ ratio decreases, implying that copper was more tightly adsorbed by high charge density soil colloids than by low charge density ones (Wu et al. 1999; Diatta et al. 2002). This is evidenced by the negative slopes (SCD_{SSA} versus $N_{kin\text{-}des}/N_{kin\text{-}ads}$) as follows:

$$SCD_{SSA} = -1.17(N_{kin\text{-}des/kin\text{-}ads}) + 11.12, \quad R^2 = 0.32, \text{ for contaminated soils}$$

$$SCD_{SSA} = -1.77(N_{kin\text{-}des/kin\text{-}ads}) + 13.13, \quad R^2 = 0.57, \text{ for uncontaminated soils}$$

For $N_{kin\text{-}des}/N_{kin\text{-}ads}$ parameters, refer to Eqs. (3.17, 3.19 and 3.20).

Therefore, if we assume the pool of high-density charges to be pH dependent, since 88 % of contaminated as well as uncontaminated soils exhibited pH > 6.0, the high-density charges could probably be limited in number.

3.8 Activation Energies for Copper Geochemical Processes

The reaction rate theory assumes that colliding molecules (e.g. reactions between solution ions and ions held in the soil sorptive complex) must be in a high energy before a reaction occurs (El-Sayed et al. 1970; Diatta et al. 2000). This energy of activation measures the magnitude of the forces that must be overcome during the process of adsorption and desorption reactions (Elkhatib et al. 1993). Energies of activation for copper adsorption (E_{aa}) and desorption (E_{ad}) are grouped in three operational ranges, respectively, for C_i levels (0.8, 1.6, 3.2 $mmol_c\,dm^{-3}$), both for contaminated and uncontaminated soils: $E_{aa} < 50$, $51 < E_{aa} < 150$, $E_{aa} > 151$ kJ mol^{-1}, namely, range I (R_I), range II (R_{II}) and range III (R_{III}), respectively, as follows:

$C_i = 0.8$ $mmol_c\,dm^{-3}$	
Contaminated	12 % of soil belong to R_I, 63 % to R_{II} and 25 % to R_{III}
Uncontaminated	25 % of soils belong to R_I, 63 % to R_{II} and 12 % to R_{III}
$C_i = 1.6$ $mmol_c\,dm^{-3}$	
Contaminated	25 % of soils belong to R_I, 63 % to R_{II} and 12 % to R_{III}
Uncontaminated	63 % of soils belong to R_I, 25 % to R_{II} and 12 % to R_{III}
$C_i = 3.2$ $mmol_c\,dm^{-3}$	
Contaminated	75 % of soils belong to R_I, 25 % to R_{II}
Uncontaminated	100 % of soils belong to R_I

According to Barrow and Whelan (1989) and Barrow (1998), the E_{aa} for investigated reactions varied from 80 to 100 kJ mol^{-1}. The activation energy of a diffusion-controlled process in a solution is about 25 kJ mol^{-1} (Sparks 1986). Activation energies for copper adsorption (E_{aa}) ranged from 11.81 to 453.81 kJ mol^{-1} and for copper desorption (E_{ad}), between 3.79 and 13.59 kJ mol^{-1} (Fig. 3.6a, b).

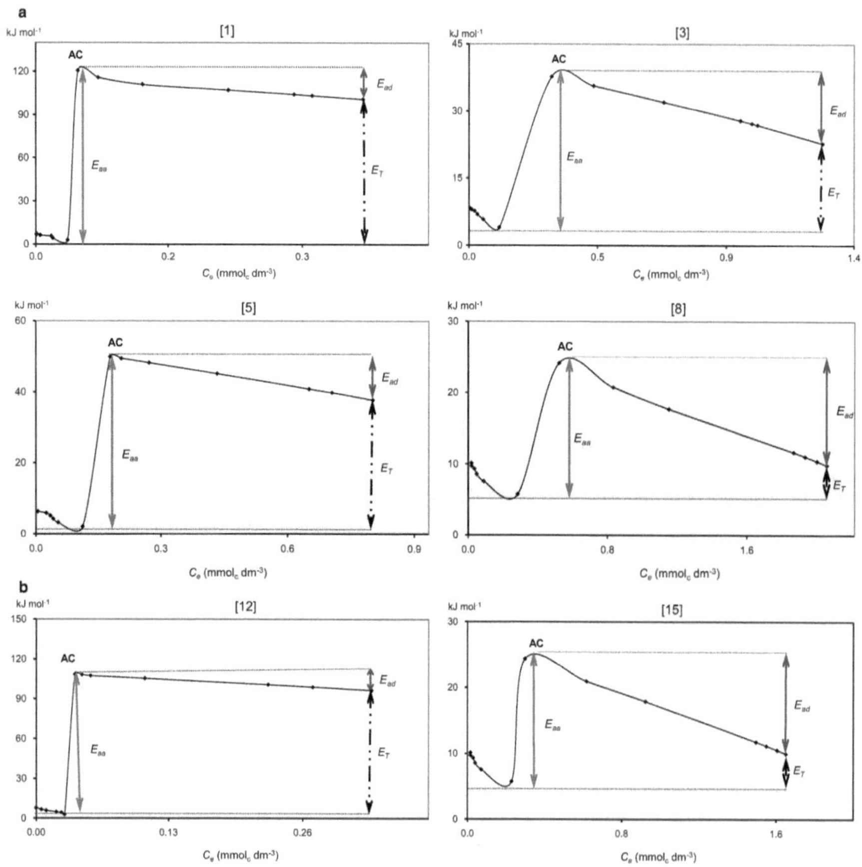

Fig. 3.6 (a) Energy flow for Cu adsorption/desorption at $C_i = 2.4$ mmol$_c$ dm^{-3}, AC = activated complex, E_{aa} and E_{ad} = adsorption and desorption activation energy, respectively. E_T = energy transfer (energy needed for Cu to move from the solution to the solid phase). (**a**) Contaminated soils (Nos. 1, 3, 5 and 8) and (**b**) Uncontaminated soils (Nos. 12 and 15)

The energy transfer (E_T) expresses a measure of the amount of energy needed to transfer copper ions from the solution to activated complex, i.e. soil solid state (Ogwada and Sparks 1986). The E_T value outcomes for soils No. 1 and No. 12 amounted for 97.78 and 93.65 kJ mol^{-1}, respectively, and were notably higher than those obtained for soils No. 3, No. 5, No. 8 and No. 15, amounting for 18.96, 39.71, 5.85 and 4.99 kJ mol^{-1}, respectively. A suggestive endothermic reaction could have been occurring in these soils. The endothermic nature of the reaction and the trend with surface coverage (a decrease of E_{aa} and E_{ad} with increasing initial copper concentration) may be attributed to the potential energy available for adsorption and the fact that most energetic sites were occupied first (Msaky and Calvet 1990). Since the latter become limited as surface coverage increases, adsorbed copper ions at lower interaction energy will be readily desorbable due to weak attraction. So relatively low energy is necessary for copper ions to transfer back to the ambient soil solution.

Investigated soils both contaminated and uncontaminated were mostly characterised by relatively weak buffering properties towards copper ions and should be managed with care: a reduction of copper inputs may increase its E_{ad} and as a result slow down copper transfer to soil solution. Mass transfer and diffusion reaction control seem to be of prime order. The multiple-phase evaluation pointed out in the current study showed the complexity of mechanisms involved in copper dynamics in soil systems.

3.9 Summary

Geochemical processes involve reactions which mediate and control the dissolution/release and (re)formation/sorption of mineral elements at geological as well as soil levels. The role of trace elements, copper among others, in soil chemical changes is still a matter of detailed studies. Copper-bearing minerals (chalcopyrite, $CuFeS_2$; chalcocite, Cu_2S, and bornite, Cu_5FeS_4) predominate in nature, although the richest Cu concentrations are related to the secondary weathering stage compounds [azurite, $Cu_3(CO_3)_2(OH)_2$, and malachite, $Cu_2CO_3(OH)_2$, for instance]. The release of Cu from these minerals may operate via CO_2 injection into deep geological structures or mining activities. The latter ones initiate slow or intensive spread of Cu compounds into the environment.

For decades, geochemists have been facing challenges related to the quantification of both natural (uncontaminated) and anthropogenic (contaminated) soil Cu concentrations. The current chapter was enriched with a case study which outlines in detail multiple steps involved in the evaluation of copper geochemical processes. The approach suggested herein takes into account the medium concentration, time dependency, model application and data fitting for Cu phase characterisation. Equilibria, exchange mechanisms, retention versus release, hysteresis as well as Cu complex formation (energy transfer) and the resulting (geo)environmental concerns were treated with substantial care.

References

Adriano DC (1986) Trace elements in the environment. Springer-Verlag, New York, p 553

Ainstworth CC, Pilon LJ, Gassman PL, Van Der Sluys W (1994) Cobalt, cadmium and lead sorption to hydrous iron oxide: residence time effect. Soil Sci Soc Am J 58:1615–1623

Allen JA, Scaife PH (1966) The Elovich equation and chemisorption kinetics. Aust J Chem 19:2015–2023

Alloway BJ (1995) Heavy metals in soils, 2nd edn. Blackie Academic & Professional, London, p 360

Amacher MC (1991) Methods of obtaining and analysing kinetic data. In: Sparks DL, Suarez DL (eds) Rates of soil chemical processes. SSSA, Madison WI, pp 19–59, Spec. Publ. Nr 27. Soil Sci. Soc. Am

Amacher MC, Kotuby-Amacher J, Selim HM, Iskandar IK (1986) Retention and release of metals by soils. Evaluation of several models. Geoderma 38:131–154

Aringhieri R, Carrai P, Petruzzelli G (1985) Kinetics of Cu^{2+} and Cd^{2+} adsorption by an Italian soil. Soil Sci 139(3):197–204

Baker DE (1990) Copper. In: Alloway BJ (ed) Heavy metals in soils. Halsted, Blackie, NY, USA, pp 151–176

Barriuso E, Laird DA, Koskinen CW, Dowdy RH (1994) Atrazine desorption from smectites. Soil Sci Soc Am J 58:1632–1638

Barrow NJ (1998) The effect of time and temperature on the sorption of cadmium, zinc, cobalt and nickel by a soil. Aust J Soil Res 36:941–950

Barrow NJ (1999) The four laws of soil chemistry: the Leeper lecture 1998. Aust J Soil Res 37:787–829

Barrow NJ, Whelan BW (1989) Testing a mechanistic model. VIII. The effect of time and temperature of incubation on the sorption and subsequent desorption of selenite and selenate by a soil. J Soil Sci 40:29–37

Basta NT, Tabatabai MA (1992) Effect of cropping systems on adsorption of metals by soils: II. Effect of pH. Soil Sci 153:195–204

Benjamin MM, Leckie JO (1981) Conceptual model for metal-ligand-surface interactions during adsorption. Environ Sci Tech 15:1050–1056

Boehringer IM (1980) Echanges cationique avec des métaux lourds sur la tourbe acide. Thèse de Doctorat Ecole Polytechnique, Fédérale, Zurich; No. 6700. p 131

Bowden JW, Posner AM, Quirk JP (1977) Ionic adsorption on variable charge mineral surfaces. Theoretical-charge development and titration curves. Aust J Soil Res 15:121–136

Brümmer GW, Gerth J, Herms U (1986) Heavy metal species, mobility and availability in soils. Z Pflanzenernehr Bodenk 149:382–398

Bunzl K, Schmidt W, Sansoni B (1976) Kinetics of ion exchange in soil organic matter. IV. Adsorption and desorption of Pb^{2+}, Cu^{2+}, Cd^{2+}, Zn^{2+} and Ca^{2+} by peat. J Soil Sci 27:32–41

Carter DL, Mortland MM, Kemper WD (1986) Specific surface. In: Klute A (ed) Methods of soils analysis, part I—physical and mineralogical methods, 2nd edn. Am Soc Agron, Madison, WI, pp 413–423

Cavallaro N, McBride MB (1978) Copper and cadmium adsorption characteristics of selected acid and calcareous soils. Soil Sci Soc Am J 42:550–556

Cavallaro N, McBride MB (1980) Activities of Cu^{2+} and Cd^{2+} in soil solutions as affected by pH. Soil Sci Soc Am J 44:729–732

Chien SH, Clayton WR (1980) Application of the Elovich equation to the kinetics of phosphate release and sorption in soils. Soil Sci Soc Am J 44:265–268

Chudziński B (1995) Assessment of heavy metals phytotoxicity on the basis of their availability, yields and quality of cereals. PhD Thesis, IOR Poznan (Poland). 166 p. (in Polish).

Cox PA (1989) The elements. Their origin, abundance and distribution. Oxford University Press, Oxford, p. 207

Dang YP, Dalal RC, Edwards DG, Tiller KG (1994) Kinetics of Zinc desorption from vertisols. Soil Sci Soc Am J 58:1392–1399

Davis JA, Leckie JO (1978) Surface ionization and complexation at the oxide-water interface. II. Surface properties of amorphous iron oxyhydroxide and adsorption of metal ions. J Colloid Interface Sci 67:90–107

de Jong E (1999) Comparison of three methods of measuring surface area of soils. Can J Soil Sci 79:345–351

Diatta JB (2002) Evaluation of adsorption parameters and charge densities of some selected soils: application to lead. EJPAU Environ Dev 5(Issue 2). http://www.ejpau.media.pl/series/volume5/issue2/environment/art-03.html

Diatta JB (2008) Mutual Cu, Fe and Mn solubility control under differentiated soil moisture status. J Elementol 13(4):473–489
Diatta JB, Kociałkowski WZ, Grzebisz W (2000) Copper distribution and quantity-intensity parameters of highly contaminated soils in the vicinity of a copper plant. Polish J Environ Stud 9(5):355–361
Diatta JB, Kociałkowski WZ, Grzebisz W (2003) Lead and zinc partition coefficients of selected soils evaluated by Langmuir, Freundlich and linear isotherms. Commun Soil Sci Plant Anal 34 (17 & 18):2419–2439
Diatta JB, Grzebisz W, Wiatrowska K (2004) Competitivity, selectivity, and heavy metals-induced alkaline cation displacement in soils. Soil Sci Plant Nutr 50(6):899–908
Diatta JB, Komisarek J, Wiatrowska K (2012) Evaluation of heavy metals competitive sorption and potential mobility on the basis of Cu/Cd and Zn/Pb binary systems. Fresenius Environ Bullet 21(5):1105–1109
Drobek L, Bukowska M, Borecki T (2008) Chemical aspects of CO_2 sequestration in deep geological structures. Gospod Surowcami Min 24(3/1):425–438
Dube A, Zbytniewski R, Kowalkowski T, Cukrowska E, Buszewski B (2001) Adsorption and migration of heavy metals in soil. Polish J Environ Stud 10(1):1–10
Duffus JH (2002) Heavy metals—a meaningless term. IUPAC technical report. Pure Appl Chem 74(5):793–807
Elkhatib EA, Elshebiny GM, Balba MA (1992) Comparison of four equations to describe the kinetics of lead desorption from soils. Z Pflanzenernahr Bodenk 155:285–291
Elkhatib EA, Elshebiny GM, Elsubruiti GM, Balba MA (1993) Thermodynamics of lead sorption and desorption in soils. Z Pflanzenernahr Bodenk 156:461–465
El-Sayed MH, Burau RG, Babcock KL (1970) Thermodynamics of copper (II)—calcium exchange on bentonite clay. Soil Sci Soc Am Proc 34:397–400
Elzinger EJ, van Grinsven MJ, Swartjes FA (1999) General of Freundlich isotherms for cadmium, copper and zinc in soils. Eur J Soil Sci 50:139–149
Eriksson JE (1988) The effects of clay, organic matter and time on adsorption and plant uptake of cadmium added to the soil. Water Air Soil Pollut 40:359–373
Evans RL, Jurinak JJ (1976) Kinetics of phosphate release from a desert soil. Soil Sci 121:205–211
Forbes EA, Posner AM, Quirk JP (1976) The specific adsorption of divalent Cd, Co, Cu, Pb and Zn on goethite. J Soil Sci 27:154–166
Forrester SD, Giles CH (1971) From manure heaps to monolayers: the earliest development of solute-solid adsorption studies. Chem Ind, 1314–1321
Freedman B, Hutchison TC (1980) Pollutant inputs from the atmosphere and accumulations in soils and vegetation near a nickel-copper smelter at Sudbury, Ontario. Canada Can J Bot 58:108–132
Fripiat JJ, Cloos P, Poncelet A (1965) Comparison entre les propriétés d'échange de la montmorillonite et d'une résine vis-à-vis des cations alcalins et alcalino-terreux. I. Réversibilité des processus. Bull Soc Chim Fr, 208–215
Gaines GL, Thomas HC (1953) Adsorption studies on clay minerals. II. A formulation of a thermodynamics of exchange-adsorption. J Chem Phys 21:714–718
Gapon EN (1933) Theory of exchange adsorption in soils. J Gen Chem (USSR) 3(2):144–152
Geiger G, Federer P, Sticher H (1993) Reclamation of heavy metal contaminated soils—field studies and germination experiments. J Environ Qual 22:201–207
Geochemical Atlas of Legnica-Głogów Copper District (1999) State Geological Institute. Scale 1:250 000. Warszawa (in Polish)
Giles CH, MacEwan TH, Nakhwa SN, Smith D (1960) Studies in adsorption. Part XI. A system of classification of solution adsorption isotherms, and its use in diagnosis of adsorption mechanisms and in measurements of specific surface areas of solids. J Chem Soc 786:3973–3993
Gregg SJ, Sing KSW (1967) Adsorption, surface area and porosity. Academic, London, UK, pp 44–50

Grzebisz W, Chudziński B (1996) Crop aided biological remediation of soils contaminated by a copper smelter. In: Gărban Z, Drăgan P (eds) Proceedings of second international symposium on metal elements in environment, medicine and biology, Publishing House "Eurobit" Timişoara—Romania, 1st edition. 1996. ISBN 973-9336-15-9, pp 63–72

Grzebisz W, Kociałkowski WZ, Chudzinski B (1997) Copper geochemistry and availability in cultivated soils contaminated by a copper smelter. J Geochem Explor 58:301–307

Harter RD (1984) Curve-fit errors in Langmuir adsorption maxima. Soil Sci Soc Am J 48:749–752

Harter R, Lehmann R (1983) Use of kinetics for the study of exchange reactions in soils. Soil Sci Soc Am J 47:666–669

Helios-Rybicka E, Jędrzejczyk B (1995) Preliminary studies on mobilisation of copper and lead from contaminated soils and readsorption on competing sorbents. Appl Clay Sci 10 (3):259–268

Hendershot WH, Duquette M (1986) A simple barium chloride method for determining cation exchange capacity and exchangeable cations. Soil Sci Soc Am J 50:605–608

Hodgson JF, Lindsay WL, Trierweiler JF (1966) Micronutrient cation complexing in soil solution. II. Complexing of zinc and copper in displacing solution from calcareous soils. Soil Sci Am Proc 30:723–726

Huijgen WJJ, Comans RNJ, Witkamp GJ (2006) Mechanisms of aqueous wollastonite carbonation as a possible CO_2 sequestration process. Chem Eng Sci 61:4242–4251

Ibsen KH, Jacobsen FL (1996) The Linde Structure, Denmark an example of a CO_2-depository with a secondary chalk cap rock. Energ Convers Manag 37(6–8):1161–1166

International Standard (1995) Soil quality—extraction of trace elements soluble in aqua regia, Geneva ISO11466

Izotermy© (1993) Wersje 1.0; 1.1; 1.2. Biuro Projektów Informatyki, A. Ratajczak, Poznań

Jopony M, Young SD (1994) The solid-solution equilibria of lead and cadmium in polluted soils. Eur J Soil Sci 45:59–70

Kabata-Pendias A (1978) The impact of copper mining and industrial activity of Lower Silesia on the chemical composition of plants. Final report—IUNG, p 173

Kabata-Pendias A (1993) Behavioural properties of trace metals in soils. Appl Geochem Suppl Issue 2, 3–9

Kabata-Pendias A, Pendias H (1992) Trace elements in soils and plants, 2nd edn. CRC Press, Boca Raton, p 365

Kabata-Pendias A, Pendias H (1999) Biogeochemia pierwiastków śladowych. Wydawnictwo Naukowe PWN, Warszawa

Kabata-Pendias A, Motowicka-Terelak T, Piotrowska M, Terelak H, Witek T (1993) Ocena stopnia zanieczyszczenia gleb i roślin metalami ciężkimi i siarką. IUNG-Puławy, str. 20

Kinniburgh DG (1986) General purpose adsorption isotherms. Environ Sci Technol 20:895–904

Kondracki J (1978) Geografia fizyczna Polski. PWN, Wyd. II, Warszawa, p 463

Kociałkowski WZ, Diatta JB, Grzebisz W (1999) Assessment of lead sorption by acid agroforest soils. Pol J Environ Stud 8(6):403–408

Konstantynowicz E (1979) Geologia surowców mineralnych. Tom 2. Złoża rud metali, Uniwersytet Śląski, Katowice

Lai TM, Mortland MM (1968) Cationic diffusion in clay minerals: I. Homogeneous and heterogeneous systems. Soil Sci Soc Am Proc 32:56–61

Laird DA, Barriuso E, Dowdy RH, Koskinen WC (1992) Adsorption of atrazine on smectites. Soil Sci Soc Am J 56:62–67

Lindsay WL, Norvell WA (1978) Development of a DTPA soil test for zinc, iron, manganese and copper. Soil Sci Soc Am J 42:421–428

Lindsay WL (1979) Chemical equilibria in soils. Wiley, New York, p. 449

Liu C, Huang PM (2000) Kinetics of phosphate adsorption on iron oxide formed under the influence of citrate. Can J Soil Sci 80:445–454

Loeppert RH, Suarez DL (1996) Carbonate and gypsum. In: Bartels JM et al (eds) Methods of soil analysis: part 3. Chemical methods, vol 5, 3rd edn. ASA and SSSA, Madison, WI, pp 437–474

Logan TJ (1990) Chemical degradation of soil. Adv Soil Sci 11:187–219
Logan EM, Pulford ID, Cook GT, McKenzie AB (1997) Complexation of Cu^{2+} and Pb^{2+} by peat and humic acid. Eur J Soil Sci 48:685–696
Low WJD (1960) Kinetics of chemisorption of gases on solids. Chem Rev 60:267–312
Maqueda C, Undabeytia T, Morrillo E (1998) Retention and release of copper on montmorillonite as affected by the presence of a pesticide. J Agric Food Chem 46(3):1200–1204
Markert B, Friese K (eds) (2000) Trace elements. Their distribution and effects in the environment. Elsevier, Amsterdam
Markert B, Breure A, Zechmeister H (eds) (2003) Bioindicators & biomonitors. Principles, concepts and applications. Elsevier, Amsterdam
Mattigod SV, Sposito G (1979) Chemical modeling of trace metal equilibria in contaminated soil solutions using the computer program GEOCHEM. Chemical modeling in aqueous systems. Henne E. A. Editions, Washington D.C. Am Chem Soc 93:837–856
McBride MB (1989) Reactions controlling heavy metal solubility in soils. Adv Soil Sci 10:1–56
McBride MB (1994) Environmental chemistry of soils. Oxford University Press, New York, p 406
McBride MB (2001) Cupric ion activity in peat soil as a toxicity indicator for maize. J Environ Qual 30:78–84
McLaren RG, Crawford DV (1973a) Studies on soil copper. I. The fractionation of copper in soils. J Soil Sci 24(2):172–181
McLaren RG, Crawford DV (1973b) Studies on soil copper. II. The specific adsorption of copper by soils. J Soil Sci 24(4):443–452
Msaky JJ, Calvet R (1990) Adsorption behaviour of copper and zinc in soils: influence of pH on adsorption characteristics. Soil Sci 150(2):513–522
Mulder J, Cresser MS (1994) Soil and soil solution chemistry. In: Moldan B, Cerny J (eds) Biogeochemistry of small catchments: a tool for environmental research. Wiley, Chichester, pp 107–131
Murali V, Aylmore LAG (1983a) Competitive adsorption during solute transport in soils. 1: mathematical models. Soil Sci 135(3):143–150
Murali V, Aylmore LAG (1983b) Competitive adsorption during solute transport in soils. 2: simulation of competitive adsorption. Soil Sci 135(4):203–213
Nelson DW, Sommers LE (1996) Total carbon, organic carbon and organic matter. In: Sparks DL (ed) Methods of soil analysis. Part 3. Chemical methods. SSSA, Madison, WI, pp 961–1010, SSA Book Ser. 5
Nriagu J, Pacyna P, Szefer P, Markert B, Wuenschmann S, Namiesnik J (eds) (2012) Heavy metals in the environment. Maralte Books, The Netherlands
O'Connor GA, Wierenga PJ, Cheng HH, Doxtader KG (1980) Movement of 2,4,5-T through large soil columns. Soil Sci 130:157–162
Ogwada RA, Sparks DL (1986) A critical evaluation on the use of kinetics for determining thermodynamics of ion exchange. Soil Sci Soc Am J 50:300–305
Pavlatou A, Polyzopoulos NA (1988) The role of diffusion in kinetics of phosphate desorption: the relevance of the Elovich equation. J Soil Sci 39:425–436
Polański A, Smulikowski K (1969) Geochemia. Wydawnictwa Geologiczne, Warszawa
Polish Standard (1994) Polish Standardisation Committee, ref. PrPN-ISO 10390 (E): soil quality and pH determination, 1st edn (in Polish)
Poonia SR, Talibudeen O (1977) Sodium-calcium exchange equilibria in salt-affected and normal soils. J Soil Sci 28:276–288
Prasa MNV (2008) Trace elements as contaminants and nutrients. Consequences in ecosystems and human health. Wiley, Hoboken, p. 777
Ram N, Verloo M (1985) Effect of various organic materials on the mobility of heavy metals in soils. Environ Pollut 10:241–248
Robert M (1996) Le sol: Interface dans l'Environnement, Ressource pour le développement. Masson, Paris, p 244

Roszyk E, Szerszeń L (1988) Accumulation of heavy metals in arable soil layers of the copper smelters sanitary zones, part II: "Głogów". Rocz Gleboz 39(4):147–158 (in Polish)
Ryan J, Estefan G, Rashid A (2001) Soil and plant analysis laboratory manual, 2nd edn. International Center for Agricultural Research in the Dry Areas (ICARDA), the National Agricultural Research Center (NARC), Aleppo, Syria, p 172
Sapek B (1976) Study on the copper sorption kinetics by peat-muck soils. In: Proceedings of the 5th international peat congress—Poznań: new recognition of peatlands and peats, vol 2. pp 236–245
Sapek B, Sapek A (1980) The application of ion-selective electrodes to investigations of binding of copper and cadmium by humic substances from peat-muck soils. Pol J Soil Sci 13(2):125–134
Sapek B, Zebrowski W (1976) Comparison of copper binding rate by peat-muck soils at various transformation stages. Pol J Soil Sci 9(2):93–99
Sarbak Z (2000) Adsorption and adsorbents. Theory and application. Wydawnictwo Nauko we UAM, Seria Chemia Nr 71., 167 str. (in Polish)
Schindler PW, Fürst B, Dick R, Wolf PU (1976) Ligand properties of surface silanol groups I. Surface complex formation with Fe^{3+}, Cu^{2+}, Cd^{2+} and Pb^{2+}. J Colloid Interface Sci 55:469–475
Schulte A, Beese F (1994) Isotherms of cadmium sorption densities. J Environ Qual 23:712–718
Schulthess CP, Dey DK (1996) Estimation of Langmuir constants using linear and nonlinear least squares regression analyses. Soil Sci Soc Am J 60:433–442
Schultz MF, Benjamin MM, Ferguson JF (1987) Adsorption and desorption of metals on ferrihydrite: reversibility of the reaction and sorption properties of the generated solid. Environ Sci Technol 21:863–869
Sillén LG, Martell AE (1971) Stability constants of metal-ion complexes. Suppl. 1, pt. I and II. The Chemical Society, Metcalfe and Cooper, Ltd, London, p 865
Skopp J (1986) Analysis of time dependent chemical processes in soils. J Environ Qual 15:205–213
Soil Taxonomy (1975) A basic system of soil classification for making and interpreting soil surveys: soil survey staff. U.S. Dept Agric. Handb. 436. U.S. Govt Print. Off. Washington, DC
Sokołowska Z (1989) The role of surface uniformity in adsorption processes occurring in soils. Problemy Agrofizyki 58:1–170, (in Polish) kinetics
Sparks DL (1986) Kinetics of reactions in pure and mixed systems. In: Sparks DL (ed) Soil physical chemistry. CRC Press, Boca Raton, FL, pp 83–178
Sparks DL (1989) Kinetics of soil chemical processes. Academic, San Diego, CA, p 210
Sparks DL (1995) Environmental soil chemistry. Academic, San Diego, CA, p 267
Sparks DL (2000) New frontiers elucidating the kinetics and mechanisms of metals and oxyanion sorption at the soil mineral/water interface. J Plant Nutr Soil Sci 163:563–570
Sposito G (1982) On the use of the Langmuir equation in the interpretation of adsorption phenomena II. The "Two-surface" Langmuir equation. Soil Sci Soc Am J 46:1147–1152
Sposito G (1984) The surface chemistry of soils. Oxford University Press, New York, p 234
Stawiński J (1977) Interrelationships between the specific surface area and some physico-chemical properties of soils. Zesz Probl Post Nauk Roln 197:229–240
Taylor RW, Kamaleldin H, Mehadi AA, Shoford WJ (1995) Kinetics of zinc sorption by soils. Commun Soil Sci Plant Anal 26(11 & 12):1761–1771
Thomas GW (1982) Exchangeable cations. In: Page AL, Miller RH, Keeney DR (eds) Methods of soil analysis, part 2. Chemical and microbial properties (No 9), 2nd edn. ASA-SSSA, Madison, WI, pp 159–165
Tiller KG, Gerth J, Brummer G (1984) The sorption of Cd, Zn and Ni by soil clay fractions: procedures for partition of bound forms and their interpretation. Geoderma 34:1–16
Van Bladel R, Laudelout H (1967) Apparent reversibility of ion-exchange reactions in clay suspension. Soil Sci 104:134–137
Van Bladel R, Halen H, Cloos P (1993) Calcium-zinc and calcium–cadmium exchange in suspensions of various types of clays. Clay Miner 28:33–38

Veith JA, Sposito G (1977) On the use of the Langmuir equation in the interpretation of "Adsorption" phenomena. Soil Sci Soc Am J 41:697–702
Verburg K, Baveye P (1994) Hysteresis in the binary exchange of cations on 2:1 clay minerals. A critical review. Clay Miner 42:207–220
Violante A, Krishnamurti GSR, Pigna M (2008) Factors affecting the sorption/desorption of trace elements in soil environments. In: Violante A, Ming Huang P, Gadd GM (eds) Biophysico-chemical processes of heavy metals and metalloids in soil environments. John Wiley & Sons, Inc., New York, pp 169–213
Weber WJ Jr (1984) Evolution of a technology. J Environ Eng Div 110:899–917
Wilk Z, Bocheńska T (2003) Hydrogeology of Polish mineral deposits and hydrological mining problems. Ed. AGH University of Science and Technology Press, Cracow, (Poland), 478 p. ISBN: 83-89388-21-9 (in Polish)
WRB (1998) World reference base for soil resources. 84 World soil resources reports. FAO-ISSS-ISRIC
Wu J, Laird DA, Thompson ML (1999) Sorption and desorption of copper on soil clay components. J Environ Qual 29(1):334–338
www.kghm.pl. Accessed 2 February 2015
Xian X, Shokohifard G (1989) Effect of pH on chemical forms and plant availability of cadmium, zinc and lead in polluted soils. Water Air Soil Pollut 45:265–273
Zehetner F, Wenzel WW (2000) Nickel and copper sorption in acid soils. Soil Sci 165(6):463–472
Zhang M, Nyborg M, Malhi SS, McKenzie RH, Solberg E (2000) Phosphorus release from coated monoammonium phosphate: effect of coating thickness, temperature, elution medium, soil moisture and placement method. Can J Soil Sci 80:127–134
Zhu H, Selim HM (2000) Hysteretic behavior of metolachlor adsorption-desorption in soils. Soil Sci 165(8):632–645

Chapter 4
Potentially Toxic Element Contamination and Its Impact on Soil Biological Quality in Urban Agriculture: A Critical Review

Avanthi Deshani Igalavithana, Sabry M. Shaheen, Jae Nam Park, Sang Soo Lee, and Yong Sik Ok

4.1 Urban Agriculture: Concept and Definitions

Zezza and Tasciotti (2010) defined "urban agriculture" as "the production of crop and livestock goods within cities and towns." The main objective of urban agriculture is to make use of neglected lands in urbanized areas to food garden (Despommier 2011). The United Nations Development Programme (UNDP) (1996) defined urban agriculture in a more advanced way by incorporating broad objectives: "Urban agriculture is an activity that produces, processes, and markets food and other products, on land and water in urban and peri-urban areas, applying intensive production methods, and (re)using natural resources and urban wastes, to yield a diversity of crops and livestock. In a broad sense, it encompasses the entire area in which a city's sphere of influence (social, ecological, or economic) comes to bear daily or directly on its population."

UNDP (1996) emphasized that urban agriculture is not a process isolated from the surrounding environment but is centered in urban areas; it is a dynamic process with very close interactions with the peri-urban areas or nearby rural areas (Martellozzo et al. 2014). Urban agriculture brings many benefits to stressed people in urban areas. Greening free spaces aids in recreational satisfaction, reduces heat accumulation, and cleans polluted air. Moreover, it provides economic benefits, social corporations, and new job opportunities (Wortman and Lovell 2013). Therefore, urban agriculture is an important concept for continuously developing and growing human populations to sustain their lives.

A.D. Igalavithana • J.N. Park • S.S. Lee • Y.S. Ok (✉)
Korea Biochar Research Center and Department of Biological Environment, Kangwon National University, Chuncheon 200-701, Republic of Korea
e-mail: soilok@kangwon.ac.kr

S.M. Shaheen
Department of Soil and Water Sciences, Faculty of Agriculture, University of Kafrelsheikh, 33 516 Kafr El-Sheikh, Egypt

I. Sherameti, A. Varma (eds.), *Heavy Metal Contamination of Soils*, Soil Biology 44, DOI 10.1007/978-3-319-14526-6_4

The urban ecosystem is characterized as a complex composite of both natural and anthropogenic factors. The local climate, geology, and geographical characteristics are the main natural factors that influence the urban environment. Anthropogenic factors may include population and settlement patterns and use/misuse of resources, and behavior of the human inhabitants is primarily governed by their socioeconomic conditions. Mega-urbanization in the form of increased population density, intense industrialization, and excessive exploitation of natural resources characterize a megacity as a unique system, but it has consequential deteriorating impacts on the overall structure and function of the ecosystem (Antrop 2000; Qureshi et al. 2010).

4.1.1 Worldwide Distribution

Urbanization is a big issue in all countries mainly due to the gathering of communities into cities due to industrialization. This community movement to city centers diminished the free land spaces while providing living rooms in skyscrapers. Although the beginning of urbanization satisfied people with a good economy, many urban communities are pressurized with life requirements beyond economic stability. In contrast, half of urban communities are facing poverty, job security, and lack of food (Drakakis-Smith et al. 1995). Hence, the importance of urban agriculture is more applicable now than the early stages.

Agriculture in urban areas has spread rapidly during the recent past. Orsini et al. (2013) reported that 100–200 million urban residents are utilizing urban spaces for commercial and subsistence agricultural production throughout the world. UNDP (1996) estimated that 800 million people worldwide are engaged in urban agriculture. Yet, all these estimations are based on judgments, and small urban areas were not considered (Martellozzo et al. 2014). However, it is clear that urban poverty leads people toward agriculture in urban areas, particularly in developing countries (Orsini et al. 2013).

4.1.2 General Introduction

Urban soils are receiving increased attention even though they are less studied, and the knowledge gap is higher than that for agricultural soils (Bullock and Gregory 1991). However, about 15 % of the world's food is produced in urban areas (USDA 2012).

Urban soils are highly spatially variable due to diverse human activities that influence its composition (Martellozzo et al. 2014). These include the addition of waste materials, pollution from numerous point and diffuse sources, landscaping, and rapid changes in land use, which can modify soil properties in an unpredictable manner (Jim 1998).

The use of vacant urban land for agriculture is increasing (Attanayake et al. 2014). However, all developed and developing countries are facing challenges in agriculture in urbanized areas (Martellozzo et al. 2014). A large and increasing population lives in urban areas, and recreational activities which often develop in urban areas enhance attention to urban soils (Abrahams 2002). Thus, environmental quality information is of crucial importance for appraising the sustainability of urban areas (Ravetz 2000; Brown 2003; van Kamp et al. 2003; Grimm and Redman 2004).

Urban soil is an important indicator of urban environmental quality as they act as sinks and sources for potentially toxic elements (PTEs) and other pollutants (Kelly et al. 1996; Mielke et al. 1999). The presence of excessive levels of PTE in urban soils, particularly in public access areas such as parks and agricultural land, is a cause for concern as pollutants can be easily transferred to humans through suspended matter, by direct contact with the soil, or via food webs (Attanayake et al. 2014). Many authors have revealed the persistence of PTE in soil for a considerable time period (Nolan et al. 2003; Wong et al. 2006; Thornton et al. 2008). Hence, soils polluted with PTE can threaten human health, the soil microbial community, water resources (surface and ground), food quality, flora, and fauna. Moreover, extensive soil pollution has significant negative effects on the quality of urban agricultural products both in developed and developing countries (Luo et al. 2012). In recent years, soils have been evaluated as a diagnostic tool of environmental conditions that influence health (Abrahams 2002; Davydova 2005). Interest in the urban geochemistry of PTE has expanded rapidly in recent years with studies conducted in several cities worldwide. Thus, PTE contamination, sources, and their significance and effect on soil microbial quality will be discussed in this chapter.

4.2 Sources of Contamination in Urban Agriculture

Frequent exposure of contamination sources is a rarely avoidable consequence in urban soils. Unfortunately, many sources of urban soil contamination are man-made, including traffic and industrial emissions, atmospheric deposits, household wastes, decay of painted walls, and industrial effluents (Banerjee 2003; Amato et al. 2009; Wei and Yang 2010). These sources have been frequently discussed by many authors as the main sources of contaminations of urban soils throughout the world (Francek 1992; Duzgoren-Aydin et al. 2006; Wei and Yang 2010). The possible sources of soil PTE pollution include atmospheric deposition of dust and aerosols, vehicle emissions (Yassoglou et al. 1987; Surthland et al. 2000), and various industrial activities (Thornton 1991; Krishna and Govil 2005; Li et al. 2007).

Human activities that add waste material to urban soils also influence element concentrations. Soil excavation and land filling at dump and construction sites with domestic or industrial wastes or with natural or disturbed soils and demolition wastes, incineration wastes, and crushed rock are also among the major human influences (Anikwe and Nwobodo 2002; Pastor and Hernandez 2012; Wang et al. 2012).

4.2.1 Traffic Emissions

One of the continuously increasing circumstances in urban areas is roads filled with vehicles (Wei and Yang 2010). PTEs found in vehicle fuels are liberated into the atmosphere during burning. Moreover, fuel tanks and other metal parts of vehicles can degrade over the time and release PTE to the atmosphere (Acosta et al. 2010; Saeedi et al. 2009; Denier van der Gon and Appelman 2009). The urban atmosphere mainly around roads has been contaminated with vehicle smoke, and the severity of contamination depends on the level of traffic (Francek 1992; Wortman and Lovell 2013). The tragedy is that all atmospheric contaminants ultimately travel into urban soil (Pivić et al. 2013).

Francek (1992) grouped road traffic into three main categories as low traffic, moderate traffic, and heavy traffic, and the range in the number of vehicles counted on a daily average was <8,000, 8,000–20,000, and >20,000, respectively. That author observed a positive correlation between soil contamination and the amount of traffic. However, the severity of contamination was reduced with respect to distance from the road, and this has also been reported by Pivić et al. (2013). Roadside soils in heavily trafficked urbanized areas are vastly contaminated with PTE (Leharne et al. 1992; Wong and Mak 1997). Urban soil contamination with Ni, Cu, Zn, Cr, and Pb has been observed by Wang and Qin (2007), and they revealed that traffic emissions were the main reason. Even though the United States government banned Pb-containing fuels for motor vehicles in 1975, alkyllead products are still contained in gasoline (Francek 1992). The European Union also banned Pb in gasoline in 1993, but it did not lead to decreasing Pb in gasoline throughout the world (Biasioli et al. 2006; Laidlaw et al. 2012).

A study conducted in Hong Kong by Li et al. (2001) reported higher Pb concentrations in urban soils in contrast to rural soils. They concluded that traffic emissions were one of the significant reasons. A comparable observation was reported by Luo et al. (2012) in a study conducted in urban areas of China. Significant roadside soil contamination with PTE has been reported in Kathmandu and Visakhapatnam in Nepal and India, respectively (Gautam et al. 2004; Goddu et al. 2004). There is no clear solution for urban traffic; hence, its influence on urban soil will not be diminished regardless of the pressure.

4.2.2 Industrial Emissions

Industrial activities play an important role in the urban and country's economy. Nevertheless, their impact on the environment has been discussed by world environmental organizations, particularly during the last half century. Industrial activities directly release detrimental amounts of PTE into the atmosphere, soil, and water in developed and developing countries (European Commission 2014). Hence, local governments and world organizations have implemented regulations to avoid and/or minimize the environmental impacts of industrial emissions but many of them are limited (USEPA 2010; European Commission 2014).

Emissions from metallurgy industries, particularly smelters, have led to a significant increase in the concentrations of PTE in neighboring soils. This may be hazardous to the environment and to the people living in the highly contaminated areas. These risks increase with the production of vegetables for human consumption.

Industrial wastewater discharge contaminates urban soils. Chromium and other PTEs containing tannery wastewater have been detected in urban soil of Damascus, Syria, as many tanneries are releasing their wastewater into the branches of the Barada River (Möller et al. 2005; Song et al. 2000). Urban soil contamination with industrially emitted Zn, Cd, and Cr was observed by Georgeaurd et al. (1997) in an urban area of Etang de Berre, France. The same PTE and Pb contaminations in urban soils by industrial emissions have been reported in Hangzhou city in China by Lu and Bai (2006).

The intensity of industrialization and a city's age are positively correlated to PTE concentrations in soils. For example, London, Hamburg, and Palermo exhibit very high PTE concentrations in soils compared to low industrialized areas such as Bangkok (Culbard et al. 1998; Lux 1993; Manta et al. 2002; Wilcke et al. 1998). Authors have described that this scenario is due to long-term industrial emissions and accumulation of PTE in old cities with respect to the new cities. Hence, industrial emissions are one of the main reasons for urban soil contamination with PTE.

4.2.3 Household Wastes

Household waste was defined by Slack et al. (2005) as "any waste produced from a domestic source." These include a range of anthropogenic sources such as paints, garden pesticides, detergents, pharmaceuticals, batteries, waste electronics and equipment, waste oils, and cosmetics and personal care products; all of these waste products could contain high or low concentrations of PTE (Slack et al. 2005). Another risk of household wastes is poor regulations by local governments for their disposal. Hence, potentially hazardous household wastes are often disposed with other wastes in urban landfills throughout the world (Schrab et al. 1993; Kjeldsen et al. 2002). The United Kingdom and some world organizations such as the European Union have paid special attention to hazardous household waste management and approved legislation to avoid urban soil contamination (Slack et al. 2009). However, humans produce widespread urban soil contamination by PTE in landfills and land dumps by disposing of hazardous household wastes (Christensen et al. 2001).

Landfills containing household waste from urban areas of Germany and Denmark leach PTE into surface and deeper soil layers. Although the concentrations are comparatively low, a high PTE concentration could develop over time in particular soils (Christensen et al. 2001). The release of PTE from landfills may vary with season (Akesson and Nilsson 1997).

4.2.4 *Atmospheric Deposition*

Avoiding atmospheric deposition of hazardous contaminants such as PTE is impracticable in urban areas (Laidlaw et al. 2012). Anthropogenic activities lead to a buildup of PTE concentrations in the urban atmosphere, particularly by smelting and burning of coal, oil, and wastes (Chen et al. 1997).

A study conducted in China clearly showed that atmospheric deposition contained considerable amounts of PTE; Cd, 0.01–0.14 mg/m^2/year; Co, 0.07–0.34 mg/m^2/year; Cr, 3.68–16.2 mg/m^2/year; Cu, 10.7–40.9 mg/m^2/year; Mn, 6.23–14.0 mg/m^2/year; Ni, 2.24–34.3 mg/m^2/year; Pb, 2.89–25.4 mg/m^2/year; and Zn 57.7–164 mg/m^2/year have been reported in Guangdong province in China (Wong et al. 2003). Atmospheric deposition mainly depends on solubility (Morselli et al. 2003).

Atmospheric particles, street dust, and surface soil collected from Nanjing, China, were analyzed by Hu et al. (2014) to determine Pb levels. The Pb content in the atmospheric particles was significantly higher than that in surface soil. Deposition of airborne contaminants is an important PTE source for urban surface soil.

4.3 Concentrations and Spatial Distribution of Potentially Toxic Elements in Urban Agricultural Land

A number of studies have described the PTE contents in the urban soils of many cities. Studies conducted in both developed and developing countries have reported the content, distribution, origin, extractability, and solid-phase speciation of PTE in urban soils (De Kimpe and Morel 2000; Govil et al. 2001; Krishna and Govil 2004, 2005, 2008; Chen et al. 2005; Lee et al. 2006; Davidson et al. 2006; Wong et al. 2006; Yay et al. 2008).

In a study by Biasioli et al. (2007), the soils of Ljubljana (Slovenia), Seville (Spain), and Turin (Italy) were analyzed. The results highlighted similarities across the cities, despite differences in geography, size, and climate. The PTE showed a wide range in concentrations revealing disperse contamination. Among the PTE, Pb and Zn exceeded the legislation threshold by 45 and 20 %, 43 and 43 %, and 11 and 2 % in Ljubljana, Turin, and Seville urban soils, respectively. The distribution of PTE was not depth dependant. Multivariate statistics showed similar associations among PTE in these three cities: group one "Cu, Pb, and Zn" and group two "Ni and Cr" signifying anthropogenic and natural origins of the PTE in the soils, respectively. Hence, Cu, Pb, and Zn distributions mostly rely on distance from the emission source.

Table 4.1 summarizes results of the concentrations of different PTE in urban agricultural soils and gardens and guidelines for different countries. The concentrations of Pb, Cu, Cr, Zn, Ni, Cd, and Hg in the urban soils increased to the maximum allowable concentration for the European community (CEC 1986) (Table 4.1). The concentrations of Cu, Zn, Ni, Cd, and Hg in urban soils of 21 Chinese cities increased to the maximum allowable concentration for the Chinese soil quality standard (Luo et al. 2012; Table 4.1). Therefore, the degree of urban soil contamination can be higher than that of agricultural soil (Douay et al. 2008).

Metal contamination in three soil groups of an industrial city in Ghaziabad district was investigated by Chabukdhara and Nema (2013). The spatial distributions of Cu, Cr, Pb, Cd, Zn, Mn, Fe, and Ni in the urban topsoil were determined, and the mean contents were 122, 288, 147, 0.4, 187, 386, 21,433, and 147 mg/kg, respectively. These were compared with mean concentrations from other cities around the world. Cu, Cr, Pb, Zn, and Ni concentrations appeared to be higher than those in many other cities in the world. Sites near commercial and industrial areas or dumpsites showed relatively higher concentrations of metals, compared to those at commercial or residential sites. These results clearly indicate significant effects of rapid urbanization and industrialization on soil PTE concentrations.

A large-scale survey of urban park soils compared to rural soils in Hong Kong showed that the mean concentrations of Cu and Zn in urban soils (24.8 and 168 mg/kg, respectively) were at least four and two times higher than those of rural soils, whereas the mean Pb concentration of urban soils (90 mg/kg) was ten times higher than that of rural soils (8.7 mg/kg) (Wong et al. 2006; Li et al. 2001). Pb represents a classic example, as concentrations in urban soils often exceed 100 mg/kg (Davidson et al. 2006). The concentrations of Cu and Zn and, in certain cases, Cd were elevated in urban soils (Lu et al. 2003; Imperato et al. 2003).

Madrid et al. (2006) compared the concentrations of Cu, Cr, Ni, Pb, and Zn in six urban parks located in different European cities with different climate and industrial histories. Broad concentration ranges were found for Cu, Pb, and Zn at most sites, but an extensive range of Cr and Ni was only observed in the Italian park, and the amount was significantly greater than that in the other cities' soils. Older cities with an intense manufacturing industry (Glasgow and Turin) showed the highest PTE,

Table 4.1 Concentrations of potentially toxic elements (PTEs) in urban soils and the guidelines of different countries

City, country		Element									References
		Pb	Cu	Cr	Zn	Ni	Cd	Mn	Hg	Co	
Palermo, Italy	Range	57–2,516	10–344	12–100	52–433	7.0–38.6	0.27–3.80	142–1,259	0.04–56	1.5–14.8	Manta et al. (2002)
	Median	253	77	39	151	19.1	0.84	566	1.85	6.5	
	SD	302	56	19	71	8.0	0.51	263	6.73	3.6	
21 cities, China[a]	Range	26.7–110	16.2–1,226	17.8–197	69.1–301.0	4.08–910	0.13–6.90	–	0.12–0.77	3.5–58.9	Luo et al. (2012)
	Median	55.2	40.4	68.5	109	24.9	0.39	–	0.31	13.6	
	Mean	61.3	99.2	76.8	133	99.6	0.88	–	0.35	25.4	
Hangzhou, China	Range	19.6–89.3	14.3–77.6	51.0–83.9	59.0–264.5	14.2–36.5	0.10–0.64	278.5–649.8	0.06–5.78	6.6–14.5	Chen et al. (2008)
	Median	46.7	34.9	60.4	109.0	19.5	0.23	411.1	0.72	9.25	
	Mean	46.2	36.6	62.2	116.1	20.0	0.26	415.3	0.69	9.25	
Suszec, Poland	Range	17.8–212.8	4.2–45.4	13.6–81.8	22.8–177.1	2.3–15.9	0.20–2.1	–	0.02–0.46	–	Loska et al. (2004)
	Median	38.56	8.08	31.47	61.40	5.56	0.77	–	0.06	–	
	Mean	39.80	8.47	34.43	62.47	6.35	0.80	–	0.07	–	
	SD	13.27	2.79	12.71	17.97	2.60	0.29	–	0.02	–	
Hong Kong, Hong Kong	Range	23–473	3.6–115	–	29–531	–	–	–	–	–	Luo et al. (2012)
	Median	81	14	–	137	–	–	–	–	–	
	Mean	108	27	–	161	–	–	–	–	–	
	SD	89	27	–	123	–	–	–	–	–	
Zagreb, Croatia	Range	1–139	4.30–183	–	15.2–277	1–282	0.25–3.85	79.2–1,282	–	–	Prohic et al. (1997)
	Median	23	17.8	–	69.7	48.7	0.50	605	–	–	
	Mean	25.9	20.8	–	77.9	49.5	0.66	613	–	–	
	SD	16.3	16.2	–	30.9	25.7	0.51	246	–	–	
Aberdeen, Scotland	Mean	172.9	44.6	22.9	113.2	15.9	–	264	–	6.2	Paterson et al. (1996)
	SD	33.9	11.0	2.8	14.6	1.7	–	24	–	0.3	

20 cities, China[b]	Range	17.1–77.3	21.2–42.5	22.7–87.7	52.2–227	15.51–38.5	0.21–2.57	–	0.08–0.73	–	Wei and Yang (2010)
	Mean	37.55	31.71	58.87	117.72	27.53	0.43	–	0.24	–	
Mt Pleasant, USA	Range	100–840	–	–	–	–	–	–	–	–	Francek (1992)
	Median	280	–	–	–	–	–	–	–	–	
	Mean	320	–	–	–	–	–	–	–	–	
Damascus, Syria	Range	<5–108	16–97	12–116	46–293	24–58	–	–	–	7–23	Möller et al. (2005)
	Median	10	30	51	84	35	–			10	
	Mean	17	34	57	103	39	–			13	
	SD	18	16	24	54	9	–			5	
Ghaziabad, India (group 1)	Range	28–341	30–444	356–2,300	42–477	206–403	0.2–4.0	105–2,613	–	–	Chabukdhara and Nema (2013)
	Mean	225	295	807	373	279	1.1	809	–	–	
	SD	123	123	684	148	61	1.2	981	–	–	
Ghaziabad, India (group 2)	Range	23–245	13–55	11–42	47–284	16–422	0.2–1.3	89–374	–	–	
	Mean	112	27	24	113	92	0.4	184	–	–	
	SD	73	12	9	78	141	0.3	88	–	–	
Ghaziabad, India (group 3)	Range	27–162	24–98	2–8	38–87	13–38	0.1–0.3	121–217	–	–	
	Mean	111	53	4	65	27	0.2	166	–	–	
	SD	60	31	2	19	10	0.1	37	–	–	
MAC in AG soils, USA, 1993		150	750	1,500	1,400	210	20	–	8	–	Chen et al. (2008)
MAC in AG soils, European community, 1986		50–300	50–140	50–150	150–300	30–75	1–3	–	1–1.5	–	CEC (1986)

(continued)

Table 4.1 (continued)

City, country		Element									References
		Pb	Cu	Cr	Zn	Ni	Cd	Mn	Hg	Co	
MAC in AG soils, Germany, 1992		100	60	100	200	50	1.5	–	1	–	Chen et al. (2008)
MAC in AG soils, Australia, 1977		100	100	100	300	100	5	–	5	50	Chen et al. (2008)
Chinese soil quality standard II		300	100	200	250	50	0.30	–	0.50	–	CEPA (1995)
Background value of Zhejiang province, China		25.61	17.76	54.60	69.0	22.31	0.17	672.0	0.15	14.0	Zhejiang Soil Survey Office (1994); Chen et al. (2007)
Dutch soil guidelines	Target value	85	36	100	140	35	0.80	–	0.30	9	VROM (2000); Department of Soil Protection, Netherlands (1994)
	Intervention value	530	190	380	720	210	12	–	10	240	

SD Standard Deviation, *MAC* maximum allowable concentration, *AG* agriculture

[a]Yangtze River Delta (Nanjing, Shanghai, and Hangzhou) in eastern China, the Pearl River Delta (Guangzhou, Shenzhen, and Hong Kong) in the south, the Bohai Rim (Beijing and Tianjin) in the north, the Old Industrial Zone (Shenyang and Dalian) in the northeast, the Chengdu–Chongqing Plain in the west, the Wei River Plain (Xi'an) in the northwest, and a few other cities

[b]Beijing, Changchun, Taicang, Fuyang, Guangzhou, Hangzhou, Luoyang, Nanjing, Shanghai, Wenzhou, Changsha, Xiangtan, Zhangzhou, Shenyang, Hongkong, Qingdao, Baoji, Jinchang, Shenzhen, Xuzhou

and the most recently established parks in the least industrialized city (Aveiro) exhibited the lowest PTE concentrations in the soil.

In the urban–rural transitional area of Hangzhou, China, topsoil samples from vegetable fields were tested for PTE contamination, and high concentrations of As, Cd, Cu, Hg, Pb, and Zn were observed (Chen et al. 2008). Moreover, these contaminating elements showed a relatively weaker correlation to inherent soil minerals but higher spatial variability, indicating that their presence and spatial heterogeneity relied on anthropogenic inputs.

Hu et al. (2013) assessed the impact of urbanization and industrialization on PTE content in surface soils of the world's most densely populated and major global manufacturing regions. They revealed that mean surface soil Cd, Cu, and Zn concentrations were two times higher than the background values and apparently contributed by anthropogenic sources. Moreover, Co, Fe, Cr, and Ni in the same surface soils were primarily due to lithogenic sources rather than anthropogenic sources.

Mitchell et al. (2014) studied distributions of some PTE in New York City community garden soils. They reported that 22 % of garden soils exceeded the health-based guideline values. Ba, Pb, Cd, and Zn were the commonly found PTE and exceeded the guideline values in many soil samples. More importantly, they revealed that PTE varied independently from organic matter and geogenic metals.

Werkenthin et al. (2014) reviewed the metals in European roadside soils and in soil solutions. The highest median values of Cr, Cu, Ni, Pb, and Zn were determined in the topsoil layer in the first 5 m beside the road and decreased with increasing soil depth and distance from the road.

The necessity for developing pollution prevention and reduction strategies to reduce PTE pollution in rapidly industrialized and urbanized areas is well documented.

4.4 Geochemical Fractions and the Mobility of PTE in Urban Soils

Despite the high PTE contents in urban soils, mobility or leachability is often not very high due to relatively strong bonding (Madrid et al. 2004). However, some studies show that the potential mobility and associated risk of PTE pollutants may be higher in urban soils compared to that in rural and agricultural soils (Mingkui et al. 2003; Zhao and Li 2013).

A sequential extraction procedure adapted by Davidson et al. (2006) revealed that Fe, Ni, and Cr in five European cities were mainly associated with the residual phase of the soil matrix. Cu was in the reducible, oxidizable, and residual fractions. Mn and Pb were strongly associated with reducible material in the same cities, and Pb showed an ability to remobilize under reduced conditions. This study also showed the highest extractable metal contents in older and more heavily industrialized cities.

Waterlot et al. (2011) studied urban kitchen garden topsoils affected by atmospheric emissions from two smelters and adjacent non-massively contaminated soils. Almost 50 % of the total Cd and Pb were strongly bound with Fe and Mn oxides and hydroxides. Zn was greatly associated with crystalline mineral matrices. Hence, the mobilities of these three metals were Cd > Zn > Pb. A similar observation was reported by Chon et al. (1998) in urban and industrialized areas of Korea.

Wastewater irrigation and its effects on PTE distribution, fractions, and mobilities in five urban garden soils in Kano, Nigeria, were studied by Abdu et al. (2011). Total Zn and Cd concentrations in the first 20 cm of the soil were 121–207 mg/kg and 0.3–2.0 mg/kg, respectively, and 50 % of that was from the weakly bound fraction, reflecting high mobility. Thus, there is a great risk of spread and translocation of Zn and Cd via food chains and aquifers in the urban and nearby areas.

Li et al. (2014) studied the binding and release kinetics of Cd, Cu, and Pb from six different urban soils in the older metropolitan area of Copenhagen. Total Cd, Cu, and Pb concentrations were elevated 5–27 times in the urban soils compared to those in an agricultural reference soil. Cd and Pb were mainly in the mobilizable pools and Cu was in the strongly bound pools. Total leachable Cu and Pb pools were higher in urban soils compared to those in the agricultural reference soil but not for Cd.

4.5 Effects of PTE on Soil Biological Properties in Urban Agricultural Lands

Numerous authors have documented that long- or short-term exposure to high concentrations of PTE diminishes the microbial biomass and activities in soils (Lasat 2002; McGrath et al. 2001). Soil microbes play vital roles in soils by recycling nutrients, maintaining soil structure, detoxifying chemicals, and controlling plant pests and plant growth (Elsgaard et al. 2001). Soil microbial parameters are considered early warning and very sensitive indicators of soil pollution (Wang et al. 2007). However, few studies have evaluated PTE concentrations and their effects on microbial parameters in urban soils (White and McDonnell 1988; Carreiro et al. 1999).

4.5.1 Microbial Biomass and Microbial Diversity

Soil microbial biomass and diversity are sensitive and negatively affected by PTE contamination. Although a limited number of studies have been conducted on urban soils contaminated with PTE and their effects on microbial biomass and diversity,

Fig. 4.1 Different levels of Pb-, Cu-, and Zn contaminated soil and microbial biomass C (adapted from Liao and Xie 2007)

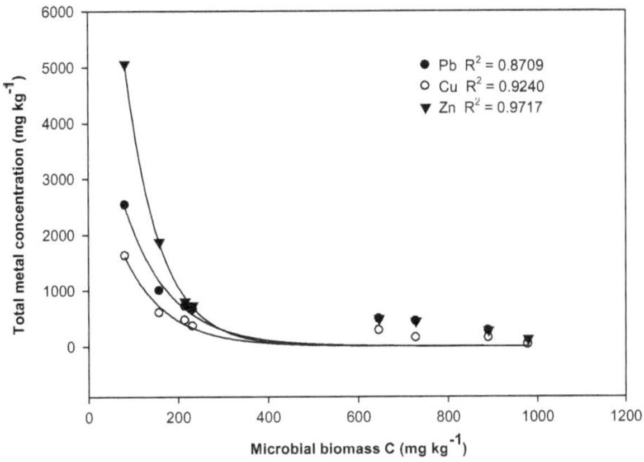

many authors have extensively discussed the adverse effects of PTE on these soil parameters. Giller et al. (1998) described the negative impacts of metals on soil microbes. Therefore, PTE contaminating urban soils would have a similar relationship with microbial biomass.

Hinojosa et al. (2005) reported reduced microbial community richness and diversity in pyrite sludge polluted soils, and this varied with the degree of PTE pollution and soil type. Similar observations of rhizosphere community shifts were reported by Müller et al. (2002) and Gremion et al. (2004) in PTE-contaminated soils. Xu et al. (2014) evaluated 16 representative Chinese cities and reported that Mn and Mg were highly correlated with variability in each city's soil microbial community. Microbial communities are highly sensitive to anthropogenic soil disturbances including PTE contamination, and this is quite prominent in urban soils which are influenced by the vast array of human activities (Zhao et al. 2013).

Poulsen et al. (2013) reported that microbial biomass was significantly affected by urban fertilizer waste applications. Moreover, these wastes were enriched with Zn and Cu, and it was the main factor causing a significant reduction in the microbial biomass in a particular soil.

Liao and Xie (2007) documented a clear reduction in microbial biomass C with increased concentrations of Pb, Cu, and Zn in the soil (Fig. 4.1) in a field study conducted at a Cu mining wasteland area in China.

4.5.2 *Enzyme Activities*

The roles of soil enzymes in biogeochemical cycles are vital. Soil enzymes are one of the most sensitive indicators of soil quality compared to physicochemical properties (Finkenbein et al. 2013). Soil enzymes have been used as a significant assessment tool to manage and improve soil quality (Pandey et al. 2014).

Most studies related to impacts of heavy metals on soil enzymes have been conducted under laboratory incubation conditions (Hinojosa et al. 2004), and Hg,

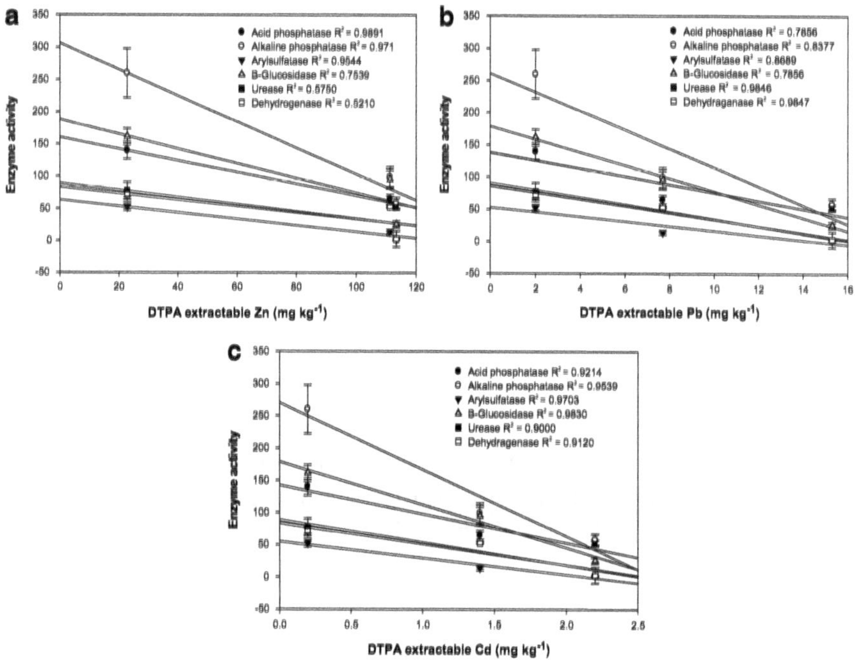

Fig. 4.2 Diethylene triamine pentaacetic acid (DTPA) extractable Zn, Pb, Cd, and enzyme activities in urban soils (adapted from Hinojosa et al. 2004)

Ag, Cr, and Cd resulted in the highest inhibition of soil enzymes (Deng and Tabatabai 1995). However, Speir and Ross (2002) reported that PTE pollution and enzyme activity under field conditions have not been studied well.

Hinojosa et al. (2004) assessed Pb, Cu, and Zn levels and enzyme activities in nonpolluted, reclaimed, and polluted soils near the Aznalcóllar mine spill and industrialized area in Spain. They revealed that acid phosphatase, alkaline phosphatase, arylsulfatase, β-glucosidase, urease, and dehydrogenase activity declined with the increase in Pb, Cu, and Zn levels in the soil (Fig. 4.2). Moreover, Zn and Cd showed comparatively higher negative impacts on all considered enzymes than that of the Cu. A few authors have reported that arylsulfatase is very sensitive to PTE contamination, whereas acid phosphatase and urease are not much affected (Bardgett et al. 1994).

Speir and Ross (2002) clearly explained that the effects of PTE content and other soil parameters on soil enzyme activities are complex. However, soil enzyme activities are profoundly influenced by PTE depending on the soil type (Hinojosa et al. 2010). Hinojosa et al. (2010) observed reduced enzyme activities with enhanced concentrations of PTE in the rhizosphere.

Bielińska et al. (2013) observed a decline in enzyme activities with higher levels of PTE (Zn, Pb, Cu, and Cd) in cities of southeastern Poland. The most influenced enzyme was dehydrogenase of the other tested enzymes such as acid phosphatase, alkaline phosphatase, urease, and protease.

Zn toxicity to arylsulfatase, protease, and urease was observed by Lessard et al. (2014), and Cu toxicity to protease activity was examined by Dussault et al. (2008) in PTE-contaminated soils. A laboratory experiment conducted by Wieczorek et al. (2014) confirmed that high PTE concentrations in soil negatively affect soil acid phosphatase activity. However, the effects changed with soil type; the greatest influence was reported in sandy soil (Wieczorek et al. 2014). Juma and Tabatabai (1977) stated that Cu, Ag, Cd, Cu, Zn, Mn, Sn, Ni, Pb, Cr, Fe, B, Al, V, and Se are significant phosphatase activity inhibitors of the PTE.

Acknowledgment This chapter is a product which is supported by the Rural Development Administration of Korea as a part of the "Evaluation of environmental change and development of management method for sustainable growth of urban agriculture" project (PJ01018204).

References

Abdu N, Agbenin JO, Buerkert A (2011) Geochemical assessment, distribution, and dynamics of trace elements in urban agricultural soils under long-term wastewater irrigation in Kano, northern Nigeria. J Plant Nutr Soil Sci 17:447–458

Abrahams PW (2002) Soils: their implications to human health. Sci Total Environ 291:1–32

Acosta JA, Faz A, Martinez-Martinez S (2010) Identification of heavy metal sources by multivariable analysis in a typical Mediterranean city (SE Spain). Environ Monit Assess 169:519–530

Akesson M, Nilsson P (1997) Seasonal changes of leachate production and quality from test cells. J Environ Eng 123:892–900

Amato F, Pandolfi M, Viana M, Querol X, Alastuey A, Moreno T (2009) Spatial and chemical patterns of PM10 in road dust deposited in urban environment. Atmos Environ 43:1650–1659

Anikwe MA, Nwobodo KC (2002) Long term effect of municipal waste disposal on soil properties and productivity of sites used for urban agriculture in Abakaliki, Nigeria. Bioresour Technol 83:241–250

Antrop M (2000) Changing patterns in the urbanized countryside of Western Europe. Landsc Ecol 15:257–270

Attanayake CP, Hettiarachchi GM, Harms A, Presley D, Martin S, Pierzynski GM (2014) Field evaluations on soil plant transfer of lead from an urban garden soil. J Environ Qual 43:475–487

Banerjee ADK (2003) Heavy metal levels and solid phase speciation in street dusts of Delhi, India. Environ Pollut 123:95–105

Bardgett RD, Speir TW, Ross DJ, Yeates GW, Kettles HA (1994) Impact of pasture contamination by copper, chromium, and arsenic timber preservative on soil microbial properties and nematodes. Biol Fertil Soils 18:71–79

Biasioli M, Barberis R, Ajmone-Marsan F (2006) The influence of a large city on some soil properties and metals content. Sci Total Environ 356:154–164

Biasioli M, Grčman H, Kralj T, Madrid F, Díaz-Barrientos E, Ajmone-Marsan F (2007) Potentially toxic elements contamination in urban soils: a comparison of three European cities. J Environ Qual 36:70–79

Bielińska EJ, Kołodziej B, Sugier D (2013) Relationship between organic carbon content and the activity of selected enzymes in urban soils under different anthropogenic influence. J Geochem Explor 129:52–56

Brown AL (2003) Increasing the utility of urban environmental quality information. Landsc Urban Plann 64:77–87
Bullock P, Gregory PJ (eds) (1991) Soils in the urban environment. Blackwell, Oxford
Carreiro MM, Howe K, Parkhurst DF, Pouyat RV (1999) Variation in quality and decomposability of red oak leaf litter along an urban-rural gradient. Biol Fertil Soils 30:258–268
CEC (1986) Council Directive of 12 June 1986 on the protection of the environment, and in particular of the soil, when sewage sludge is used in agriculture. Official Journal of the European Communities L 181, pp 6–12
CEPA (1995) Environmental quality standards for soils (GB 15816–1995) (in Chinese). Chinese Environmental Protection Administration
Chabukdhara M, Nema A (2013) Heavy metals assessment in urban soil around industrial clusters in Ghaziabad, India: probabilistic health risk approach. Ecotoxicol Environ Saf 87:57–64
Chen J (2007) Rapid urbanization in China: a real challenge to soil protection and food security. Catena 69:1–15
Chen TB, Wong JWC, Zhou HY, Wong MH (1997) Assessment of trace metal distribution and contamination in surface soils of Hong Kong. Environ Pollut 96:61–68
Chen TB, Zheng YM, Lei M, Huang ZC, Wu HT, Chen H, Fan KK, Yu K, Wu X, Tian QZ (2005) Assessment of heavy metal pollution in surface soils of urban parks in Beijing, China. Chemosphere 60:542–551
Chen JM, Tan MG, Li YL, Zheng J, Zhang YM, Shan Z, Zhang GL, Li Y (2008) Characteristics of trace elements and lead isotope ratios in PM2.5 from four sites in Shanghai. J Hazard Mater 156:36–43
Chon HT, Ahn JS, Jung MC (1998) Seasonal variations and chemical forms of heavy metals in soils and dusts from the satellite cities of Seoul, Korea. Environ Geochem Health 20:77–86
Christensen TH, Kjeldsen P, Bjerg PL, Jensen DL, Christensen JB, Baun A, Albrechtsen HJ, Heron G (2001) Biogeochemistry of landfill leachate plumes. Appl Geochem 16:659–718
Culbard EB, Thornton I, Watt J, Wheatley M, Moorcroft S, Thompson M (1988) Metal contamination in British urban dusts and soils. J Environ Qual 17:226–234
Davidson CM, Urquhart GJ, Ajmone-Marsan F, Biasioli M, Duarte AC, Díaz- Barrientos E, Grcman H, Hossack I, Hursthouse AS, Madrid L, Rodrigues S, Zupan M (2006) Fractionation of potentially toxic elements in urban soils from five European cities by means of a harmonised sequential extraction procedure. Anal Chim Acta 565:63–72
Davydova S (2005) Heavy metals as toxicants in big cities. Microchem J 79:133–316
De Kimpe CR, Morel JL (2000) Urban soil management: a growing concern. Soil Sci 165:31–40
Deng SP, Tabatabai MA (1995) Cellulase activity of soils: effect of trace elements. Soil Biol Biochem 27:977–979
Denier van der Gon H, Appelman W (2009) Lead emissions from road transport in Europe. A revision of current estimates using various estimation methodologies. Sci Total Environ 407:5367–5372
Department of Soil Protection, Netherlands (1994) The Netherlands Soil Contamination Guidelines, Netherlands Intervention Values for Soil Remediation, Reference #DBO/07494013
Despommier D (2011) The vertical farm: controlled environment agriculture carried out in tall buildings would create greater food safety and security for large urban populations. J Verbr Lebensm 6:233–236
Douay F, Pruvot C, Roussel H, Ciesielski H, Fourrier H, Proix N, Waterlot C (2008) Contamination of urban soils in an area of Northern France polluted by dust emissions of two smelters. Water Air Soil Pollut 188:247–260
Drakakis-Smith D, Bowyer-Bower T, Tevera D (1995) Urban poverty and urban agriculture: an overview of the linkages in harare. Habitat Int 19:183–193
Dussault M, Bécaert V, François M, Sauvé S, Deschênes L (2008) Effect of copper on soil functional stability measured by relative soil stability index (RSSI) based on two enzyme activities. Chemosphere 72(5):755–762

Duzgoren-Aydin NS, Wong CSC, Aydin A, Song Z, You M, Li XD (2006) Heavy metal contamination and distribution in the urban environment of Guangzhou, SE China. Environ Geochem Health 28:375–391

Elsgaard L, Petersen SO, Debosz K, Kristiansen IB (2001) Effects of linear alkylbenzene sulfonates (LAS) on soil microbiology. Tenside Surfact Det 38:94–97

European Commission (2014) http://ec.europa.eu/environment/industry/stationary/studies.htm

Finkenbein P, Kretschmer K, Kuka K, Klotz S, Heilmeier H (2013) Soil enzyme activities as bioindicators for substrate quality in revegetation of a subtropical coal mining dump. Soil Biol Biochem 56:87–89

Francek MA (1992) Soil lead levels in a small town environment: a case study from Mt pleasant, Michigan. Environ Pollut 76:251–257

Gautam P, Blaha U, Appel E, Neupane G (2004) Environmental magnetic approach towards the quantification of pollution in Kathmandu urban area, Nepal. Phys Chem Earth 29:973–984

Georgeaud VM, Rochette P, Ambrosi JP, Vandamme D, Williamson D (1997) Relationship between heavy metals and magnetic properties in a large polluted catchment: The Etang de Berre (south of France). Phys Chem Earth 22(1–2 Spec Iss):211–214

Giller KE, Witter E, McGrath SP (1998) Toxicity of heavy metals to microorganisms and microbial processes in agricultural soils: a review. Soil Biol Biochem 30:1389–1414

Goddu SR, Appel E, Jordanova D, Wehland F (2004) Magnetic properties of road dust from Visakhapatnam (India)-relationship to industrial pollution and road traffic. Phys Chem Earth 29:985–995

Govil PK, Reddy JLN, Krishna AK (2001) Contamination of soil due to heavy metals in the Patancheru industrial development area, Andhra Pradesh, India. Environ Geol 41:461–469

Gremion F, Chatzinotas A, Kaufman K, Von Sigler W, Harms H (2004) Impacts of heavy metal contamination and phytoremediation on a microbial community during a twelve-month microcosm experiment. FEMS Microbiol Ecol 48:273–283

Grimm N, Redman CL (2004) Approaches to the study of urban ecosystems: the case of Central Arizona-Phoenix. Urban Ecosyst 7:199–213

Hinojosa MB, Carreira JA, García-Ruíz R, Dick RP (2004) Soil moisture pre-treatment effects on enzyme activities as indicators of heavy metal-contaminated and reclaimed soils. Soil Biol Biochem 36:1559–1568

Hinojosa MB, Carreira JA, García-Ruıź R, Dick RP (2005) Microbial response to heavy metal-polluted soils: community analysis from phospholipid-linked fatty acids and ester-linked fatty acids extracts. J Environ Qual 34:1789–1800

Hinojosa BM, Carreira JA, García-Ruíz R, Rodríguez-Maroto JM, Daniell TJ, Griffiths BS (2010) Plant treatment, pollutant load, and soil type effects in rhizosphere ecology of trace element polluted soils. Ecotoxicol Environ Saf 73:970–981

Hu Y, Liu X, Bai J, Shih K, Zeng EY, Cheng H (2013) Assessing heavy metal pollution in the surface soils of a region that had undergone three decades of intense industrialization and urbanization. Environ Sci Pollut Res 20:6150–6159

Hu X, Sun Y, Ding Z, Zhang Y, Wu J, Lian H, Wang T (2014) Lead contamination and transfer in urban environmental compartments analyzed by lead levels and isotopic compositions. Environ Pollut 187:42–48

Imperato M, Adamo P, Naimo D, Arienzo M, Stanzione D, Violante P (2003) Spatial distribution of heavy metals in urban soils of Naples city (Italy). Environ Pollut 124:247–256

Jim CY (1998) Urban soil characteristics and limitations for landscape planting in Hong Kong. Landsc Urban Plann 40:235–249

Juma NG, Tabatabai MA (1977) Effects of trace elements on phosphatase activity in soils. Proc Soil Sci Soc Am 41:343–346

Kelly J, Thornton I, Simpson PR (1996) Urban geochemistry: a study of the influence of anthropogenic activity on the heavy metal content of soils in traditionally industrial and non-industrial areas of Britain. Appl Geochem 11:363–370

Kjeldsen P, Barlaz MA, Rooker AP, Baun A, Ledin A, Christensen TH (2002) Present and long-term composition of MSW landfill leachate: a review. Crit Rev Environ Sci Tech 32:297–336
Krishna AK, Govil PK (2004) Heavy metal contamination of soil around Pali Industrial Area, Rajasthan, India. Environ Geol 47:38–44
Krishna AK, Govil PK (2005) Heavy metal distribution and contamination in soils of Thane Belapur industrial development area, Mumbai, Western India. Environ Geol 47:1054–1061
Krishna AK, Govil PK (2008) Assessment of heavy metal contamination in soils around Manali industrial area, Chennai, southern India. Environ Geol 54:1465–1472
Laidlaw MAS, Zahran S, Mielke HW, Taylor MP, Filippelli GM (2012) Re-suspension of lead contaminated urban soil as a dominant source of atmospheric lead in Birmingham, Chicago, Detroit and Pittsburgh, USA. Atmos Environ 49:302–310
Lasat MM (2002) Phytoextraction of toxic metals: a review of biological mechanisms. J Environ Qual 31:109–120
Lee CS, Li XD, Shi WZ, Cheung SC, Thornton I (2006) Metal contamination in urban, suburban, and country park soils of Hong Kong: a study based on GIS and multivariate statistics. Sci Total Environ 356:45–61
Leharne S, Charlesworth D, Chowdhry B (1992) A survey of metal levels in street dusts in an inner London neighbourhood. Environ Int 18:263–270
Lessard I, Sauvé S, Deschênes L (2014) Enzymatic functional stability of Zn-contaminated field-collected soils: an ecotoxicological perspective. Sci Total Environ 484:1–9
Li X, Poon C, Liu PS (2001) Heavy metal contamination of urban soils and street dusts in Hong Kong. Appl Geochem 16:13610–13618
Li FR, Kang LF, Gao XQ, Hua W, Yang FW, Hei WL (2007) Traffic-related heavy metal accumulation in soils and plants in Northwest China. J Soil Contam 16:473–484
Li L, Holm PE, Marcussen H, Hansen HCB (2014) Release of cadmium, copper and lead from urban soils of Copenhagen. Environ Pollut 187:90–97
Liao M, Xie XM (2007) Effect of heavy metals on substrate utilization pattern, biomass, and activity of microbial communities in a reclaimed mining wasteland of red soil area. Ecotoxicol Environ Saf 66:217–223
Loska K, Wiechula D, Korus I (2004) Metal contamination of farming soils affected by industry. Environ Int 30:159–165
Lu SG, Bai SQ (2006) Study on the correlation of magnetic properties and heavy metals content in urban soils of Hangzhou city, China. J Appl Geophys 60:1–12
Lu Y, Gong ZT, Zhang GL, Zhang GL, Burghardt W (2003) Concentrations and chemical speciations of Cu, Zn, Pb and Cr of urban soils in Nanjing, China. Geoderma 115:101–111
Luo XS, Yu S, Zhu YG, Li XD (2012) Trace metal contamination in urban soils of China. Sci Total Environ 421–422:17–30
Lux W (1993) Long-term heavy metal and As pollution of soils, Hamburg, Germany. Appl Geochem 2:135–143
Madrid L, Diaz-Barrientos E, Reinosa R, Madrid F (2004) Metals in urban soils of Sevilla: seasonal changes and relations with other soil components and plant contents. Eur J Soil Sci 55:209–217
Madrid L, Diaz-Barrientos E, Ruiz-Cortés E, Reinoso R, Biasioli M, Davidson CM, Duarte AC, Grcman H, Hossack I, Hursthouse AS, Kralj T, Ljung K, Otabbong E, Rodrigues S, Urquhart GJ, Ajmone-Marsan F (2006) Variability in concentrations of potentially toxic elements in urban parks from six European cities. J Environ Monit 8:1158–1165
Manta DS, Angelone M, Bellanca A, Neri R, Sprovieri M (2002) Heavy metals in urban soils: a case study from the city of Palermo (Sicily), Italy. Sci Total Environ 300:229–243
Martellozzo F, Landry JS, Plouffe D, Seufert V, Rowhani P, Ramankutty N (2014) Urban agriculture: a global analysis of the space constraint to meet urban vegetable demand. Environ Res Lett 9:064025
McGrath SP, Zhao FJ, Lombi E (2001) Plant and rhizosphere processes involved in phytoremediation of metal-contaminated soils. Plant and Soil 232:207–214

Mielke HW, Gonzales CR, Smith MK, Mielke PW (1999) The urban environment and children's health: soils as an integrator of lead, zinc, and cadmium in New Orleans, Louisiana, USA. J Environ Res 81:117–129
Mingkui Z, Meiqing W, Xingmei L, Hong J, Jianming X (2003) Characterization of soil quality under vegetable production along an urban-rural gradient. Pedosphere 13:173–180
Mitchell R, Spliethoff HM, Ribaudo LN, Lopp DM, Shayler HA, Marquez-Bravo LG, Lambert VT, Ferenz GS, Russell-Anelli JM, Stone EB, McBride MB (2014) Lead (Pb) and other metals in New York City community garden soils: factors influencing contaminant distributions. Environ Pollut 187:162–169
Möller A, Müller HW, Abdullah A, Abdelgawad G, Utermann J (2005) Urban soil pollution in Damascus, Syria: concentrations and patterns of heavy metals in the soils of the Damascus Ghouta. Geoderma 124:63–71
Morselli L, Olivieri P, Brusori B, Passarini F (2003) Soluble and insoluble fractions of heavy metals in wet and dry atmospheric depositions in bologna, Italy. Environ Pollut 124:457–469
Müller AK, Westergaard K, Christensen S, Sørensen SJ (2002) The diversity and function of soil microbial communities exposed to different disturbances. Microb Ecol 44:49–58
Nolan AL, Lombi E, McLaughlin MJ (2003) Metal bioaccumulation and toxicity in soils—why bother with speciation? Aust J Chem 56:77–91
Orsini FS, Kahane R, Nono-Womdim R, Gianquinto G (2013) Urban Agriculture in the developing world: a review. Agron Sustain Dev 33:695–720
Pandey D, Agrawal M, Bohra JS (2014) Effects of conventional tillage and no tillage permutations on extracellular soil enzyme activities and microbial biomass under rice cultivation. Soil Till Res 136:51–60
Pastor J, Hernandez AJ (2012) Heavy metals salts and organic residues in old solid urban waste landfills and surface waters in their discharge areas: determinants for restoring their impact. J Environ Manage 95:42–49
Paterson E, Sanka M, Clark L (1996) Urban soils as pollutant sinks—a case study from Aberdeen, Scotland. Appl Geochem 11:129–131
Pivić RN, Stanojković Sebić AB, Jošić DL (2013) Assessment of soil and plant contamination by select heavy metals along a major European highway. Pol J Environ Stud 22:1465–1472
Poulsen PHB, Magid J, Luxhøi J, de Neergaard A (2013) Effects of fertilization with urban and agricultural organic wastes in a field trial—waste imprint on soil microbial activity. Soil Biol Biochem 57:794–802
Prohic E, Davis JC, Hansberger G (1997) Geochemical patterns in soils of the karst region, Croatia. J Geochem Explor 60:139–155
Qureshi S, Breuste JH, Lindley SJ (2010) Green space functionality along an urban gradient in Karachi, Pakistan: a socio-ecological study. Hum Ecol 38(2):283–294
Ravetz J (2000) Integrated assessment for sustainability appraisal in cities and regions. Environ Impact Assess 20:31–64
Saeedi M, Hosseinzadeh M, Jamshidi A, Pajooheshfar SP (2009) Assessment of heavy metals contamination and leaching characteristics in highway side soils, Iran. Environ Monit Assess 151:231–241
Schrab GE, Brown KW, Donnelly KC (1993) Acute and genetic toxicity of municipal landfill leachate. Water Air Soil Pollut 69:99–112
Slack RJ, Gronow JR, Voulvoulis N (2005) Household hazardous waste in municipal landfills: contaminants in leachate. Sci Total Environ 337:119–137
Slack RJ, Gronow JR, Voulvoulis N (2009) The management of household hazardous waste in the United Kingdom. J Environ Manage 90:36–42
Song Z, Williams CJ, Edyvean RGJ (2000) Sedimentation of tannery wastewater. Water Resour 34:2171–2176
Speir TW, Ross DJ (2002) Hydrolytic enzyme activities to assess soil degradation and recovery. In: Burns RG, Dick RP (eds) Enzymes in the environment. Activity, ecology and applications. Marcel Dekker, New York, pp 407–431

Surthland RA, Tolosa CA, Tack FMG, Verloo MG (2000) Characterization of selected element concentrations and enrichment ratios in background and anthropogenically impacted roadside areas. Arch Environ Contam Toxicol 38:428–438
Thornton I (1991) Metal contamination of soils in urban areas. In: Bullock P, Gregory PJ (eds) Soils in the urban environment. Blackwell, Oxford, pp 47–75
Thornton I, Farago ME, Thums CR, Parrish RR, McGill RAR, Breward N, Fortey NJ, Simpson P, Young SD, Tye AM, Crout NMJ, Hough RL, Watt J (2008) Urban geochemistry: research strategies to assist risk assessment and remediation of brownfield sites in urban areas. Environ Geochem Health 30:565–576
UNDP (1996) Urban agriculture—food, jobs and sustainable cities, vol 1. United Nations Development Programme Publication Series for Habitat II, New York
USDA (2012) Urban agriculture. Alternative farming systems information center. National Agricultural Library, U.S. Department of Agriculture. http://afsic.nal.usda.gov/farms-and-community/urban-agriculture
USEPA (2010) The history of the clean air act. http://www.epa.gov/air/caa/caa_history.html
van Kamp I, Leidelmeijer K, Marsman G, de Hollander A (2003) Urban environmental quality and human well-being: towards a conceptual framework and demarcation of concepts; a literature study. Landsc Urban Plann 81:5–18
VROM (2000) Circular on target values and intervention values for soil remediation Annex A: target values, soil remediation intervention values and indicative levels for serious contamination. Dutch Ministry of Housing, Spatial Planning and Environment
Wang XS, Qin Y (2007) Relationships between heavy metals and iron oxides, fulvic acids, particle size fractions in urban roadside soils. Environ Geol 52:63–69
Wang Y, Shi J, Wang H, Lin Q, Chen X, Chen Y (2007) The influence of soil heavy metals pollution on soil microbial biomass, enzyme activity, and community composition near a copper smelter. Ecotoxicol Environ Saf 67:75–81
Wang Y, Sikora S, Kim H, Dubey B, Townend T (2012) Mobilization of iron and arsenic from soil by construction and demolition debris landfill leachate. Waste Manag 32:925–932
Waterlot C, Pruvot C, Ciesielski H, Douay F (2011) Effects of a phosphorus amendment and the pH of water used for watering on the mobility and phytoavailability of Cd, Pb and Zn in highly contaminated kitchen garden soils. Ecol Eng 37:1081–1093
Wei B, Yang L (2010) A review of heavy metal contaminations in urban soils, urban road dusts and agricultural soils from China. Microchem J 94:99–107
Werkenthin M, Kluge B, Wessolek G (2014) Metals in European roadside soils and soil solution—a review. Environ Pollut 189:98–110
White CS, McDonnell MJ (1988) Nitrogen cycling processes and soil characteristics in an urban versus rural forest. Biogeochemistry 5:243–262
Wieczorek K, Wyszkowska J, Kucharski J (2014) Influence of zinc, copper, nickel, cadmium and lead in soils on acid phosphatase activity. Fresen Environ Bull 23:274–284
Wilcke W, Muller S, Kanchanakool N, Zech W (1998) Urban soil contamination in Bangkok: heavy metal and aluminium partitioning in topsoils. Geoderma 86:211–228
Wong JWC, Mak NK (1997) Heavy metal pollution in children playgrounds in Hong Kong and its health implications. Environ Technol 18:109–115
Wong CSC, Li XD, Zhang G, Qi SH, Peng XZ (2003) Atmospheric deposition of heavy metals in the pearl river delta, China. Atmos Environ 37:767–776
Wong CSC, Li X, Thornton I (2006) Review. Urban environmental geochemistry of trace metals. Environ Pollut 142:1–16
Wortman SE, Lovell ST (2013) Environmental challenges threatening the growth of urban agriculture in the United States. J Environ Qual 42:1283–1294
Xu HJ, Li S, Su JQ, Nie S, Gibson V, Li H, Zhu YG (2014) Does urbanization shape bacterial community composition in urban park soils? A case study in 16 representative Chinese cities based on the pyrosequencing method. FEMS Microbiol Ecol 87:182–192
Yassoglou N, Kosmas C, Asimakopoulos J, Kallianou C (1987) Heavy metal contamination of roadside soils in the greater Athens area. Environ Pollut 47:293–304

Yay OD, Alagha O, Tuncel G (2008) Multivariate statistics to investigate metal contamination in surface soil. J Environ Manage 86:581–594

Zezza A, Tasciotti L (2010) Urban agriculture, poverty, and food security: empirical evidence from a sample of developing countries. Food Policy 35:265–273

Zhao H, Li X (2013) Risk assessment of metals in road-deposited sediment along an urban-rural gradient. Environ Pollut 174:297–304

Zhao D, Li F, Yang Q, Wang R, Song Y, Tao Y (2013) The influence of different types of urban land use on soil microbial biomass and functional diversity in Beijing, China. Soil Use Manag 29:230–239

Zhejiang Soil Survey Office (1994) Zhejiang soils (in Chinese). Zhejiang Technology Press, Zhejiang, China, pp 556–562

Part II
Monitoring Heavy Metals in Soils

Chapter 5
The Adsorption Equilibrium of Pb and Cd on Several Soils in Various pH Solutions or in the Presence of Dissolved Humic Substances

Takeshi Kobayashi, Yuyu Liu, Yukari Takahashi, Takashi Kameya, and Yuko Yamada

5.1 Introduction

Soil contamination by hazardous metals such as lead and cadmium is a serious environmental problem. To understand this problem, an investigation into heavy metal contamination of soil and possible countermeasures was carried out. In Japan, soil contamination is typically assessed by elution tests, which evaluate the extent to which hazardous metals pollute groundwater after permeating through soil, and by content tests, which evaluate how hazardous metals are absorbed by humans through direct ingestion of soil or via inhalation of airborne dust. There is often considerable variation in the elution tests, because the ease with which hazardous metals elute from soil is greatly influenced by the states of metals in soil. Insoluble lead compounds such as lead oxides and lead carbonate seldom cause groundwater contamination. However, water-soluble lead can be dissolved in pore water and attains the adsorption equilibrium with the soil particles. And the elute property is greatly changed by the various environmental factors, such as pH, ORP, and coexisting substances.

This article investigates the adsorption equilibrium of hazardous metals, including lead, and examines how various environmental factors (pH, humic substances, and salt concentration) influence the adsorption and elution of hazardous metals from contaminated soil.

T. Kobayashi (✉) • T. Kameya • Y. Yamada
Graduate School of Environment and Information Sciences, Yokohama National University, Yokohama, Japan
e-mail: koba@ynu.ac.jp

Y. Liu
Graduate School of Environmental Studies, Tohoku University, Sendai, Japan

Y. Takahashi
Toyama University of International Studies, Toyama, Japan

I. Sherameti, A. Varma (eds.), *Heavy Metal Contamination of Soils*, Soil Biology 44, DOI 10.1007/978-3-319-14526-6_5

5.2 The Adsorption Equilibrium

5.2.1 The Adsorption Equilibrium Equation

The relationship between the metal concentration in groundwater and the amount of metals adsorbed onto the soil is often analyzed by adsorption isotherms such as the Freundlich and Langmuir equations (Anghinoni et al. 1996; Atenassova and Okazaki 1997; Houng and Lee 1988; Ramachandran and Souza 1999; Kinniburgh et al. 1999; Limousin et al. 2007; Febrianto et al. 2009). However, these adsorption isotherms are only applicable over a limited concentration range and cannot account for the influence of pH and coexisting substances, which differ according to soil type. Therefore, parameters for the adsorption isotherms have to be determined for each environmental condition.

Urano et al. demonstrated experimentally that the adsorption equilibrium of a metal ion to chelating resin, when the system contains another coexisting ion, can be represented by an adsorption isotherm shown in Eq. (5.1) (a competitive ion-exchange equation) (Urano et al. 1981). As described in the next section, this equation holds for the adsorption of metals to soil:

$$Q_1 = \frac{Q_{\max}}{1 + \frac{C_2^{n_2}}{K_2^1 \cdot C_1^{n_1}}} \tag{5.1}$$

where C_1 and C_2 are the concentrations of metal ion 1 and coexisting ion 2 in the adsorption equilibrium, respectively; Q_1 is the amount of ion 1 exchanged in equilibrium (the adsorbed amount of the ion 1 in equilibrium); n_1 and n_2 are ion-exchange constants, which represent slopes of the ion isotherms; K_2^1 is the selectivity coefficient of ion 1 to ion 2; and $Q_{\max}$ is the maximum ion-exchange capacity of the soil, which is equivalent to the cation-exchange capacity of soil when the metal ion is cationic. Each substance is expressed by equivalent weights.

When multiple coexisting ions are involved, a multicomponent adsorption isotherm expressed by Eq. (5.2) can be applied:

$$Q_1 = \frac{Q_{\max}}{1 + \sum_{k=2}^{\infty} \left(\frac{C_k^{n_k}}{K_k^1} \right) \cdot \frac{1}{C_1^{n_1}}} \tag{5.2}$$

Here, C_k is the equilibrium concentration of the coexisting ion k, n_k is the ion-exchange constant of the coexisting ion k, and K_k^1 is the selectivity coefficient of the target ion to the coexisting ion. When C_1 is small enough, Eq. (5.2) can be approximated by Eq. (5.3). This means that the relationship between the amount exchanged in equilibrium, Q_1, and the equilibrium concentration, C_1, can be linearly approximated on a logarithmic scale, which approaches a Freundlich-type adsorption isotherm:

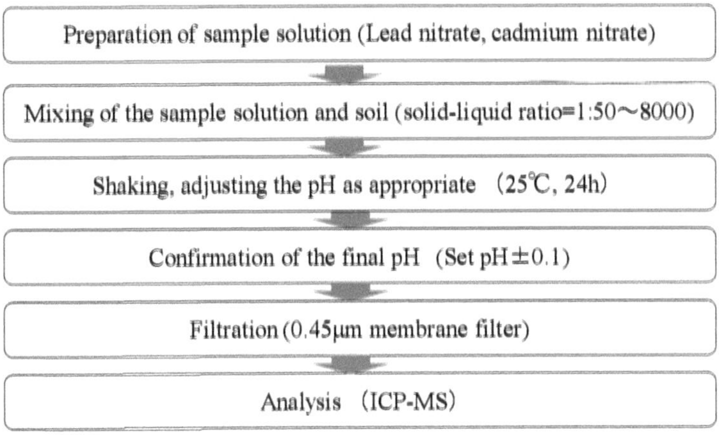

Fig. 5.1 Experimental method of adsorption equilibrium

$$Q_1 = Q_{\max} \cdot \sum_{k=2}^{\infty} \left(\frac{K_k^1}{C_k^{n_k}} \right) \cdot C_1^{n_1} \tag{5.3}$$

5.2.2 *The Adsorption Equilibrium Experiment*

Figure 5.1 outlines the adsorption isotherm experiments (Kameya et al. 2004; Liu et al. 2011). Wet soil and sample solution were prepared at a given solid–liquid ratio and shaken at 25 °C for 24 h, adjusting the pH as appropriate. After checking the final pH, the liquid was filtered through a 0.45 μm membrane filter to separate solid and liquid phases. The metal concentration of the filtrate was quantified by inductively coupled plasma mass spectrometry (ICP-MS). The amount adsorbed in equilibrium was calculated by subtracting the equilibrium concentration of the filtrate from the initial concentrations of the metals. Diluted nitric acid was added to the filtrate to prevent deposition and precipitation of the metals.

5.3 The Examples of Adsorption Equilibrium of Cd

Figure 5.2 is an example of adsorption isotherms of metals. This particular example is an adsorption equilibrium experiment performed with cadmium and andosol at pH 5–7. The abscissa represents the concentration of an equilibrated metal ion aqueous solution, and the ordinate represents the amount of the ions exchanged in equilibrium. Solid lines are obtained through curve fitting by varying the selectivity coefficient K in Eq. (5.1). In this experiment, the cation-exchange capacity (CEC) was used as $Q_{\max}$, and n_{Cd} and n_H were parameters independent of the soil type. The parameters for andosol, Kanto loam, and sandy soil are summarized in Table 5.1. It

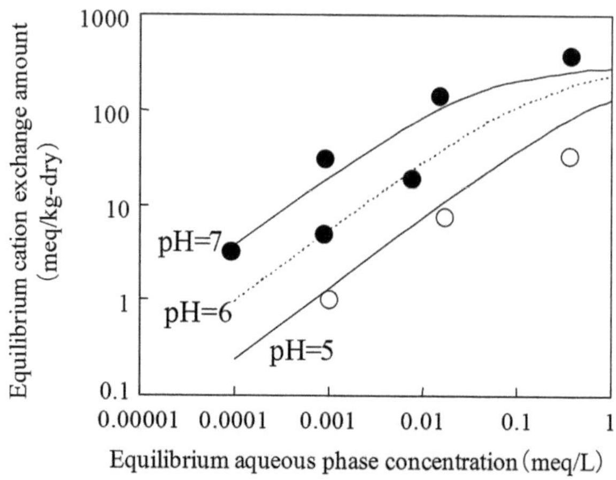

Fig. 5.2 The example of adsorption isotherms of metals (Cd, Andosol)

Table 5.1 The parameters of ion-exchange isotherms

Soil	Andosol	Kanto loam	Sand
Q_{max}	300	200	73
K_H^{Cd}	0.05	0.015	0.20
K_H^{Pb}	0.20	0.15	0.15
$n_H = 0.60$	$n_{Cd} = 0.75$	$n_{Pb} = 060$	

was found that pH variation, which can be regarded as coexistence of hydrogen ions, can be analyzed by Eq. (5.1). It is apparent that Q_{max} and the selectivity coefficient to hydrogen ions vary greatly according to soil type.

5.4 The Presence of Dissolved Humic Substances

Figure 5.3 shows an example of adsorption isotherms of lead on andosol at pH 7 and 5. In the adsorption experiment at pH 7, humus extract solution was added to the sample solution to obtain adsorption isotherms at various organic carbon concentrations (Liu et al. 2011, 2013). When the total organic carbon (TOC), that is, concentration of humic substances, is low, it was found that the adsorption isotherm can be expressed by Eq. (5.1). The parameters are summarized in Table 5.1. At lower equilibrium concentrations (approx. 1×10^{-3} mEq/L or lower) at pH 7 (neutral range), soils with high organic carbon content tended to show slightly higher lead concentrations in the liquid phases. When andosol is used in the experiment, humic substances are eluted into the filtered sample solutions, which are clear, pale-yellow solutions, at pH 7 or higher. Because lead exists in a form that is bound to the humic substances, and given the relative abundance of humic substances in soil, lead bound to the humic substances and dissolved in the water phase becomes impossible to ignore. Accordingly, when adsorption isotherms were obtained with

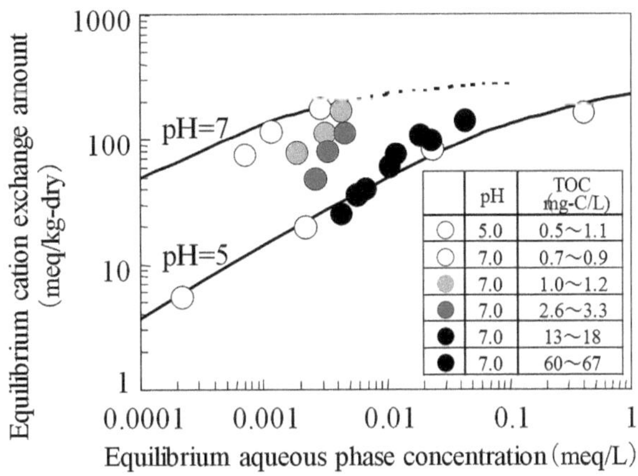

Fig. 5.3 The example of adsorption isotherms at pH 7 and 5, with humic substances (Pb, Andosol)

addition of humic substances, the higher the TOC, the higher the lead concentration was in the liquid phase. It is thought that in soil with high levels of organic carbon, lead is bound to humic substances and elutes from the soil with the humic substances when the pH is high.

This is critical to understanding the underground permeation behavior in basic soils. In addition, lead bound to humic substances can, upon solid–liquid separation, exist in both solid and liquid phases, which contributes to the variation of the measured elution test values.

5.5 The Adsorption Equilibrium of Pb(II) in the Presence of Dissolved Salts

Figure 5.4 shows an example of adsorption isotherms of lead with high salt concentration. The salt concentration was adjusted with artificial seawater. EC = 4,500 indicates a salt concentration similar to that of seawater. This experiment demonstrated that the coexistence of hydrogen ions and lead ions can be analyzed by Eq. (5.2). In addition, when the concentration of equilibrated ions is low and the salt concentration is high, the adsorbed amount in equilibrium will be small. This is because sodium ions act as coexisting ions and compete with the lead ions. This influence is significant at higher pH. Moreover, the influence is greater in sandy soil than in andosol, because the Q_{max} of sandy soil is smaller than that of andosol.

Moreover, the underground permeation of lead was investigated in areas of Japan affected by tsunami. Although there was very little data, the underground permeation of lead into soil with high salt concentration tended to be high.

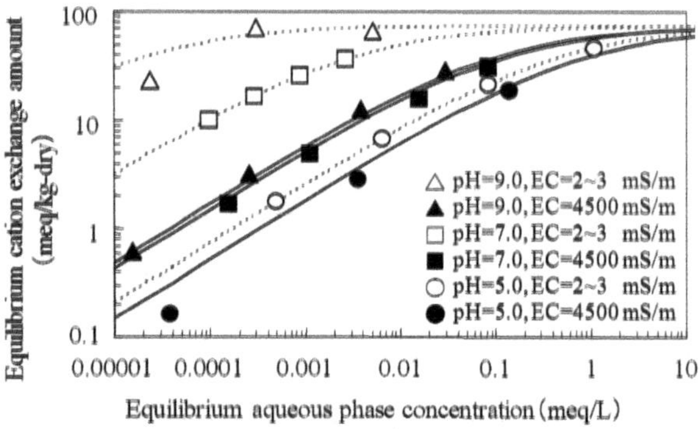

Fig. 5.4 The examples of the effects of salt concentration to the adsorption isotherms (Pb, sand)

5.6 Conclusion

This article described how the adsorption of metal ions such as lead and cadmium onto soil can be represented by the ion-exchange isotherms. The influence of pH, coexisting humic substances, and salt concentration on the solubility of aqueous metal ions is also described. Although this article only addresses cations, anions such as chromium and arsenic can be analyzed in a similar way. Preparing a database of Q_{max} and selectivity coefficients for hydrogen ions and hydroxide ions of various soils will be helpful in analyzing the permeability of contaminated groundwater and understanding the contamination states.

Acknowledgment This work was partially supported by the Environmental Research and Technology Development Fund (B-0914) of the Ministry of the Environment, Government of Japan, and JSPS KAKENHI Grant Number 25550077.

References

Anghinoni I, Baligar VC, Wright RJ (1996) Phosphorus sorption isotherm characteristics and availability parameters of Appalachian acidic soils. Commun Soil Sci Plant Anal 27:2033–2048

Atenassova I, Okazaki M (1997) Adsorption-desorption characteristics of high levels of copper in soil clay fractions. Water Air Soil Pollut 98:213–228

Febrianto J, Kosasih AN, Sunarso J, Ju YH, Indraswati N, Ismadji S (2009) Equilibrium and kinetic studies in adsorption of heavy metals using biosorbent: a summary of recent studies. J Hazard Mater 162:616–645

Houng KH, Lee DY (1988) Comparisons of linear and nonlinear langmuir and Freundlich curve-fit in the study of Cu, Cd, and Pb adsorption on Taiwan soils. Soil Sci 163:115–121

Kameya T, Ohashi Y, Kobayashi T, Urano K (2004) Exchange isotherm of toxic metal cations on soils in consideration of coexistence in wide concentration range for metal contamination survey. J Jpn Soc Civil Eng 776(VII-33):29–37

Kinniburgh DG, van Riemsdijk WH, Koopal LK, Borkovec M, Benedetti MF, Avena MJ (1999) Ion binding to natural organic matter: competition, heterogeneity, stoichiometry and thermodynamic consistency. Colloids Surf A 151:147–166

Limousin G, Gaudet JP, Charlet L, Szenknect S, Barthes V, Krimissa M (2007) Sorption isotherms: a review on physical bases, modeling and measurement. Appl Geochem 22:249–275

Liu Y, Kobayashi T, Kameya T, Takahashi Y, Ohashi Y (2011) Simulation of adsorption equilibrium of heavy metal cations on soils in circumneutral aqueous solution: influences of solution pH and dissolved humus substances. Adv Mater Res 287–290:2822

Liu Y, Kobayashi T, Takahashi Y, Kameya T, Urano K (2013) A simple simulation of adsorption equilibrium of Pb(II) on Andosols in the presence of dissolved humic substances for monitoring soil contamination. J Environ Sci Health A 48:1694–1699

Ramachandran VD, Souza TJ (1999) Adsorption of cadmium by Indian soils. Water Air Soil Pollut 111:225–234

Urano K, Chanyasak V, Fujii N (1981) Adsorption isotherm equation of single heavy metal ion in acidic solution on chelating resin. J Colloid Interface Sci 82(1):251–253

Chapter 6
Arsenic in Soil: Availability and Interactions with Soil Microorganisms

Silvia Rita Stazi, Rosita Marabottini, Ruxandra Papp, and Maria Cristina Moscatelli

6.1 Arsenic a Worldwide Alert

> "All people, regardless of ethnicity, gender, or income from their geographical origin, have the right to a healthy environment, balanced and sustainable." Senator Nelson, 1970 first Earth Day.

Soil is the foundation of the world's ecosystems. A healthy soil supports life.

According to the European Environment Agency (EEA), 250,000 soils have been identified as contaminated across the EU, mainly by metals and mineral oils (EEA 2007). The quality of soil, especially the degree of pollution, can affect human health and ecosystems in many different ways (Malik et al. 2010; Zhang et al. 2011). Soil contaminated by metals and metalloids constitutes a major environmental problem, which may result in significant human exposure (Zhuang et al. 2009). How toxic these exposures are can be subject to a number of variables, including route of exposure, absorption, metabolism, and distribution in the human body (Ajmone-Marsan et al. 2008; Madrid et al. 2008). Paradoxically, As is classified as both an important micronutrient to some animals, including humans, and highly toxic to the same animals, as well as plants, at higher concentration (Agency for Toxic Substance and Disease Registry 2007). Chronic exposure to As has been shown to cause a variety of cancers (Wu et al. 1989; Smith et al. 1992) and is also associated with a variety of other human ailments (Cullen and Reimer 1989; Jain and Ali 2000).

Due to its nature and behavior in the environment, As is persistent, does not degrade, and can bioaccumulate and biomagnify in the trophic chain (Huq et al. 2006). Background concentrations in soil range from 1 to 40 $\mu g\ g^{-1}$ (Agency for Toxic Substance and Disease Registry 2007). Total As concentration in soils can

S.R. Stazi (✉) • R. Marabottini • R. Papp • M.C. Moscatelli
Department of Biological, Agrifood and Forest Systems, DIBAF, University of Tuscia, Viterbo, Italy
e-mail: srstazi@unitus.it

I. Sherameti, A. Varma (eds.), *Heavy Metal Contamination of Soils*, Soil Biology 44, DOI 10.1007/978-3-319-14526-6_6

reach levels up to more than 1,000 mg kg^{-1} (Mukhopadhyay et al. 2002) and in waters may reach concentrations as high as 5,000 μg L^{-1} (Smedley and Kinniburgh 2002).

The nature of the soil parent material is a major factor determining the As concentration in soils, even though soils are enriched in As compared with their parent rocks.

Although As is a naturally occurring element in the Earth's crust and is commonly found as a trace constituent of rocks, soils, sediments, water, and biota (Vaughan 2006), both natural and anthropogenic processes can elevate concentrations in ore bodies, groundwater, selected natural surface water bodies, soils, and sediments to toxic levels.

There are various natural As sources in soil and many of them are being currently analyzed and others are to be found. The hot spots of As contaminations are concentrated in the earth's crust associated with sulfide minerals and hydrothermal fluids (Henke 2009). They present a continuous risk due to erosion and weathering of the As containing rocks that enter in a complicated cycle characterized mainly by oxido-reduction processes. Hydrothermal fluids also contain a high concentration of soluble inorganic As, compounds incorporated from minerals and rocks. Anthropogenic sources include mining, smelting of nonferrous metals, and burning of fossil fuels. Other contributors to the anthropogenic As deposition in agricultural soils are use of arsenical herbicides, insecticides, and wood preservatives, application of phosphate fertilizers, and irrigation with As-contaminated groundwaters.

Arsenic occurs naturally in the environment in inorganic and organic forms; it has four oxidation states, which are denoted −3, 0, +3, and +5, such as arsenic acid (H_3AsO_4) and arsenous acid (H_3AsO_3) and their dissociation derivatives $H_2AsO_4^{-}$, $HAsO_4^{-2}$, $H_2AsO_3^{-}$, and $HAsO_3^{-2}$, arsenites, arsenates, monomethylarsenic acid $(CH_3)_2AsO(OH)$ (MMA), and dimethylarsenic acid (DMA) $(CH_3)AsO(OH)_2$.

In soil pH range 4–8 the most thermodynamically stable species are As(III) and As(V) (Smith et al. 1998), from which the pentavalent form comes in the highest proportion, in oxic conditions. As in water bonds with oxygen and forms *arsenites* As_2O_3 (As(III)) and *arsenates* As_2O_5 (As(V)).

As(III) is the predominant species under anaerobic conditions, and it is more soluble, mobile, and phytotoxic than As(V). As(V) is the main As species in aerobic soils. Many As-bearing minerals associated with sulfides have been identified; in general, arsenide, arsenic-sulfide, and arsenic-rich sulfide minerals are associated with anoxic hydrothermal ore deposits and *metamorphic* and *intrusive igneous rocks*. Once these minerals are exposed to oxygen and water under surface or in near-surface conditions, they weather to arsenite and arsenate minerals (Henke 2009).

Because of the chemical similarity between As(V) and phosphate, application of phosphate causes the desorption of As(V) from soil colloids and increases its concentration in the soil solution, enhancing As phytoavailability. Biological organisms can also create less volatile and more complex organoarsenicals, such as arsenic sugars and methyl arsenates (e.g., DMA(V) and MMA(V)). Organo

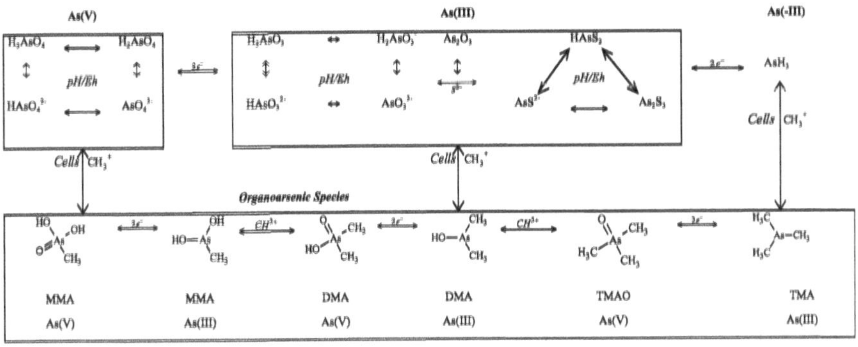

Fig. 6.1 Main As species and inter-transformation (adapted from Wang and Mulligan 2006)

arsenicals are sometimes detected in water, sediments, soils, and rocks, especially in geologic materials that are rich in biological carbon. However, their concentrations are usually higher in biological samples, including algae, plant parts, urine, hair, and animal carcasses (Francesconi et al. 2002).

The following description refers to the most important As natural and artificial compounds that are found in water and soil environment. Figure 6.1 shows the main forms of As in nature and how they transform according to different influences.

6.2 Fate of Arsenic in Soil

The fate and mobility of As in soil are mainly controlled by several factors such as redox potential (*Eh*), pH, dissolved organic carbon (DOC), the contents of iron/aluminum/manganese oxides/hydroxides, clay contents, and chemical speciation that play a crucial role in these processes (Martin et al. 2007). Consequently, the occurrence, distribution, and mobility of As are dependent on the interplay of these geochemical and biological factors that can differ markedly between soil types, growth seasons, and land use (Cheng et al. 2009).

As(V)/As(III) reactions (equilibrium constants) can be predicted from thermodynamic properties of the redox couples involved in them both in soils and groundwater environments. For example, Fe(III) (hydro)oxide minerals can oxidize arsenite, but only when the pH is below 4–5; at pH 8 this oxidation reaction does not take place. Redox reactions can control aqueous As concentrations by their effects on As speciation and consequently As adsorption and desorption.

Arsenate and arsenite can be adsorbed on mineral surface forming two different complexes: outer-sphere (nonspecific adsorption) and inner-sphere (specific adsorption) complexes. Outer-sphere complexes are due to (strong or weak) nonspecific interactions of the external shell of the solvated anion containing As and the surface, for example, between the water molecules solvating the arsenate ion in solution and the surface. Inner-sphere complexes, instead, are due to the direct interaction of the ion and the surface, for example, the interaction of an oxygen of the arsenate and a metal of the surface. As(V) binds more strongly on positively charged surfaces like the (hydro)oxides of Fe, Mn, Al, and Ca and these may form precipitates or coprecipitates (Mohan and Pittman 2007), as compared to the As(III) species (Bissen and Frimmel 2003). However, the binding mechanisms are dependent on the pH and redox potential of the environment. In fact, *Eh* and pH play a key role in controlling speciation and distribution of As species between solid phases and aqueous phases. The strong adsorption of As(V) by Fe and Al (hydro)oxides at low pH and that of As(III) at higher pH values have been observed in many experimental studies (Dixit and Hering 2003). As species may compete with inorganic anions (similar or with higher charge densities) such as phosphate (PO_4^{3-}), sulfate (SO_4^{2-}), and carbonate (CO_3^{2-}). The competition is for the same sorption sites on mineral surfaces affecting arsenate adsorption and desorption reactions. In particular, phosphate, which has similar chemical properties and behavior as arsenate, strongly competes with arsenate for sorption sites on metal oxides surfaces (Violante and Pigna 2002). Organic species may interfere with As adsorption from minerals and soils (Chen et al. 2008) and/or may chemically mobilize As by promoting redox reaction (Bauer and Blodau 2006; Mohapatra et al. 2007). In natural environments, arsenate may form precipitates or coprecipitates with Al, Fe, Mn, and Ca (Sadiq 1997) forming different arsenic-containing salts. Also microorganisms play a significant role in the biogeochemistry of As in the environment; they interact with As through a variety of mechanisms, including sorption, mobilization, precipitation, and redox and methylation transformation (Huang 2014).

Total concentration of As in soils does not necessarily represent its biological availability or potential toxicity measures which are much more important for assessing possible environmental impacts. Most risk from As is, in fact, associated with As forms biologically available for absorption, or "bioavailable" to plants, soil microbes, and finally humans. Bioavailability is a function of the abundance, chemical form (i.e., oxidation state), the nature of its binding to soil particles, and biological factors (Marabottini et al. 2013). In Italy, a complex study, performed by (Cubadda et al. 2010), reported the amount of phytoavailable As fractions and all data were mapped and can be used as guideline for future agricultural activities.

Chemical fractionation methods, which consist in sequential extraction procedures (SEP) with solvent of growing extractive power, allow to determine the bioavailability of As (Marabottini et al. 2013): As extracted in "milder" solvents is more bioavailable, while that extracted in "stronger" ones is less bioavailable. Thus, chemical fractionation methods allow a more correct evaluation of the potential health and environment risks, supporting the development of more accurate strategies for waste disposal. The fractions of As in soil may be classified according to the different availability. As is adsorbed strongly to the organic, and inorganic, colloidal constituents of soil. The resulting moderation of mobility and bioavailability of As in the environment must rank as one of the most important "ecosystem services" attributable to soils with implications for both micronutrient availability and potential toxicity to terrestrial and aquatic organisms. The lowest

As concentrations are found in sandy soils and those derived from granites, whereas higher concentrations are found in alluvial and organic soils.

6.3 Arsenic and Soil Microorganisms

6.3.1 Effects on Soil Microbial Biomass

There exists a considerable body of literature on the effects of metals and metalloids on soil microbial biomass and microbial-mediated processes related to nutrient cycling, C sequestration, and organic matter decomposition (Tripathy et al. 2014). Accumulation of metals and metalloids may modify microbial biomass and activity and affect the structure and diversity of soil microbial communities (Bååth 1989). The role of soil microorganisms, mainly fungi and bacteria, and the different processes they perform are fundamental for soil functioning. Moreover soil microbial biomass is one of the most important indicators of soil quality and may thus serve as a sensitive tool to assess the impact of heavy metal pollution on soil resource.

High heavy metal loads may modify soil microorganisms in total biomass amount, activity, and their genetic/functional diversity (Gans et al. 2005). As a consequence, a decrease in diversity may lead to less resistance to additional disturbances (Degens et al. 2001). Decreased size of microbial biomass can be partially explained by physiological causes such as a decrease in the microbial substrate utilization efficiency and an increased maintenance energy requirement (Giller et al. 2009). Pollution leads to a decrease in microbial diversity in terms of species richness due to the extinction of species lacking sufficient tolerance or to shifts toward more tolerant groups.

As far as arsenic is concerned, the current literature, actually, reports positive, negative, or no effects on soil microorganisms and their processes (Lorenz et al. 2006; Speir et al. 1999). This can be due to the actual toxicity level, soil type and physicochemical properties, the fraction of soil the metal is associated with, and synergic effects with other pollutants. The ability of a microbial community to perform equally well under polluted and non-polluted situations could also be determined by the extent to which the community can acquire tolerance to the pollutant in question.

The time factor is another important factor in this context; chronic pollution conditions and spike addition of pollutants in laboratory experiments can produce different effects on soil biological properties (Nordgren et al. 1988).

Recent ecotoxicological studies, performed in the Netherlands, reported a ranking for 17 metals/metalloids in relation to their potential hazard to soil biota: Se > Tl > Sb > Cd > V > Hg > Ni > Cu > Cr > As > Ba > Zn > Co > Sn > Ce > Pb > Mo (Vodyanitskii 2013). However, the author reports that this ranking does not take into account the diversity of mineralogical and chemical forms of pollutants. In fact,

Table 6.1 Pearson's correlation coefficient between total As in soil and different bioavailable As forms—(WS: water soluble, NSS: not specifically sorbed) to microbial parameters: As MBC = microbial incorporated As, *q*mic = microbial quotient, qCO_2 = metabolic quotient

As forms	AsMBC	*q*mic	Bas. Resp	Cum. Resp	qCO_2
Total	ns	ns	ns	ns	ns
WS	−0.547*	ns	−0.529*	−0.697**	−0.523*
NSS	ns	ns	ns	−0.540*	ns
Total bioav.	ns	ns	−0.521*	−0.667**	ns

Significance of Pearson's correlation coefficient is reported (ns not significant; $^{*}p < 0.05$; $^{**}p < 0.01$)

potential risk of heavy metals is strictly related to their bioavailability, which in such a complex matrix as soil may be negligible or worrying in relation to the soil fraction they are associated with (Das et al. 2013; Marabottini et al. 2013; Ghosh et al. 2004); all the above authors reported, indeed, a close relationship between most microbial indicators and water-soluble heavy metals in soil, this fraction being the one mostly influencing microbial metabolism. Table 6.1 shows correlation between microbial parameters and As in soil: microbial parameters, except the microbial quotient, were all inversely correlated to the different forms of bioavailable As (water soluble, WS; not specifically sorbed, NSS; and total bioavailable), WS being the fraction most influencing microbial respiration and indexes.

Tolerance and adaptation of microorganisms to As contamination are common phenomena and the presence of tolerant fungi and bacteria in polluted environments has frequently been observed. The increased abundance of tolerant organisms in polluted environments can be due to genetic changes, physiological adaptations involving no alterations in the genotype, or replacement of metal sensitive species with species that already are tolerant to the heavy metal.

Soil microorganisms may be involved in As pollution issues in an active and passive way at the same time. They can induce environmental As bioavailability due to chemical–biochemical processes, microbial mediated, that favor its release in water-soluble forms or more mobile (As III). Conversely, soil microbial biomass can be deeply affected in its size, functions, and diversity by the concentration of this ion in soil solution and impact in this way soil functioning.

However there is still much research to be carried out about this environmental threat affecting human and environmental health at the same time. In particular, a search performed on SCOPUS database (July 11th, 2014), using as keywords "arsenic," "soil," "microbial biomass," and "diversity" combined together consecutively, sorted the following numbers of articles: 6662 "As" and "soil," 65 "As" and "soil" and "microbial biomass," and only 8 "As" and "soil" and "microbial biomass" and "diversity."

Considerable variation in tolerance to As compounds applied to the soil has been shown by the soil microflora. Maliszewska et al. (1985) reported that As(III) compounds were more toxic than As(V) compounds to microorganisms important to maintaining soil fertility. These effects were strictly dependent on soil type (e.g., texture) and different microbial groups, fungi being more tolerant than bacteria.

6.3.2 *Effects on Soil Enzymes*

Most studies published in the last decades have focused on the effects of only four metals, two of which are toxic (Cd and Pb) and two of which are essential elements or micronutrients (Cu and Zn), as these metals are often found as soil contaminants. Other metals such as Cr and Ni have been studied on some occasions, whereas studies involving V, As, Co, Hg, Ag, Mn, and Se are relatively scarce. The studies also vary widely in regard to the enzyme(s) used as indicators of the effects of the metals (Bello et al. 2014). Ions of metalloids like As are assumed to inactivate enzymes by reacting with sulfhydryl groups, a reaction analogous to the formation of As sulfide. Sulfhydryl groups in enzymes serve as integral parts of the catalytically active sites or as groups involved in maintaining the correct structural relationship of the enzyme protein (Tabatabai 1977). As can also reduce enzyme activity by interacting with enzyme–substrate complex, by denaturing the enzyme protein, or by interacting with the protein active groups (Dick 1997). Arsenic is toxic to almost all bacteria by inhibiting basic cellular functions, which are linked with energy metabolism.

Most enzymes exhibit significant decrease under As pollution; however, several authors report a specific behavior of acid phosphatase showing no significant effect due to As presence in soil (dos Santos et al. 2013; Lyubun et al. 2013; Stazi personal communication), preventing this hydrolase from being a good indicators of soil quality under As pollution.

Table 6.2 reports a synthesis of the effects of diverse As concentrations in soil on several microbial parameters contaminated by natural and anthropic sources.

6.3.3 *Tolerance and Adaptation Mechanisms*

An important help to clear the interactions of arsenic forms in soil and soil microorganisms was given by the multidisciplinary approach combining the knowledge of different fields of studies from chemistry, geochemistry, and microbiology with the contribution of genetics.

Bacteria isolated from natural polluted sites together with contaminated soils from mine tailings are a reservoir of important information on microbial adaptation mechanisms. A brief summary of these behaviors will be presented encompassing both arsenite As(III) and arsenate As(V).

It is already known that As is a bizarre metalloid and, especially in soil, it can be influenced by various factors from pH, phosphorus content, redox potential, organic matter composition, Fe oxides, etc. The same complex trend is reflected in the

Table 6.2 Arsenic effects on microbial parameters in soils contaminated by natural and anthropic sources

As pollution level (ppm)	Microbial biomass size	Diversity	Fungi	Bacteria	Basal respiration	Enzyme activities	Source of pollution	Reference
0–50	+	n.d.	n.d.	n.d.	+	–	Lab exp	Speir et al. (1999)
10–40	n.d.	–	n.d.	n.d.	n.d.	–	Anthropic	Lorenz et al. (2006)
200–4,000	x	n.d.	n.d.	n.d.	–	n.d.	Anthropic	Marabottini et al. (2013)
200–4,000	–	–	–	–	n.d.	–	Anthropic	Stazi personal communication
10–40	–	n.d.	n.d.	n.d.	–	–	Natural	Ghosh et al. (2004)
10–40	n.d.	n.d.	n.d.	n.d.	n.d.	–	Natural	Bhattacharyya et al. (2008)
10–400	–	n.d.	–	x	n.d.	–	Natural	Das et al. (2013)
2,400	–	–	n.d.	+As resistant	n.d.	–x	Natural	Garau et al. (2011)
34–2,940	–	–	–	–	–	n.d.	Natural	Edvantoro et al. (2003)
0–600	–	–	–	–	–	x	Anthropic	dos Santos et al. (2013)

+: increase, –: decrease, *x*: *no effect*, *n.d.*: not determined

behavior of soil microbes. As everything in the microbial world has to do with survival even in hostile polluted environments, there is wide a resistance to As found at the soil microbial biomass especially bacteria (Silver et al. 2002). The presence of oxygen and therefore the oxidative states of As seem to be the main trigger of the metabolic mechanisms. They can be divided into defense mechanisms, modifying the valence of the metalloid or using As as an energy source.

6.3.4 Arsenic Reducing Organisms

Anaerobic conditions support the growth of dissimilatory arsenate reducing prokaryotes (DAPS); therefore, As breathing bacteria, usually G+ Eubacteria and G− Proteobacteria, are thermophiles and can survive in extreme environments with high pH and salinity (Oremland and Stolz 2003) where As(V) is used as an electron acceptor. Importantly, DARPs can reduce either aqueous or solid-phase As (V) found in minerals like scorodite (Oremland and Stolz 2005). The key enzyme in these processes is As reductase.

In aerobic conditions microbes developed a defense mechanism encoded in operons containing ARS genes. These microorganisms are called Arsenic-Resistant Microbes (ARMs). As(V) is transported by proteins in the cell, reduced to As(III) by cytoplasmic reductase (Ars C), and then released out of the cell by specific efflux pumps (ArsB) able to extrude As from the cell cytoplasm (Silver et al. 2002). In some cases an ATP-ase is bound to Ars B to facilitate As (III) efflux (Yamamura and Amachi 2014).

The basic biochemical systems are Ars R, B, C found in the *E. coli* genome and on *Staphylococcus aureus* (Kruger et al. 2013); the process is not generating any energy. Strangely, the product of this defense strategy is 50 times more toxic than As(V); it is presumed to be an evolved mechanism under anoxic conditions where As(V) wasn't present. When the atmosphere became oxic and As(V) became abundant, arsenate reductases are believed to have evolved quickly from phosphatase to complement the arsenic resistance pathway (Kruger et al. 2013).

There is much more information found for the ARM bacteria as they were discovered first. It is thought that only DARP and Fe (III) reducers are involved in mineral dissolution and bioreduction of solid-phase As(V) in aquifer materials (Oremland and Stolz 2005). Furthermore, in anoxic environments bacteria like *Clostridium* and *Desulfovibrio* have the Ars defense system (Oremland and Stolz 2003).

6.3.5 Arsenic Oxidating Organisms

Under anoxic conditions As(III) is oxidized by phototrophic bacteria using arsenite oxidase enzyme; this transformation may have occurred in oxygen-free arsenic-rich environments as a partial detoxification mechanism (Cavalca et al. 2013).

In both aerobic and anaerobic conditions, organisms (e.g. Eubacteria) called chemo-litho autotrophic arsenite oxidizers (CAO) oxidize As(V) for energy (using arsenite as an electron donor and CO_2 or HCO_3^- as a carbon source). At the same time with As oxidation, the reduction of oxygen or nitrate occurs and the energy is used to fix CO_2 into organic cellular material and achieve growth (Oremland and Stolz 2003).

Another group are heterotrophic arsenite oxidizers (HAO) bacteria that cannot develop without organic matter. HAO have been isolated including *Alcaligenes faecalis*, *Agrobacterium albertimagni*, and three *Thennus* sp. The genes involved in prokaryotic aerobic As(III) oxidation were recently unified under the name *aio* (Lett et al. 2012) and they were seen even in soils and sediments containing even low As levels (Ellis et al. 2001). By transforming the more toxic As(III) into less toxic As(V) and at the same time gaining energy, these bacteria may have an ecological advantage over other microorganisms (Cavalca et al. 2013). The oxidation of As(III) encountered on the cell's outer membrane into the less toxic form perhaps makes it less likely to enter the cell (Oremland and Stolz 2005).

6.3.6 Bioremediation Techniques

The mechanisms previously presented can hold the key to a better understanding of the arsenic's sources and sinks in the biogeochemical cycle. Knowing the microbial community structure and the dominant taxa in the studied zones can simplify choosing between specific bioremediation methods.

Unfortunately classic soil As bioremediation is time- and money-consuming activities. The polluted soils are mainly replaced by other soils and the contaminated ones are further solidified/stabilized in special containers (Yamamura and Amachi 2014). An answer comes from new techniques as precipitating Fe oxides in the contaminated soil by adding ferrous sulfate and lime that control the sorption of dissolved As (Warren et al. 2003). As for microorganisms, a special environmental importance is given to DARPs due to their reductive reactions fueled by the oxidation of organic substrates and release of As in groundwaters and soil (Cavalca et al. 2013); while ARMS are involved in reducing As(V) present only in water. The DARP mechanism of reduction to a less adsorptive As(III) can promote the removal from solid to the aqueous phase applicable for soil remediation (Yamamura and Amachi 2014). Moreover, Fe-oxidizing microbes release As(V) that will be further reduced to As(III) as a possible factor in increasing the concentrations of the metalloid. For water purification, there are scientific studies using oxidizing bacteria in different pack-bed columns, some of them having a 95 % efficiency in removal of As(III) (Cavalca et al. 2013).

Soil bioremediation techniques are well reviewed by Rahaman et al. (2014) describing a wide range of living organisms used. According to the study, the following strategies are adopted by the organisms: active extrusion of As, intracellular chelation (eukaryotes) by several metal-binding peptides, transformation onto organic forms with reduced toxicity, and compartmentalization. Types of organisms used in As removal are prokaryotes, especially DARPs, eukaryotes with a special attention to ferns like *Pteris vittata*, maize plants, gymnosperms, *Eucalyptus cladocalyx* (especially for mine tailings), etc.

Lately, genetically engineered microbes are produced because they can convert As(III) into its less harmful form (enzyme-mediated redox reactions); they can promote the cellular accumulation of As by highly selective As binding ligands (Singh et al. 2008). An important aspect is that they can be associated both with GM plants and with naturally occurring tolerant plants, synergistically removing As and its toxic effects. In the same paper by Rahman et al. (2014) As detoxification mechanisms of bio-organisms are described as being:

(a) enzyme-mediated redox reactions, As oxidase and/or reductase, mediated by both aerobic and anaerobic bacteria; (b) bio-volatilization, a natural process conducted by genetically engineered organisms with methylating capabilities; (c) intracellular chelation, where As(III) is converted into inactive complexes by binding itself to many types of intracellular chelating proteins or peptides; (d) extrusion or the suppression of As cytotoxicity, seen in yeast species, by the extrusion pumps that transport As(III) outside the cell membrane; and (e) transformation of As into an organic form by methylation and further volatilization.

There are many aspects still to be elucidated in order to assess the complex and diverse interactions of soil microorganisms and As. The difficulty is mainly given by the metalloid's high sensitivity to biotic and abiotic factors. It is time that health and environmental problems linked to As are found and strictly monitored in time using all the available scientific approaches.

References

Agency for Toxic Substance and Disease Registry (2007) Toxicological profile for arsenic. US Department of Health and Human Services, Agency for Toxic Substances and Disease Registry, Atlanta, GA, p 193

Ajmone-Marsan F, Biasioli M, Kralj T, Grčman H, Davidson CM, Hursthouse AS, Madrid L, Rodrigues S (2008) Metals in particle-size fractions of the soils of five European cities. Environ Pollut 152(1):73–81, http://dx.doi.org/10.1016/j.envpol.2007.05.020

Bååth E (1989) Effects of heavy metals in soil on microbial processes and populations (a review). Water Air Soil Pollut 47(3–4):335–379. doi:10.1007/BF00279331

Bauer M, Blodau C (2006) Mobilization of arsenic by dissolved organic matter from iron oxides, soils and sediments. Sci Total Environ 354:179–190

Bello D, Trasar-Cepeda C, Gil-Sotres F (2014) Enzymes and environmental contaminants significant to agricultural sciences in enzymes in agricultural sciences. OMICS Group International, Foster City, CA

Bhattacharyya P, Tripathy S, Chakrabarti K, Chakraborty A, Banik P (2008) Fractionation and bioavailability of metals and their impacts on microbial properties in sewage irrigated soil. Chemosphere 72(4):543–550, http://dx.doi.org/10.1016/j.chemosphere.2008.03.035

Bissen M, Frimmel FH (2003) Arsenic, a review. Part I: occurrence, toxicity, speciation, mobility. Acta Hydrochim Hydrobiol 31:9–18

Cavalca L, Corsini A, Zaccheo P, Andreoni V, Muyzer G (2013) Microbial transformations of arsenic: perspectives for biological removal of arsenic from water. Future Microbiol 8 (6):753–768. doi:10.2217/fmb.13.38

Chen Z, Cai Y, Liu G, Solo-Gabriele H, Snyder GH, Cisar JL (2008) Role of soil-derived dissolved substances in arsenic transport and transformation in laboratory experiments. Sci Tot Environ 406:180–189

Cheng H, Hu Y, Luo J, Xu B, Zhao J (2009) Geochemical processes controlling fate and transport of arsenic in acid mine drainage (AMD) and natural systems. J Hazard Mater 165:13–26

Cubadda F, Ciardullo S, D'Amato M, Raggi A, Aureli F, Carcea M (2010) Arsenic contamination of the environment-food chain: a survey on wheat as a test plant to investigate phytoavailable arsenic in Italian agricultural soils and as a source of inorganic arsenic in the diet. J Agric Food Chem 58(18):10176–10183

Cullen WR, Reimer KJ (1989) Arsenic speciation in the environment. Chem Rev 89(4):713–764. doi:10.1021/cr00094a002

Das S, Jean J-S, Kar S, Chakraborty S (2013) Effect of arsenic contamination on bacterial and fungal biomass and enzyme activities in tropical arsenic-contaminated soils. Biol Fertil Soils 49(6):757–765. doi:10.1007/s00374-012-0769-z

Degens BP, Schipper LA, Sparling GP, Duncan LC (2001) Is the microbial community in a soil with reduced catabolic diversity less resistant to stress or disturbance? Soil Biol Biochem 33 (9):1143–1153, http://dx.doi.org/10.1016/S0038-0717(01)00018-9

Dick RP (1997) Soil enzyme activities as integrative indicators of soil health. In: Pankhurst CE, Doube BM, Gupta VVSR (eds) Biological indicators of soil health. CAB International, New York, NY, pp 121–156

Dixit S, Hering JG (2003) Comparison of arsenic(V) and arsenic(III) sorption onto iron oxide minerals: implications for arsenic mobility. Environ Sci Technol 37:4182–4189

dos Santos JV, de Melo Rangel W, Azarias Guimarães A, Duque Jaramillo P, Rufini M, Marra L, Varón López M, Pereira da Silva M, Fonsêca Sousa Soares C, de Souza Moreira F (2013) Soil biological attributes in arsenic-contaminated gold mining sites after revegetation. Ecotoxicology 22(10):1526–1537. doi:10.1007/s10646-013-1139-9

Edvantoro BB, Naidu R, Megharaj M, Singleton I (2003) Changes in microbial properties associated with long-term arsenic and DDT contaminated soils at disused cattle dip sites. Ecotoxicol Environ Saf 55(3):344–351, http://dx.doi.org/10.1016/S0147-6513(02)00092-1

European Environment Agency (EEA) (2007) Progress in management of contaminated sites (CSI 015). European Environment Agency, Copenhagen

Ellis PJ, Conrads T, Hille R, Kuhn P (2001) Crystal structure of the 100 kDa arsenite oxidase from alcaligenes faecalis in two crystal forms at 1.64 Å and 2.03 Å. Structure 9(2):125–132, http://dx.doi.org/10.1016/S0969-2126(01)00566-4

Francesconi K, Visoottiviseth P, Sridokchan W, Goessler W (2002) Arsenic species in an arsenic hyperaccumulating fern, Pityrogramma calomelanos: a potential phytoremediator of arsenic-contaminated soils. Sci Total Environ 284:27–35

Gans J, Wolinsky M, Dunbar J (2005) Computational improvements reveal great bacterial diversity and high metal toxicity in soil. Science 309(5739):1387–1390. doi:10.1126/science.1112665

Garau G, Silvetti M, Deiana S, Deiana P, Castaldi P (2011) Long-term influence of red mud on As mobility and soil physico-chemical and microbial parameters in a polluted sub-acidic soil. J Hazard Mater 185(2–3):1241–1248, http://dx.doi.org/10.1016/j.jhazmat.2010.10.037

Ghosh AK, Bhattacharyya P, Pal R (2004) Effect of arsenic contamination on microbial biomass and its activities in arsenic contaminated soils of Gangetic West Bengal, India. Environ Int 30 (4):491–499, http://dx.doi.org/10.1016/j.envint.2003.10.002

Giller KE, Witter E, McGrath SP (2009) Heavy metals and soil microbes. Soil Biol Biochem 41 (10):2031–2037, http://dx.doi.org/10.1016/j.soilbio.2009.04.026

Henke K (2009) Arsenic: environmental chemistry, health threats and waste treatment. Wiley, Hoboken, NJ

Huang J (2014) Impact of microorganisms on arsenic biogeochemistry: a review. Water Air Soil Pollut 225:1548

Huq SMI, Joardar JC, Parvin S, Correll R, Naidu R (2006) Arsenic contamination in food-chain: transfer of arsenic into food materials through groundwater irrigation. J Health Popul Nutr 24:305–316

Jain CK, Ali I (2000) Arsenic: occurrence, toxicity and speciation techniques. Water Res 34 (17):4304–4312. doi:10.1016/S0043-1354(00)00182-2

Kruger M, Bertin P, Heipieper H, Arsène-Ploetze F (2013) Bacterial metabolism of environmental arsenic—mechanisms and biotechnological applications. Appl Microbiol Biotechnol 97 (9):3827–3841. doi:10.1007/s00253-013-4838-5

Lett MC, Muller D, Lievremont D, Silver S, Santini J (2012) Unified nomenclature for genes involved in prokaryotic aerobic arsenite oxidation. J Bacteriol 194(2):207–208. doi:10.1128/JB.06391-11

Lorenz N, Hintemann T, Kramarewa T, Katayama A, Yasuta T, Marschner P, Kandeler E (2006) Response of microbial activity and microbial community composition in soils to long-term arsenic and cadmium exposure. Soil Biol Biochem 38(6):1430–1437, http://dx.doi.org/10.1016/j.soilbio.2005.10.020

Lyubun YV, Pleshakova EV, Mkandawire M, Turkovskaya OV (2013) Diverse effects of arsenic on selected enzyme activities in soil–plant–microbe interactions. J Hazard Mater 262:685–690, http://dx.doi.org/10.1016/j.jhazmat.2013.09.045

Madrid F, Biasioli M, Ajmone-Marsan F (2008) Availability and bioaccessibility of metals in fine particles of some urban soils. Arch Environ Contam Toxicol 55(1):21–32. doi:10.1007/s00244-007-9086-1

Malik R, Jadoon W, Husain S (2010) Metal contamination of surface soils of industrial city Sialkot, Pakistan: a multivariate and GIS approach. Environ Geochem Health 32(3):179–191. doi:10.1007/s10653-009-9274-1

Maliszewska W, Dec S, Wierzbicka H, Woźniakowska A (1985) The influence of various heavy metal compounds on the development and activity of soil micro-organisms. Environ Pollut Ser A Ecol Biol 37(3):195–215, http://dx.doi.org/10.1016/0143-1471(85)90041-8

Marabottini R, Stazi SR, Papp R, Grego S, Moscatelli MC (2013) Mobility and distribution of arsenic in contaminated mine soils and its effects on the microbial pool. Ecotoxicol Environ Saf 96:147–153, http://dx.doi.org/10.1016/j.ecoenv.2013.06.016

Martin M, Violante V, Barberis E (2007) Fate of arsenite and arsenate in flooded and not flooded soils of southwest Bangladesh irrigated with arsenic contaminated water. J Environ Sci Health 42:1775–1783

Mohan D, Pittman CU Jr (2007) Arsenic removal from water/wastewater using adsorbents - a critical review. J Hazard Mater 142:1–53

Mohapatra D, Mishra D, Rout M, Chaudhry GR (2007) Adsorption kinetics of natural dissolved organic matter and its impact on arsenic(V) leachability from arsenic-loaded ferrihydrite and Al-ferrihydrite. J Environ Sci Health 42:81–88

Mukhopadhyay R, Rosen BP, Phung LT, Silver S (2002) Microbial arsenic: from geocycles to genes and enzymes. FEMS Microbiol Rev 26:311–325

Nordgren A, Baath E, Soderstorn B (1988) Evaluation of soil respiration characteristics to assess heavy metal effect on soil microorganism using glutamic acid as a substrate. Soil Biol Biochem 20:949–954

Oremland RS, Stolz JF (2003) The ecology of arsenic. Science 300(5621):939–944. doi:10.1126/science.1081903

Oremland RS, Stolz JF (2005) Arsenic, microbes and contaminated aquifers. Trends Microbiol 13 (2):45–49. doi:10.1016/j.tim.2004.12.002

Rahman S, Kim KH, Saha SK, Swaraz AM, Paul DK (2014) Review of remediation techniques for arsenic (As) contamination: A novel approach utilizing bio-organisms. J Environ Manage 134:175–185. doi:10.1016/j.jenvman.2013.12.027

Sadiq M (1997) Arsenic chemistry in soils: an overview of thermodynamic predictions and field observations. Water Air Soil Pollut 93:117–136

Silver S, Phung LT, Rosen BP (2002) Arsenic metabolism: resistance, reduction, and oxidation. In: Jr WTF (ed) Environmental chemistry of arsenic. Marcel Dekker, Inc., New York, NY, pp 247–272

Singh S, Mulchandani A, Chen W (2008) Highly selective and rapid arsenic removal by metabolically engineered Escherichia coli cells expressing Fucus vesiculosus metallothionein. Appl Environ Microbiol 74(9):2924–2927

Smedley PL, Kinniburgh DG (2002) A review of the source, behaviour and distribution of arsenic in natural waters. Appl Geochem 17(5):517–568, http://dx.doi.org/10.1016/S0883-2927(02)00018-5

Smith AH, Hopenhayn-Rich C, Bates MN, Goeden HM, Hertz-Picciotto I, Duggan HM, Wood R, Kosnett MJ, Smith MT (1992) Cancer risks from arsenic in drinking water. Environ Health Perspect 97:259–267

Smith E, Naidu R, Alston AM (1998) Arsenic in the soil environment: a review. Adv Agron 64:149–195

Speir TW, Kettles HA, Parshotam A, Searle PL, Vlaar LNC (1999) Simple kinetic approach to determine the toxicity of AS[V] to soil biological properties. Soil Biol Biochem 31 (5):705–713, http://dx.doi.org/10.1016/S0038-0717(98)00169-2

Tabatabai MA (1977) Effects of trace elements on urease activity in soils. Soil Biol Biochem 9 (1):9–13, http://dx.doi.org/10.1016/0038-0717(77)90054-2

Tripathy S, Bhattacharyya P, Mohapatra R, Som A, Chowdhury D (2014) Influence of different fractions of heavy metals on microbial ecophysiological indicators and enzyme activities in century old municipal solid waste amended soil. Ecol Eng 70:25–34, http://dx.doi.org/10.1016/j.ecoleng.2014.04.013

Vaughan DJ (2006) Arsenic. Elements 2(2):71–75. doi:10.2113/gselements.2.2.71

Violante A, Pigna M (2002) Competitive sorption of arsenate and phosphate on different clay minerals and soils. Soil Sci Soc Am J 66(6):1788–1796

Vodyanitskii YN (2013) Contamination of soils with heavy metals and metalloids and its ecological hazard (analytic review). Euras Soil Sci 46(7):793–801, ISSN 1064-2293, Pleiades Publishing, Ltd

Warren GP, Alloway BJ, Lepp NW, Singh B, Bochereau FJ, Penny C (2003) Field trials to assess the uptake of arsenic by vegetables from contaminated soils and soil remediation with iron oxides. Sci Total Environ 311(1–3):19–33. doi:10.1016/S0048-9697(03)00096-2

Wang S, Mulligan CN (2006) Effect of natural organic matter on arsenic release from soils and sediments into groundwater. Environ Geochem Health 28:197–214

Wu MM, Kuo TL, Hwang YH, Chen CJ (1989) Dose-response relation between arsenic concentration in well water and mortality from cancers and vascular diseases. Am J Epidemiol 130 (6):1123–1132

Yamamura S, Amachi S (2014) Microbiology of inorganic arsenic: from metabolism to bioremediation. J Biosci Bioeng 118(1):1–9

Zhang J, Pu L, Peng B, Gao Z (2011) The impact of urban land expansion on soil quality in rapidly urbanizing regions in China: Kunshan as a case study. Environ Geochem Health 33 (2):125–135. doi:10.1007/s10653-010-9326-6

Zhuang P, Zou B, Li NY, Li ZA (2009) Heavy metal contamination in soils and food crops around Dabaoshan mine in Guangdong, China: implication for human health. Environ Geochem Health 31(6):707–715. doi:10.1007/s10653-009-9248-3

Chapter 7
Egg as a Biomonitor of Heavy Metals in Soil

Muhammad Zaffar Hashmi, Naeem Akhtar Abbasi, Xianjin Tang, and Riffat Naseem Malik

7.1 Introduction

The use of avian tissues for biomonitoring of contaminants has widely been used. Avian tissues are usefully used indicator because the concentration level of heavy metals in their tissues reflects their exposure and may also be indicative for trend of the past exposure. Initially, internal organs were collected after sacrificing a live bird, but more recently the demand of noninvasive technique leads to the use of feathers, eggs, and feces for biomonitoring of heavy metals (Burger 1993). Feathers and eggs are very special for metal accumulation studies not only because they can be collected without affecting the life of a bird but also for they have certain important information which is not possible to acquire otherwise (Dauwe et al. 1999). The use of feathers and eggs for heavy metal monitoring in soil has their own respective limitations, but eggs have certain several advantages over the other noninvasive biomonitoring matrices because they are easier to collect and can be preserved for a long time. In addition, eggs are relevant to a specific portion of the population, called as egg-laying female birds; they are formed during a specific duration of the life of a female and hence proven as a very good indicator for local exposure. Prior to egg laying, females of all the bird species from temperate and

M.Z. Hashmi
Department of Environmental Engineering, College of Environmental and Resource Sciences, Zhejiang University, Hangzhou 310058, People's Republic of China

N.A. Abbasi • R.N. Malik (✉)
Environmental Biology and Ecotoxicology Laboratory, Department of Environmental Sciences, Faculty of Biological Sciences, Quaid-i-Azam University, Islamabad PO 45320, Pakistan
e-mail: r_n_malik2000@yahoo.co.uk

X. Tang
Institute of Soil and Water Resources and Environmental Sciences, Zhejiang University, 866 Yuhangtang Road Hangzhou 310058, P.R. China

I. Sherameti, A. Varma (eds.), *Heavy Metal Contamination of Soils*, Soil Biology 44, DOI 10.1007/978-3-319-14526-6_7

tropical zones spend a lot of time in their breeding grounds feeding on local resources for egg production (Dauwe et al. 2005). Unlike feathers, the composition of an egg is advantageous because its composition is highly consistent which does not change with size, age, or the position of the body. Eggs have been widely used for many of the recent biomonitoring studies, and it has been a proven fact that collecting a single egg in favorable circumstance has produced negligible effects on population parameters of the species (Furness 1993).

Soil is an important medium to support the life on earth. It is an essential part of biosphere which provides space for animal and plant anchorage and growth and regulates the cycles of nutrients and water. Since the advent of industrial revolution, the concentration of heavy metals increases alarmingly in the soil. As a result of rampant discharge of agricultural and industrial waste, the concentration level of heavy metals like most of the pollutants suddenly touched to the threatening threshold. Municipal waste, ore smelting, and the use of pesticide and fertilizers are the major anthropogenic sources through which toxic heavy metals are introduced in the soil (Meli et al. 2013). Heavy metals are carried from the source to the distant areas through water currents or land/soil movement. Soil is the ultimate medium which receives heavy metals from different sources and deposits them for a long time. Being the largest supporting medium for heavy metal deposition, soil and sediments serve as an archive and reservoir for metal deposition and their subsequent dissemination in plants and animal body parts through their feed. Most of the heavy metals particularly mercury, cadmium, and lead readily adsorbed on dead organic matter, waxy plant material, animal's membranes, fats, and oppositely charged soil particles. After adsorption, the bioavailability of these toxic heavy metals for an organism will either depend on the properties of the metals and/or the physical, chemical, and biological properties of the medium into which it is released (Gamberg et al. 2005). Elevated concentration of heavy metals in soil greatly affects the capability of the soil to sustain its biological productivity, regulate the local environment, and provide the best condition for the growth of animals and plants and betterment of human health.

Heavy metals present in soil are taken by plants and animals through absorption and ingestion, respectively. Birds being the agile group of vertebrates receive metals during their lifetime exposure to soil. The concentration level in eggs of a species is a very good indicator for local exposure to soil. This is because of the reason that an egg is the only development in birds which is produced during a particular period of time by an egg-laying female fraction of a population. For breeding, birds are either sedentary (remain in their same habitat) or migratory (migrate toward their breeding grounds for breeding). Before egg laying, females build an actively defended breeding/nesting territory and exploit the local resources to the balance of nutrients during egg-laying season. Burger and Gochfeld (1995) reported that the concentration of metals in eggs represents the circulation level of respective metals in the blood which in turn reflects the recent exposure. Metals usually accumulate in the eggshell and/or egg contents during metal sequestration by the female (Mora 2003). Similarly, Burger (1993) said that eggs are the most vulnerable part to be affected with heavy metals and it exhibits the local exposure of the adult female which has laid them. Different studies reported the direct relation between the heavy metal concentration in soil and the level recorded in eggs of different bird species. The concentration of metals such as Pb and Cd concentration was found to be correlated in the soil and egg which suggests that soil is an important source for these elements (Waegeneers et al. 2009a, b). Similarly, the concentrations of Pb, Cd, and Cu in the eggs of gray heron (*Ardea cinerea*) and black-crowned night heron (*Nycticorax nycticorax*) were found indicative for their respective concentration in sediments of Nallihan Bird Paradise, Turkey (Ayaş 2007). In this way, an egg has been proven as a very useful biomonitoring tool for heavy metal contamination in soil.

The concentration of heavy metals in eggs in relation to soil may be affected by several factors. As this relation is much dependent on food of the organism, difference in the food chain may exist. It also depends on the dietary preferences of the species during breeding period. Interspecies differences and spatiotemporal differences were recorded in several studies (Akearok et al. 2010; Burger 2002; Dauwe et al. 2004). The difference between the eggshell and egg contents also gives useful information. Usually, for toxic trace metals such as Pb, Cd, Se, and Hg, the eggshell has been suggested to be a better indicator, while egg contents could be used as a best indicator for essential elements such as Fe, Mn, Zn, etc. (Dauwe et al. 1999). Moreover, it is suggested that data for egg metals for the same species should be collected for a period of consecutive years to identify the addition or deletion of any point or nonpoint pollution source and also to record the past trends of exposure (Burger and Gochfeld 1993). This chapter relates the metal concentration in eggs which is indicating the pollution level at different habitats, factors which affect the soil to egg metal ratio and the inexplicability of egg to soil metal relation. Further, this chapter also discusses the pros and cons for an egg as a biomonitoring tool for metal contamination and its scope for future biomonitoring research.

7.2 Soil Characteristics and Metal Bioavailability for Birds

Depending on the ecosystem, either terrestrial or aquatic, pollutant dynamics are rather different. In terrestrial ecosystems, the ambient environment is usually soil (Burger et al. 2004; Lam et al. 2005; Shahbaz et al. 2013), while in an aquatic one, it is usually the water or the sediment (Boncompagni et al. 2003; Mora 2003; Mora et al. 2007). Once metals get into soil or sediment, they have long residence times before they are eluted to other compartments. Acidity has a marked effect on the solubility of metals in soils and water. The mobility of metals in soils is dictated largely by the clay contents, amount of organic matter, and pH. In general, the higher the clay and/or organic content and pH are, the more firmly bound the metals are and the longer their residence time is in soil. In water, the solubility of metals is strongly pH dependent. In aquatic systems, metals may become "blocked up" in bottom sediments, where they may remain for many years. However, if the pH falls, metal solubility increases, and they become more mobile and hence more bioavailable (Cotín Martínez 2012). As the acidity changes, the oxidation state changes accordingly, and as a result, heavy metals can generate or break bounds with organic compounds generating specific compounds such as methylmercury. Methylated compounds are taken up more readily than the unmethylated form, due to its high lipoaffinity (Cotín Martínez 2012).

Unlike organochlorides (another known family of pollutants), which are distributed widely among body tissues, largely in relation to their lipid contents, heavy metals tend to be held in one particular tissue at much higher levels than the others. The site specificity of metals has an important influence on the choice of tissue for monitoring studies. For example, although females can excrete mercury into eggs, the amount they shed in this way is usually small compared to the amount put into feathers during molting (Honda et al. 1986). There are several different approaches for measuring bioavailability. In vivo tests use an animal to measure absolute bioavailability and toxicity. In vitro tests are performed outside the organism (Furman et al. 2006).

7.3 Egg as a Biomonitor for Soil Heavy Metals from Different Ecosystems

Governmental agencies, tribal nations, policy makers, managers, and the general public are concerned about the health of different ecosystems of the environment and require bioindicators that assess the levels of contaminants in individual organisms, populations, and communities. Chemical use is increasing in our environment and may pose a threat to some species and populations. The levels of many chemicals have been elevated in aquatic ecosystems (wetland, islands, marine land, and coastal areas) and terrestrial ecosystems (agriculture, forest, rural and urban) by using eggs (Table 7.1) as a biomonitor (Burger et al. 2009; Hashmi et al. 2013; Lam et al. 2005; Mora et al. 2011; Shahbaz et al. 2013), because of the influx from rivers, as well as runoff and direct pollution (Furness and Rainbow 1990), and from atmospheric transport and deposition.

7.3.1 Terrestrial Ecosystem

Terrestrial ecosystem exists only on landforms which may cover agricultural areas, forest areas, urban areas, and rural areas. Hence, terrestrial ecosystems are major hotspot for the chemical deposition from anthropogenic activities (agriculture and industrial) and natural processes. These anthropogenic activities and natural processes have not only affected terrestrial ecosystem services but also other living things (human, animals, etc.). Biomonitoring approaches using bird's eggs as a bioindicator tool to probe out the integrity of different areas of terrestrial ecosystem have been used a few decades ago. Several research articles reported eggs as a suitable bioindicator of agricultural areas (Hashmi et al. 2013; Shahbaz et al. 2013),

Table 7.1 Biomonitoring of heavy metals using eggs from different areas and bird species

Authors	Study area	Species
Shahbaz et al. (2013)	Islam Headworks, Pakistan	Little egret
		Cattle egret
	Trimmu Headworks, Pakistan	Little egret
		Cattle egret
Ruuskanen et al. (2014)	Eurasia	*Ficedula hypoleuca*
Jackson et al. (2011)	Virginia	Carolina wren
Beltrame et al. (2010)	Argentina	Burrowing crab
Páez-Osuna et al. (2010)	Mexico	Sea turtle (albumen)
		(yolk)
Burger et al. (2009)	Aleutian Islands, Alaska	Glaucous-winged Gull
Burger and Gochfeld (2009)	New York, USA	Common tern
Custer et al. (2008)	Lostwood, North Dakota	Tree swallows
Tsipoura et al. (2008)	New Jersey, USA	Red-winged blackbird
		Marsh wren
Roodbergen et al. (2008)	Netherlands	Black-tailed godwit
Custer et al. (2007a)	Agassiz National Wildlife Refuge, Minnesota	Franklin's gull
		Black-crowned night heron
		Eared grebe
		Pied-billed grebe
Bostan et al. (2007)	Gujranwala, Pakistan	Cattle egret
Burger and Gochfeld (2007)	Aleutians Islands, Alaska	Common eider
		Glaucous-winged Gull
Custer et al. (2007b)	Mississippi, USA	Tree swallows
	Mexico	Mourning dove
		Burrowing owl
		Marsh wren
Swaileh and Sansur (2006)	Birzeit, Palestine	House sparrows
Ikemoto et al. (2005)	Torishima Island, Japan	Short-tailed albatross
		Black-footed albatross
Dauwe et al. (2005)	Antwerp, Belgium	Great tits
		Great tits
		Great tits
Lam et al. (2005)	Hong Kong	Black-crowned night heron
		Little egret
		Bridled tern
Burger et al. (2004)	Florida	Scrub jay

(continued)

Table 7.1 (continued)

Authors	Study area	Species
Mora (2003)	Arizona, USA	Yellow-breasted chat
		Yellow-breasted chat
		Yellow-breasted chat
		Bell's vireo
		Lesser goldfinch
Boncompagni et al. (2003)	Haleji Lake, Pakistan	Little egret
		Intermediate egret
	Karachi, Pakistan	Little egret
	Taunsa, Pakistan	Little egret
		Cattle egret
Hui (2002)	USA	Rock dove
Burger (2002)	New Jersey, USA	Herring gull
		Great black-backed gull
		Common tern
		Forster's tern
		Skimmer

urban areas (Burger et al. 2004; Burger and Gochfeld 1995, 2004; Dev and Bhattacharjee 2010; Ek et al. 2004; Evers et al. 2003; Fu et al. 2014; Gochfeld 1997; Hui 2002; Lam et al. 2005; Mora 2003; Mora et al. 2007, 2011; Orłowski et al. 2010; Sekeroglu et al. 2013; Swaileh and Sansur 2006), rural areas (Kitowski et al. 2013; Orłowski et al. 2010), and industrial areas (Burger et al. 2004; Burger and Gochfeld 2004; Dauwe et al. 1999, 2005; Gochfeld 1997; Mora et al. 2011; Orłowski et al. 2010; Scheuhammer et al. 2001).

7.3.2 Aquatic Ecosystem

Aquatic ecosystems or environments are particularly sensitive to heavy metal contamination because aquatic physicochemical conditions can facilitate the conversion of unavailable forms of heavy metals into more bioavailable forms, which accumulate very efficiently in top predators. The life histories and ecological niches occupied by many aquatic birds predispose them to accumulating heavy metals, and the toxicology and merits/limitations of using eggs for monitoring heavy metals have been studied extensively. Aquatic ecosystems such as wetlands, islands, marine lands, and coastal areas provide breeding habitats for egrets, herons, and waterfowls and receive untreated effluents from agriculture and industries and cities from point and nonpoint sources of pollution. Hence, wetlands, islands, marine lands, and coastal areas act as a depository reservoir of organic as well as inorganic pollutants. Species that forage in aquatic environments are particularly vulnerable because of the potential for the rapid movement of contaminants in water, compared to the movement in terrestrial environments, and because chemicals can be stored in sediments in intertidal environments, providing a pool for years to come.

A variety of toxic substances enter the water from surface runoff, industrial effluent, air pollution, wave action, the movement of organisms, and the high traffic of marine shipping entering these ports. Organisms that live in these estuarine systems can bioaccumulate organic and certain inorganic substances over time and are at risk from both lethal and sublethal effects, as their body burdens increase. Several studies used eggs as a biomonitor of heavy metal contamination in the soil and sediments of wetland ecosystems (Shahbaz et al. 2013; Hashmi et al. 2013; Jackson et al. 2011; Ayaş 2007; Kertész and Fáncsi 2003; Burger and Gochfeld 2004; Mora 2003; Mora et al. 2007 Boncompagni et al. 2003; Bischoff et al. 2002; Goutner et al. 2001), island ecosystems (Xu et al. 2011; Burger et al. 2009; Pereira et al. 2009; Burger et al. 2008; Ayaş et al. 2008; Agusa et al. 2005; Burger 2002; Gochfeld 1997; Sanpera et al. 2000, 1997; Burger and Gochfeld 1995), marine ecosystems (Akearok et al. 2010; Day et al. 2006), and coastal ecosystems (Pereira et al. 2009; Burger and Gochfeld 2003; Mateo et al. 2003). However, most of the published literature focused on the island and wetland ecosystems to monitor the heavy metal contamination by using eggs than the other ecosystems.

7.4 Factors Influencing Accumulation of Soil Heavy Metals in Eggs

7.4.1 Interspecific Variations

Metal accumulation in eggs is different for different species of birds. Several factors such as age and health of the female, diet preferences during breeding season, taxonomic affiliation, and the trophic level of the species and metal concentration level in the surrounding environment either singly or synergistically influence the overall metal burden in egg of different bird species. Heavy metal accumulation in eggs of different species largely depends upon the metal availability, metal intake, and physiological mechanisms for metal circulation, accumulation in the parts of the body, regulation, and subsequent excretion. All of these factors vary for different species and/or within the species and hence cause intra- and interspecific variation for metal accumulation in eggs. Differences of heavy metal concentrations in the eggs of different species have been reported in several studies (Akearok et al. 2010; Boncompagni et al. 2003; Burger 2002). There are several possible reasons for interspecific metal variation in eggs. One of the most important is the difference of the habitats/locations which different species prefer for foraging during breeding season. As the choice and requirement of nutrients differ for species, they accumulate metals in different quantities in their eggs (Ayaş 2007). A study reported by Boncompagni et al. (2003) concluded that the dissimilar concentration level of heavy metals in the eggs of three egret species foraging in the same area was because the female may have to visit different habitats for foraging prior to or during breeding and egg formation. So the phenomenon of visiting different locations for the sake of fulfilling its nutrient requirement has proved to be the basic factor influencing the ratio of soil and egg concentration level of heavy metals for different species. The next question in this regard may be that why eggs of different species exhibit different concentrations of metals inhabiting the same location? There is no straightforward answer yet being provided through research. But it is described by the physiological mechanism differences for the circulation, regulation, excretion, and replenishment of metals for different species. In general, one can say that both of the abovementioned factors, i.e., the difference of location and difference of the regulation of body metal burden, markedly influence the metal concentration level in eggs.

7.4.2 Influence of the Trophic Levels

Concentrations of metals in eggs are mainly attributed to the diet of the organism. Shahbaz et al. (2013) reported that elevated concentrations of toxic metals such as Cd, Cr, and Pb and some essential elements, viz, Mn, Fe, and Zn, in the egg contents of cattle egret are directly related to the higher concentrations of these metals in prey and soil of their foraging sites. They also deduced that unlike feathers, the concentration of heavy metals in eggs is more influenced by the environmental condition and food items. Similarly, elevated concentrations of Pb and Cd in the diet and sediments were found to be directly associated with the higher concentration of these metals in herons of Korea (Kim et al. 2010). The higher concentration of strontium in the eggshell of passerine birds from Arizona was probably associated with the higher deposition of Sr in the surrounding of the nesting sites of these birds. These passerine birds feed on certain invertebrates and insects which were associated with the recycling of Sr in the soil (Mora 2003). While studying the accumulation of Hg in different species, Burger (2002) found that larger fish-eating birds tend to accumulate more metals in their eggs when compared to small fish eaters. Similarly, in comparison with carnivores, omnivores and invertebrate-eating species accumulate lower metal accumulation in their eggs and showed that for birds, certain metal concentration increases with the increase in the trophic position. But this is not true for all the metals nor for all species. Being higher on the food web, the concentration of Hg in the eggs of common tern (*Sterna hirundo*) was found twice than those of duck species (Akearok et al. 2010). Boncompagni et al. (2003) reported that the concentration level of different heavy metals in the eggs and different other tissues of egret species was indicative of the concentration of the respective metals in sediments and prey items. So the birds which feed on the prey which largely relies on metal-rich soil for their food should have more metal concentration in their eggs. In general, one can say that the concentration level of heavy metals increases with the increase in the trophic position of the species, but one has to keep in mind the respective food chain and food web to get a meaningful result.

7.4.3 Temporal Variations (Inter-clutch and Intra-clutch Differences)

Temporal variation in egg metal content describes the time-based changes in the level of metal concentration and hence the difference in the local exposure. Temporal variation in the egg metal level provides information that can be useful for understanding the sudden shift of metals which in turn can be used to find any new addition in the metal pollution source (Burger and Gochfeld 2003). It is usually recommended to collect year-wise data for the same species for consecutive years to get a better understanding about the trends of metal accumulation. Significant differences in the concentration of heavy metals such as Pb, Cd, Se, and Hg were recorded in eggs of herring gulls (*Larus argentatus*) between the years 1998 and 1994. But the concentration level of metals in eggs rarely remains consistent for any metals between the years because of the changing metal pollution level for the location which the female is visiting for foraging prior to egg laying (Burger 1994). Furthermore, seasonal variation in the availability of heavy metals is obvious, so the exposure varies between different species and seasons. This is because the onset of breeding period is not the same for all the species, so the female of the different species may be exposed to the varying levels of heavy metals and hence accumulate different levels of metals in their egg. Seasonal bioavailability of heavy metals in soil is largely based on soil characteristics and local environmental condition and weather attributes. Besides inter-clutch variation in metal concentration, intra-clutch variation has also been observed for several metals. Custer et al. reported that sometimes systematic trends of metal concentration could be observed with egg-laying order, but it is not consistent for all the species nor for all the metals. Usually, the early laid egg has higher concentration of metals than later ones. Dauwe et al. (2004) reported that the time difference for laying eggs in any one breeding season has no significant effect on the metal accumulation in the female egg and hence on the number of sperm stored in perivitelline layer of the egg. Conversely, the concentration of several trace metals varied significantly between the home-produced chicken eggs in autumn and spring, but this temporal discrepancy of metals has not solely depended upon the concentration of these metals in soil (Waegeneers et al. 2009a). Similarly, temporal trends were recorded as spatial and temporal variations of Hg in the egg contents of gannet (*Morus bassanus*) between the years 1974 and 2004 from Bass Rock (North Sea) and Ailsa Craig (eastern Atlantic) (Pereira et al. 2009). Intra- and inter-clutch variations in the Hg level in the eggs of marine birds were reported and attributed to the diet switching of these marine species during their clutch periods. It is also proposed that an early clutch could be a better representative of the environmental state because it exhibits maximum metal load in females, metal burden which may be transferred to the offspring, and metals available for the predator (Akearok et al. 2010). The selenium level increases more than twofolds in the eggs of Florida scrub jays (*Aphelocoma coerulescens*) from 1996 to 2001 (Burger et al. 2004). No intra-clutch variation in the egg Hg level was reported in some piscivore and insectivore birds. The reason for this could be the body replenishment of Hg during egg-laying season mainly through ingestion and/or partly from the stored methylmercury in tissues and muscles (Brasso et al. 2010). Burger et al. (2008) reported that the clutches with more number of eggs exhibit variation than those with less number of eggs. In other words, the metal concentration varies within and among clutches, but no consistent trend has yet been observed. The lack of inconsistency among clutches of consecutive years is primarily attributed to the difference in the metal pollution level at the site where females visit during breeding season and partly may be due to species-specific responses toward metal burden in the consecutive years. Females may replenish the metal deficiency after laying the first egg and make up the metal deficiency effectively from the surrounding environment after each egg laid.

7.4.4 Geographical Difference

Metal contamination in egg is also influenced by the metal distribution level in the surrounding environment. Although not true for all the cases, the metal concentration level in egg is greatly influenced by the overall metal pollution burden of the surrounding environment and produces certain impairments in the egg. Differential exposure of species to soil heavy metals affects the total metal burden of the eggs and may produce certain abnormalities in the egg which in turn may affect reproduction. The exposure to the extreme level of heavy metals did not affect the egg of blue tit (*Parus caeruleus*) (Dauwe et al. 2002), while substantial impairments were observed in the eggs of pied flycatcher (*Ficedula hypoleuca*) and great tit (*Parus major*). It was also concluded that poor forest soil with less nutrient in combination with acidification affects the metal accumulation in egg and hence effects on egg quality (Eeva and Lehikoinen 1996). The soils with greater heavy metal pollution have been reported with lower sperm quality in males, poor mobility, and subsequent trapping in the perivitelline layer of the egg (Dauwe et al. 2002). Roodbergen et al. (2008) noted that the concentration of heavy metals such as Hg, Pb, and Cd was greater in soil, in earthworms, and subsequently in the egg of black-tailed godwit (*Limosa limosa*) at a polluted site which is contrasting to the reference site and suggested a direct transport of metals from the soil to the eggs through earthworm which is the primary prey of black-tailed godwit. Sekeroglu et al. (2013) reported that the metal concentration in eggs from the village chicken of Turkey is directly proportional to the distance from the roadside. Similarly, the concentration of toxic metals such as Pb, Cu, and Zn increases in the eggs of black kite (*Milvus migrans*) and decreases in distance from the solid waste incinerator (Blanco et al. 2003). Concentrations of heavy metals in the eggs of egret species were found higher in cosmopolitan areas where the metal pollution level was quite higher than relatively pristine areas (Boncompagni et al. 2003). Locational differences of Hg and Cd and Pb, As, Se, and Mn contamination in common tern (*Sterna hirundo*) eggs were recorded from five different locations of Barnegat Bay, New

Jersey (Burger and Gochfeld 2003). It is evident from the above discussion that soil from different locations has the varying capacities of heavy metals which it can offer to plants, small invertebrates, and higher organisms. The greater the metal pollution sources, the higher the concentration in soil, and consequently more metals reached to the avian tissues and sequestered in egg which truly represents the local environmental state.

7.4.5 Eggshell or Egg Content Differences

Eggshell is thought to be associated with the sequestration of most of inorganic elements (Dauwe et al. 1999) and usually exhibits greater metal concentration of most of the inorganic elements when compared with egg contents. Mora (2003) reported that the concentration of most of the studied metals including Ba, Ni, Sr, Zn, Mn, and V was 2–35 times higher in eggshells of yellow-breasted chats (*Icteria virens*) and willow flycatchers (*Empidonax traillii extimus*) when compared with egg contents. Similar results for higher eggshell/egg content ratio were recorded in black-tailed gulls (*Larus crassirostris*) in Japan (Agusa et al. 2005) and for little egret (*Egretta garzetta*) and black-crowned night heron (*Nycticorax nycticorax*) in Hong Kong (Lam et al. 2004). Metals sequestered through eggs depend upon the requirement of the respective metals for embryonic development and need for metal regulation and excretion. Ayaş (2007) reported that the metal concentration in the eggshell itself is a very good indicator for the metal concentration level in egg content. The level of metal contents in the eggshell and egg content is different for different metals. Dauwe et al. (1999) draw an important conclusion that there is a discrepancy for metal sequestration between the egg contents and eggshell. They reported that essential metals such as Zn and Cu were greater in egg contents of great tit (*Parus major*) and blue tit (*Parus caeruleus*), whereas the nonessential metals such as Cd, Pb, and As were higher in the eggshell. No significant difference between the polluted and reference site was recorded. Further, they deduced that the eggshell being highly associated with sequestered nonessential metals is a good indicator as compared to egg content. Similarly, the concentration of Zn, Cu, Hg, and Mn was found 80–99 % lower in the eggshell of Audouin's gulls (*Ichthyaetus audouinii*) than in egg contents (Morera et al. 1997). It is reported that eggshells are important for the excretion of heavy metals in birds (Burger 1994; Hashmi et al. 2013). The greater concentration level of essential nutrients in egg content is evidently associated with the regulation mechanism of these metals in the egg contents, whereas nonessential metals are sequestered in the eggshell for excretion. But the elevated concentration of these nonessential metals in the respective soil would tend to accumulate more metals in different bird tissues and subsequently sequester in the eggshell at a higher rate and affect egg quality and hence reproduction.

7.5 Heavy Metals in Soil and Eggs: An Equivocal Relation

Egg laying may be an additional excretory pathway for heavy metals available to laying female birds. Contaminant levels in eggs are useful bioindicators of exposure because they potentially represent different periods of exposure. The concentrations of metals in eggs are derived from females and represent the recent exposure in waters close to the breeding grounds, as well as mobilization of stored metals from the past intake (Burger 1994; Burger and Gochfeld 1996). However, the contaminant concentration in the eggs is also influenced by the presence of contaminants in nonbreeding areas (Burger 2002). Eggshell formation and egg laying are recognized as a means of excreting environmental contaminants (e.g., metals) by oviparous females (Burger and Gochfeld 1995; Gochfeld and Burger 1998; Lam et al. 2004, 2005), and thus contaminant concentrations in eggs can potentially be used as a surrogate for contaminant monitoring in egg-laying wildlife. Besides metals, bird eggs are also useful for evaluating lipophilic persistent organic pollutants (Connell et al. 2003). However, the contaminant concentrations in the eggs can also be influenced by contaminants present in areas utilized by these birds outside the breeding seasons (Burger 2002). Further, heavy metal content in the bird's egg reflects local pollution levels (Burger 2002); because hens live in close contact with the outside environment, they are exposed to contaminants (Furness et al. 1993). Also, numerous studies have related the heavy metal contents of bird's eggs and internal organs come from environmental pollution (Burger 2002; Burger et al. 2004; Hui 2002; Mora 2003; Sekeroglu et al. 2013). Hence, eggs are not influenced directly by soil heavy metal contamination, and there are no reports which suggest a direct relationship between soil heavy metal contamination and eggs (Day et al. 2006). However, several reports suggested eggs as bioindicator of soil heavy metal contamination by food chain (Boncompagni et al. 2003; Ikemoto et al. 2005; Mora et al. 2011; Shahbaz et al. 2013) or by intake of heavy metals from air into birds. The literature published so far on eggs as indicator of soil heavy metal contamination considered the food chain. Then, the birds after the regulation of heavy metals in the internal body excrete or maternally transfer heavy metals in the eggs or eggshell. Scheuhammer (1987) suggests that the transfer to the eggs only takes place if the concentration in the tissues of the female is high. Egg heavy metal contamination suggested that maternal transfer, food chain, and inhalation of metals could be responsible. However, some reports suggested that eggshells (Burger 1994) and egg contents (Lewis et al. 1993; Monteiro and Furness 1995; Monteiro et al. 1995) are important for the excretion of some trace elements in birds. Hence, eggshells and egg contents may be suitable to assess burdens of trace elements in mother birds during the period of egg development (Fossi et al. 1994). Hence, the published literature discussed here indicated that heavy metals in soil and eggs exhibited an equivocal relation.

7.6 Scope of Egg as a Biomonitor of Soil Heavy Metals

The egg in birds is a self-sustained life support system which comprised of the egg contents and eggshell. The bird's egg virtually serves as a good indicator for heavy metal pollution in the environment. Heavy metals are sequestered by the females during egg production (Burger 2002). But the sequestration of heavy metals is not consistent; cadmium, for example, is reported to be transmitted to the egg in a very limited quantity (Burger 1993). So, the use of egg for biomonitoring of those metals which are not sequestered properly by females will be limited. Scheuhammer (1987) suggested that the transfer of heavy metals to eggs takes place only if the concentration of the respective metals is very high in body tissues. In this way, females predominately sequester only those elements which are present in higher concentration in their body tissues, and the information about the other elements would be masked. So far, very limited information on the maternal transfer rate of heavy metals is available which means that the percentage of heavy metals in the egg to the whole body burden of metals has rarely been studied. To fill this gap of knowledge, extensive research on the maternal transfer rate of heavy metals to the eggs should be studied which will not only enhance the pool of existing knowledge on metal soil to egg metal ratio but will also provide the information about the extent to which eggs could be useful for biomonitoring of heavy metals in soil. Moreover, the concentration level of most of the essential metals such as Cu, Zn, and Fe in the eggs could partly be explained by the internal regulatory mechanism of the body. Some metals may be mobilized by the female body during the egg production from the stored metals in different tissues in the body which the organism has ingested elsewhere not from the local soil. For example, Pb sequestered in egg may reflect the mobilization of lead from accumulated mass during the previous breeding season from the same area. Therefore, concentrations of certain metals are not necessarily linked to the soil metal level. For a meaningful interpretation of the results, one has to take care about all these possible discrepancies. As the egg provides information only about the concentration in the soil of breeding grounds, the information about the metal addition at wintering grounds would be missing. Before using egg as a biomonitoring tool for heavy metal accumulation through soil, one must know the characteristics and history of the local soil, especially the turnover rate of sediments. By using these information and egg-related discrepancies, one could give the evocative description on soil to egg metal transfer.

References

Agusa T et al (2005) Body distribution of trace elements in black-tailed gulls from Rishiri Island, Japan: age-dependent accumulation and transfer to feathers and eggs. Environ Toxicol Chem 24:2107–2120

Akearok JA et al (2010) Inter-and intraclutch variation in egg mercury levels in marine bird species from the Canadian Arctic. Sci Total Environ 408:836–840

Ayaş Z (2007) Trace element residues in eggshells of grey heron (*Ardea cinerea*) and black-crowned night heron (*Nycticorax nycticorax*) from Nallihan Bird Paradise, Ankara-Turkey. Ecotoxicology 16:347–352

Ayaş Z, Celikkan H, Aksu ML (2008) Lead (Pb) and copper (Cu) concentration in the eggshells of Audouin's gulls (Larus audouinii) in Turkey. Turk J Zool 32:379–384

Beltrame MO, De Marco SG, Marcovecchio JE (2010) Influences of sex, habitat, and seasonality on heavy-metal concentrations in the burrowing crab (Neohelice granulata) from a coastal lagoon in Argentina. Arch Environ Contam Toxicol 58:746–756

Bischoff K, Pichner J, Braselton W, Counard C, Evers D, Edwards W (2002) Mercury and selenium concentrations in livers and eggs of common loons (Gavia immer) from Minnesota. Arch Environ Contam Toxicol 42:71–76

Blanco G, Frías O, Jiménez B, Gómez G (2003) Factors influencing variability and potential uptake routes of heavy metals in black kites exposed to emissions from a solid-waste incinerator. Environ Toxicol Chem 22:2711–2718

Boncompagni E et al (2003) Egrets as monitors of trace-metal contamination in wetlands of Pakistan. Arch Environ Contam Toxicol 45:399–406

Bostan N, Ashraf M, Mumtaz AS, Ahmad I (2007) Diagnosis of heavy metal contamination in agro-ecology of Gujranwala, Pakistan using cattle egret (Bubulcus Ibis) as bioindicator. Ecotoxicology 16:247–251

Brasso RL, Abdel Latif MK, Cristol DA (2010) Relationship between laying sequence and mercury concentration in tree swallow eggs. Environ Toxicol Chem 29:1155–1159

Burger J (1993) Metals in avian feathers: bioindicators of environmental pollution. Rev Environ Toxicol 5:203–311

Burger J (1994) Heavy metals in avian eggshells: another excretion method. J Toxicol Environ Health 41:207–220

Burger J (2002) Food chain differences affect heavy metals in bird eggs in Barnegat Bay, New Jersey. Environ Res 90:33–39

Burger J et al (2004) Metal and metalloid concentrations in the eggs of threatened Florida scrub-jays in suburban habitat from south-central Florida. Sci Total Environ 328:185–193

Burger J, Gochfeld M (1993) Lead and cadmium accumulation in eggs and fledgling seabirds in the New York Bight. Environ Toxicol Chem 12:261–267

Burger J, Gochfeld M (1995) Heavy metal and selenium concentrations in eggs of herring gulls (Larus argentatus): temporal differences from 1989 to 1994. Arch Environ Contam Toxicol 29:192–197

Burger J, Gochfeld M (1996) Heavy metal and selenium levels in Franklin's gull (*Larus pipixcan*) parents and their eggs. Arch Environ Contam Toxicol 30:487–491

Burger J, Gochfeld M (2003) Spatial and temporal patterns in metal levels in eggs of common terns (*Sterna hirundo*) in New Jersey. Sci Total Environ 311:91–100

Burger J, Gochfeld M (2004) Metal levels in eggs of common terns (*Sterna hirundo*) in New Jersey: temporal trends from 1971 to 2002. Environ Res 94:336–343

Burger J, Gochfeld M (2007) Metals and radionuclides in birds and eggs from Amchitka and Kiska Islands in the Bering Sea/Pacific Ocean ecosystem. Environ Monit Assess 127:105–117

Burger J, Gochfeld M (2009) Comparison of arsenic, cadmium, chromium, lead, manganese, mercury and selenium in feathers in bald eagle (Haliaeetus leucocephalus), and comparison with common eider (Somateria mollissima), glaucous-winged gull (Larus glaucescens), pigeon guillemot (Cepphus columba), and tufted puffin (Fratercula cirrhata) from the Aleutian Chain of Alaska. Environ Monit Assess 152:357–367

Burger J, Gochfeld M, Jeitner C, Snigaroff D, Snigaroff R, Stamm T, Volz C (2008) Assessment of metals in down feathers of female common eiders and their eggs from the Aleutians: arsenic, cadmium, chromium, lead, manganese, mercury, and selenium. Environ Monit Assess 143:247–256

Burger J et al (2009) Mercury and other metals in eggs and feathers of glaucous-winged gulls (Larus glaucescens) in the Aleutians. Environ Monit Assess 152:179–194

Connell D et al (2003) Risk to breeding success of fish-eating Ardeids due to persistent organic contaminants in Hong Kong: evidence from organochlorine compounds in eggs. Water Res 37:459–467

Cotín Martínez J (2012) Birds as bioindicators of pollution in aquatic and terrestrial environments. Thesis, Universitat de Barcelona. http://hdl.handle.net/10803/101099

Custer TW, Custer CM, Johnson KM, Hoffman DJ (2008) Mercury and other element exposure to tree swallows (Tachycineta bicolor) nesting on Lostwood National Wildlife Refuge, North Dakota. Environ Pollut 155:217–226

Custer TW, Custer CM, Eichhorst BA, Warburton D (2007a) Selenium and metal concentrations in waterbird eggs and chicks at Agassiz National Wildlife Refuge, Minnesota. Arch Environ Contam Toxicol 53:103–109

Custer TW et al (2007b) Water level management and contaminant exposure to tree swallows nesting on the Upper Mississippi River. Environ Monit Assess 133:335–345

Dauwe T et al (1999) Are eggshells and egg contents of great and blue tits suitable as indicators of heavy metal pollution? Belg J Zool 129:439–447

Dauwe T et al (2005) Heavy-metal concentrations in female laying great tits (*Parus major*) and their clutches. Arch Environ Contam Toxicol 49:249–256

Dauwe T et al (2004) The effect of heavy metal exposure on egg size, eggshell thickness and the number of spermatozoa in blue tit *Parus caeruleus* eggs. Environ Pollut 129:125–129

Dauwe T et al (2002) Great and blue tit feathers as biomonitors for heavy metal pollution. Ecol Indic 1:227–234

Day RD et al (2006) Murre eggs (Uria aalge and Uria lomvia) as indicators of mercury contamination in the Alaskan marine environment. Environ Sci Technol 40:659–665

Dev B, Bhattacharjee P (2010) Heavy metals in egg shells of six species of Ardeidae (Aves) from Barak Valley, Assam. Ass Univ J Sci Technol 5:48–52

Eeva T, Lehikoinen E (1996) Growth and mortality of nestling great tits (*Parus major*) and pied flycatchers (*Ficedula hypoleuca*) in a heavy metal pollution gradient. Oecologia 108:631–639

Ek KH et al (2004) Platinum group elements in raptor eggs, faeces, blood, liver and kidney. Sci Total Environ 334:149–159

Evers D et al (2003) Common loon eggs as indicators of methylmercury availability in North America. Ecotoxicology 12:69–81

Fossi MC et al (1994) Blood esterase inhibition in birds as an index of organophosphorus contamination: field and laboratory studies. Ecotoxicology 3:11–20

Fu J et al (2014) Monitoring of non-destructive sampling strategies to assess the exposure of avian species in Jiangsu Province, China to heavy metals. Environ Sci Pollut Res 21:2898–2906

Furman O et al (2006) Risk assessment test for lead bioaccessibility to waterfowl in mine-impacted soils. J Environ Qual 35:450–458

Furness R (1993) Birds as monitors of pollutants. In: Furness RW, Greenwood JJD (eds) Birds as monitors of environmental change. Springer, New York, NY, pp 86–143

Furness R et al (1993) Can birds be used to monitor the environment? In: Furness RW, Greenwood JJD (eds) Birds as monitors of environmental change. Springer, New York, NY, pp 1–41

Furness RW, Rainbow PS (1990) Heavy metals in the marine environment. CRC Press, Boca Raton, FL

Gamberg M et al (2005) Spatial and temporal trends of contaminants in terrestrial biota from the Canadian Arctic. Sci Total Environ 351:148–164

Gochfeld M (1997) Spatial patterns in a bioindicator: heavy metal and selenium concentration in eggs of Herring gulls (*Larus argentatus*) in the New York Bight. Arch Environ Contam Toxicol 33:63–70

Gochfeld M, Burger J (1998) Temporal trends in metal levels in eggs of the endangered roseate tern (*Sterna dougallii*) in New York. Environ Res 77:36–42

Goutner V, Furness R, Papakostas G (2001) Mercury in feathers of Squacco Heron (Ardeola ralloides) chicks in relation to age, hatching order, growth, and sampling dates. Environ Pollut 111:107–115
Hashmi MZ et al (2013) Heavy metals in eggshells of cattle egret (*Bubulcus ibis*) and little egret (*Egretta garzetta*) from the Punjab province, Pakistan. Ecotoxicol Environ Saf 89:158–165
Honda K et al (1986) Distribution of heavy metals and their age-related changes in the eastern great white egret, Egretta alba modesta, in Korea. Arch Environ Contam Toxicol 15:185–197
Hui CA (2002) Concentrations of chromium, manganese, and lead in air and in avian eggs. Environ Pollut 120:201–206
Ikemoto T, Kunito T, Tanabe S, Tsurumi M, Sato F, Oka N (2005) Non-destructive monitoring of trace element levels in short-tailed albatrosses (Phoebastria albatrus) and black-footed albatrosses (Phoebastria nigripes) from Torishima Island, Japan using eggs and blood. Mar Pollut Bull 51:889–895
Jackson AK et al (2011) Mercury exposure affects the reproductive success of a free-living terrestrial songbird, the Carolina wren (Thryothorus ludovicianus). Auk 128:759–769
Kertész V, Fáncsi T (2003) Adverse effects of (surface water pollutants) Cd, Cr and Pb on the embryogenesis of the mallard. Aquat Toxicol 65:425–433
Kim J et al (2010) Effects of age on heavy metal concentrations of black-crowned night herons Nycticorax nycticorax from Korea. J Environ Monit 12:600–607
Kitowski I et al (2013) Trace elements in eggshells of the Grey Heron (*Ardea cinerea*) from the colony in the Roztocze Hills (South East Poland). Zool Ecol 23:240–244
Lam JC et al (2005) Risk to breeding success of waterbirds by contaminants in Hong Kong: evidence from trace elements in eggs. Environ Pollut 135:481–490
Lam JC et al (2004) Trace element residues in eggs of Little Egret (*Egretta garzetta*) and Black-crowned Night Heron (*Nycticorax nycticorax*) from Hong Kong, China. Mar Pollut Bull 48:390–396
Lewis S et al (1993) Mercury levels in eggs, tissues, and feathers of herring gulls *Larus argentatus* from the German Wadden Sea Coast. Environ Pollut 80:293–299
Mateo R, Beyer WN, Spann J, Hoffman D, Ramis A (2003) Relationship between oxidative stress, pathology, and behavioral signs of lead poisoning in mallards. J Toxic Environ Health A 66:1371–1389
Meli M et al (2013) Population-level consequences of spatially heterogeneous exposure to heavy metals in soil: an individual-based model of springtails. Ecol Model 250:338–351
Monteiro L, Furness R (1995) Seabirds as monitors of mercury in the marine environment. Water Air Soil Pollut 80:851–870
Monteiro L et al (1995) Mercury levels in seabirds from the Azores, mid-north Atlantic Ocean. Arch Environ Contam Toxicol 28:304–309
Mora MA (2003) Heavy metals and metalloids in egg contents and eggshells of passerine birds from Arizona. Environ Pollut 125:393–400
Mora MA et al (2011) Regional and interspecific variation in Sr, Ca, and Sr/Ca ratios in avian eggshells from the USA. Ecotoxicology 20:1467–1475
Mora MA et al (2007) Potential ecotoxicological significance of elevated concentrations of strontium in eggshells of passerine birds. Condor 109:199–205
Morera M, Sanpera C, Crespo S, Jover L, Ruiz X (1997) Inter-and intraclutch variability in heavy metals and selenium levels in Audouin's gull eggs from the Ebro Delta, Spain. Arch Environ Contam Toxicol 33:71–75
Orłowski G et al (2010) Geographical and habitat differences in concentrations of copper, zinc and arsenic in eggshells of the Rook Corvus frugilegus in Poland. J Ornithol 151:279–286
Páez-Osuna F, Calderón-Campuzano M, Soto-Jiménez M, Ruelas-Inzunza J (2010) Lead in blood and eggs of the sea turtle, Lepidochelys olivacea, from the Eastern Pacific: concentration, isotopic composition and maternal transfer. Mar Pollut Bull 60:433–439
Pereira MG, Walker LA, Best J, Shore RF (2009) Long-term trends in mercury and PCB congener concentrations in gannet (Morus bassanus) eggs in Britain. Environ Pollut 157:155–163

Roodbergen M, Klok C, van der Hout A (2008) Transfer of heavy metals in the food chain earthworm Black-tailed godwit (Limosa limosa): comparison of a polluted and a reference site in The Netherlands. Sci Total Environ 406:407–412
Ruuskanen S et al (2014) Large-scale geographical variation in eggshell metal and calcium content in a passerine bird (Ficedula hypoleuca). Environ Sci Pollut Res 21:3304–3317
Sanpera C, Morera M, Crespo S, Ruiz X, Jover L (1997) Trace elements in clutches of yellow-legged gulls, Larus cachinnans, from the Medes Islands, Spain. Bull Environ Contam Toxicol 59:757–762
Sanpera C, Morera M, Ruiz X, Jover L (2000) Variability of mercury and selenium levels in clutches of Audouin's gulls (Larus audouinii) breeding at the Chafarinas Islands, southwest Mediterranean. Arch Environ Contam Toxicol 39:119–123
Scheuhammer A (1987) The chronic toxicity of aluminium, cadmium, mercury, and lead in birds: a review. Environ Pollut 46:263–295
Scheuhammer A et al (2001) Mercury, methylmercury, and selenium concentrations in eggs of common loons (*Gavia immer*) from Canada. Environ Monit Assess 72:79–94
Sekeroglu A et al (2013) Effect of distance from the roadway on heavy metal content and egg quality of village laying hen's egg along roadsides of Tokat-Turhal, Turkey. J Agr Sci 50 (2):299–304
Shahbaz M et al (2013) Relationship between heavy metals concentrations in egret species, their environment and food chain differences from two Headworks of Pakistan. Chemosphere 93:274–282
Swaileh K, Sansur R (2006) Monitoring urban heavy metal pollution using the House Sparrow (Passer domesticus). J Environ Monit 8:209–213
Tsipoura N, Burger J, Feltes R, Yacabucci J, Mizrahi D, Jeitner C, Gochfeld M (2008) Metal concentrations in three species of passerine birds breeding in the Hackensack Meadowlands of New Jersey. Environ Res 107:218–228
Waegeneers N et al (2009a) Transfer of soil contaminants to home-produced eggs and preventive measures to reduce contamination. Sci Total Environ 407:4438–4446
Waegeneers N et al (2009b) Trace elements in home-produced eggs in Belgium: levels and spatiotemporal distribution. Sci Total Environ 407:4397–4402
Xu X, Li Y, Wang Y, Wang Y (2011) Assessment of toxic interactions of heavy metals in multi-component mixtures using sea urchin embryo-larval bioassay. Toxicol in Vitro 25:294–300

Chapter 8
Enzyme Activities in Soils Contaminated with Heavy Metals in Varying Degrees

Krystyna Ciarkowska

8.1 Introduction: Soil Enzymes and Their Role in Soil

The biological activity of the soil, which is determined, among other things, by estimating the enzyme activity, is an important component indicating the regularity of soil processes. Soil enzymes catalyze processes related to the breakdown of organic matter released into the soil during plant growth. They also stimulate the processes connected with the formation, decomposition of soil humus, and a distribution and release of mineral substances. Moreover, enzymes are responsible for making those substances available to plants and for binding of molecular nitrogen, detoxification of xenobiotics, as well as nitrification and denitrification processes (Das and Varma 2011; Friedlova 2010; Hang et al. 2013; Khan et al. 2007; Pang and Yu 2011). Soil enzyme activity results from the activity of the enzymes accumulated in soil (extracellular enzymes) and from the enzyme activity of reproducing microorganisms (intracellular enzymes). Cells of microorganisms, as well as remains of plants and animals, are sources of enzymes accumulated in the soil (Mukhopadhyay and Maiti 2010). The enzyme activity of the soil is most often assessed on the basis of the effect of five enzymes: dehydrogenase, phosphatase, urease, invertase, and protease. The activity of soil enzymes is also known as a sensitive indicator of natural and anthropogenic changes in ecosystems, which is used to assess the impact of various pollutants including heavy metals in the soil, both in a long and short period of time (Gulser and Erdogan 2008; Januszek 1999; Kuziemska 2012; Yang et al. 2007). Moreover, the activity of enzymes can be used to indicate the effectiveness of rehabilitation processes of ecosystems or to reflect the quality of the soil following the restoration

K. Ciarkowska (✉)
Soil Science and Soil Protection Department, University of Agriculture, 31-120 Krakow, Poland
e-mail: rrciarko@cyf-kr.edu.pl

I. Sherameti, A. Varma (eds.), *Heavy Metal Contamination of Soils*, Soil Biology 44, DOI 10.1007/978-3-319-14526-6_8

of soil environment which has been destroyed by industrial processes (Ciarkowska et al. 2014; Finkenbein et al. 2013; Mukhopadhyay and Maiti 2010).

8.2 Characterization of Selected Soil Enzymes

Dehydrogenases catalyze redox processes. Under aerobic conditions, they are transferred by a series of intermediaries onto the components of the respiratory chain, and ultimately onto oxygen, while in anaerobic conditions, oxidized inorganic forms such as NO^{3-}, Mn (IV), Fe (III), SO_4^{2-}, and CO_2 serve as acceptors of organic compounds, in the fermentation process. Thus, dehydrogenases are indicators of the respiratory metabolism of soil microorganisms—mainly of bacteria and actinomycetes (Burns et al. 2013; Hang et al. 2013; Kumar et al. 2013; Kuziemska 2012; Wolińska and Stępniewska 2012). They constitute an integral part of living cells and do not accumulate extracellularly in the soil. They belong to constitutive enzymes, that is, those which are present in almost constant amounts in the cell, independently of the amount of substrate (Das and Varma 2011; Friedlova 2010; Stępniewska and Wolińska 2005). It has been stated that the optimal soil pH value for the activity of dehydrogenases is 6.6–7.2. These enzymes are located in the cytoplasm or in specific structures developed from the cytoplasmic membrane. Dehydrogenase activity depends on the total population of microorganisms living in the soil, and that is why they are considered to be an indicator of overall soil microbial activity (Mukhopadhyay and Maiti 2010).

Phosphatases belong to a large group of enzymes that catalyze the hydrolysis of organic phosphorus compounds and are used to assess a potential rate of mineralization of these compounds in the soil (Gulser and Erdogan 2008; Hang et al. 2013; Yang et al. 2007). They play an essential role in the phosphorus circulation and are responsible for the management of phosphorus in the plant. They belong to periplasmic enzymes which means that they are deposited on the membrane or between the cytoplasmic membrane and the cell wall. These are the following types of phosphatases: acidic, neutral, and alkaline ones. Acid reaction, which corresponds to pH value ranging from 4 to 6, is optimal for acid phosphatase, whereas alkaline pH (pH 8–10) is accurate for alkaline phosphatase. The neutral phosphatase shows optimal activity at a pH value of 6.5–7.0. Microorganisms are the main source of phosphatases in the soil, but they are also produced by roots of plants and soil fauna. Phosphatase activity in soil reflects the activity of enzymes associated with soil colloids and humic substances (Das and Varma 2011; Śliwińska-Wyrzychowska and Nadgórska-Socha 2011).

Urease catalyzes the hydrolysis of urea in the soil to carbon dioxide and ammonia, which is one of the sources of nitrogen for plants. The activity of this enzyme reflects the ability to transform organic nitrogen to effective nitrogen and the ability to provide inorganic nitrogen (Ciarkowska et al. 2014; Gulser and Erdogan 2008; Friedlova 2010; Ianelli et al. 2012; Li et al. 2012; Yang et al. 2007). The rate of decomposition of urea depends on the soil reaction. The optimal value of soil pH for urease activity is between 6 and 7. This enzyme is present in the cells of plants and many microorganisms, especially bacteria as both intra- and extracellular enzyme. Urease is bound in the soil mainly by humic substances but also by the clay minerals on a smaller scale (Das and Varma 2011).

Proteases catalyze the hydrolysis of proteins in the soil environment, breaking down the peptide bonds (CO-NH) to amino acids. They are produced mainly by aerobic and anaerobic bacteria, fungi, and actinomycetes. Most proteases produced by bacteria are most active at pH = 7–8. Proteases produced by the fungi are less sensitive to acid reaction because they are active even at pH = 4. Extracellular proteases are enzymes whose activity indicates not only the biological activity of the soil in the enzymatic transformations of substrates, which is independent of the activity of microorganisms, but also play a role in microbial ecology of ecosystems. These enzymes are bound by the mineral and organic colloids (Das and Varma 2011; Mocek-Płóciniak 2011).

Invertase is an enzyme which is widely distributed in soils. It plays a significant role in increasing the content of soluble nutrients in the soil and carbon transformations. The activity of this enzyme reflects the ability of the soil to break down sucrose and release simple sugars, which are the main energy sources of soil microorganisms (Ciarkowska et al. 2014; Li et al. 2012; Yang et al. 2007).

8.3 Natural Environmental Factors Influencing the Activity of Soil Enzymes

Research findings on the variability of enzyme activity in soils indicate the impact of many environmental factors such as soil type, soil profile depth, composition and diversity of soil microorganisms, soil pH, organic matter content, and temperature on the enzyme activity. Many studies have shown that the activity of enzymes, such as dehydrogenases, urease, protease, phosphatase, and invertase, is proportional to the content of organic matter (organic content of C and total N) and is much higher in the rhizosphere than in the deeper horizons of the soil profile (Chodak and Niklińska 2010; Januszek 1999; Mocek-Płóciniak 2011). According to many authors (Friedlova 2010; Mukhopadhyay and Maiti 2010; Wolińska and Stępniewska 2012), the factors which have a significant positive effect on the activity of dehydrogenases are the water content and cations, Ca, Mg, K, and Fe, whereas a negative impact is exerted by the oxidizing-reducing potential and aerating of the soil. Under conditions of the humid temperate climate of Poland, it is observed that in most cases, enzymes are most active in summer. Nevertheless, high levels of activity have also been noted in spring or even in winter, which shows the large impact of freezing and thawing effects on the soil enzyme activity (Januszek 1999). The rate of enzymatic reaction is also conditioned by the concentration of the enzyme involved in the reaction, the substrate concentration, as well as the presence of activators and inhibitors (Kuziemska 2012).

The enzymes released from litter, living and dead microorganisms, as well as plant roots, are washed down to lower horizons. Their further fate, stability, and activity in the soil depend on the soil texture, mineral composition, as well as physical and chemical properties of soils. Extracellular enzymes secreted by living or dying cells can be present on the surface of cell walls and cell membranes, as well as fragments of cell organelles or plasma. They can also be accumulated in the soil, where they form labile enzyme-substrate compounds, are adsorbed onto the surface of mineral particles, or enter into complex compounds with colloids of humic substances. They can even remain partially, and on a short-term basis, in the soil solution (Burns et al. 2013). Free enzymes are very active but short lived, and they quickly undergo proteolysis. There are not only enzymes bound to organic colloids as a result of copolymerization during the process of the formation of humus and accumulated in humic complexes which may survive for long periods in the soil but also, to a smaller extent, enzymes bound to mineral colloids (Januszek 1999). In comparison with free enzymes, bound enzymes usually display lower levels of activities. It is because the complexation blocks the access to the substrate by occluding the active centers. Therefore, bound enzymes constitute a potential reservoir of enzymes. They can also be the source of enzymes in case of substrate deficiency or stressful conditions. One theory states that the enzyme spread increases as the enzyme-substrate availability decreases, which allows the producer of enzyme to access more distant substrate (Burns et al. 2013). The level of enzyme activity in the soil can also depend on vegetation and its succession, as extracellular enzymes are often released by plant roots. Research findings of Yang et al. (2007) indicate that the urease activity varies depending on the species of a plant or a combination of plants.

8.4 Mechanisms Connected with the Influence of Heavy Metals on Soil Properties

The increasing interest in the environmental pollution caused by heavy metals results from their toxic effect on plants, animals, and humans. They suppress the activity of enzymes, damage nucleic acid chains, and tend to bioaccumulate in tissues of living organisms. The consequences of their occurrence in the environment are mutagenic and carcinogenic changes in the metabolism or limitation of photosynthesis (Lock et al. 2003; Mocek-Płóciniak 2011; de Mora et al. 2005).

The effect of heavy metals on soil enzyme activity may be direct or indirect. The direct impact refers to the extracellular free enzymes, while the indirect one is manifested through the influence on the biosynthesis of enzymes performed by microorganisms. The indirect impact is also manifested by the effect of heavy metals on the composition of the population of microorganisms in the soil, as well as root exudates production and the release of enzymes from dead cells (Wyszkowska et al. 2006). The toxicity of discussed elements results from their possible influence on the growth and metabolism of soil microorganisms by disturbing the integrity of cell membranes or adversely affecting their functioning. Heavy metals also reduce soil enzyme activities by interacting with the complex enzyme-substrate and masking catalytically active groups, which produces the effect of denaturation toward active proteins of enzymes (Gianfreda et al. 2005; Zaborowska et al. 2006). Heavy metals affect the number, diversity, and activity of microorganisms. In the soil polluted with heavy metals, there prevail less diverse, slowly growing microorganisms which are more resistant to metals, but are characterized by a lower biological activity (Friedlova 2010). Lock and Janssen (2005) claim that the zinc contamination of soil causes microorganisms sensitive to this metal to die out. This can increase the number of resistant microorganisms.

Many authors think that as a result of the heavy metal contamination of the soil, the activity of intracellular soil enzymes produced by soil microorganisms is inhibited to a greater extent than the extracellular soil enzyme activity (Ciarkowska et al. 2014; Gulser and Erdogan 2008; Khan et al. 2007; Pang and Yu 2011). The influence of the heavy metal pollution of the soil on the activity of enzymes depends on the degree of contamination, the type of metal, time of exposure, and environmental factors, including mainly the soil reaction, which plays the key role in the solubility and the activity of metals in the soil (Januszek 1999; Oliveira and Pampulha 2006; Wyszkowska et al. 2006). Neutral or alkaline soil reaction contributes to the formation of metal compounds which are hardly soluble and therefore unavailable to plants (Ciarkowska and Gambuś 2004). The enzyme activity inhibited by heavy metals can also be associated with decreased contents of C, N, and P, which results in a lack of balance in the nutrient content and the soil degradation (Zhang et al. 2010). However, heavy metals which are present in small quantities in soils may stimulate the activity of enzymes. Not until they exceed the threshold content do they play a role in the reduction of microorganism activity and the production of extracellular enzymes (Wyszkowska et al. 2006).

8.5 Effects of the Short-Term Influence of Heavy Metals on the Activity of Soil Enzymes Observed During the Incubation and Pot Experiments

Incubation laboratory experiments enable us to trace the relationship between the content of heavy metals in soil and the level of activity of soil enzymes. The results of several studies show that as more time passes since the start of incubation and as the concentration of metals in the soil grows bigger, the enzyme activity gets lower. The extent of inhibition and the time after which the greatest reduction of enzyme activity is observed depend on the enzyme, the type of metal, their combination, and doses.

8.5.1 Cadmium and Lead

Khan et al. (2007) and Pang and Yu (2011) incubated the soil with the additional amounts of lead and cadmium salts. The researchers were progressively increasing amounts of those metals treating them both separately and as a combination. They observed that while they were increasing the doses of metals, the activity of urease, acid phosphatase, and dehydrogenase was getting lower proportionally to the dose of introduced metal. The enzyme activity was at its lowest after the scientists applied a combination of salts and the highest dose of metals. In the Pang and Yu (2011) experiment, the addition of 100 mg of Cd + 500 mg Pb kg^{-1} of soil resulted, after 10 weeks, in 50 % reduction of urease and dehydrogenase activity, whereas the acid phosphatase activity was reduced by 30 % after 2 weeks. In their incubation experiment, Khan et al. (2007) observed that the strongest inhibition of the activity of dehydrogenases and acid phosphatase occurred after 2 weeks and was followed by an increase in those enzyme activities. The authors explained that the decline in the enzyme activity which occurred over incubation time had been caused by the lack of nutrients which would be easily available for microorganisms. They also claim that the collapse of activity immediately after the application of metals and the increase in activity over time were both results of the adaptation of microorganisms to the conditions of contamination. Many authors (Khan et al. 2007; Mukhopadhyay and Maiti 2010; Pang and Yu 2011) unanimously state that a particularly negative effect on the activity of enzymes is exerted by cadmium and a combination of cadmium and lead. The impact of Cd on the enzyme activity is more toxic than Pb due to the greater mobility of the former element and a lower affinity with soil colloids. Mukhopadhyay and Maiti (2010) report that a significant inhibition of dehydrogenase activity occurs after applying very high doses of lead (5,000 mg of metal kg^{-1} of soil), while lower doses of the metal do not always reduce, but sometimes even slightly stimulate dehydrogenase activity. The effects lead has on the activity of enzymes can also be modified by the type of the plant cover. It was indicated by the result of the pot experiment in which lead had been introduced into the soil at a dose of 600 mg kg^{-1}. Such an amount of Pb led to the decline in the activity of urease activity in both cases: the soil covered with a combination of several plants and monoculture. However, the activity of this enzyme was greater in soils covered with a combination of several plants than in those with the monoculture plant cover. This may indicate that the presence of several plant species can reduce the negative influence of lead (Yang et al. 2007). The implemented dose of lead did not alter the activity of dehydrogenases and acid phosphatase and even resulted in an increase in the activity of alkaline phosphatase. The increased activity of alkaline phosphatase was explained by the increase in the soil pH value which had been caused by the introduction of alkalizing lead compounds (Yang et al. 2007). In her incubation experiment, Ciarkowska (2010) observed the varied impact of lead and cadmium on the activity of dehydrogenases. She compared the influence of low and high doses of lead and cadmium on the soils of sandy and silty textures, which were adjusted to the value of pH equal to 4.5, 5.5, and 7.0. An increased lead content in the soil, which resulted from low and high doses of this metal introduced into both acid soils, slightly inhibited the activity of dehydrogenases. Artificial lead contamination at a low and medium level of a sandy soil with a pH of 5.5 led to the inhibition of dehydrogenase activity in the range of 7.2–11.7 %, while adding a small amount of lead to a silty soil of the same pH value increased the enzyme activity by more than 26 %. Under the same conditions, contamination with lead at a medium level slightly reduced the dehydrogenase activity. Increasing the concentration of lead in both soils with a neutral soil reaction, regardless of the amount of the metal added, generally stimulated the activity of dehydrogenases to a limited extent. A small amount of cadmium (2.25 mg kg^{-1}) incorporated into the silty soil stimulated enzyme activity in the range of 2.1–23.4 %. Such an impact of a small cadmium dose (1.5 mg kg^{-1}) on a sandy soil was observed only in the soil with pH = 7. Cadmium contamination of soil at a medium level (addition of 4.5 mg Cd to the light soil and 6.75 mg Cd kg^{-1} to the medium soil) had generally a clearly toxic effect in both soils and at each level of pH value, except for the soil with silty texture and pH = 5.5, in which a slight increase in the activity of dehydrogenases was recorded.

8.5.2 Zinc

The results of incubation experiments, in which different doses of zinc had been introduced into the soil, showed that this metal was a stronger inhibitor of the activity of enzymes than lead and cadmium. In the experiment of Ciarkowska (2010), the introduction of an increased amount of zinc into the soil strongly suppressed the activity of dehydrogenases from the very beginning. In the sandy soil, a negative effect of additional amounts of zinc was clearly seen since the first day after the introduction of the metal. In the silty soil, which has a greater sorption capacity than the sandy soil, the toxic effect of small additional amounts of zinc was observed only after 7–14 days since the start of the experiment. The introduction of zinc played a role in reducing the dehydrogenase activity in the range of 5.8–46.1 % in relation to the control soil. The amounts added were the following: 200 mg Zn kg^{-1} into the light soil and 300 mg Zn kg^{-1} into the medium soil. The enzyme activity decreased to the range of 34.5–74.5 % when 600 mg Zn kg^{-1} was introduced into the light soil and 750 mg Zn kg^{-1} was added into the medium soil.

Wyszkowska et al. (2006) showed a similar inhibiting effect of zinc on enzyme activity in the incubation experiment, in which they added increasing doses of zinc into the soil to test their effect on the activity of dehydrogenases, urease, as well as acid and alkaline phosphatases. The dose of 5 g Zn kg^{-1} decreased dehydrogenase by 3.5 %, and a dose of Zn, which was 400 times higher, brought the dehydrogenase activity down by 89 %. Excessive doses of zinc led to the decline of urease activity, but the influence was much weaker than in the case of dehydrogenases. The highest dose of zinc made the activity of the discussed enzyme 4.3 times smaller. Phosphatase activities were even less affected by the increased zinc content. The highest dose of zinc used in this experiment (2,000 g Zn kg^{-1}) resulted in a more than twofold decrease in alkaline phosphatase activity and only a 1.5-fold decrease in acid phosphatase activity.

8.5.3 Combination of Zinc, Lead, and Cadmium

Yang et al. (2006) studied the effect that the increasing amounts of Zn, Pb, and Cd, individually or as a whole, had on catalase, urease, alkaline phosphatase, and invertase activities. They found out that, in case of all objects of experiment, lead, either as a single agent or in combination with the other two metals, stimulated the activity of catalase. The authors explain the fact by a reaction of Pb with the functional groups of catalase. The influence of heavy metal contamination of soil on the activity of urease was always inhibitory and proportional to the dose of introduced metals. It was stronger when they were applying the combination of metals rather than each one individually. A particularly strong synergistic interaction was observed for Zn and Cd. The activity of alkaline phosphatase was inhibited the most by the addition of Cd, while the effect of lead was not significant. Invertase was most strongly inhibited by a combination of three metals. Each of the enzymes responded differently to the single metal contamination and a combination of three metals. Various effects of individual metals on the enzyme activity resulted from different reactions into which enzyme functional groups and metals entered. In the experiment of Yang et al. (2006), cadmium was a strong inhibitor of all analyzed enzyme activity. Whereas zinc inhibited urease and catalase activities, lead proved to have the least inhibitory effect on enzymes and even a stimulating effect on catalase activity. The inhibitory effect of Cd on catalase and urease was strongly intensified by the addition of Zn. As far as the studied enzymes are concerned, urease has proven to be the most sensitive to the combinations of heavy metals. According to the authors, this enzyme can be a good biochemical indicator for the evaluation of Zn, Pb, and Cd contamination of soils.

8.5.4 Chromium

It was also observed that chromium (Cr (III) and Cr (VI)) was another element which had a negative influence on the activity of dehydrogenases in various types of soils. The results of the incubation experiment showed a rapid and significant decrease in the activity of dehydrogenases in each of the soils after the addition of higher doses of chromium, whereas small doses of chromium (2 mg kg^{-1}) increased the enzyme activity. The authors explained that the small additional amounts of Cr compensated for deficiencies of this element in the soils. The inhibitory effect of chromium on the activity of dehydrogenases depends on the type of soil, especially on the content of the organic substance. The activity of dehydrogenases was more sensitive to the effects of chromium in *Haplic Luvisol* and *Eutric Cambisol* than in *Mollic Gleysol* (Stępniewska and Wolińska 2005).

8.5.5 Nickel

The purpose of the 4-year pot experiment of Kuziemska (2012) was to determine the effect which the increased amounts of nickel introduced into the soil after bean cultivation had on the dehydrogenase activity. The results showed a significant change in the enzyme activity in each year of the experiment. After nickel was introduced at the lowest dose of 50 mg Ni kg^{-1}, the scientists observed, during the three following years a significantly increased activity of dehydrogenases, whereas higher doses of the metal (100 and 150 mg Ni kg^{-1}) strongly decreased their activity. In the fourth year of the experiment, only the highest dose of nickel (150 mg kg^{-1}) lowered the activity of these enzymes. Liming of the acid soils reduced the negative impact of nickel.

8.5.6 Summary

The results of research connected with the impact of heavy metals on the enzyme activity in the soils contaminated with heavy metals for a short period of time are very ambiguous. This fact implies that various soil microbial populations have different sensitivities to heavy metals, even if the soils have similar physical and chemical properties. The differences result from the diverse structures of the microbial population, as well as from the various interactions of metals and their forms with the soil. They are also caused by the chemical and exchangeable sorption of heavy metals by the organic substance, clay minerals, metal oxides and hydroxides, as well as free amorphous aluminosilicates (Yang et al. 2007).

The abovementioned results of experiments can be summarized by stating that:

- The heavy metals discussed above, such as zinc, lead, cadmium, nickel, and chromium, tend to have a negative effect on enzyme activities in the soils which have not been contaminated for a long time.
- The inhibitory effect of metals on the activity of enzyme depends on the type of enzyme, soil properties, and the dose of metal.
- The inhibitory effect of zinc, lead, and cadmium can be arranged in the following order: $Cd > Zn > Pb$ or $Zn > Cd > Pb$. Lead has a greater affinity for soil organic matter than zinc and is less mobile than cadmium. That is why it is complexed with soil organic matter to a greater degree and has a smaller inhibitory impact on the activity of enzymes.
- Combination of metals has a stronger negative impact on the activity of enzymes than a single metal because of a synergistic interaction of metal combinations.

- The toxic effect of metal interaction decreases over time.
- The results of studies do not unanimously point to one of the enzymes (dehydrogenase, urease, phosphatase, invertase) as an indicator of biochemical assessment of heavy metal contamination of soil.

8.6 Effects of the Long-Term Influence of Heavy Metals on the Activity of Soil Enzymes

Effects of long-term heavy metal pollution on the soil enzyme activity were studied in soil samples taken from the mining areas and/or metallurgy areas connected with nonferrous metals (copper, zinc, lead). Most authors observed a decrease in the activity of enzymes such as alkaline and acid phosphatases, urease, and dehydrogenase in soils which were in the direct vicinity of the emitter (Bielińska and Mocek-Płóciniak 2010; Castaldi et al. 2004; Januszek 1999). There was a close connection between the distance from metal smelters and enzyme activities, which indicates a significant negative effect of heavy metals on the activity of soil enzymes. The authors emphasize the particularly close relationship between the activity of enzymes and the content of bioavailable forms of zinc and cadmium (extracted by 0.01 M $CaCl_2$ or 1 M NH_4NO_3) (Śliwińska-Wyrzychowska and Nadgórska-Socha 2011). The negative effect of heavy metals on enzyme activity is also confirmed by the results of the research analyzing the activity of urease and alkaline phosphatase in soils which are located at different distances from roads with heavy traffic. The research indicated that the enzyme activity was higher as the distance from a road was growing. It also showed a significant negative correlation between the activity of enzymes (urease and alkaline phosphatase) and the content of Mn, Cr, and Pb. This suggests that heavy metal contamination of the soils located along roads inhibited the activity of the enzymes discussed above (Gulser and Erdogan 2008).

Friedlova (2010) showed a high inactivation of dehydrogenases in soils under conditions of long-term industrial emissions when she carried out the study of the enzyme activity of soils contaminated with emissions of lead industry, which had been functioning since the eighteenth century. The activity of dehydrogenases significantly decreased when the pollution increased, whereas changes in the urease activity were rather ambiguous and correlated with the content of nitrogen. However, one can also cite results of research on dehydrogenase activity in soils strongly contaminated with heavy metals over centuries. These soils are characterized by the significant activity of dehydrogenases. This can be explained by the fact that microbial populations which are exposed to long-term impact of toxic metals are increasingly tolerant to metals (Ciarkowska and Gambuś 2004; Januszek 1999; Zaborowska et al. 2006).

According to some researchers, extracellular enzyme activity, as compared with the activity of dehydrogenases, can be more useful to evaluate changes in the soil environment. Bielińska and Mocek-Płóciniak (2010) believe that the phosphatase and urease activities are sensitive indicators of changes in the soil environment exposed to a long-term impact of metals. Mocek-Płóciniak (2011) also described the activity of phosphatase as a good indicator of heavy metal pollution of the soil, but the results of her study did not demonstrate the negative impact of a high content of heavy metals (Cu and Pb) on urease activity. Studies of Castaldi et al. (2004) confirm that the urease activity is not significantly correlated with the content of heavy metals in the soil. Resistance of this enzyme to heavy metal pollution of the soil apparently results from the specific properties of urease, which is resistant to external factors, and can even increase its activity in extreme conditions. The only limiting factor is the availability of urea because urease, as an extracellular enzyme, is synthesized only in its presence. Results of analyses of urease and invertase activities which were conducted by Ciarkowska et al. (2014) on the land that had been subjected to the long-term effects of strong pollution by zinc, lead, and cadmium showed that long-term and significant accumulation of metals in the soils had no reducing effect on the activity of enzymes and the invertase activity was even higher in polluted soils than in unpolluted forest soil. The authors explained that fact by arguing that the growth of metal-resistant microbial population and/or organic matter accumulation, as well as neutral reaction of the soil, caused heavy metals to occur mainly in biologically inactive forms. Also, according to Ciarkowska et al. (2014) and Finkenbein et al. (2013), the increased activity of enzymes in soils suffering from long-term stress, which resulted from heavy metal contamination, was the effect of changes in the composition of the microbial population. However, the decrease in the enzymatic activity may be the result of the consumption of energy in the process of the physiological adaptation of the microorganisms by which they develop the tolerance of heavy metals. This adaptation consists of the following processes: synthesis of intracellular and extracellular proteins which sequester metals as well as biochemical reactions of metal precipitation or capture which occur on the surfaces of microbial cells (Zhang et al. 2010). Śliwińska-Wyrzychowska and Nadgórska-Socha (2011) explain also that the lack of relationship between the enzymatic activity of certain enzymes (dehydrogenases, urease, and phosphatases) and the content of heavy metals in soils characterized by various degrees of heavy metal contamination, including those located near metallurgical plants of nonferrous metals, results from a high content of organic matter and a low concentration of soluble forms of heavy metals.

Drawing conclusions from the above-presented research results, we can observe that a variety of enzyme reactions to long periods of heavy metal pollution may result from the various levels of sensitivity of microorganisms to metal toxicity, as well as from the development of mechanisms of resistance to metals and changes in the structure of microbial population, which can compensate for losses in more sensitive populations. Therefore, it seems that a reliable assessment of the quality of the soil environment can be achieved only by studying a series of soil enzymes.

8.6.1 Determining the Level of Degradation and the Progress of Ecosystem Restoration

The determination of enzyme activity is used as an indicator of both the rate of degradation of ecosystems and the improvement in the quality of mining areas as a result of rehabilitation procedures (Ciarkowska et al. 2014; Schimann et al. 2012). The reclaimed soil has to be microbiologically active, so it is very important to study the activity of microorganisms by, for example, testing the enzyme activity (Finkenbein et al. 2013). Habitats created after the opencast mining of nonferrous metals, apart from strong contamination by heavy metals, are often characterized by very unfavorable soil conditions such as extreme pH values, as well as scarcity of organic matter and nutrients for plants. The activity of many enzymes, both intra- and extracellular, is determined by the organic matter content or composition of the vegetation growing on degraded lands. Soil texture and pH value are other important properties, which determine the enzyme activity, of the soil forming on degraded areas. They often are characteristics affecting the enzyme activity even more strongly than the vegetation or organic matter content. In strongly acid soils, the presence of organic carbon and total nitrogen may not stimulate microbial activity (Chodak and Niklińska 2010; Finkenbein et al. 2013).

An important and difficult issue is to determine the point at which the soil can be considered completely reclaimed. For this reason, indicators which reflect the state of the soil and its functions, such as the activity of enzymes which provide information about the soil microbiology as well as its physical and chemical parameters, are very useful (Antunes et al. 2011; Zhang et al. 2010). According to Ciarkowska et al. (2014), the activity of urease and invertase can be such an indicator. It has been noted that the activity of these enzymes in the soils of zinc and lead ores mining areas which have been restored in recent years is significantly lower than in the soils of the same area but reclaimed decades ago. This suggests that the newly reclaimed soils undergo a slow rehabilitation process. This process, however, is not yet completed, and microbial communities are still under the influence of mining activities (Ciarkowska et al. 2014). Similarly, Mukhopadhyay and Maiti (2010), who were examining the activity of dehydrogenases in reclaimed soils which were degraded after opencast mining of metal ores, found out that the activity of these enzymes in reclaimed soils was lower than in referential soils. The lowest activity was found in the soils where the period of time since the completion of the rehabilitation was the shortest. The activity of dehydrogenase in 3-year heaps was several times lower than in the soils of 20-year heaps, the latter level of activity being similar to the one noted in nondegraded soils of this area.

The results support the opinion that enzymatic activity is reduced immediately after the introduction of metals into the soil, but usually, as time passes, it returns to its original level (Ciarkowska and Gambuś 2004). The use of a biological parameter allows us to specify the ecologically negative effects of mining and metal ore processing as well as destabilization of the soil ecosystem.

8.7 Conclusions

Despite the fact that the results of various authors' research on the relationship between the content of heavy metals in soil and the level of enzyme activity are not fully unequivocal, one can draw some general conclusions, according to which the soil enzyme activity:

- Reflects the degree and size of heavy metal pollution soil
- Is a good indicator of both short- and long-term heavy metal contamination of the environment
- Allows to estimate the effects of reclamation of degraded ecosystems

References

Antunes SC, Pereira R, Marques SM, Castro BB, Gonçalves F (2011) Impaired microbial activity caused by metal pollution: a field study in a deactivated uranium mining area. Sci Total Environ 410–411:87–95. doi:10.1016/j.scitotenv.2011.09.003

Bielińska EJ, Mocek-Płóciniak A (2010) Impact of ecochemical soil conditions on selected heavy metals. Pol J Environ Stud 19(5):895–900

Burns RG, DeForest JL, Marxsen J, Sinsabaugh RL, Stromberger ME, Wallenstein MD, Weintraub MN, Zoppini A (2013) Soil enzymes in a changing environment: current knowledge and future directions. Soil Biol Biochem 58:216–234, http://dx.doi.org/10.1016/j.soilbio.2012.11.009

Castaldi S, Rutigliano FA, Virzo de Santo A (2004) Suitability of soil microbial parameters as indicators of heavy metal pollution. Water Air Soil Pollut 158:21–35

Chodak M, Niklińska M (2010) Effect of texture and tree species on microbial properties of mine soils. Appl Soil Ecol 46:268–275. doi:10.1016/j.apsoil.2010.08.002

Ciarkowska K (2010) Impacts of agro-technology and industrial pollution on the biological activity and physicochemical and micromorphological properties of soils. Zesz Nauk 458:335, ss.128, (in Polish)

Ciarkowska K, Gambuś F (2004) Activity of dehydrogenase in heavy metal contaminated soils in the area of Olkusz Zesz. Probl Post Nauk Rol 501:79–87 (in Polish)

Ciarkowska K, Sołek-Podwika K, Wieczorek J (2014) Enzyme activity as an indicator of soil rehabilitation processes at a zinc and lead ore mining and processing area. J Environ Manage 132:250–256, http://dx.doi.org/10.1016/j.jenvman.2013.10.022

Das SK, Varma A (2011) Role of enzymes in maintaining soil health. In: Shukla G, Varma A (eds) Soil enzymology, vol 22, Soil biology. Springer, Berlin. doi:10.1007/978-3-642-14225-3_2, Chapter 2

Finkenbein P, Kretschmer K, Kuka K, Klotz S, Heilmeier H (2013) Soil enzymatic activities as bioindicators for substrate quality in revegetation of a subtropical coal mining dump. Soil Biol Biochem 56:87–89

Friedlova M (2010) The influence of heavy metals on soil biological and chemical properties. Soil Water Res 5(1):21–27

Gianfreda L, Rao MA, Piotrowska A, Palumbo G, Colombo C (2005) Soil enzymes activities as affected by anthropogenic alterations: intensive agricultural practices and organic pollution. Sci Total Environ 341:265–279

Gulser F, Erdogan E (2008) The effects of heavy metal pollution on enzyme activities and basal soil respiration of roadside soils. Environ Monit Assess 145:127–133. doi:10.1007/s10661-0007-0023-7

Hang X., Li F., Liu T., Xu C., Duan D., Peng C., Zhu S., Shi J. 2013. The variations in the soil enzyme activity, protein expression, microbial biomass and community structure of soil contaminated by heavy metals. ISRN Soil Sci 2013, Article ID: 803150, http://dx.doi.org/10.1155/2013/803150

Ianelli R, Bianchi V, Macci C, Peruzzi E, Chiellini C, Petroni G, Masciendaro G (2012) Assessment of pollution impact on biological activity and structure of seabed bacterial communities in the Port of Livorno (Italy). Sci Total Environ 426:56–64

Januszek K (1999) Enzyme activity of selected forest soils of southern Poland, in the light of field and laboratory studies. Zesz Nauk AR Kraków 250: ss. 132 (in Polish)

Khan S, Cao Q, Hesham A, Xia Y, He J (2007) Soil enzymatic activities and microbial community structure with different application rates of Cd and Pb. J Environ Sci 19:834–840

Kumar S, Chaudhuri S, Maiti SK (2013) Soil dehydrogenase activity in natural and mine soil – a review. Mid E J Sci Res 13(7):898–906. doi:10.5829/idosi.mejsr.2013.13.7.2801

Kuziemska B (2012) The activity of dehydrogenases in the soil contaminated with nickel. Ochr Środ i Zasob Natur 52:103–112

Li H, Shao H, Li W, Bai Z (2012) Improving soil enzyme activities and related quality properties of reclaimed soil by applying weathered coal in opencast mining areas of the Chinese Loess Plateau. Clean Soil Air Water 40:233–238. doi:10.1002/clean.201000579

Lock K, Janssen CR (2005) Influence of soil zinc concentrations on zinc sensitivity and functional diversity of microbial communities. Environ Pollut 136:275–281

Lock K, Janssens F, Janssen CR (2003) Effects of metal contamination on the activity and diversity of springtails in an ancient Pb-Zn mining area at Plombieres Belgium. Eur J Soil Biol 39(1):25–29

Mocek-Płóciniak A (2011) Impact of heavy metals on microorganisms and the soil enzymatic activity. Soil Sci Ann 62(4):211–220 (in Polish)

de Mora AP, Ortega-Calvo JJ, Cabrera F, Madejón E (2005) Changes in enzyme activities and microbial biomass after "in situ" remediation of a heavy metal contaminated soil. Appl Soil Ecol 28(2):125–137

Mukhopadhyay S, Maiti SK (2010) Dehydrogenase activity in natural and mine soil - a review. Ind J Environ Prot 30(11):921–933

Oliveira A, Pampulha ME (2006) Effects of long-term heavy metal contamination on soil microbial characteristics. J Biosci Bioeng 102(3):157–161

Pang J, Yu L (2011) Effects of Cd or/and Pb on soil enzyme activities and microbial community structure. Ecol Eng 37:1889–1894

Schimann H, Petit-Jean K, Guitet S, Reis T, Domenach AM, Roggy J-C (2012) Microbial bioindicators of soil functioning after disturbance: the case of gold mining in tropical rain forests of French Guiana. Ecol Indic 20:34–41. doi:10.1016/j.ecolind.2012.01.021

Stępniewska Z, Wolińska A (2005) Soil dehydrogenase activity in the presence of chromium (III) and (VI). Int Agrophys 19:79–83

Śliwińska-Wyrzychowska A, Nadgórska-Socha A (2011) The effect of heavy metal pollution and presence of *Lycopodium annotinum* on soil enzyme activity. Chem Met Alloys 4:38–45

Wolińska A, Stępniewska Z (2012) Dehydrogenase activity in the soil environment. In Tech, Rijeka, http://dx.doi.org/10.5772/48294

Wyszkowska J, Zaborowska A, Kucharski J (2006) Activity of enzymes in zinc contaminated soil. EJPAU Environ Develop 9(1): 06, www.ejpau.media.pl/volume9/issue1/art-06.html

Yang Z, Liu S, Zheng D, Feng S (2006) Effects of cadmium, zinc and lead on soil enzyme activities. J Environ Sci 18(6):1135–1141

Yang R, Tang J, Chen X, Hu S (2007) Effects of coexisting plant species on soil microbes and soil enzymes in metal lead contaminated soils. Appl Soil Ecol 37:240–246. doi:10.1016/j.appsoil.2007.07.004

Zaborowska M, Wyszkowska J, Kucharski J (2006) Microbiological activity of zinc-contaminated soils. J Elem 11(4):543–557

Zhang FP, Li CF, Tong LG, Yue LX, Li P (2010) Response of microbial characteristics to heavy metal pollution of mining soils in central Tibet, China. Appl Soil Ecol 45:144–151. doi:10.1016/j.apsoil.2010.03.006

Chapter 9
Heavy Metals in Native Mediterranean Grassland Species Growing at Abandoned Mine Sites: Ecotoxicological Assessment and Phytoremediation of Polluted Soils

Mª J. Gutiérrez-Ginés, J. Pastor, and A.J. Hernández

9.1 Introduction

The closing of several mines in the central Iberian Peninsula has left behind a bleak scenario of areas of soils highly polluted with heavy metals and trace elements. The plant communities that thrive at these sites are mainly comprised of grassland species, and it is of major concern that these plants are consumed by livestock or wild animals. These grasslands have been the focus of several years of study by our research group, both because of the impacts of their polluted soils on ecosystems and because of their possible remediation role (Pastor and Hernández 2008).

Based on the results of numerous plant surveys, the dynamics of these communities exposed to elevated levels of trace elements (Hernández and Pastor 2008a) can be summarised as follows. The most evident impact of pollutants on plant communities is a loss of species diversity. The presence of a pollutant in a habitat affects either the area occupied by each species or the resources they use. These effects depend on the tolerance level of species or their sensitivity to a given pollutant (Pastor and Hernández 2007). As a consequence, effects of pollutants on competition may upset the balance among the community's components, which could wipe out some of the more sensitive populations. This may be observed at the sites examined here; although the dominance of grassland species is evident, legume species are hardly present, probably due to the heavy metals in their soils (Hernández and Pastor 2005). Thus, species frequencies in an ecosystem will vary along a gradient of chronic pollution.

MªJ. Gutiérrez-Ginés (✉) • J. Pastor
Department of Biogeochemistry and Microbial Ecology, National Museum of Natural Sciences, CSIC, C/Serrano, 115 bis, 28006 Madrid, Spain
e-mail: mjesus.gutierrezg@uah.es

A.J. Hernández
Department of Life Sciences, University of Alcalá, Ctra. Madrid-Barcelona, km 33.6, 28871 Alcalá de Henares, Spain

I. Sherameti, A. Varma (eds.), *Heavy Metal Contamination of Soils*, Soil Biology 44, DOI 10.1007/978-3-319-14526-6_9

Knowledge of soil-plant processes and interrelations in terrestrial ecosystems is essential to understand how these systems work and how they are affected by any perturbation. Processes such as those controlling heavy metal dynamics are slow and complex. When toxic elements are incorporated into food webs, they may have direct effects on ecosystem health, human health or population welfare (Hernández et al. 2009). These problems need to be understood in detail to avoid their consequences through the design of remediation measures.

This study is a part of a research line that focuses on the ecological restoration of sites with heavy metal-polluted soils (Hernández and Pastor 2008b). Based on our knowledge of the behaviour of plant communities at these abandoned mine sites, the present study was designed to determine the heavy metal, trace element and Al contents of both the topsoil layers and plant species growing at these sites with the objectives: (1) to assess the remediation capacity of the native species forming these communities and, perhaps more importantly, (2) to assess the ecological risks of the accumulating behaviour of these plants given that the grasslands are grazed by wild and domestic animals.

9.2 Materials and Methods

9.2.1 Mine Sites and Sampling

The eight abandoned mine sites (Fig 9.1) examined are named according to their location (town or village) in central Spain; the main element mined and Universal Transverse Mercator (UTM) coordinates for the site are indicated in parentheses. Five sites are found in the Comunidad de Madrid: Garganta de los Montes (Cu, $X = 443297$, $Y = 4529867$, 30 T, henceforth *Garganta*), Bustarviejo (Ag, $X = 438423$, $Y = 4524252$, 30 T), Navas del Rey (Ba, $X = 393455$, $Y = 4474405$, 30 T, henceforth *Navas*), Colmenar del Arroyo (Pb, $X = 395182$, $Y = 4474723$, 30 T, henceforth *Colmenar*) and Colmenarejo (Cu, $X = 415799$, $Y = 4486912$, 30 T). Their main characteristics are described in Fernández-Rubio (2007), Jiménez et al. (2004) and Jiménez-Ballesta et al. (1990). The other three sites are located in the Comunidad de Castilla-La Mancha; two in the Toledo province, Mazarambroz (Ag, $X = 406730$, $Y = 4397755$, 30 S) and Buenasbodas (Au, $X = 332410$, $Y = 4392147$, 30 S); and the other in the Guadalajara province, Hiendelaencina (Ag, $X = 500459$, $Y = 4548926$, 30 T) (Libro Blanco de la Minería de Castilla-La Mancha 2004).

Soil and plant sampling was conducted in different geomorphologic units at each site (tailings, slopes and valleys affected by mining activities) since heavy metals spread beyond tailings (Encabo et al. 1987; Gutiérrez-Maroto et al. 1989; Lacal et al. 1995). In each geomorphologic unit, sampling was random. The number of points sampled depended on the unit size and its visual heterogeneity. At each point, an average soil sample was collected of the topsoil layer (0–10 cm) using a hoe. Plant sampling was performed across 1-m^2 squares to include the above-

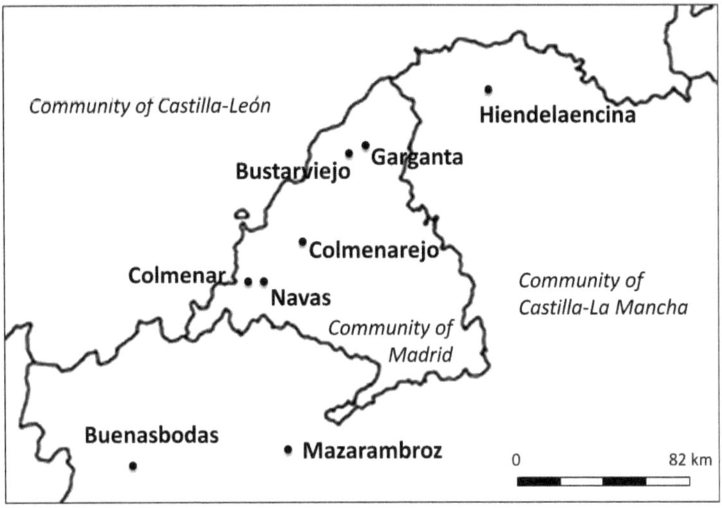

Fig. 9.1 Localization of the abandoned mine sites

ground mass of dominant species or species contributing to the food web (171 grass species total). Additionally, we collected samples of 15 woody species because of their structural role. Species were identified according to the Flora de Andalucía and Flora Ibérica although the nomenclature used is that accepted in January 2013 by Anthos (Sistema de Información sobre las plantas de España, www.anthos.es). The name of the author describing each species is not included in the result tables. Species are classified into the three most abundant Mediterranean grassland families (Poaceae or grasses, Fabaceae or legumes and Asteraceae or composites).

9.2.2 Chemical and Data Analysis

Soil samples were dried at room temperature for 1 week and sieved through a 2 mm mesh. Tests conducted were pH in slurry, organic matter by potassium dichromate reduction (according to protocols described in Hernández and Pastor 1989), pseudo-total metal contents by inductively coupled plasma-optical emission spectroscopy (ICP-OES, PerkinElmer Optima 4300DV), following HNO_3 and $HClO_4$ digestion (Walsh and SSSA 1971), and As and Ba contents by X-ray fluorescence (Siemens SRS 300). All plant samples of above-ground parts were washed with tap water and rinsed twice in deionised water, oven-dried at 70 °C for 48 h and ground in an IKA Werke Yellow Line A10 grinder. The procedures used to determine metals were those described for the soil samples. The metals determined in plants were those that may enter the food chain (Cu, Zn, Pb, Cd, Ni, Cr, Al and Mn).

Results are provided as minimum and maximum concentrations.

The index Relative Deviation to Background (RDB) described by Kabata-Pendias and Mukherjee (2007), relating the average heavy metal contents of all samples of the same species to a reference value, was used as a measure of the accumulation capacity of native plants.

9.3 Results and Discussion

9.3.1 Concentrations of Heavy Metals and Trace Elements in the Mine Soils

In total, 191 soil samples from the eight abandoned mine sites were analysed (see Table 9.1). At the Garganta and Mazarambroz sites, two sampling sessions were needed due to their large size and widely varying metal concentrations.

The prevailing soil pH of the eight sites is acidic, determining a risk of high metal bioavailability. Soils from valleys and slope breaks showed the higher organic matter (OM) contents. In contrast, tailing samples had the least OM. Since it is difficult to find permissible reference values for Al and Mn, values obtained for these elements were compared to the mean for the Earth's crust and light sandy soils (as are most of the soils examined) as described in Kabata-Pendias and Mukherjee (2007). Soil Cr levels were always under their reference values. Ni was detected in a concentration higher than the reference just in two sampling points (one in Navas and one in Mazarambroz). Ba was only determined at the mine where this element was exploited (Navas). The following is a summary of the different sites in terms of the metals that mostly contribute to their polluted soils.

Practically all the mine soils were found to contain several metals at higher concentrations than reference levels. The Garganta mine is mainly polluted by Cu. Although one of the samples examined showed substantial Pb and Cd concentrations (3,750 mg kg^{-1} and 340 mg kg^{-1}, respectively), these metals normally appeared at concentrations of around 100 mg kg^{-1} and $<$10 mg kg^{-1}, respectively. Arsenic is the main pollutant of the Bustarviejo mine, but in some areas Cu, Zn, Pb and Cd levels were also worrying. Only two soil samples from the Navas site featured higher element levels than reference values. Similarly, only one soil sample from the Colmenar mine contained high Cu, Zn and Pb concentrations. Mazarambroz was the mine with the largest area affected by heavy metals, mainly Zn and Pb, but also As, Cu and Cd at several points. None of the samples collected from the Hiendelaencina mine showed polluting levels of metals as determined by XRF; thus, these samples were not subjected to ICP-OES tests.

Table 9.1 Number of soil samples (*N*) collected from each mine site and soil pH, organic matter (OM %), pseudo-total contents of Al (%) and metals (mg kg^{-1}) and total content (XRF) of As (mg kg^{-1})

Mine	*N*	pH	OM	Ba	Al	Mn	Cu	Zn	Pb	Cd	As	Cr	Ni
Garganta	50	4.6–7.6	0.3–17.9	–	1.6–5.4	70–1,450	30–3,500	40–540	5–3,750	0.1–340	21–45	2.0–35	8.0–26
Bustarviejo	15	4.0–6.2	0.3–9.0	–	1.2–4.1	200–690	45–2,070	140–11,010	40–3,953	4.5–263	205–15,436	2.0–12	1.0–6.0
Colmenarejo	12	5.7–7.0	0.9–8.0	–	1.2–3.0	266–449	22–10,909	87–248	11–178	0.0–6.0	9.0–224	0.0–10	1.0–5.1
Navas	11	4.7–6.7	2.0–7.1	42–1,630	1.7–2.9	135–605	4.5–888	69–1,130	10–249	4.0–5.5	–	5.6–21	3.5–32
Colmenar	4	6.2–7.6	3.7–7.5	–	2.0–3.8	58–609	1.5–435	131–2,641	209–8,584	0.0–16	n.d.	1.0–10	1.5–15
Mazarambroz	79	1.6–8.0	0.3–9.7	–	0.4–5.2	45–1,625	8.5–201	62–10,430	20–6,542	0.0–35	36–531	0.0–54	3.0–41
Buenasbodas	12	5.0–6.9	1.4–10.0	–	1.6–5.2	41–1,315	5–25	16–69	1–25	0.0–0.5	–	2.5–10	2.0–24
Hiendelaencina[a]	8	–	–	–			26–63	91–420	18–516	–	–		
Reference level	pH < 7[b]			160[c]	8, 4[d]	1,200, 2,000[d]	50	150	50	1	29[c]	100	30
	pH > 7[b]						210	450	300	3		150	112

[a]Samples only subjected to XRF due to their low pollutant levels
[b]Reference according to Spanish law RD1310/1990 (BOE 1990)
[c]Dutch reference
[d]Concentration in Earth's crust—concentration in light sandy soils, Kabata-Pendias and Mukherjee (2007)
n.d.: not detected
–: not analysed

9.3.2 Heavy Metals, Trace Elements and Al in the Shoots of Grass Species Growing at the Mine Sites

One hundred and seventy-one native grass species of the eight mine sites plus 16 woody species were tested. Most species had metals and trace elements at higher concentrations than considered normal. Tables 9.2, 9.3 and 9.4 show the most relevant species according to their accumulation capacity. The chosen ones were those with concentrations higher than that considered toxic for plants (Table 9.5) of at least two metals.

Interestingly, the Poaceae family (Table 9.2) contained a great number of species with high metal concentrations. Among all those tested, *Agrostis castellana* was the species showing the highest levels of Cu, Zn and Pb. This species was also broadly distributed across all sites. Many species of this family featured outstanding concentrations of Zn, but species with high levels of Cu, Pb, Cd or As were not that frequent. Of interest, Zn, Pb, Cd and As concentrations were high in *Arrhenatherum elatius* subsp. *bulbosum*, *Holcus annuus* subsp. *setiglumis* and *Vulpia myuros*. Apart from *Agrostis castellana*, *A. stolonifera*, *Corynephorus canescens* and *Lolium multiflorum* showed the greatest Cu contents.

The metal contents of the legumes were unremarkable (Table 9.3) and always lower than the concentrations detected in the grasses, as also reported by Kabata-Pendias and Mukherjee (2007). However, metal levels were high in some species such as *Trifolium scabrum* (maximum 496 mg Zn kg^{-1}, 100 mg Pb kg^{-1}) or *T. campestre* (maximum 8 mg Cd kg^{-1}). None of the legumes were found to contain As in their shoots, so this element is not shown in the table. In contrast, Jana et al. (2012) detected an As concentration of 208 mg kg^{-1} in the leaves of *Trifolium pratense*.

Species of the Asteraceae family (Table 9.3) are well known for their high element accumulation capacity. However, although some of the species showed significant metal concentrations, these were never comparable to those detected in the grasses. The highest Zn concentration (1,087 mg Zn kg^{-1}) was observed in *Leontodon saxatilis*. *Carduus pycnocephalus* and *Crepis vesicaria* featured high Pb concentrations, and marked Cd concentrations were observed in all of this family's species (up to 44 mg kg^{-1} in *Andryala ragusina*), even exceeding those of the grasses. Once again, Arsenic was not detected in any species.

In the remaining families (Table 9.4), several species showed a notable accumulation capacity. *Scirpoides holoschoenus* was the only species containing As (up to 57 mg kg^{-1}), although similar species have been described by Jana et al. (2012) to take up this metal: *Juncus effusus* (29 ± 0.3 mg As kg^{-1}) and *Plantago major* (7 mg kg^{-1}). Species of the *Rumex* genus featured marked overall contents of Cu, Zn, Pb and Cd. *Spergularia rubra*, species of the genus *Plantago* and those of the Brassicaceae family are also interesting accumulators, though their Cu accumulation capacity is limited. The maximum Cd concentration in all the tested plants was found in *Jasione montana* (64 mg Cd kg^{-1}), but species such as *Andryala ragusina* or *Corrigiola telephiifolia* were also observed to accumulate

Table 9.2 Heavy metal and As concentrations (mg kg^{-1}) detected in the shoots of native Poaceae species collected from the mine sites

Poaceae species		Cu	Zn	Pb	Cd	As	Al	Mn
Agrostis	***A. castellana***	0.0–424	17–9,065	0.0–1,572	0.0–12	0.0–135	11–3,059	40–2,465
	A. stolonifera	2.1–941	3.6–123	0.0–12	0.0–2.8	0.0	1,247	9.5–118
Aira	*A. caryophyllea*	0.8–93	39–140	0.0	0.0–2.0	0.0	189–626	49–231
Arrhenatherum	*A. album*	0.0–39	43–268	0.0–8.3	0.0	–	40–171	0.0–218
	A. elatius	0.0–98	20–790	0.0–43	0.0–16	–	139–1,078	20–373
	A. elatius* subsp. *bulbosum	0.5–16	0.6–1,740	0–198	0–10	0.0–51	1.8–325	1.1–848
Avena	***A. barbata***	0.7–14	30–423	0.0–249	0.0–5.5	–	54–2,511	32–442
	A. sterilis	4.8–11	14–320	0.0–44	0.0–2.6	0.0	2.1–77	28–484
Bromus	*B. diandrus*	2.3–45	0.0–194	0.0	0.0	–	28–362	10–130
	B. hordeaceus	2.0–64	14–210	0.0–18	0.0–1.2	–	31–150	25–253
	B. madritensis	1.3–26	131–720	3–74	0.0–3.1	–	61–335	8.2–78
	B. rubens	1.2–19	17–677	0.0–309	0.0–1.9	0.0	44–2,205	5.1–97
	B. tectorum	0.0–47	33–487	0.0–78	0.0–4.7	0.0	25–389	1.9–278
Corynephorus	***C. canescens***	2.2–659	54–108	0.0	0.0–0.4	0.0	249–2,419	30–198
Cynodon	***C. dactylon***	7.1–86	24–462	0.0–98	0.0–6.0	0.0	53–571	13–834
Dactylis	*D. glomerata*	0.0–24	21–424	0.0–50	0.0–0.8	–	12–362	7.2–327
Holcus	***H. annuus* subsp. *setiglumis***	0.4–34	24–1,220	0.0–262	0.0–4.6	0.0–51	18–456	4.1–695
	H. lanatus	0.0–84	20–840	0.0–3.5	0.0–8.6	0.0	24–2,531	72–1,140
Lolium	***L. multiflorum***	0.3–662	2.6–209	0.0	0.0–4.9	0.0	0.0–133	1.5–117
	L. rigidum	4.3–38	48–147	0.0–6.0	0.0	0.0–195	41–1,510	135–295
Melica	*M. ciliata*	1.9–10	24–206	0.0–4.6	0.0–4.1	–	79–94	6.2–35
Micropyrum	***M. tenellum***	2.0–21	78–781	0.0–104	0.0–2.6	–	13–1,171	32–206
Molineriella	*M. laevis*	5.2–89	33–748	0.0	0.0–7.8	0.0	31–431	74–524
Periballia	***P. involucrata***	3.2–33	111–304	0.0–21	0.0–1.4	0.0–200	37–716	82–147
Poa	*P. trivialis*	1.5–46	12–339	0.0	0.0–2.0	–	–	29–82

(continued)

Table 9.2 (continued)

Poaceae species		Cu	Zn	Pb	Cd	As	Al	Mn
Stipa	*S. lagascae*	0.7–16	15–301	0.0–75	0.0–1.1	0.0	14–286	10–129
Taeniatherum	*T. caput-medusae*	1.2–152	0.0–21	0.0	0.0	0.0–57	41–100	18–26
Trisetum	*T. ovatum*	1.9–7.2	25–109	0.0	0.0–0.3	–	–	47–400
Vulpia	***V. bromoides***	3.6–49	88–892	0.0–58	0.0–2.4	0.0	56–1,998	63–394
	V. ciliata	4.3–8.1	133–254	0.0–107	0.0–0.7	0.0	480–3,608	0.0–398
	V. myuros	0.0–47	6.3–3,165	0.0–771	0.0–20	0.0–139	22–2,118	2.5–602

Table 9.3 Heavy metal and As concentrations (mg kg^{-1}) detected in the shoots of native Fabaceae and Asteraceae species collected from the mine sites

Fabaceae species		Cu	Zn	Pb	Cd	Al	Mn
Lotus	*L. corniculatus*	2.0–6.9	34–324	0.0	0.0–1.3	–	63–91
	L. pedunculatus	2.1–7.6	1.0–80	0.0	0.0–2.7	–	21–236
Ornithopus	*O. compressus*	4.4–13	27–209	0.0	0.0	32–275	138–353
	O. perpusillus	1.5–6.8	2.1–234	0.0	0.0–3.6	–	20–113
Trifolium	*T. campestre*	7.0–38	84–214	0.0	0.0–8.1	111–3,294	43–171
	T. gemellum	3.0–5.0	62–348	0.0–32	0.0	78–311	65–151
	T.ochroleucon	8.6–23	50–220	0.0	0.0–0.5	18–205	7.0–69
	T. scabrum	13–17	114–496	0.0–100	0.0–0.1	154–308	45–223
	T. strictum	4.0–79	34–120	0.0	0.0–1.5	50–114	22–69
	T. subterraneum	7.0– 21	93–214	0.0	0.0–1.2	66–168	131–198
	T. tomentosum	2.6–15	21–295	0.0–76	0.0–2.4	65–552	14–29
Asteraceae species							
Anacyclus	*A. clavatus*	5.9–66	43–79	0.0	0.3–4.6	25–432	155–334
Andryala	***A. integrifolia***	2.2–51	12–172	0.0	0.0–20	26–637	7.5–63
	A. laxiflora	7.4–15	191–285	2.1–8.2	0.0–10	48–96	71–206
	A. ragusina	6.3–27	167–824	0.0–57	4.8–44	158–811	32–60
Artemisia	*A.* sp.	0.8 – 1.0	12–206	0.0	0.0–2.0	–	9.4–14
Carduus	***C. pycnocephalus***	3.7–20	27–472	0.0–115	0.0–5.1	21–75	0.0–44
Centaurea	*C. paniculata*	8.9–13	14–94	0.0	0.0–2.5	20–57	4.8–14
	C. melitensis	1.3–8.0	71–224	0.0–52	0.0–6.1	41–128	0.0–96
Chondrilla	***C. juncea***	1.0–19	0.4–294	0.0–6.4	0.0–14	48–1,019	0.0–1,312
Cirsium	***C.* sp.**	4.5–19	14–82	0.0	0.0–6.5	28–239	2.3–9.1
Crepis	*C. capillaris*	6.2–39	48–264	0.0–8.6	0.0–8.7	74–1,060	25–248
	C. vesicaria	2.4–25	34–603	0.0–176	0.0–20	23–534	3.9–346
Filago	*F. arvensis*	1.5–74	12–459	0.0	0.0–4.0	–	21–281
	F. gallica	11–25	85–461	9.0–97	1.7–13	62–425	37–51

(continued)

Table 9.3 (continued)

Fabaceae species		Cu	Zn	Pb	Cd	Al	Mn
Hypochaeris	***H. radicata***	2.4–73	36–724	0.0–0.2	0.0–17	90–1,734	20–327
Leontodon	***L. saxatilis***	4.8–54	91–1,087	0.0–45	0.0–24	70–199	9.3–433
Pilosella	***P. officinarum***	5.0–25	15–84	0.0–1.9	0.0–16	99–1,144	10–75
Santolina	***S. rosmarinifolia***	5.0–52	15–193	0.0	0.0–9.3	85–1,537	0.5–107
Scolymus	*S. hispanicus*	0.0–23	40–579	0.0–23	0.0–8.5	41–173	0.0–86
Sonchus	***S. asper***	0.9–3.5	12–380	0.0–43	0.0–13	–	1.6–7.9

large quantities of this metal. These species, however, do not show this metal in their shoots as frequently as *J. montana*, which took up this metal in all the samples tested. Despite being highly bioavailable (Kabata-Pendias and Mukherjee 2007), Cd concentrations in shoots are not usually high. In a study examining Cd accumulation in native species from mine sites, Zhang et al. (2012) also detected scarce amounts of Cd in vascular plants. In general, the metal concentrations detected here in native species from eight mine sites in central Spain were higher than those reported by other authors (Conesa et al. 2006, 2007; García-Salgado et al. 2012; Jana et al. 2012; Massa et al. 2010; Moreno-Jiménez et al. 2009; Pratas et al. 2013; Zhang et al. 2012), though for different plant communities.

The main woody species growing at the 8 mine sites were also tested (Table 9.4) since they are often consumed by goats. However, appreciable metal concentrations were only detected in the leaves of *Thymus zygis*.

Other analysed species that did not show very relevant concentration of metals and As were *Aegilops geniculata*, *A. triuncialis*, *Anthoxanthum aristatum*, *Avena sativa*, *Bromus sterilis*, *Cynosurus cristatus*, *C. echinatus*, *Festuca rothmaleri*, *Hordeum murinum*, *Koeleria vallesiana*, *Phalaris canariensis*, *P. minor*, *Phleum pratense*, *Poa annua*, *P. bulbosa*, *Stipa parviflora*, *Vulpia membranacea* and *V. unilateralis* of Poaceae family; *Anthyllis vulneraria*, *Lathyrus angulatus*, *Medicago minima*, *M. sativa*, *M. turbinate*, *Melilotus officinalis*, *Ononis pusilla*, *O. spinosa*, *Trifolium angustifolium*, *T. arvense*, *T. cherleri*, *T. dubium*, *T. glomeratum*, *T. repens*, *T. striatum*, *T. suffocatum*, *T. sylvaticum*, *Vicia angustifolia*, *V. benghalensis* and *V. hirsuta* of Fabaceae family; *Carthamus lanatus*, *Centaurea alba*, *C. calcitrapa*, *Chamaemelum mixtum*, *Cichorium intybus*, *Mantisalca salmantica*, *Pallenis spinosa*, *Picnomon acarna*, *Podospermum laciniatum and Tragopogon crocifolius* of Asteraceae family; *Anarrhinum bellidifolium*, *Campanula rapunculus*, *Carum verticillatum*, *Centranthus calcitrapae*, *Clinopodium vulgare*, *Crucianella angustifolia*, *Eryngium campestre*, *Foeniculum vulgare*, *Juncus effusus*, *Lomelosia simplex*, *Malva neglecta*, *Marrubium vulgare*, *Papaver rhoeas*, *Parentucellia latifolia*, *Paronychia argentea*, *Plantago holosteum*, *Rumex induratus*, *R. pulcher*, *Salvia verbenaca*, *Sanguisorba minor*, *Sesamoides purpurascens*, *Silene colorata*, *Torilis nodosa* and *Veronica arvensis* of other herb families; and *Cistus ladanifer*, *Crataegus* sp., *Daphne gnidium*, *Genista cinerascens*, *Helianthemum caput-felis*, *Lavandula pedunculata*, *Olea europaea*, *Quercus ilex* ssp. *ballota*, *Salvia verbenaca* and *Sambucus nigra* of woody species.

9.3.3 Ecotoxicological Assessment and Implications for Phytoremediation

For an ecotoxicological assessment, it is necessary to know which species can tolerate high soil metal concentrations since these could enter the food webs of the

Table 9.4 Heavy metal and As concentrations (mg kg^{-1}) in the shoots of native herb species of other families and woody species collected from the mine sites

Species of other families		Cu	Zn	Pb	Cd	Al	Mn
Armeria	***A. arenaria***	10–156	34–219	0.0	0.0–2.2	125–3,831	15–74
Bartsia	*B. trixago*	2.8–21	39–385	0.0–102	0.0–2.0	59–1,986	15–99
Cerastium	*C. glomeratum*	2.0–4.4	22–185	0.0–3.2	0.0–0.3	41–207	34–122
Convolvulus	*C. arvensis*	10–122	56–106	0.0	0.0–2.3	138–814	18–85
Coronopus	***C.* sp.**	10–125	21–100	0.0	0.0	–	21–73
Corrigiola	***C. telephiifolia***	4.6–465	31–554	0.0	0.0–57	53–630	10–575
Daucus	*D. carota*	1.0–20	22–179	0.0–7.3	0.0–6.2	54–1,253	9.0–200
Digitalis	***D. purpurea***	6.9–17	211–276	0.0–8.5	5.4–12	–	19–303
Diplotaxis	*D. catholica*	3.0–14	63–1,375	0.0–96	0.0–10	39–115	50–88
Echium	*E. vulgare*	2.8–25	59–581	0.0–57	0.0–5.1	35–653	20–520
Galium	***G. aparine***	7.0–8.0	134–481	10–66	0.0–1.9	94–186	33–459
Geranium	***G. molle***	5.5–29	84–216	0.0	0.0–18	349–558	3.9–130
Heliotropium	*H. europaeum*	9.0–13	171–248	20–52	0.0	26–107	33–133
Hirschfeldia	***H. incana***	0.0–6.6	29–827	0.0–142	0.0–11	31–1,094	0.0–34
Hypericum	***H. humifusum***	3.6–15	14–242	0.0–0.1	0.0–15	92–118	31–651
	H. perforatum	2.1–12	12–94	0.0	0.0–25	39–312	28–71
Jasione	***J. montana***	1.7–26	35–1,549	0.0–43	0.0–64	51–508	43–654
Marrubium	*M. supinum*	1.9–11	74–352	0.0–80	0.0–0.7	49–320	0.0–32
Mentha	*M. pulegium*	4.1–18	70–354	0.0–24	0.0–15	32–393	5.0–91
Petrorhagia	***P. nanteuilii***	6.2–31	12–305	0.0	0.0–7.1	98–687	25–111
	P. prolifera	1.0–37	38–184	0.0	0.0–3.2	106–261	12–172
Plantago	***P. afra***	3.7–15	294–1,105	2.1–256	0.0–12	23–164	15–410
	P. coronopus	5.4–31	72–799	0.0–239	0.0–8.0	43–420	8.9–423
	P. lagopus	1.8–97	21–1,622	0.0–292	0.0–14	46–3,037	3.6–985
	P. lanceolata	3.1–39	10–417	0.0–41	0.0–4.5	49–824	7.5–322

	P. ovata	4.0–11	27–400	0.0–31	0.0–3.5	–	13–45
Reseda	*R. lutea*	3.3–20	60–173	0.0–7.9	0.0–1.3	–	14–62
Rumex	***R. acetosella***	0.5–447	5.3–1,114	0.0–108	0.0–11	65–1,036	0.0–998
	R. bucephalophorus	2.6–177	17–3,077	0.0–525	0.0–25	54–664	0.0–564
	R. papillaris	6.3–597	12–242	0.0	0.0–6.4	61–322	1.2–89
Scandix	***S. pecten-veneris***	3.0–6.3	180–461	37–78	0.0	86–194	58–389
Scirpoides	***S. holoschoenus***	0.0–36	11–1,747	0.0–231	0–20	6.4–401	4.5–2,901
Silene	*S. vulgaris*	5.4–18	24–127	0.0	0.0–4.3	84–681	1.6–302
Sinapis	*S. arvensis*	2.4–7.3	17–1,656	0.0–83	0.0–8.8	10–166	20–59
Spergularia	*S. rubra*	3.0–82	69–2,775	0.0–805	0.0–11	51–1,914	22–590
Verbascum	*V. pulverulentum*	5.8–12	669–726	26–38	0.4–2.0	109–145	11–42
Xolantha	*X. guttata*	14–20	68–69	0.0	3.2–5.5	105–2,518	140–426
Woody species							
Adenocarpus	*A. complicatus*	15	103	0.0	2.4	161	100
Cytisus	*C. scoparius*	15	225	0.0	2.7	–	234
Helichrysum	***H. stoechas***	31	64	0.0	13	2,189	143
Lavandula	*L. stoechas*	6.9	192	0.0	1.7	–	245
Thymus	*T. zygis*	40	431	275	1.5	625	166
	T. mastichina	37	73	0.0	2.6	999	66

Table 9.5 Element concentrations (mg kg^{-1}) in plant shoots considered normal versus toxic for the plants themselves, permissible levels for forage plants and maximum levels tolerated by livestock

Element	Normal for plants[a]	Toxic for plants[a]	Maximum levels permissible for forage[b]	Maximum levels tolerated by livestock[c]	
				Cattle	Sheep
Cu	5–30	20–100	2.9	100	25
Zn	25–150	100–400	45	500	300
Pb	5–10	30–300	1.3	30	30
Cd	0.01–0.2	5–30	0.23	0.5	0.5
As	1–1.5	5–20	1	50	50
Mn	30–300	400–1,000	32.5	1,000	1,000

[a]Kabata-Pendias and Mukherjee (2007)
[b]WHO (1992, 1995), FAO (2000)
[c]Adapted from Madejón et al. (2006)

ecosystems where they are found. For phytoremediation purposes, we need to know which species are particularly good at accumulating a trace metal. In ecology, both these issues have been interesting from a perspective of species indicators of polluted sites.

9.3.3.1 Bioaccumulation Capacity

Here, we used the RDB index described by Kabata-Pendias and Mukherjee (2007) as a measure of the average metal accumulation capacity of species that grow in polluted soils. The species found to show a good accumulation capacity (RDBs above 100 % for two or more heavy metals or high indices for highly toxic elements such as Cd or As) are shown in bold at Tables 9.2, 9.3 and 9.4.

The maximum metal concentration found in a species indicates its maximum accumulation capacity. In contrast, the RDB index refers to an average concentration such that it better reflects the general or mean accumulation capacity of a species. As an example, the aforementioned Cd accumulation capacity of *Corrigiola telephiifolia* and *Jasione montana* was revealed by their similar maximum Cd concentrations, yet according to their RDBs, *J. montana* (RDB = 2320) is a better accumulator than *C. telephiifolia* (RDB = 487).

Despite the great metal accumulation capacity of many species collected from the mine sites, only *Agrostis castellana* showed higher or similar Zn and Pb concentrations to the limits generally used to describe a species as a hyperaccumulator (10,000 mg kg^{-1} for Mn and Zn, 1,000 mg kg^{-1} for the rest of the metals, except Cd for which the limit is 100 mg kg^{-1}). According to McGrath and Zhao (2003), these limits are rather arbitrary, and these authors argue that hyperaccumulators share the following common characteristics: a bioaccumulation index (BI) > 1 but in some cases as high as 50–100; a transfer factor (shoot/root ratio) > 1, meaning good metal transport to shoots; and a hypertolerance to metals in the soil and inside the plant, indicating good internal detoxification.

Although we consider these criteria reasonable, they are not easy to apply. The accumulation capacity of species (BI) depends on the soil concentration of the pollutant (Gutiérrez-Ginés et al. 2012). In effect, when we calculated the BI for the present species, values > 1 were observed in samples collected from soils with insignificant metal concentrations. Similarly, it was not frequent to obtain values > 1 in plants collected from the most polluted sampling points. Total or pseudo-total soil metal concentrations are generally used to calculate this index, yet bioavailable contents more realistically reflect the amount of metal that in fact can be taken up by plants. For this reason, we did not use this criterion to classify the present species.

Although *Agrostis castellana* was the only species classified as a hyperaccumulator, the following species emerged as able to accumulate large amounts of Zn and Pb together (at the sites Mazarambroz, Colmenar del Arroyo and Bustarviejo): the grasses *Avena barbata*, *Arrhenatherum elatius* subsp. *bulbosum*, *Bromus rubens*, *Holcus annuus* subsp. *setiglumis*, *Micropyrum tenellum*, *Vulpia bromoides* and *V. myuros*; the composite *Crepis vesicaria*; the Brassicaceae species *Diplotaxis catholica*, *Hirschfeldia incana* and *Sinapis arvensis*; and most species of the genera *Plantago* and *Rumex*, as well as *Spergularia rubra* and *Scirpoides holoschoenus*.

Though many species were able to tolerate the extremely high Cu levels found in Colmenarejo and Garganta, few species were good accumulators of this metal. The species showing the highest Cu concentrations were *Agrostis stolonifera*, *Corynephorus canescens*, *Lolium multiflorum*, *Armeria arenaria*, *Convolvulus arvensis*, *Corrigiola telephiifolia* and *S. holoschoenus* and the species of the *Rumex* genus. Among these, *C. telephiifolia* accumulated also the highest concentration of Cd.

9.3.3.2 Tolerance, Resistance and Toxicity

Although mine sites have been ignored for decades, they have aroused some interest regarding plant selection and adaptation over time (Dickinson et al. 1991; Shaw 1990). Thus, the study of metallophytes—plants adapted to heavy metal-enriched soils—has provided knowledge of the natural mechanisms of adaptation to heavy metal stress. Barceló and Poschenrieder (1992) summarised the complexity of stress resistance and behaviour models of plants exposed to metal toxicity: plants can either avoid/exclude the metal or tolerate it.

Since all the species examined here were collected from mine sites, they may be described as tolerant to the heavy metal concentrations of the soils that sustain them. The fact that legumes do not seem to have evolved as tolerant to heavy metals, as described by Ernst (1996), may be why they were practically absent from our study sites. In contrast, grasses (Poaceae) and composites (Asteraceae) are heavy metal accumulators, indicating the tolerance of these families to the metal conditions of the soils where they grew.

Growth within a plant community can increase the tolerance of some species to pollutants. Some species grow in the most polluted areas, where they would never be found growing alone. Plant interrelations can also reduce the pollutant tolerance of a species when the effort required by one species to tolerate high metal levels reduces its capacity to compete against other species.

Many of the tested species, even those growing in highly polluted soils, did not reflect such high metal concentrations in their shoots. Some plants are excluders and hardly take up metals from soils. The main tolerance mechanism of excluders is thought to be reduced metal transport from roots to shoots (Schat et al. 2000).

In contrast, other species featured such high metal concentrations in their shoots (even higher than those considered toxic) that they must have some sort of internal detoxification mechanism. The main symptom of heavy metal and Al toxicity in plant species growing in polluted soils appears to be their reduced root growth (Barceló and Poschenrieder 1992; Hernández 1986; Hernández et al. 2007).

However, of greater concern is the toxicity produced in animals consuming these plants. Table 9.5 shows some approximate values that may be compared with those in Tables 9.2, 9.3 and 9.4. Many of the species examined here could be harmful to the livestock that graze on these sites (cattle in Garganta or sheep in Mazarambroz or Colmenarejo). Indeed, 70 % of all the analysed species contained at least one metal at a concentration that could cause health problems in these animals.

Many studies have shown toxic element concentrations in animals grazing on polluted soils (Madejón et al. 2006; Morcombe et al. 1994; Petersson et al. 1997; Ronneau and Cara 1984). However, this type of data is still rather scarce (Hapke 1996; Madejón et al. 2009), especially with respect to native species. Accumulating native species are a health risk to both primary consumers and those that feed on them, thus affecting the entire food webs (Kabata-Pendias and Mukherjee 2007). The need to control heavy metals in terrestrial ecosystems prompted the Food and Agriculture Organization of the United Nations (FAO) (2000) to define maximum limits for Pb and Cd levels, and the EU has subsequently adopted such limits.

In prior reports, we presented our results regarding the behaviour of Poaceae species (maize, sorghum, oats and ray grass) when grown in soils obtained from abandoned Cu, Ag and Al mines. Results indicated that the shoots of these species, especially maize, accumulated high quantities of metals (Gutiérrez-Ginés et al. 2010, 2011; Hernández et al. 2007; Pastor and Hernández 2009; Pastor et al. 2012). These species could therefore be candidates for "induced phytorestoration", as their biomass is much greater than that of native accumulator species. However, if they are employed as forage, their use for the phytoremediation of heavy metal-polluted soils is not recommended unless their consumption can be avoided.

9.3.3.3 Phytoremediation Versus Ecotoxicology

Collectively, the results of the present study provide useful information for the phytoremediation of old mine sites. The native grass species of these sites are a valuable resource for this type of planned action. These species are adapted to the environmental conditions in which they grow, which in many cases (such as in the Toledo province with its summer drought) could be too harsh for other introduced species. They are also able to tolerate the high metal concentrations of the soils that sustain them even in situations of competition with other species of the community. Further, by growing on pronounced slopes, such as those of landfills or mountainous regions (such as the Madrid sites), they can thwart erosion, which is among the worse problems facing these sites. These benefits and the stable nature of these communities also translate to economic incentives when designing remediation measures.

As a drawback, we should mention that these native species with the remediation potential do not show a regular distribution within a site itself. This determines a need for more detailed studies at each site before designing any specific remediation protocol for each one. In addition, many of the species showing the greater accumulator capacity are small sized such that the amount of metal they can extract from the soil is limited. In situations in which a site is in urgent need of phytoremediation or in zones where the native vegetation is insufficient, rapid-growth cultivated plants such as maize could be a good complementary option. However, we recommend that the use of cultivated species should not be the main remediation strategy for this type of site. Indeed, conserving the autochthonous communities of abandoned mine sites should be a priority of any form of restoration.

Despite the vast amount of literature focusing on the use of cultivated or allochthonous species, many scientists are starting to appreciate the benefits of native communities adapted to these environments and have centred their studies on this strategy (Conesa et al. 2006, 2007; Jana et al. 2012; Moreno-Jiménez et al. 2009; Zhang et al. 2012).

Finally but not least importantly, we should take into account that many of the native species examined here and those employed in our past trials are forage species or base components of food chains. Thus, their assessment for remediation purposes should also consider this aspect. These consumable populations require monitoring to assess possible risks for human and ecosystem health (Hernández et al. 2009), given that any species that may be used for phytoremediation according to their accumulating capacity can also be a risk factor for introducing metals in food webs.

9.4 Conclusions

The heavy metal and trace element pollution of soils (Cu, Zn, Pb, Cd, As, Cr, Ni, Mn, Al and Ba) was examined at eight abandoned mine sites in central Spain. Altogether, six sites showed worrying levels of at least five metals. However, plant communities (Mediterranean grasslands) growing on tailings and surrounding soils seem unaffected by this pollution. After testing the shoots of 171 native species from these communities, our results indicate the plant accumulation capacity for one or more of these metals. Accumulation capacity depends on the potential uptake by each species, as well as the edaphic environment (thus metal bioavailability). As a consequence, classification according to metal tolerance or hyperaccumulation is not easy. The Poaceae family, however, showed the largest number of species with Cu, Zn, Pb and As accumulation capacity. *Agrostis castellana* emerged as a hyperaccumulator of Zn and Pb that can also take up Cu, As, Al and Mn.

Our data reveal the metal accumulation capacity of a large number of grassland species growing at abandoned mine sites. Although they provide important information concerning the candidate species for phytoremediation, we cannot ignore the toxicity risks for the animals that consume these plants.

Acknowledgements This study was funded by project grants CTM 2008-04827/TECNO from the Spanish Ministry of Science and Innovation and P2009/AMB-1478 from the EIADES Program of the Comunidad de Madrid. MJGG holded a FPU fellowship (AP2008-02934) awarded by the Spanish Ministry of Education.

References

Barceló J, Poschenrieder C (1992) Respuesta de las plantas a la contaminación por metales pesados. Suelo y Planta 2:345–361

BOE (Boletín Oficial del Estado) (1990) RD 130/1990. Real Decreto 1310/1990 de 29 de octubre, por el que se regula la utilización de los lodos de depuración en el sector agrario. Boletín Oficial del Estado del 1 de noviembre de 1990

Conesa H, Faz A, Arnaldos R (2006) Heavy metal accumulation and tolerance in plants from mine tailings of the semiarid Cartagena-La Union mining district (SE Spain). Sci Tot Environ 366 (1):1–11

Conesa H, Faz A, Arnaldos R (2007) Initial studies for the phytostabilization of a mine tailing from the Cartagena-La Union Mining District (SE Spain). Chemosphere 66(1):38–44

Dickinson NM, Turner AP, Lepp NW (1991) How do trees and other long-lived plants survive in polluted environments? Funct Ecol 199:5–11

Encabo C, Cala V, Gutierrez-Maroto A (1987) Evaluación de la dispersión de metales pesados en suelos del entorno de una mina, mediante el método de especiación secuencial química. Boletín Geológico y Minero 108:57–68

Ernst WHO (1996) Phytotoxicity of heavy metal. In: Rodríguez-Barrueco CR (ed) Fertlilizers and environment. Kluwer Acad Publishers, Dordrecht, pp 423–430

FAO (2000) Infections and intoxications of farm livestock associated with feed and forage. www.fao.org/es/ESN/animal/animapdf7annex-4.pdf

Fernández-Rubio R (ed) (2007) Activos Ambientales de la Minería Española. Consejo Superior de Colegios de Ingenieros de Minas de España, Madrid

García-Salgado S, García-Casillas D, Quijano-Nieto MA, Bonilla-Simón MM (2012) Arsenic and heavy metal uptake and accumulation in native plant species from soils polluted by mining activities. Water Air Soil Poll 223:559–572

Gutiérrez-Ginés MJ, Pastor J, Hernández AJ (2010) Effect of heavy metals from mine soils on Avena sativa L. and education strategies. Fresen Environ Bull 19:2083–2086

Gutiérrez-Ginés MJ, Hernández AJ, Pastor J (2011) Estudio del comportamiento de Lolium perenne L. en suelos del centro de España contaminados por metales pesados. In: Ortiz

Silla R, Sánchez Navarro A (eds) Control de la degradación y uso sostenible del suelo. Universidad de Murcia, Murcia, pp 177–180

Gutiérrez-Ginés MJ, Pastor J, Hernández AJ (2012) Integrated approach to assessing the effects of soils polluted with heavy metals on a plant population. Ecotoxicology 21:1965–1978

Gutiérrez-Maroto A, Sobrados L, Jimenez-Ballesta R, Morcillo E, Álvarez JB (1989) Dispersión de elementos pesados y su incidencia en el medio natural. Boletín Geologico y Minerero 100–105:886–896

Hapke HJ (1996) Heavy metals transfer in the food chain to humans. In: Rodríguez-Barrueco CR (ed) Fertilizers and environment. Kluwer Acad Publishers, Dordrecht, pp 431–436

Hernández AJ (1986) Acción del aluminio del suelo sobre los vegetales, microflora y microfauna edáficas. Anal Edafol Agrobiol 9–10:1369–1388

Hernández AJ, Pastor J (1989) Técnicas analíticas para el estudio de las interacciones suelo-planta. Henares Revista de Geología 3:67–102

Hernández AJ, Pastor J (2005) Incidencia conjunta de metales pesados en pastos de vacuno ubicados en el entorno de una mina abandonada en la Sierra de Guadarrama. In: De La Roza B, Martínez-Fernández A, Carballal A (eds) Producciones agroganaderas: Gestión eficiente y conservacion del medio natural. SERIDA, Gijón, pp 955–963

Hernández AJ, Pastor J (2008a) Relationship between plant biodiversity and heavy metal bioavailability in grasslands overlying an abandoned mine. Environ Geochem Health 30 (2):127–133

Hernández AJ, Pastor J (2008b) La restauración ecológica de emplazamientos con suelos contaminados: evaluación de metodologías para la acción. In: Millán R, Lobo C (eds) Contaminación de Suelos. Tecnologías para su recuperación. CIEMAT, Madrid, pp 61–82

Hernández AJ, Alexis S, Gutiérrez-Ginés MJ, Pastor J (2007) Acción conjunta de metales pesados del suelo en cultivos (maíz y sorgo) para alimentación humana. In: Libro de Actas IX Congreso Nacional de Sanidad Ambiental, Sevilla

Hernández AJ, Gutiérrez-Ginés MJ, Pastor J (2009) Ecology and health in risk analysis of polluted soils. In: Brebbia CA (ed) WIT Press, Environmental Health Risk Southampton, pp 257–268.

Jana U, Chassany V, Bertrand G, Castrec-Rouelle M, Aubry E, Boudsocq S, Laffray D, Repellin A (2012) Analysis of arsenic and antimony distribution within plants growing at an old mine site in Ouche (Cantal, France) and identification of species suitable for site revegetation. J Environ Manage 110:188–193

Jiménez R, Jordá L, Jordá R, Prado P (2004) La minería metálica en Madrid. Bocamina 14:50–89

Jiménez-Ballesta R, Cala V, García R, Patino M (1990) Diferenciación textural en suelos de la cuenca de Madrid. Alteración y génesis mineral Boletín Geológico y Minero 101:593–599

Kabata-Pendias A, Mukherjee AB (2007) Trace elements from soil to human. Springer-Verlag, Berlin

Lacal M, Encabo C, Jiménez-Ballesta R, Gutiérrez-Maroto A (1995) Contaminación ambiental provocada por una escombrera en Garganta de los Montes (CM). In: Actas VI Congreso de Geoquímica. pp 5061–5065

Libro Blanco de la Minería de Castilla-La Mancha (2004) Consejería de Industria y Tecnología

Madejón P, Murillo JM, Marañón T, Cabrera F (2006) Bioaccumulation of trace elements in a wild grass three years after the Aznalcóllar mine spill (South Spain). Environ Monit Assess 114:169–189

Madejón P, Domínguez M, Murillo JM (2009) Evaluation of pastures for horses grazing on soils polluted by trace elements. Ecotoxicology 18:417–428

Massa N, Andreucci F, Poli M, Aceto M, Barbato R, Berta G (2010) Screening for heavy metal accumulators amongst autochtonous plants in a polluted site in Italy. Ecotoxicol Environ Saf 73:1988–1997

McGrath SP, Zhao FJ (2003) Phytoextraction of metals and metalloids from contaminated soils. Curr Opin Biotechnol 14:277–282

Morcombe PW, Petersson DS, Masters HG, Ross PJ, Edwards JR (1994) Cd concentrations in kidney of sheep and cattle in Western Australia. I. Regional distribution. Aust J Agr Res 45:851–862

Moreno-Jiménez E, Peñalosa JM, Manzano R, Carpena-Ruiz RO, Gamarra R, Esteban E (2009) Heavy metals distribution in soils surrounding an abandoned mine in NW Madrid (Spain) and their transference to wild flora. J Hazard Mater 162:845–859

Pastor J, Hernández AJ (2007) La restauración en sistemas con suelos degradados: estudio de casos en vertederos, escombreras y emplazamientos de minas abandonadas. In: Millán R, Lobo C (eds) Procesos de Degradación y Recuperación de suelos. CIEMAT, Madrid, pp 529–550

Pastor J, Hernández AJ (2008) Multi-functional role of grassland systems in the ecological restoration of mines, landfills, roadside slopes and agroecosystems. Options Mediterranéennes 79:103–107

Pastor J, Hernández AJ (2009) Acción Conjunta de Al y Metales Pesados del Suelo en la Nutrición Mineral del Maíz: Bioensayos en Microcosmos y Reflexiones con Fines de Fitorremediación. In: Romero L (ed) Presente y Futuro de la Nutrición Mineral de las Plantas. Granada, INDEMA, pp 661–673

Pastor J, Gutiérrez-Ginés MJ, Hernández AJ (2012) Respuesta del maíz (*Zea mays*) en suelos contaminados por metales pesados después de crecer una comunidad de pasto. In: Canals RM, San Emeterio L (eds) Nuevos retos de la ganadería extensiva: un agente de conservación en peligro de extinción. Sociedad Española para el Estudio de los Pastos. pp 177–184

Petersson KP, Thierfelder T, Jorhem L, Oskarsson A (1997) Cd levels in kidneys from Swedish pigs in relation to environmental factors-temporal and spatial trends. Sci Tot Environ 208:111–122

Pratas J, Favas PJC, D'Souza R, Varun M, Paul MS (2013) Phytoremedial assessment of flora tolerant to heavy metals in the contaminated soils of an abandoned Pb mine in Central Portugal. Chemosphere 90:2216–2225

Ronneau C, Cara J (1984) Correlations of element deposition on pastures with analysis of cows' hair. Sci Tot Environ 39:135–142

Schat H, Llugany M, Bernhard R (2000) Metal-specific patterns of tolerance, uptake, and transport of heavy metals in hyperaccumulating and nonhyperaccumulating metallophytes. In: Terry N, Bañuelos G (eds) Phytoremediation of contaminated soil and water. CRC, Boca Ratón, pp 171–188

Shaw AJ (ed) (1990) Heavy metal tolerance in plant: evolutionary aspects. CRC, Boca Raton

Walsh LM, SSSA (Soil Science Society of America) (1971) Instrumental methods for analysis of soils and plant tissue. VII, 334, Soil Science Society of America, Wisconsin, USA.

WHO (World Health Organization) (1992) Cadmium environmental aspect. Environmental Health Criteria 135. Geneve

WHO (World Health Organization) (1995) Inorganic lead. Environmental Health Criteria 165. Geneve

Zhang S, Li T, Huang H, Zou T, Zhang X, Yu H, Zheng Z, Wang Y (2012) Cd accumulation and phytostabilization potential of dominant plants surrounding mining tailings. Environ Sci Pollut Res Int 19:3879–3888

Part III
Adaptation Strategies of Plants and Bacteria in Response to Heavy Metals

Chapter 10
Heavy Metal Uptake and Transport in Plants

Tayyaba Komal, Midhat Mustafa, Zeshan Ali, and Alvina G. Kazi

10.1 Introduction

Plants have developed different biochemical systems to overcome the heavy metal-induced stresses. An increase in the metal ion concentration in soil, metallothioneins, stress proteins, etc. results into reactive oxygen species (ROS) production that ultimately leads to programmed cell death. To deal with such problems, plants have developed certain defense mechanisms or adaptation strategies including restriction of metal ion uptake, metal export from the plant, chelation and compartmentalization, etc. These processes involve metal transporters, i.e., copper transporter family, ZIP (ZRT-IRT-like protein) family, NRAMP (natural resistance-associated macrophage protein) family of transporters, MATE (multidrug and toxic compound extrusion) protein transporters, HMA (heavy metal ATPase) transporters, oligopeptide transporters, ABC (ATP-binding cassette) family of transporters, and cation-diffusion facilitator family of transporters. These transporters act through a series of signaling events like phosphorylation cascades, hormones, mitogen-activated protein kinases, and calcium-calmodulin systems, ultimately leading to the balance of nutrients in the plant necessary for its survival.

Trace quantities of some heavy metals are essential for plant metabolism; however, at higher concentration they are potentially toxic to plants and soil ecosystem (Nagajyoti et al. 2010). Applying inorganic fertilizers leads to the buildup of metals in soils (Li et al. 2010b). Soil and sediments obtained from

T. Komal • M. Mustafa • A.G. Kazi (✉)
Atta-ur-Rahman School of Applied Biosciences, National University of Sciences & Technology (NUST), Islamabad, Pakistan
e-mail: alvina_gul@yahoo.com

Z. Ali
Ecotoxicology Research Institute, National Agricultural Research Centre, Park Road, Islamabad 45500, Pakistan

I. Sherameti, A. Varma (eds.), *Heavy Metal Contamination of Soils*, Soil Biology 44, DOI 10.1007/978-3-319-14526-6_10

domestic water sources are found to possess radioactive elements along with heavy metals (Muhammad et al. 2012). In addition to other heavy metals, polonium-210 is found in the vicinity of phosphate fertilizer industry as a major contaminant source (Aoun et al. 2010). Mining, smelting, and coal-fired power plants also contribute to the accretion of heavy metals in soils and plants (Okedeyi et al. 2013). High altitude points with comparatively less anthropogenic activities are usually less infested with these metals (Cheng et al. 2013). Plains are largely affected due to extensive human activities.

Over a period of time, the accumulation of heavy metals in soil affects the quality of agricultural soil, which leads to transmission of toxic metals to the human diet as a consequence of increased crop uptake (Reis et al. 2012). Bioaccumulation of these metals in the edible parts of plants increases the crop risk to human health as observed in *Brassica chinensis* and other vegetables (Li et al. 2012). High salt contents in the soils result into reduction of alkalinity, inhibition of seed germination, and unavailability of Mn that results into the depletion of vegetation. Heavy metal accumulation in soils leads to reduction in the faunal diversity. As described in a study ore-concentrating factory caused 1.5–6.3-fold decrease of microorganisms and 3–4-fold decrease of streptomycetes, hence decreasing their number and diversity as well (Hryshko and Syshchykova 2010). Guo et al. (2011) studied that the metabolic activities of various microbial communities inhabiting polluted soils are also affected. The utilization of carbon sources by soil microbes was activated in slightly polluted soils where as in heavily polluted soils the carbon source utilization was inhibited (Guo et al. 2011). It can affect plants with the decreased chlorophyll and increased peroxidase activity resulting in the damage to plants (Li et al. 2010a). In the soil, metabolic characteristics of different microbial organisms vary with the pollution intensity. With the increase in pollution, the decrease in carbon source utilization and the decrease in the activity of enzymes like urease, alkaline phosphatase, catalase, sucrase, protease, and cellulose takes place that results into damaging effects (Guo et al. 2012). Risks from the intake of polluted vegetables are of great concern as the presence of metals is observed in the edible parts of selected *Brassica* species (Khan et al. 2010). The distribution of these metals in surface top soil like zinc, lead, and copper can directly affect the health of workers besides other ecological concerns (Fujimori and Takigami 2014). When this polluted soil was decontaminated using clean water, the stored acidity was reduced after a long period of time showing the difficulty in reducing the contamination (Chen et al. 2010). Heavy metals can pose a serious risk to food chains and freshwater. Composts are well suited for revegetation of contaminated soil, but care should be taken in order to avoid any complication (Farrell et al. 2010).

10.2 Restriction of Metal Ion Uptake from Soil

Metal availability and transport in the rhizosphere are influenced by microorganisms and root exudates. Root exudates are described as the substances which are released into the medium, present in the vicinity of healthy and intact plant roots (Bais et al. 2006). Literature survey reveals a variety of compounds released by intact roots, which include enzymes, peptides, amino acids, sugars, vitamins, nucleotides, organic acids, fungal stimulators, attractants, inhibitors, and other miscellaneous compounds (Bais et al. 2006; Dakora and Phillips 2002).

In higher plants, well-organized systems for obtaining inorganic nutrients as well as metal ions from the soil are present. Such systems are based on a few transport mechanisms. It is known that heavy metals are cotransported in the roots through the plasma membrane (Krämer et al. 2007). Metals enter into the cells through cation transporters with a wide range of substrate specificity. Since toxic heavy metals such as Pb and Cd with no known biological functions are also transported by same mechanisms, specific transporters do not exist (Silver et al. 2002).

10.3 Metal Ion Binding to Extracellular Exudates

The major function of root exudates is to prevent metal uptake inside the cells and also their chelation. For example, histidine and citrate are present in root exudate which chelate Ni and reduce its uptake from the soil (Hooper et al. 2010). The release of citrate in root exudates of maize and that of malate in sorghum (*Sorghum bicolor*) helps to prevent Cd from entering the plant (Pinto et al. 2008). Moreover, release of certain oxidants and oxygen in the rhizosphere results in iron plaque formation on the root surface which has been suggested as a mechanism to prevent heavy metal toxicity. This plaque prevents Cd from entering into the roots by adsorbing and sequestering Cd onto root surface in rice plant (Liu et al. 2008). Although excluding excess heavy metals from plant is helpful in preventing heavy metal stress, some tolerant and hyperaccumulator plants have a higher uptake level as compared to sensitive plants, so other mechanisms also exist (Rascio and Navari-Izzo 2011).

10.4 Metal Ion Binding to the Cell Wall

The exclusion of metal ions from the cell is an important heavy metal tolerance strategy in plants. The metal content of root cells is also controlled by binding of metal ions such as Zn and Cu in the apoplast. Large portions of heavy metals are accumulated in plasma membrane-cell wall interface (Hossain et al. 2012). The

CEC (cation exchange capacity) of a plant is essentially controlled by the available exchange sites on the cell walls. Cation binding sites for metal exchange are present on the root cell walls. These influence greatly the availability of ions for uptake (Morley 2010). The cell wall plays a key role in the immobilization as well as uptake of toxic heavy metal ions into the cytosol by providing histidyl groups, pectic sites, and extracellular carbohydrates such as mucilage and callose. Composition of cell wall is the characteristic of a specific plant genotype (DalCorso et al. 2013). Thus, different plant genotypes possessing chemically distinct root cell walls have dissimilar sensitivities to a specific metal toxicity. The cell wall has limited number of binding sites and it is in direct contact with metal ions; this suggests that it has minor impact on metal tolerance (YANG et al. 2005; Clemens 2006). The function of the cell wall in metal tolerance is, however, still unclear.

10.5 Heavy Metal Transport Through the Plasma Membrane in the Root

Various families of plasma membrane transporters have evolved in plant cell membranes which are significantly involved in the uptake and homeostasis of metal ions (Wang et al. 2011). Metal transporters on plasma membrane and tonoplast maintain physiological concentrations of heavy metals in a cell, but they may also contribute to heavy metal stress responses. Such transporters belong to the heavy metal P_{1B}-ATPase, the NRAMP, the CDF, and the ZIP families (Lee et al. 2010a; Maestri et al. 2010). The biological function, cellular location, and metal specificity of most of these transporters are still unknown. These metal ion transporters were discovered by complementation in mutant *Saccharomyces cerevisiae* organisms which were defective in metal uptake (Manara 2012).

10.6 ZIP Family

This is one of the principal metal transporter families involved in metal uptake. The first identified proteins of this family are ZRT and IRT-like protein, and this is where the name ZIP comes from. The first ZIP transporter discovered by expression cloning in yeast double-mutant strain (*fet3fet4*, which have impaired iron transport) was the *A. thaliana*, IRT1. Certain ZIP proteins are expressed in *A. thaliana* shoots and roots in response to Zn or Fe loading, as a part of stress response (Nishida et al. 2008). Up to now over 25 ZIP family transporters have been identified. On the basis of amino acid similarities, the respective genes fall into two subfamilies (Grotz et al. 1998; Claus and Chavarría-Krauser 2012).

The ZIP families of transporters are involved in the translocation of divalent cations across cellular membranes, and they are found in many plants, bacteria, fungi, and animal species. IRT1 functions in Fe, Zn, Cd, and Mn transport, so, in plants with Fe deficiency once IRT1 is expressed, it can also transport other divalent metals. It was observed that Cd competes with Fe for uptake in IRT1-expressing yeast. So in order to overcome Cd toxicity, one must increase the Fe concentration in soil. In *A. thaliana*, Ni excess induces the expression of IRT1, which is involved in accumulation and uptake of Ni (Thomine et al. 2000).

The Zn transporters of yeast, ZRT1 and ZRT2, are two other members of ZIP family. They were discovered on the basis of their similarity with IRT1 which is 54–65 %. Studies have demonstrated that normally yeast has two systems of Zn uptake (Claus and Chavarría-Krauser 2012). One system has high affinity and is activated only in Zn deficit cells, and the other has low affinity and is induced in cells with enough zinc. The transporter protein of the former is encoded by the ZRT1 gene, whereas ZRT2 gene encodes transporter of the latter (Guerinot 2000; Manara 2012).

Similar method was used in the identification of IRT1 and ZIP1, ZIP2, and ZIP3 genes of the *A. thaliana*. It has been found that these three proteins are not involved in Fe transport, but they have significant role in Zn transport and are the first zinc transporter genes to be cloned (Lin and Aarts 2012). Another *A. thaliana* ZIP identified via analysis of the genomic sequence is ZIP4, which is induced in shoots and roots by Zn deficiency (Küpper and Kochian 2010). In another study on rice, the OsZIP8 gene which encodes a plasma membrane Zn transporter was characterized, and it was found that it is upregulated in roots and shoots under Zn deficiency (Lee et al. 2010b).

In the study on maize (*Zea mays*), nine ZIP-encoding genes were recognized. Transgenic studies showed that all ZmZIP (*Zea mays* ZIP) proteins are present on plasma membrane and ER (endoplasmic reticulum); a variable region between TM-3 and TM-4 and a conserved transmembrane domain are present. During different developmental stages of embryo and endosperm, the expression patterns of all genes were studied and variations were observed. It was found that the buildup of ZmZIP6 and ZmIRT1 was greater in the late developmental stages of embryo, whereas ZmZIP4 was upregulated in early stages. Moreover the expression of ZmZIP5 was found correlated with the middle stage of the development of endosperm and embryo (Li et al. 2013). In Fabaceae family the Zn homeostasis was studied in legume *Medicago truncatula* model; Zn transporters from ZIP family were characterized. Three of the six ZIP studies were found to be involved in Zn transport which are MtZIP1, MtZIP6, and MtZIP5 (Stephens et al. 2011).

10.7 NRAMP Family

NRAMP family has been demonstrated to transport a variety of heavy metal ions like Mn, Cu, Fe, Zn, Cd, Ni, and Co in eukaryotes, including plants, fungi, animals, and the bacteria also. NRAMP is an acronym of natural resistance-associated macrophage protein (Sasaki et al. 2012). The *A. thaliana* NRAMPs 1, 3, and

4 are found to be the cause of Fe, Mn, and Cd transport. When studied in yeast, NRAMP 6 transported Cd only and the function of NRAMP 2 is still unknown although it showed no transport activity for Fe. In rice (*Oryza sativa*), there are six and in *A. thaliana* genome, there are seven members (Sasaki et al. 2012). NRAMP 1 expressed in the cell membrane of root cells has high affinity for Mn. NRAMP 3 and 4 are located in tonoplast and function in Fe transport from vacuole (Cailliatte et al. 2010). Recently it has been reported that NRAMP 3 and 4 are also involved in intracellular transport of Mn like Fe. In rice, only 2/7 NRAMP genes have been sequenced and characterized. In yeast NRAMP 1 transported Cd and Fe but not Mn. NRAMP 1 is predicted to have a role in the uptake and transport of Cd within plants, but exact function of OsNRAMP1 (rice NRAMP 1) is unknown (Cailliatte et al. 2010; Sasaki et al. 2012). NRAMP 4 gene was isolated from *Thlaspi japonicum*, which is a Ni hyperaccumulator species. It showed its capability for Ni transport but in yeast it didn't transport Mn, Cd, or Zn (Lin and Aarts 2012).

10.8 Copper Transporter Family

The copper transporter (COPT/Ctr) is found in plants, eukaryotes, animals, and fungi. The *A. thaliana* copper transporter was identified by functional expression mutation of the yeast mutant strain ctr 1-3, which is copper uptake defective (Sancenón et al. 2004; Klaumann et al. 2011). All proteins of this family contain three TM (transmembrane) regions, TM1, TM2, and TM3. There is a short connecting sequence between TM2 and TM3 which is reported to have an essential function, and it binds the two regions tightly at the extracellular side of the membrane (Yuan et al. 2011). COPTR/Ctr proteins can form homodimer or heterocomplex with themselves or other proteins which are also associated with copper transport. Such heterocomplex is predicted to provide a channel for Cu passage through lipid bilayer. The presence of COPTR on lysosomal membrane shows that they are involved in both intracellular and extracellular transports of Cu ions. Their expression is upregulated in conditions of Cu deficiency and is downregulated when copper is in excess (Yuan et al. 2014).

In rice, COPTR family comprises of seven members, COPTR1–7. The rice COPTR family has all conserved features. COPTR 1 and COPTR 5 make homodimers and heterodimers also. In *A. thaliana* COPTR 5 is found important for Cu transport, mainly efflux from the vacuole at cellular level, and in the plant as a whole it is involved in the interorgan transportation of Cu ions from the root to important organs like reproductive organs (Klaumann et al. 2011). In some cases the COPTRs have also been found to be involved in the uptake of other substances also. Nonetheless, in a study it is demonstrated via mutant yeast strains that rice COPTRs cannot transport Zn and Fe alone. The yeast mutants (zrt1zrt2ZHY3) are missing in the zrt1 and zrt2 proteins required for Zn uptake. When the mutant cells are transformed with any one of the rice COPTRs, they are unable to grow in selective medium without Zn supplement (Yuan et al. 2011).

10.9 Plant Responses to Heavy Metals at Root-Shoot Level

Several types of proteins contribute in the transport of metals from the root to shoot. Soon after the metals are taken by the roots, metal ions move toward the xylem to be transported to the shoots in the form of complexes with various chelators (Migeon et al. 2010). Organic acids, e.g., citrates and malates, and various amino acid derivatives, e.g., histidine and nicotianamine (NA), are chelators of Fe, Ni, and Cu chelators in the xylem (Manara 2012). As the phloem translocates from source to sink, metal ions move along. Hence, phloem sap contains metals obtained from source tissues, like Fe, Cu, Mn, and Zn. In phloem only NA is present as chelator (Curie et al. 2009; Hossain et al. 2012).

10.10 Heavy Metal Transporting P-Type ATPase (HMAs) Transporter Family

The HMAs are involved in the transport of metal ions between cellular compartments, cytoplasm and xylem. They are basically involved in the efflux of heavy metal ions from the cytoplasm by transport across the plasma membrane or into organelles in *A. thaliana* and some hyperaccumulator species (Rascio and Navari-Izzo 2011). In a study it was found that in rice (*Oryza sativa*), the HMA2, i.e., OsHMA2, preferentially delivers Zn to developing tissues. It was found that OsHMA2 is expressed mainly in the mature zone of the roots and nodes. Its expression remains unaltered by Zn deficiency or excess (Yamaji et al. 2013). An efficient and well-organized transportation of metals from the root to shoot is a hallmark of hyperaccumulator plants. HMA4 is predicted to be located in the root and plays an important role in loading metals into the xylem, not only in hyperaccumulator plant *T. caerulescens* but also in *A. thaliana*. HMA3 is present on tonoplast and is Cd specific. It's responsible for confiscation of Cd in leaf vacuoles (Xu et al. 2010). In a study using transgenic strategies, in hyperaccumulating rice species (*Oryza sativa*), OsHMA3 gene was identified that controls root to shoot Cd transport. It basically encodes a P_{1B}-ATPase transporter and found to be localized to tonoplast (Miyadate et al. 2011). Overexpression of HMA4 and HMA3 generates tolerance in *A. thaliana* against Zn and Cd. HMA5 of *A. thaliana* is also expressed in roots and other organs but in response to Cu (Xu et al. 2010).

10.11 Multidrug and Toxic Compound Extrusion (MATE) Protein Family of Efflux Proteins

The MATE proteins have been identified in many eukaryotes and prokaryotes including plants. These are basically the efflux proteins involved in extrusion of toxic compounds from cell (Buer et al. 2010). In recent studies it has been demonstrated that a cluster of genes in plants encodes the citrate transporter, which plays a role in detoxification of aluminum and translocation of Fe from roots to shoots (Yokosho et al. 2010). They participate in the loading of Fe and citrate into the vascular tissue in the roots. In order to transport Fe from the root to shoots, ferric citrate complexes must form (Durrett et al. 2007). In another study with rye plant, the two mate genes, ScFRDL1 and ScFRDL2, of rye were characterized. The former is involved in translocation of Fe and the latter is activated by Al (Maron et al. 2010). It is found that both of them are localized to root tip endodermal cells and play a role in efflux of citrate into the xylem (Magalhaes 2010; Durrett et al. 2007).

10.12 Oligopeptide Transporter Family

The oligopeptide transporters (OPT) are a superfamily of oligopeptides found in animals, plants, and fungi. They are predicted to have 13–16 transmembrane domains (Xiang et al. 2013). These transporters are capable of transporting wide range of substrates. The YSL, i.e., yellow stripe 1 protein is specific for plants, takes its name from maize yellow stripe 1 protein (YS1). The maize YS1 transporter is involved in translocation of Fe, Ni, Cu, Zn, and, to some extent, Cd and Mn. These ions are chelated prior to translocation either by NA or phytosiderophores (Lubkowitz 2011). In a study nine OPTs belonging to peptide transporter group of genes in rice (*Oryza sativa*) were identified based on some similarity with maize YS1 gene. They were named OsOPT1 to OsOPT9 and are expressed in roots, stems, hulls, leaves, embryo, and pedicles, but not in every tissue. It was demonstrated that OsOPT1, OsOPT3, OsOPT4, OsOPT5, and OsOPT7 are capable of transporting ferrous or ferric iron chelated to NA (Vasconcelos et al. 2008). Similarly eight YSL transporters were identified in *A. thaliana*, AtYSL1 to AtYSL8. These are also widely expressed and involved in translocation of chelated iron. AtYSL1 is expressed in leaf xylem parenchyma and pollen as well. AtYSL2 is expressed in shoot and root tissues (Cao et al. 2011; Stacey et al. 2006).

10.13 Complex Formation and Compartmentalization of Heavy Metals in Cytosol

If the heavy metal ion is not to be utilized in any metabolic activity inside the cell immediately, it may rise to toxic concentration. Plants have developed mechanisms to prevent toxicity caused by them. If the concentration of that metal increases a specific threshold, mechanism to sequester this condition activates (DalCorso et al. 2008). One such strategy is to transport heavy metals out of the cell into the apoplast as described above or to store them in vacuole, in clearing the cytosol from metal. Another strategy used by most of the plants is phyto-chelation by various compounds (DalCorso et al. 2008, 2010).

10.14 Metal Sequestration in the Vacuole

Cytosol and other cellular compartments are to be cleared of the toxic metals because sensitive metabolic activities take place there (DalCorso et al. 2010). Therefore central vacuole is the most appropriate region for this. The main storage compartment for metal ions in plant cell is vacuole (Marschner and Marschner 2011). Several families of intracellular transporters found on tonoplast have been identified in plants involved in this compartmentalization.

10.15 ABC Family of Transporters

ABC transporters, i.e., ATP-binding cassette transporters, are capable of transporting xenobiotics and chelated metals into the vacuole. The subfamily MRP is now redefined as the class C of ABC proteins, i.e., ABCC transporters, and they are found in all eukaryotes. In *A. thaliana* 15 ABCC genes have been described, and these proteins are localized on tonoplasts, even though some may also be found on plasmalemma (Wanke and Kolukisaoglu 2010). Of these, ABCC11 and ABCC2 are best studied. These two proteins (ABCC11 and ABCC2) act together in PC transport response to heavy metal ions. ABCC1 and ABCC2 have been demonstrated to transport PC-Cd into the vacuole, thereby having a role in Cd sequestration, whereas the role of ABCC3 remains unclear in Cd transport (Noctor et al. 2013).

10.16 Cation-Diffusion Facilitator Family of Transporters

Proteins of the CDF family proteins play a role in homeostasis of Cd, Zn, Fe, and Co in plants, animals, and some other microbes. The CDF family proteins comprise of six transmembrane domains, a conserved sequence between transmembrane domains 1 and 2, histidine-rich cytosolic regions at both termini; this histidine-rich region acts as a sensor of metal concentration (Kawachi et al. 2008). In most bacterial cells, CDF proteins are found to cause metal tolerance by exporting the cations out of the cell or inside in the case of Zn for storage and detoxification (Haney et al. 2005).

10.17 HMA and NRAMP Family of Transporters

As mentioned above P_{1B}-ATPases (HMAs) are the cause of efflux of metal ions from the cytosol. The *A. thaliana* HMA has a major role in detoxification of various metals including Pb, Zn, Co, and Cd through storage in the vacuole. The NRAMP transporters such as *A. thaliana* NRAMP3 and NRAMP4 are found in the tonoplast and their function is still unknown. But the overexpression of NRAMP3 increases Cd sensitivity (Sasaki et al. 2012; Rascio and Navari-Izzo 2011).

10.18 Conclusion

Presence of toxic heavy metals in the environment leads to numerous health hazards. Environmental pollution is an outcome of promptly growing population and increasing anthropogenic activities. This is an alarming and serious matter as soil contamination leads to plant toxicity ultimately resulting in food chain contamination. The main pollutants present in the soil are Zn, Cu, Ni, Cd, Al, As, Fe, Co, Se, and Cr. These metals when exceed threshold concentrations become hazardous. Plants adopt different molecular and cellular adaptations in order to combat heavy metal toxicity through various tolerance responses. By this, plants ultimately respond and reduce the health risks associated with heavy metals. Plants also shield themselves from heavy metal toxicity by activating different antioxidant systems which may be enzymatic or nonenzymatic. Different plants elicit symptoms against particular heavy metal toxicity through different mechanisms. Similarly the response and tolerance mechanisms also vary. Therefore, it is difficult to state a common tolerance mechanism against different heavy metals. Significant progresses have been achieved in understanding heavy metal responses, tolerance, and toxicity mechanisms. Many important components that ensure heavy metal tolerance have been identified but still many key questions are unanswered. GSH is the most significant nonenzymatic antioxidant molecule utilized by plants against heavy metal stress. The reviewed literature confirms the crucial role of glutathione (GSH) defense and tolerance against various heavy metals. Soil remediation strategies on the other hand can be used to protect plants against heavy metal toxicity. It is known as a challenging process by researchers but has numerous advantages. Phytoremediation, phytoextraction, rhizofiltration, and phytostabilization are the currently known types of soil remediation.

References

Aoun M, El Samrani AG, Lartiges BS, Kazpard V, Saad Z (2010) Releases of phosphate fertilizer industry in the surrounding environment: investigation on heavy metals and polonium-210 in soil. J Environ Sci (China) 22(9):1387–1397

Bais HP, Weir TL, Perry LG, Gilroy S, Vivanco JM (2006) The role of root exudates in rhizosphere interactions with plants and other organisms. Annu Rev Plant Biol 57:233–266

Buer CS, Imin N, Djordjevic MA (2010) Flavonoids: new roles for old molecules. J Integr Plant Biol 52(1):98–111

Cailliatte R, Schikora A, Briat J-F, Mari S, Curie C (2010) High-affinity manganese uptake by the metal transporter NRAMP1 is essential for Arabidopsis growth in low manganese conditions. Plant Cell 22(3):904–917

Cao J, Huang J, Yang Y, Hu X (2011) Analyses of the oligopeptide transporter gene family in poplar and grape. BMC Genomics 12(1):465

Chen A, Lin C, Lu W, Ma Y, Bai Y, Chen H, Li J (2010) Chemical dynamics of acidity and heavy metals in a mine water-polluted soil during decontamination using clean water. J Hazard Mater 175(1–3):638–645. doi:10.1016/j.jhazmat.2009.10.055

Cheng F, Cheng JP, Sang HC, Yu JL, Xi L, Pi SS (2013) Assessment and correlation analysis of heavy metals pollution in soil of Dajinshan Island. Huan Jing Ke Xue 34(3):1062–1066

Claus J, Chavarría-Krauser A (2012) Modeling regulation of zinc uptake via ZIP transporters in yeast and plant roots. PLoS One 7(6):e37193

Clemens S (2006) Toxic metal accumulation, responses to exposure and mechanisms of tolerance in plants. Biochimie 88(11):1707–1719

Curie C, Cassin G, Couch D, Divol F, Higuchi K, Le Jean M, Misson J, Schikora A, Czernic P, Mari S (2009) Metal movement within the plant: contribution of nicotianamine and yellow stripe 1-like transporters. Ann Bot 103(1):1–11

Dakora FD, Phillips DA (2002) Root exudates as mediators of mineral acquisition in low-nutrient environments. Plant Soil 245(1):35–47

DalCorso G, Farinati S, Furini A (2010) Regulatory networks of cadmium stress in plants. Plant Signal Behav 5(6):663–667

DalCorso G, Farinati S, Maistri S, Furini A (2008) How plants cope with cadmium: staking all on metabolism and gene expression. J Integr Plant Biol 50(10):1268–1280

Dalcorso G, Manara A, Furini A (2013) An overview of heavy metal challenge in plants: from roots to shoots. Metallomics 5(9):1117–1132. doi:10.1039/c3mt00038a

Durrett TP, Gassmann W, Rogers EE (2007) The FRD3-mediated efflux of citrate into the root vasculature is necessary for efficient iron translocation. Plant Physiol 144(1):197–205

Farrell M, Perkins WT, Hobbs PJ, Griffith GW, Jones DL (2010) Migration of heavy metals in soil as influenced by compost amendments. Environ Pollut 158(1):55–64. doi:10.1016/j.envpol.2009.08.027

Fujimori T, Takigami H (2014) Pollution distribution of heavy metals in surface soil at an informal electronic-waste recycling site. Environ Geochem Health 36(1):159–168. doi:10.1007/s10653-013-9526-y

Grotz N, Fox T, Connolly E, Park W, Guerinot ML, Eide D (1998) Identification of a family of zinc transporter genes from Arabidopsis that respond to zinc deficiency. Proc Natl Acad Sci U S A 95(12):7220–7224

Guerinot ML (2000) The ZIP family of metal transporters. Biochim Biophys Acta 1465 (1):190–198

Guo H, Zhu J, Zhou H, Sun Y, Yin Y, Pei D, Ji R, Wu J, Wang X (2011) Elevated CO2 levels affects the concentrations of copper and cadmium in crops grown in soil contaminated with heavy metals under fully open-air field conditions. Environ Sci Technol 45(16):6997–7003. doi:10.1021/es2001584

Guo XL, Gu J, Chen ZX, Gao H, Qin QJ, Sun W, Zhang WJ (2012) Effects of heavy metals pollution on soil microbial communities metabolism and soil enzyme activities in coal mining area of Tongchuan, Shaanxi Province of Northwest China. Ying Yong Sheng Tai Xue Bao 23 (3):798–806

Haney CJ, Grass G, Franke S, Rensing C (2005) New developments in the understanding of the cation diffusion facilitator family. J Ind Microbiol Biotechnol 32(6):215–226

Hooper AM, Tsanuo MK, Chamberlain K, Tittcomb K, Scholes J, Hassanali A, Khan ZR, Pickett JA (2010) Isoschaftoside, a *C*-glycosylflavonoid from *Desmodium uncinatum* root exudate, is an allelochemical against the development of *Striga*. Phytochemistry 71(8):904–908

Hossain MA, Piyatida P, da Silva JAT, Fujita M (2012) Molecular mechanism of heavy metal toxicity and tolerance in plants: central role of glutathione in detoxification of reactive oxygen species and methylglyoxal and in heavy metal chelation. J Bot 2012:37

Hryshko VM, Syshchykova OV (2010) Changes in quantity and species composition of streptomycetes associations in soil polluted by heavy metals. Mikrobiol Z 72(3):20–28

Kawachi M, Kobae Y, Mimura T, Maeshima M (2008) Deletion of a histidine-rich loop of AtMTP1, a vacuolar Zn2+/H+ antiporter of Arabidopsis thaliana, stimulates the transport activity. J Biol Chem 283(13):8374–8383

Khan S, Rehman S, Khan AZ, Khan MA, Shah MT (2010) Soil and vegetables enrichment with heavy metals from geological sources in Gilgit, northern Pakistan. Ecotoxicol Environ Saf 73 (7):1820–1827. doi:10.1016/j.ecoenv.2010.08.016

Klaumann S, Nickolaus SD, Fürst SH, Starck S, Schneider S, Ekkehard Neuhaus H, Trentmann O (2011) The tonoplast copper transporter COPT5 acts as an exporter and is required for interorgan allocation of copper in Arabidopsis thaliana. New Phytol 192(2):393–404

Krämer U, Talke IN, Hanikenne M (2007) Transition metal transport. FEBS Lett 581 (12):2263–2272

Küpper H, Kochian LV (2010) Transcriptional regulation of metal transport genes and mineral nutrition during acclimatization to cadmium and zinc in the Cd/Zn hyperaccumulator, Thlaspi caerulescens (Ganges population). New Phytol 185(1):114–129

Lee K, Bae DW, Kim SH, Han HJ, Liu X, Park HC, Lim CO, Lee SY, Chung WS (2010a) Comparative proteomic analysis of the short-term responses of rice roots and leaves to cadmium. J Plant Physiol 167(3):161–168

Lee S, Kim SA, Lee J, Guerinot ML, An G (2010b) Zinc deficiency-inducible OsZIP8 encodes a plasma membrane-localized zinc transporter in rice. Mol Cells 29(6):551–558

Li FL, Yuan J, Sheng GD (2012) Altered transfer of heavy metals from soil to Chinese cabbage with film mulching. Ecotoxicol Environ Saf 77:1–6. doi:10.1016/j.ecoenv.2011.10.019

Li Q, Cai S, Mo C, Chu B, Peng L, Yang F (2010a) Toxic effects of heavy metals and their accumulation in vegetables grown in a saline soil. Ecotoxicol Environ Saf 73(1):84–88. doi:10.1016/j.ecoenv.2009.09.002

Li S, Zhou X, Huang Y, Zhu L, Zhang S, Zhao Y, Guo J, Chen J, Chen R (2013) Identification and characterization of the zinc-regulated transporters, iron-regulated transporter-like protein (ZIP) gene family in maize. BMC Plant Biol 13(1):1–14

Li SH, Zeng XB, Li LF, Bai LY, Wang DL (2010b) Distribution characteristics of heavy metals in soil profile of facility vegetable fields. Ying Yong Sheng Tai Xue Bao 21(9):2397–2402

Lin Y-F, Aarts MG (2012) The molecular mechanism of zinc and cadmium stress response in plants. Cell Mol Life Sci 69(19):3187–3206
Liu H, Zhang J, Christie P, Zhang F (2008) Influence of iron plaque on uptake and accumulation of Cd by rice (*Oryza sativa* L.) seedlings grown in soil. Sci Total Environ 394(2):361–368
Lubkowitz M (2011) The oligopeptide transporters: a small gene family with a diverse group of substrates and functions? Mol Plant 4(3):407–415
Maestri E, Marmiroli M, Visioli G, Marmiroli N (2010) Metal tolerance and hyperaccumulation: costs and trade-offs between traits and environment. Environ Exp Bot 68(1):1–13
Magalhaes JV (2010) How a microbial drug transporter became essential for crop cultivation on acid soils: aluminium tolerance conferred by the multidrug and toxic compound extrusion (MATE) family. Ann Bot 106(1):199–203
Manara A (2012) Plant responses to heavy metal toxicity. In: Furini A (ed) Plants and heavy metals. Springer, New York, pp 27–53
Maron LG, Piñeros MA, Guimarães CT, Magalhaes JV, Pleiman JK, Mao C, Shaff J, Belicuas SN, Kochian LV (2010) Two functionally distinct members of the MATE (multi-drug and toxic compound extrusion) family of transporters potentially underlie two major aluminum tolerance QTLs in maize. Plant J 61(5):728–740
Marschner H, Marschner P (2011) Marschner's mineral nutrition of higher plants, vol 89. Elsevier, Amsterdam, Online access
Migeon A, Blaudez D, Wilkins O, Montanini B, Campbell MM, Richaud P, Thomine S, Chalot M (2010) Genome-wide analysis of plant metal transporters, with an emphasis on poplar. Cell Mol Life Sci 67(22):3763–3784
Miyadate H, Adachi S, Hiraizumi A, Tezuka K, Nakazawa N, Kawamoto T, Katou K, Kodama I, Sakurai K, Takahashi H (2011) OsHMA3, a P1B-type of ATPase affects root-to-shoot cadmium translocation in rice by mediating efflux into vacuoles. New Phytol 189(1):190–199
Morley PS (2010) Calcium. In: Pilbeam DJ (ed) Handbook of plant nutrition. CRC, Boca Raton, p 121
Muhammad BG, Jaafar MS, Abdul Rahman A, Ingawa FA (2012) Determination of radioactive elements and heavy metals in sediments and soil from domestic water sources in northern peninsular Malaysia. Environ Monit Assess 184(8):5043–5049. doi:10.1007/s10661-011-2320-3
Nagajyoti P, Lee K, Sreekanth T (2010) Heavy metals, occurrence and toxicity for plants: a review. Environ Chem Lett 8(3):199–216
Nishida S, Mizuno T, Obata H (2008) Involvement of histidine-rich domain of ZIP family transporter TjZNT1 in metal ion specificity. Plant Physiol Biochem 46(5):601–606
Noctor G, Mhamdi A, Queval G, Foyer CH (2013) Regulating the redox gatekeeper: vacuolar sequestration puts glutathione disulfide in its place. Plant Physiol 163(2):665–671
Okedeyi OO, Dube S, Awofolu OR, Nindi MM (2013) Assessing the enrichment of heavy metals in surface soil and plant (Digitaria eriantha) around coal-fired power plants in South Africa. Environ Sci Pollut Res Int. doi:10.1007/s11356-013-2432-0
Pinto A, Sim es I, Mota A (2008) Cadmium impact on root exudates of sorghum and maize plants: a speciation study. J Plant Nutr 31(10):1746–1755
Rascio N, Navari-Izzo F (2011) Heavy metal hyperaccumulating plants: how and why do they do it? And what makes them so interesting? Plant Sci 180(2):169–181
Reis A, Patinha C, da Silva EF, Sousa A (2012) Metal fractionation of cadmium, lead and arsenic of geogenic origin in topsoils from the Marrancos gold mineralisation, northern Portugal. Environ Geochem Health 34(2):229–241
Sancenón V, Puig S, Mateu-Andrés I, Dorcey E, Thiele DJ, Peñarrubia L (2004) The Arabidopsis copper transporter COPT1 functions in root elongation and pollen development. J Biol Chem 279(15):15348–15355
Sasaki A, Yamaji N, Yokosho K, Ma JF (2012) Nramp5 is a major transporter responsible for manganese and cadmium uptake in rice. Plant Cell 24(5):2155–2167

Silver S, Phung LT, Rosen BP (2002) Arsenic metabolism: resistance, reduction, and oxidation. In: Frankenberger WT Jr (ed) Environmental chemistry of arsenic. Marce Denkker, New York, pp 247–272

Stacey MG, Osawa H, Patel A, Gassmann W, Stacey G (2006) Expression analyses of Arabidopsis oligopeptide transporters during seed germination, vegetative growth and reproduction. Planta 223(2):291–305

Stephens BW, Cook DR, Grusak MA (2011) Characterization of zinc transport by divalent metal transporters of the ZIP family from the model legume Medicago truncatula. Biometals 24 (1):51–58

Thomine S, Wang R, Ward JM, Crawford NM, Schroeder JI (2000) Cadmium and iron transport by members of a plant metal transporter family in Arabidopsis with homology to Nramp genes. Proc Natl Acad Sci U S A 97(9):4991–4996

Vasconcelos MW, Li GW, Lubkowitz MA, Grusak MA (2008) Characterization of the PT clade of oligopeptide transporters in rice. Plant Genome 1(2):77–88

Wang P, Kinraide TB, Zhou D, Kopittke PM, Peijnenburg WJ (2011) Plasma membrane surface potential: dual effects upon ion uptake and toxicity. Plant Physiol 155(2):808–820

Wanke D, Kolukisaoglu HU (2010) An update on the ABCC transporter family in plants: many genes, many proteins, but how many functions? Plant Biol 12:15–25. doi:10.1111/j.1438-8677.2010.00380.x

Xiang Q, Wang Z, Zhang Y, Wang H (2013) An oligopeptide transporter gene family in Phanerochaete chrysosporium. Gene 522(2):133–141. doi:10.1016/j.gene.2013.03.069

Xu J, Yin H, Liu X, Li X (2010) Salt affects plant Cd-stress responses by modulating growth and Cd accumulation. Planta 231(2):449–459

Yamaji N, Xia J, Mitani-Ueno N, Yokosho K, Ma JF (2013) Preferential delivery of zinc to developing tissues in rice is mediated by P-type heavy metal ATPase OsHMA2. Plant Physiol 162(2):927–939

Yang XE, Jin XF, Feng Y, Islam E (2005) Molecular mechanisms and genetic basis of heavy metal tolerance/hyperaccumulation in plants. J Integr Plant Biol 47(9):1025–1035

Yokosho K, Yamaji N, Ma JF (2010) Isolation and characterisation of two MATE genes in rye. Funct Plant Biol 37(4):296–303

Yuan M, Li X, Xiao J, Wang S (2011) Molecular and functional analyses of COPT/Ctr-type copper transporter-like gene family in rice. BMC Plant Biol 11(1):69

Yuan M, Zhao J, Huang R, Li X, Xiao J, Wang S (2014) Rice MtN3/saliva/SWEET gene family: evolution, expression profiling, and sugar transport. J Integr Plant Biol 56:559–570

Chapter 11
Adaptation Mechanisms of *Pinus sylvestris* L. in Industrial Areas

Ewa Chudzińska, Aleksandra Wojnicka-Półtorak, Wiesław Prus-Głowacki, Konrad Celiński, Jean B. Diatta, and Leszek Drobek

11.1 Introduction

In spite of a positive direction of changes occurring in the contaminated environment, the local industry still appears to exert a negative influence on plant vegetation. Forests which grow in many highly industrialized zones enable research on the influence of anthropopression on the natural population and are one of the best models for the study of plant adaptation to heavy metals in soil. In some cases, it is possible to follow processes of re-naturalization occurring on post-industrial areas in situ. Adaptive genetic diversity reflects differences in the survival capabilities of individuals exposed to stress and shows the selective pressure against trees with specific genotypes. This chapter emphasizes on the Scots pine (*Pinus sylvestris* L.) as one of the most frequently used bioindicators in the European forests and their application in the study of microevolutionary processes in tree populations. It may enhance a better understanding of how the soil pollution can change the genetic structure of important forest species.

E. Chudzińska (✉) • A. Wojnicka-Półtorak • W. Prus-Głowacki • K. Celiński
Department of Genetics, Institute of Experimental Biology, Adam Mickiewicz University, ul. Umultowska 89, 61-614, Poznań, Poland
e-mail: evpell@amu.edu.pl

J.B. Diatta
Department of Agricultural Chemistry and Environmental Biogeochemistry, Poznan University of Life Sciences, ul. Wojska Polskiego 71F, 60-625, Poznań, Poland

L. Drobek
Department of Environment Engineering, Central Mining Institute, Plac Gwarków 1, 40-166, Katowice, Poland

I. Sherameti, A. Varma (eds.), *Heavy Metal Contamination of Soils*, Soil Biology 44, DOI 10.1007/978-3-319-14526-6_11

11.2 *Pinus sylvestris* as a Bioindication of Tree Species

A critical role in the adaptive potential of a species plays the presence of substantial heritable genetic variation. The evolutionary response of a population to a changing environment is the intensity of 'the best' phenotype (Rice and Emery 2003). Thus, looking for an adaptation, the following questions arise: (1) what kind of features characterize the most invulnerable individuals, (2) is it reflected in the structure of the population, and (3) is it a matter of a special adaptation process or just a matter of chance?

One of the best models for the study of adaptation to the industrial areas is that associated with the impact of heavy metals (HM). According to multiple sources (Linhart and Grant 1996; Markert et al. 2012), only a complex approach to the problems connected with the adaptation of living organisms to a highly polluted environment allows a better understanding of evolutionary mechanisms developed by the observed individuals. Despite many years of research, still very little is known about the mechanisms of conifer response to heavy metal toxicity.

In nature, the process of selection may lead to distinct differences between resistant and sensitive individuals and, finally, create genetic differentiation with respect to metal tolerance. Different concepts of plant adaptation in stressful environments can be found in many articles (e.g. Dickinson et al. 1991; Godbold 1998; Kozlowski and Pallardy 2002; Kozlov et al. 2009). Such a wide range of tolerance patterns has been reflected in comprehensive studies conducted using methods of statistical meta-analysis. Surprisingly, only two regularities were registered concerning the effects of pollution (including contamination by HMs)—namely, (1) the diversity of organisms with respect to decreases in size due to a strong dose of pollutants and (2) larger plants (trees) suffered more from pollution than smaller (herbaceous) plants (Kozlov et al. 2009; Kozlov and Zvereva 2011). In the case of trees, where heritable changes take place at a slow rate, selection for individuals with a higher tolerance may take many decades before a sufficient resistant population is built up. Thus it has been suggested that only phenotypic plasticity may provide a mechanism that improves the survival of the long-lived trees in metal-contaminated environments (Dickinson et al. 1991; Korshikov et al. 2002).

A positive direction of changes occurring in the environment is observed in many highly industrialized zones. Despite this, the local industry is still appearing to exert a negative influence on forest areas. Upper Silesia (South Poland) is a region where a massive extinction of forests, particularly affecting the pine ecosystems, was observed in the early 1980s. Currently, natural regeneration of forests in this region enables research on the influence of anthropopression on the Scots pine population. In this case, it is possible to follow processes of re-naturalization occurring on post-industrial areas in situ. Research already undertaken in this region showed an interesting phenomenon of differentiation among the Scots pine populations with respect to health status. In a homogeneous age group, there are two types of trees: gradually dying, with distinct damage, and healthy, with no signs of damage. Observed differences affect the ability to pass genetic material to next generations and the process of natural selection can induce reproductive success as a measure of competitive effectiveness. The ongoing transformation of tree populations is a classic example of local adaptation.

Simultaneous verification of the number of different parameters allows more precise diagnoses of tree fitness and indicates with the highest probability in which stress factor is the most responsible for the occurrence of visible damage. The complexity of responsiveness of an organism to stress is the reason that a combination of the specific type of stress with its particular result in terms of cause and effect is often very difficult. The study of species used in bioindication is particularly important in this context due to large amounts of standardized data available on the topic.

Bioindicators are defined as organisms which carry information about the quality of the environment. Trees are considered sensors that record the environmental disturbances, as they live for a long time in the same place and are widespread geographically. Conifers capture pollutants from the atmosphere and deposit them directly on the bark and needles or take them up from the soil by the roots. Generally, coniferous species have much higher heavy metal content than deciduous species. *Pinus sylvestris* as a species very sensitive to industrial pollution is frequently used for monitoring contamination by HM (Nieminen and Helmisaari 1996; Derome and Saarsalmi 1999; Harju et al. 2002; Diatta et al. 2011; Ivanov et al. 2011). It is one of the most common native forests species in Europe and fulfils a series of conditions which make it useful for bioindication. The bioindicative properties of *P. sylvestris*, such as its wide geographical range, easily recognizable taxonomic range and low tolerance to toxic agents expressed in a simple-to-analyse reaction, allow not only a detailed description of the current health of forest but also identification of the direction of changes which affect them. In *P. sylvestris*, some metals, especially Pb, are accumulated in roots, probably due to physiological barriers against metal transport to aerial parts, while others, such as Cd, are easily transported (Kabata-Pendias 2004). Heavy metals enter the plant mainly by roots where they are bound to the carboxy groups of mucilage uronic acids. Close to the pollution source, concentrations of HM in all the pine compartments are higher as compared to their background levels. In conditions of severe threat by industrial emissions, pines are characterized by an unusual dwarf habit. The reduction in shoot growth and its thickness leads to changes in crown conformation. Branches and, especially, stem bark are usually richer in heavy metals than other tissues and, in addition to wood, contributed significantly to the accumulation of metals in the biomass of the polluted pine forests. However, most often in bioindication, pine needles are used.

11.2.1 Pinus sylvestris*: Morphological Alteration and Heavy Metals*

Like most coniferous trees, the Scots pine holds needles for several growing seasons, which allows for long-term observations. Taking into account that the first effects of pollution in pine stands are changes connected with leaf degradation, morphology and anatomy of needles are frequently used as a diagnostic tool for the identification of pollution. Research carried out in Upper Silesia identified several dissimilarities between the needles of trees growing under conditions of high stress and needles of trees from an ecologically clean area.

The changes in morphological and anatomical characteristics found in Scots pine needles can be a result of an exposure to elevated levels of HMs in the soil. The structure of leaves differed in relation to the control population with respect to a significant decrease in needle length correlated with an increase in needle height. Needles from the contaminated area are distinctly shorter and thicker. This is consistent with reports which showed that the reduction in leaf growth under conditions of heavy metal pollution is an expression of adaptation, because it reduces transpiration and thus limits the uptake of metals (Potters et al. 2007). The higher pollutant accumulation is correlated with the density of stomata on needles close to the smelter, and an increase of stomatal density is noted on both adaxial and abaxial needle surfaces (Chudzińska and Urbaniak 2008). The number of resin canals, another basic feature counted during pine needle observation (Lin et al. 2001), is regarded as one of the better characteristics which allow one to distinguish between resistant and sensitive trees growing in conditions of severe contamination. Alterations of the resin–canal system are closely linked to various types of stress. Damaged trees in our results show a low number of resin canals (Fig. 11.1c).

Some of the studied needle features are associated with higher resistance to the adverse conditions. Resistant trees had longer needles with a greater density of stomata and significantly more resin canals (Fig. 11.1a-c). Many authors emphasize the dependence of variation in needle area on growing conditions and explain the reduction in these parameters with the adverse effects of heavy metals on cell elongation and expansion, microtubule formation and reduction of water content in tissues (Novikova and Milyutin 2006 and literature cited herein). Therefore, longer needles have better adaptation to the high level of HMs in the soil. Low concentrations of heavy metals in leaves can cause an increase in the number of stomata resulting from the reduction of leaf area (Klánowá et al. 2009; Prasad and Hagemeyer 2004). In contrast, a high concentration of metals may lead to the reduction in the number of stomata, which was detected in sensitive trees (Chudzińska and Urbaniak 2008). Our results show also that trees with better fitness have more resin canals than trees that are close to dying.

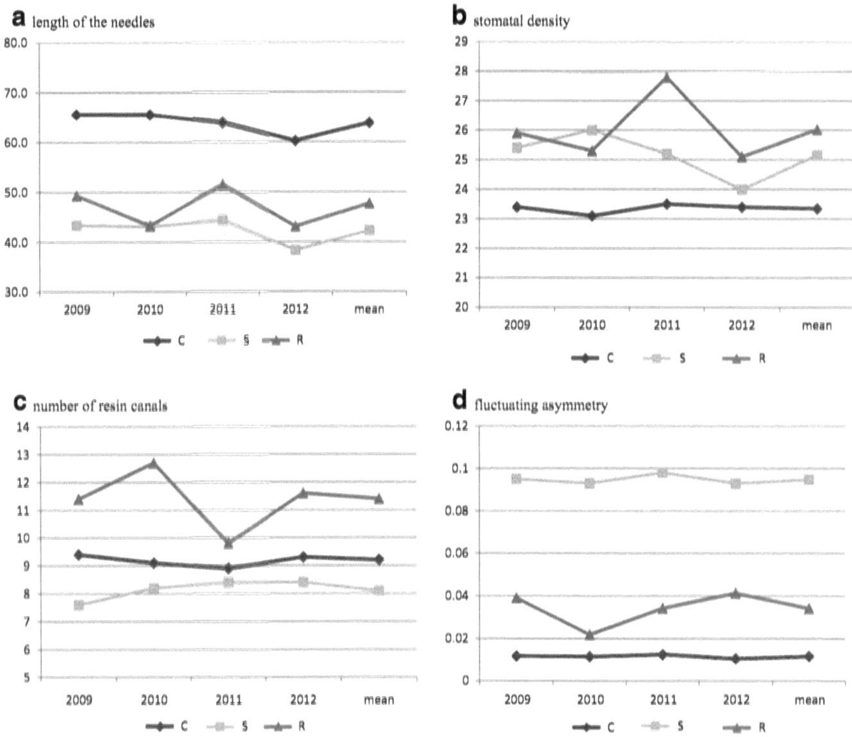

Fig. 11.1 Differences related to morphological features of *P. sylvestris* needles; (**a**) length of the needles, (**b**) stomatal density, (**c**) number of resin canals, (**d**) fluctuating asymmetry. Needles of trees from resistant (R) and sensitive (S) groups exposed to HM contamination and needles of trees from an unpolluted area (C)

Double-needled fascicles of *P. sylvestris* are additionally an ideal object for testing fluctuating asymmetry (FA), one of the parameters demonstrating an inability to control development under genetic and environmental stress (Kozlov et al. 2002; Chudzińska et al. 2013a, b). Deviation from needle symmetry as a result of disruption to normal growth may be caused by the allocation of resources from developmental quality to growth in morphogenesis under stress. Therefore, a significant increase in the value of the FA is commonly observed among the pine trees growing in the environment heavily contaminated with HMs as in Upper Silesia where the higher fitness of trees resistant to contamination was confirmed by the results of the FA measurements (Fig. 11.1d). For the two-needle Scots pine, FA measurement gives a simple answer to whether the tree is growing in unfavourable term because the variation among pine needles within a fascicle is analogous to variation between right and left sides of a bilaterally symmetrical trait (Chudzińska et al. 2013a, b).

11.2.2 Pinus sylvestris*: Cytology and Reproduction*

Disturbances occurring under the influence of HMs in cell divisions are another important parameter of bioindication. This negative impact of industrial emissions on the Scots pine lowers yields and quality of seeds (Oleksyn et al. 1994). Palowski (2000) found considerable accumulation of heavy metals in pine seeds which significantly reduced the ability and energy of seed germination in the vicinity of 'Huta Katowice' smelter. The elevated level of HMs in the soil causes an increasing number of abnormally germinating seeds, which is associated with its high mortality.

In Upper Silesia faster germination of seeds belonging to the group vulnerable to pollution than in group of resistant tree seeds was noted, and percentage of germinated seeds differ significantly (averaged 63.02 % for resistant, 46.32 % for sensitive and 90.17 % for control tree seeds). One of the reasons for these differences may be a low resistance of Scots pine seedlings to elevated concentrations of zinc, the metal present in the highest concentration in the studied area (Miasteczko Śląskie). In experimental conditions, Zn influence was manifested in reduced seed germinability, an inhibition of root growth, root development and numerous cytological aberrations (Ivanov et al. 2011). The level of Scots pine tolerance to an elevated zinc concentration is lower than that of commonly used model plants: *Arabidopsis thaliana*, *Nicotiana tabacum*, *Brassica napus* and others (Ivanov et al. 2011).

Similarly, the presence of Pb, Cu and Cd in soil induces several abnormalities as aggregation of chromosomes during metaphases, production of binucleate cells and stickiness, which increase with increasing levels of those metals (Kim et al. 2003). The increased level of aberration frequency in meristematic cells of specimens from the polluted region, which differ significantly from those recorded in the control population, indicates that the majority of these irregularities result from the impact of HM contamination on the tree stands studied. There may be various reasons for this result. First of all, the complex life cycle of conifers and the long duration of the generative phase cause a strong sensitivity of the reproductive organs to the harmful effects of a wide range of pollutants. During such a long period of maturation, seeds can accumulate numerous clearly visible DNA damages, which can be observed in their meristematic tissues. After germination, seedling roots are in the place of first contact with contaminated soil and are exposed to the highest concentrations of HMs. A lot of plants prevent or reduce uptake into root cells by restricting metal ions to the apoplast, binding them to the cell wall or to cellular exudates, or by inhibiting long-distance transport. The major role of root exudates is to chelate metals and to prevent their uptake inside the cell (Marschner 1995). The cell wall plays a key role in the immobilization of heavy metal ions and prevents their uptake into the cytosol (Manara 2012). Pine seedlings inoculated with metal-tolerant ecotypes of ectomycorrhizal fungi *Suillus* have lower metal concentrations in their needles than seedlings inoculated with sensitive strains, indicating that the metal-tolerant isolates restrict metal transfer more effectively. The influence of Zn and Cu causes a reduction in the number of lateral roots in pine seedlings. High concentrations of both metals, along with lead, can completely destroy protoplast cells of the rootcap meristem zone, by inhibiting mitotic divisions, as a result of the damage to cytokinesis (Schützendübel et al. 2001).

In general, coniferous species do not exhibit a high protective ability against contaminants taken up from the atmosphere, and an assessment of cytogenetic anomalies in the intercalary meristem of young needles appears to be a promising test system. However, in the case of HM soil contamination, it seems to be less convenient. In fact, for all analysed specimens, a greater number of abnormal cell divisions were observed in the root apical meristems than in the meristems of needles. Chemical analysis of pine seedlings (Diatta et al. 2008) showed that the content of HMs in roots was significantly higher than in shoots.

The most commonly used indicator of stress—mitotic index (MI)—in all cases also pointed to a higher number of cell divisions in the root apical meristems than in the meristems of pine needles. The differences of the mitotic index (MI) between the seedlings from the Miasteczko Śląskie and from the NPW can be considered an adaptation, giving organisms under extreme conditions more time for repair (Prus-Głowacki et al. 2006). This was verified by the dominance of cells in prophase, when 'arrested' mitosis gives an opportunity to repair genomic damages, in Scots pines from the Khrenovskoy stand (Butorina et al. 2001). The highest MI observed in the meristematic cells of seedling roots from the control trees seems to confirm this suggestion.

DNA damage observed in the form of chromosomal aberrations mainly related to the first phase of mitosis. In Scots pine, we observed mainly such abnormal mitosis types as: asymmetric mitosis, chromosomal stickiness, chromosome bridges, chromosome lagging, fragments of chromosomes and retarded chromosomes. These abnormalities results from spontaneous mutations induced by HMs, while in Russia (Voronezh), similar types of mitotic abnormalities were apparently related to a high level of gaseous air pollution (Butorina et al. 2001). This indicates that, like evidence from Voronezh, soil from MŚL also contains hazardous substances capable of disturbing mitotic divisions. Interesting data obtained in studies of *Syringa vulgaris* L. and *Armeniaca vulgaris* Lam. demonstrated that the increasing yield of abnormal mitoses was related to contamination with heavy metals from the soil in which the affected populations of these plants grew (in control samples there were no such aberrations). Similarly, a broad spectrum of cytogenetic aberrations was detected in three species of the genus *Pinus* (including *P. sylvestris*) growing in the vicinity of two integrated metallurgical works in Slovakia (Mičieta and Murín 1998). According to Kalashnik (2008), HMs rarely have a direct effect on DNA, but they are strong inducers of various mitotic abnormalities. Current data confirm that heavy metal soil pollution strongly affects Scots pine populations from the Miasteczko Śląskie region, and noted differences in the fitness of specimens are correlated with the level of cytological disturbances, especially with abnormal mitosis indices. The stand includes trees resistant and sensitive to growth conditions which vary in their range of chromosomal aberrations. The increased frequencies of abnormal mitoses in cells of *P. sylvestris* from polluted plots, which significantly differ from those recorded in control plots, indicate that the majority of these aberrations result from the impact of industrial pollution on the tree stands studied.

Mostly the chromosome aberrations are fixed by the reparative system of the cell, while the remaining defects lead to the formation of micronuclei. The formation of micronuclei results from chromosome lagging during mitosis and from multipolar mitoses. Therefore, micronuclei reflect disturbances in the genetic apparatus and are used to assess cytotoxicity and genotoxicity. For example, in the seedlings grown from the seeds collected in radioactively contaminated zones, its level was significantly elevated (Oficerov and Igonina 2008). Also the nucleolar activity (NA) is used in biomonitoring, because nucleoli are responsible for RNA synthesis in the cell, and the proportion of interphase cells with a specified number of nucleoli is an index of cell metabolism. Cytological data from multiple coniferous species suggests that pines can have many nucleolar chromosomes in their genomes. The NA is highly variable; therefore, *P. sylvestris* cells may contain from 1 to 15 nucleoli. Their number is higher under extreme conditions, e.g. in a pine growing in a bog (Chudzińska 2013). The number of nucleoli counted in silver-stained interphase pine-seedling cells varied from 1 to 15 in data from Miasteczko Śląskie. Cells with three to six nucleoli correspond to the three pairs of chromosomes with functioning nucleolar organizers, which prevailed in all studied samples. The nucleolar activity according to literature (Butorina et al. 2001; Butorina and Vostrikova 2006) is the most sensitive criterion of cytogenetic monitoring as an expression of the transcription activity of ribosomes. The number of nucleoli in a cell indicates the presence of ribosomal genes. If so, the increase of their number in trees from contaminated regions can be possibly regarded as response to stress induced by pollution of the environment. The increase nucleoli number in interphase cells and the decrease nucleolus–nucleus ratio under stressful conditions of growth have been marked in coniferous trees by Sedelnikova and Muratova (Sedelnikova and Muratova 2001). In the opinion of the authors, this phenomenon is an adaptive sign of coniferous species (as well as *P. sylvestris*) to stressful conditions. The number of nucleoli (up to 12 per cell) increases in conditions of severe stress, such as high background radiation, air pollution or soil contamination with HMs (Oficerov and Igonina and literature cited herein).

The level of chromosome aberrations, nucleolar activity and mitotic index studied in the seed and needle meristems of three Scots pine populations growing in areas differing in the degree of soil pollution provide evidence for the adverse effect of environmental stress on the chromosomal structure of these trees. With respect to environmental quality assessment, the Scots pine turned out to be a sensitive indicator species, and the analysis of chromosome aberrations in anaphase–telophase cells was most effective among the cytogenetic methods used. The differences observed between the trees growing in the same conditions cannot be explained by the influence of the environment. Their source must be sought in the varying capability of single individuals to adapt.

11.3 Sensitivity and Resistance: Genetic Basis

Organisms adapt to their environments by developing heritable tolerance mechanisms. According to Macnair et al. (2000), tolerance to heavy metals in plants may be defined as the ability to survive in a soil that is toxic to other plants and is manifested by an interaction between a genotype and its environment. Metal tolerance is a complex process by which populations respond to long-term environmental stresses by permanent genetic or epigenetic changes (Kovalchuk et al. 2004; Mehes-Smith et al. 2013). Adaptability is composed of two components: the physiological adaptability of individual trees (depending on the individual genotype) and the genetic adaptability of tree populations (based on the genetic variation within populations), which represents the evolutionary potential (Bergmann and Hosius 1996).

11.3.1 Protein Build-Up and Metabolism Control

Plants have evolved detoxification mechanisms, and consequently metal tolerance, to minimize heavy metal exposure and impact. In the cell, metals are under mechanisms of storage and detoxification strategies including transport, chelation and sequestration in the vacuole. Several types of proteins are involved in the metal chelation and in their root-to-shoot transport (Rauser 1999). The best known is the role of phytochelatins (PCs) and metallothioneins (MTs). The first (PCs) are small, heavy metal-binding, cysteine-rich polypeptides. They form complexes with toxic metal ions in the cytosol and then transport them into the vacuole, protecting plants from the deleterious effect of heavy metals (Salt and Rauser 1995). Glutathione (GSH), low-molecular-weight thiol, is a substrate for phytochelatin synthesis catalysed by the enzyme phytochelatin synthase (PCS) (Freeman et al. 2004). It is commonly believed that glutathione has a fundamental significance in determining plants' heavy metal tolerance (Yadav 2010). The second—metallothioneins (MTs)—are low-molecular-weight proteins which contain cysteine-rich domains at the amino- and carboxy-terminal regions. As they contain mercaptide groups, they are able to bind metal ions (Manara 2012). Transcription of MTs is typically induced by the same metal ion that binds to the protein, thus providing a direct activation of their protective function (Waalkes and Goering 1990). Some researches hypothesize that metallothioneins can protect plants and their DNA against oxidative damage by removing free radicals (Xue et al. 2009).

Other proteins are involved in the activation of oxidative stress mechanisms. When the concentration of heavy metal ions in the cell exceeds defence mechanisms, then the plant suffers oxidative stress due to the inhibition of metal-dependent antioxidant enzymes and production of reactive oxygen species (ROS) (Schützendübel and Polle 2002). They are very reactive molecules with an unpaired electron which can provoke oxidation and modification of cellular amino acids, proteins, membrane lipids and DNA (Ogawa and Iwabuchi 2001). The very important role glutathione plays in the mechanisms of heavy metal stress tolerance involved detoxifying ROS through ascorbate–glutathione cycle (Yadav 2010).

Plants activate their stress responses, including the induction of many enzymes, which may have different variants called allozymes (because they are encoded by different *alleles* at an enzyme gene locus). Studies (e.g. Prus-Głowacki et al. 2006; Chudzińska et al. 2013a, b) demonstrate differential survival and reproduction of individuals with different allozyme genotypes living under stress conditions. These results would imply that allozymes may be direct sites of action for toxicity to certain toxicants and in particular heavy metals (Van Straalen and Timmermans 2002). One of the mechanisms suggested explaining the relationship between pollution tolerance and allozyme genotype was that heavy metals competitively inhibit magnesium-dependent allozymes differentially (Guttman 1994). It is suspected that the degree of resistance to a highly contaminated environment may be linked to enzyme systems involved in various metabolic pathways, whose activity is modified under the influence of heavy metals. For example, enzyme 6-phosphogluconate dehydrogenase (6PGD) plays a critical role in the oxidative pentose pathway, which is physiologically very important under stressful conditions (Bergmann and Hosius 1996). A significant role for the 6PGD was also emphasized by these authors in the adaptation of Norway spruce to heavy metal-contaminated soils. Shikimate dehydrogenase (ShDH) is connected to detoxification mechanisms in environments polluted by heavy metal ions as well as by SO and NO (Tomsett and Thurman 1988). As numerous works have shown, cadmium intoxication strongly depresses glutamate oxaloacetate transaminase (GOT) activity (Benavides et al. 2005 and literature cited herein). Measurement of enzyme capacity of GOT and GDH (glutamate dehydrogenase) is recommended as a useful criterion for the evaluation of the phytotoxicity of soils contaminated by zinc and/or cadmium (Van Assche et al. 1998). Glutamate dehydrogenase also plays an essential role in the adaptation of plants to ammonia assimilation, SO_2 and other toxic gas detoxification (Van Assche et al. 1998). Studies on phosphoglucomutase (PGI) showed its relevance to oxygen availability, soil type and pH, and according to Van Assche et al. (1998), it is likely to be under the control of selection as an enzyme involved in the process of glycolysis. Research conducted on the Scots pine with regard to the above-described allozyme data revealed clear differences between genotypes of trees resistant and susceptible to heavy metals (Chudzińska 2013; Chudzińska et al. 2013a, b).

11.3.2 Mechanisms and Stability of DNA: Heavy Metal Impact

Most of the cellular and molecular aspects of metal toxicity in trees are unknown, even though deleterious effects and the mortality of trees have been recognized.

According to Macnair (Macnair 1993) and Ernst (Ernst 2006), genes for the tolerance of metals are pre-existing at low frequency in non-tolerant populations of certain plant species. In a number of genetic studies, adaptive metal tolerance has been shown to be governed by a small number of major genes with possible contributions from some minor modifier genes (Macnair et al. 2000; Schat et al. 2000). The mechanisms of genetic control and the major genes controlling heavy metal tolerance in trees have not been identified; thus, genetic variation can only be demonstrated indirectly by measuring the responses of different genotypes. The adaptive processes include changes in gene expression patterns, which ultimately lead to biochemical, cellular and physiological changes. To understand the process of plant response to environmental stresses, it is necessary to know the function of crucial genes and phytohormones and their regulation during different phases of the life cycle (Liu et al. 1998; Mukhopadhyay et al. 2004).

Transcriptional factors regulating the stress-responsive gene expression play important roles in stress adaptation (Stockinger et al. 1997; Liu et al. 1998). Tang et al. (2005) found that overexpression of an ERF/AP2 pepper transcription factor (CaPF1) in transgenic *Pinus virginiana* Mill. confers tolerance to heavy metals. Antioxidant enzymes, ascorbate peroxidase (APOX), glutathione reductase (GR) and superoxide dismutase (SOD), demonstrated higher levels of enzyme activity in transgenic Virginia pine plants overexpressing the CaPF1 gene, which may protect cells from the oxidative damage caused by stresses. Kovalchuk et al. (2004) found that the genomic DNA of pine trees exposed to radiation was considerably hyper-methylated. Hyper-methylation appeared to be dependent upon the radiation dose absorbed by the trees and may be viewed as a defence strategy of plants that prevents genome instability and reshuffling of the hereditary material, allowing survival in an extreme environment.

The study of DNA damage and repair is crucial, as DNA is the fundamental unit of inheritance and reproduction, so that perturbations in its structure and function could lead to changes in population dynamics or demography (Theodorakis 2001). The ideal molecular marker for the study of adaptive variations should comply with the following criteria: (1) has an identified DNA sequence, (2) is directly involved in the genetic control of adaptive traits and (3) has easily identifiable allelic variation (González-Martínez et al. 2006). Among the molecular markers that have been used to study populations of metal-tolerant plants are RAPD (random amplified polymorphic DNA), SSR (simple sequence repeats) or ISSR (inter-simple sequence repeat) (Nkongolo et al. 2001; Deng et al. 2007). Due to its hyper-variability, microsatellite DNA (simple sequence repeats, nuSSRs) is often used in studies of coniferous trees (Nowakowska 2006). The application of nuclear microsatellites as codominant markers in plant population studies is related to their ability to define genotypes unambiguously and makes it possible to assess genetic variation of populations growing under strong stress conditions. Microsatellites, found both in coding and non-coding regions in eukaryotic genomes, are markers whose usage is mainly based on the assumption that they are selectively neutral (Ellegrin 2004 and literature cited herein). However, some reports indicate that natural selection can control the level of variability in the microsatellite loci, although a general function of these markers is not yet known. Chambers and MacAvoy (2000) discuss in detail the potential tasks of microsatellite DNA, such as chromatin organization or regulation of gene activity, demonstrating that microsatellites may be under selection. Observed SSR diversity associated with such different ecological factors as elevated CO_2 level, climate changes or soil contamination suggests that these simple sequence repeats are not necessarily biologically neutral (Rocha et al. 2002; Badri et al. 2008). Mutation rates in nuSSRs are highly compared to mutations at coding gene loci, which makes these markers useful in studying adaptive responses of plants to stress. González-Martínez et al. (2006) worked on the polymorphism pattern of 18 candidate gene responses in *Pinus taeda* for drought stress and identified microsatellites in all those genes. Similar results were obtained by Eveno et al. (2008) in *Pinus pinaster* and by Fluch et al. (2011) in *Picea abies*. In this context, the study of variation in SSRs is a suitable method, but it should be complemented by other molecular techniques, e.g. differential-display reverse transcriptase PCR, in order to identify expressed genes which are involved in various stress responses and in the adaptive process.

11.3.3 Population Genetics: Heavy Metal Impact

Heavy metal impact on plant genetic structure, besides somatic effects (e.g. DNA damage), may include population genetic effects (changes in genetic diversity or gene frequencies). At the population level, concordant responses between changes in population genetic structure and elevated levels of DNA damage may indicate that the population genetic changes are influenced by exposure to toxic substances. If the frequencies of alleles or other genetic parameters differ between contaminated and reference populations, connections between DNA damage and genotype may provide evidence that these changes are due to toxicant-induced selection (Theodorakis 2001). Environmental contamination may act as a stress-selective factor affecting the genetic structure of the population and result in the extinction of the most vulnerable individuals or, in extreme cases, extinction of entire population. Among the conifer populations exposed to different pollutants, impoverishment of the gene pool is often observed (Geburek et al. 1987; Korshikov et al. 2002). Enzymatic studies of *Picea abies* revealed genetic differences between groups of sensitive trees in polluted areas (Scholz 1991). Observations of higher heterozygosity in tolerant plants of European beech (Müller-Starck 1985), Scots pine and *Populus tremuloides* (Geburek et al. 1987) have been reported.

Another parameter of interest is population diversity. Effects of contaminants on genetic diversity within the population are important because the level of genetic biodiversity affects a population's ability to adapt to natural changes in the environment (Hartl and Clark 1997). Many coniferous species with high genetic diversity are more viable and have better adaptation capacities to unfavourable conditions of an industrially polluted environment (Scholz 1991; Prus-Głowacki and Godzik 1991). This is in accordance to Lerner's classical hypothesis (1954) of heterozygotic adaptive advantage (as compared to homozygotes), due to the greater capacity of heterozygotes to maintain genetic and physiological homeostasis. Lerner's hypothesis leads to the conclusion that high genetic variability or extensive genetic richness (expressed in the high number of various alleles present in heterozygotic combination) is required for the progress of adaptive processes. Studies on Scots pine populations from natural regeneration and experiments on seedling survival in the vicinity of high levels of heavy metals showed that surviving individuals manifested a high level of homozygosity, particularly evident in some isoenzymatic loci, such as SHDH, MDH, GDH and ADH. A lower number of alleles have also been observed (Prus-Głowacki and Godzik 1991). In studies of *P. sylvestris* growing in areas surrounding operating metallurgical plants, Korshikov et al. (2002) reported a decrease in parameters such as the number of genotypes, alleles per locus and observed and expected heterozygosity. Prus-Głowacki et al. (2006) observed that the presence of heavy metals in soils, combined with SO_2 air pollution, resulted in a potential reduction in genetic variation in *P. sylvestris* populations. This type of adaptive strategy, linked to high level of environmental stress (i.e. heavy metals as pollutants), which drastically narrow the ecological niche, involves selection of the individuals of the specific genetic structure which is suitable in the specific, altered environment. In such cases, high genetic variability no longer presents an adaptive advantage and may be even disadvantageous for the population. This type of adaptation of forest trees to a polluted environment was observed in Norway spruce, European silver fir, common beech and black and Scots pine (Longauer et al. 2001; Korshikov et al. 2002; Schaberg et al. 2008). A decrease in heterozygosity within individuals can be associated with decreased resistance to diseases and might affect population growth. Interesting results are presented in the work of Chudzińska et al. (2013a, b) and Chudzińska (2013) on naturally regenerated trees (35 years old) of *P. sylvestris* and *P. nigra* existing in a highly heavy metal-contaminated site. Sensitive and resistant trees, which differed in fitness, exhibited differences in their genetic structures exhibited in isoenzyme markers and microsatellite loci of nuclear DNA. The greatest differentiation between the two subpopulations was observed in the loci of enzymes involved in metabolic pathways, whose activity is modified by heavy metal ions (Fig. 11.2). Sharma and Dubey (2005) listed over 20 of such enzymes. The Ewens–Watterson test for neutrality and the test of Vitalis et al. (2001) confirmed that allele frequencies at locus GOT B, as well as GDH, MDH C, NDH and SHDH B, of *P. nigra* have been influenced by selection, acting either directly on this locus or on adjacent genomic regions. Among the analysed microsatellite loci of nuclear DNA, two (PtTX 4001 and SPAC 11.4) appeared to be under balancing selection (Chudzińska 2013). SSR markers showed lower or higher genetic diversity of sensitive and resistant subpopulations with respect to certain loci. Two explanations for this phenomenon are most probable: first, those microsatellites lie in close proximity of the gene and are selected because of hitchhiking, and second, microsatellites themselves can be under selection (Meyer et al. 2010). The adaptation strategy of resistant Scots and black pine trees is the result of two processes acting in different loci: (1) strong directional

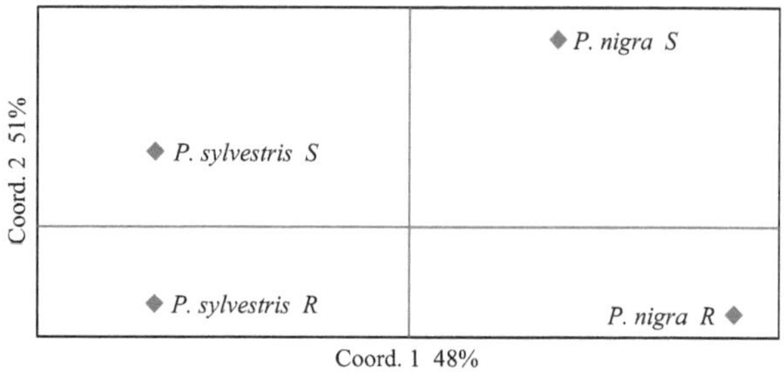

Fig. 11.2 Principal Coordinates Analysis via Nei's genetic distance (isozyme codominant markers) between sensitive (S) and resistant (R) subpopulations of *P. sylvestris* and *P. nigra* from Miasteczko Śląskie experimental trial in Poland

selection favouring strictly specialized homozygous genotypes (reduction in the level of genetic variation) and (2) stabilizing selection increasing heterozygote frequency (elimination of homozygotes) (Chudzińska et al. 2013a, b).

Both types of adaptive strategy, selection against homozygotes or heterozygotes, result in narrowing of the population gene pool due to elimination of sensitive individuals, which can carry unique alleles. The loss of genetic diversity in populations subjected to anthropogenic stress may be regarded as a factor of concern in risk assessment of heavy metals. This may diminish the genetic variability and fitness of affected populations and make them more susceptible to extinction following a subsequent stress (Guttman 1994).

11.4 Conclusions and Future Research

The heavy metal stress affects plants as a complex and dynamic system of highly variable factors; therefore, plants have many kinds of integrated mechanisms of resistance. It is generally accepted that heavy metals can affect DNA organization, function or gene expression at the individual tree level as well as on the genetic structure at the population level. The occurrence of heavy metal tolerance in plants is connected with the process of selection and microevolution.

Although our knowledge of heavy metal impact on plants has expanded in the last years, mechanisms of tolerance of forest trees are still unclear. To clarify the function of genes of forest trees and knowledge of mechanisms of metal tolerance, further studies are needed. Issues, such as which genes coordinate the improvement of resistance to heavy metal stress, how effectively are contaminated soils able to recultivate or in which way natural forests revive, still remain unanswered. However, the increased use of genetic molecular techniques should have a significant impact on our understanding of heavy metal tolerance and be helpful in the use of plants to remove contaminants from the environment by dendroremediation. This method offers a cost-effective alternative for large areas of unused lands that have been contaminated by heavy metals, especially in the context of tree biotechnology (Nevo et al. 2001). The improvement of the phytoremediation properties of trees can be achieved through modification of their primary and secondary metabolism

(Davison 2005). The example of hybrid poplar, engineered with a modified mercuric ion reductase gene, shows that trees are able to sustain high ionic mercury concentrations (Nevo et al. 2001). Transgenic poplars with higher peroxidase activity showed increased tolerance to zinc due to an enhanced ability to detoxify reactive oxygen species (Bittsanszky et al. 2005). Still, this technology is still in the developmental stage, and the field testing of transgenic plants for phytoremediation is very limited (Hur et al. 2011).

An equally important aim is connected with reintroduction of natural forests on soil contaminated by heavy metals. Mycorrhiza fulfils an important role here. Pine seedlings inoculated with metal-tolerant ecotypes of ectomycorrhizal fungi *Suillus* have lower metal concentrations in their needles than seedlings inoculated with sensitive strains, indicating that the metal-tolerant isolates restrict metal transfer more effectively. The evolutionary adaptation in *Suillus* species contributes to the survival of host trees in metalliferous soils and might be exploited in phytostabilization strategies for heavy metal-contaminated soils (Colpaert et al. 2011). Because of the key role of the ectomycorrhizal symbiosis in tree fitness, the evolution of Cd tolerance in ectomycorrhizal *S. luteus* can be of major importance for the establishment of pine forests on Cd-contaminated soils (Krznaric et al. 2009).

The capacity of plant populations to adapt to the environmental stress is influenced by their genetic potential. Adaptation can only take place if the population possesses or generates appropriate genetic variation. It should be understood that organism adaptation to stress involves genetic costs, namely, gene pool erosion and reduction of genetic variation. Tolerant individuals (genotypes) have to spend a significant part of their life energy to 'the defence' at different levels, which affects their viability and fertility.

After so many years of studies and thousands of well-documented experiments and observations, only one thing is known for sure: in order to describe the effects of pollution effectively, as many aspects as possible should be considered at the same time. For this reason, the value of integrating knowledge of plants in heavy metal-polluted environment should be underlined for making informative management and conservation implications.

References

Badri M, Zitoun A, Ilahi H, Huguet T, Aouani M (2008) Morphological and microsatellite diversity associated with ecological factors in natural populations of *Medicago laciniata* Mill. (Fabaceae). J Genet 87(3):241–255

Benavides MP, Gallego SM, Tomaro ML (2005) Cadmium toxicity in plants. Bras J Plant Physiol 17(1):21–34

Bergmann F, Hosius B (1996) Effects of heavy-metal polluted soils on the genetic structure of Norway spruce seedlings populations. Water Air Soil Poll 89:363–373

Bittsanszky A, Kömives T, Gullne G, Gyulai G, Kiss J, Heszky L, Radimszky L, Rennenberg H (2005) Ability of transgenic poplars with elevated glutathione content to tolerate zinc (2+) stress. Environ Int 31:251–254

Butorina AK, Kalaev VN, Mazurova SA, Doroshev EV (2001) Cytogenetic variation in populations of Scots pine. Russ J Ecol 32(3):198–202

Butorina AK, Vostrikova TV (2006) Cytogenetic responses of birch to stress factors. Biol Bull 33 (2):85–190

Chambers GK, MacAvoy ES (2000) Microsatellites: consensus and controversy. Comp Biochem Physiol B126:455–476

Chudzińska E (2013) Genetic diversity of Scots pine (*Pinus sylvestris* L.) as an expression of adaptation to heavy industrial pollution: a case study of the population from Miasteczko Śląskie. Wydawnictwo Naukowe UAM, Seria Biologica, Poznań

Chudzińska E, Diatta J, Półtorak W (2013a) Adaptation strategies and referencing trial of Scots and black pine populations subjected to heavy metal pollution. Environ Sci Pollut Res 21:2165–2177. doi:10.1007/s11356-013-2081-3

Chudzińska E, Pawlaczyk EM, Celiński K, Diatta J (2013b) Response of Scots pine (*Pinus sylvestris* L.) to stress induced by different types of pollutants—testing the fluctuating asymmetry. Water Environ J 28:533–539. doi:10.1111/wej.12068

Chudzińska E, Urbaniak L (2008) *Pinus sylvestris* L. response to heavy metal contamination express in anatomical traits of needles. Manag Environ Protect Forests 2:72–84

Colpaert JV, Wevers JHL, Krznaric E, Adriaensen K (2011) How metal-tolerant ecotypes of ectomycorrhizal fungi protect plants from heavy metal pollution. Ann Forest Sci 68:17–24

Davison J (2005) Risk mitigation of genetically modified bacteria and plants designed for bioremediation. J Ind Microbiol Biotechnol 32:639–650

Deng DM, Shu WS, Zhang J, Zou HL, Ye ZH, Wong MH, Lin Z (2007) Zinc and cadmium accumulation and tolerance in populations of *Sedum alfredii*. Environ Pollut 147:381–386

Derome J, Saarsalmi A (1999) The effect of liming and correction fertilization on heavy metal and macronutrients concentrations in soil solutions in heavy metal polluted Scots pine stands. Environ Pollut 104:249–259

Diatta JB, Chudzinska E, Wirth S (2008) Assessment of heavy metal contamination of soils impacted by a zinc smelter activity. J Elementol 13(1):5–16

Diatta JB, Wirth S, Chudzinska E (2011) Spatial distribution of Zn, Pb, Cd, Cu, and dynamics of bioavailable forms at a Polish metallurgical site. Fresen Environ Bull 20(4):976–982

Dickinson NM, Turner AP, Lepp NW (1991) How do trees and other long lived plants survive in polluted environments. Funct Ecol 5:5–11

Ellegrin H (2004) Microsatellites: simple sequences with complex evolution. Nat Rev Genet 5:435–445

Ernst WHO (2006) Evolution of metal tolerance in higher plants. Forest Snow Landscape Res 80:251–274

Eveno E, Collada C, Guevara MA, Léger V, Soto A et al (2008) Contrasting patterns of selection at *Pinus pinaster* Ait. drought stress candidate genes as revealed by genetic differentiation analyses. Mol Biol Evol 25(2):417–437

Freeman JL, Persans MW, Nieman K, Albrecht C, Peer W, Pickering IJ, Salt DE (2004) Increased glutathione biosynthesis plays a role in nickel tolerance in *Thlaspi* nickel hyperaccumulators. Plant Cell 16:2176–2191

Fluch S, Burg A, Kopecky D, Homolka A, Spiess N, Vendramin G (2011) Characterization of variable EST SSR markers for Norway spruce (*Picea abies* L.). BMC Res Notes 4:401–412

Geburek T, Scholz F, Knabe W, Vornweg A (1987) Genetic studies by isoenzyme gene loci on tolerance and sensitivity in an air polluted *Pinus sylvestris* field trial. Silv Genet 36:49–53

Godbold DL (1998) Stress concepts and forest trees. Chemosphere 35(4–5):859–864

González-Martínez SC, Ersoz E, Brown GR, Wheeler NC, Neale DB (2006) DNA sequence variation and selection of tag single nucleotide polymorphisms at candidate genes for drought-stress response in *Pinus taeda* L. Genetics 172:1915–1926

Guttman SI (1994) Population genetic structure and ecotoxicology. Environ Health Perspect 102 (Suppl 12):97–100

Hartl DL, Clark AG (1997) Principles of population genetics. Sinauer Associates, Inc., Sunderland, MA
Harju L, Saarela KE, Rajander J, Lill JO, Lindroos A, Heselius SJ (2002) Environmental monitoring of trace elements in bark of Scots pine by thick—target PIXE. Nucl Instrum Methods 189:163–167
Hur M, Kim Y, Song H, Kim JM, Choi YI, Yi H (2011) Effect of genetically modified poplars on soil microbial communities during the phytoremediation of waste mine tailings. Appl Environ Microbiol 77:7611–7619
Ivanov YV, Savochkin YV, Kuznietzov V (2011) Scots pine as a model plant for studying the mechanisms of conifers adaptation to heavy metal action: 1. Effects of continuous zinc presence on morphometric and physiological characteristics of developing pine seedlings. Russ J Plant Physiol 58(5):871–878
Kabata-Pendias A (2004) Soil–plant transfer of trace elements: an environmental issue. Geoderma 122:143–149
Kalashnik NA (2008) Chromosome aberration as indicator of technogenic impact of Conifer stands. Russ J Ecol 39(4):261–271
Kim GH, Bell JN, Power SA (2003) Effects of soil cadmium on *Pinus sylvestris* L. seedlings. Plant Soil 257:443–449
Klánowá J, Čupr P, Baráková D, Šeda Z, Anděl P, Holoubek I (2009) Can pine needles indicate trends in the air pollution levels at remote sites? Environ Pollut 157(12):3248–3254
Korshikov II, Velikorydko TI, Butilskaya IA (2002) Genetic structure and variation in *Pinus sylvestris* L. populations degrading due to pollution-induced injury. Silv Genet 51(2–3):45–49
Kovalchuk I, Abramov V, Pogribny I, Kovalchuk O (2004) Molecular aspects of plant adaptation to life in the Chernobyl zone. Plant Physiol 135:357–363
Kozlov MV, Niemelä P, Junttila J (2002) Needle fluctuating asymmetry is a sensitive indicator of pollution impact on Scots pine (*Pinus sylvestris*). Ecol Indic 1:271–277
Kozlov MV, Zvereva E (2011) A second life for old data: global patterns in pollution ecology revealed from published observational studies. Environ Pollut 159:1067–1075
Kozlov MV, Zvereva E, Zverev V (2009) Impacts of point polluters on Terrestrial Biota. Environ Pollut 15:197–224
Kozlowski TT, Pallardy SG (2002) Acclimation and adaptive responses of woody plants to environmental stresses. Bot Rev 68:270–334
Krznaric E, Verbruggenc N, Wevers JHL, Carleer R, Vangronsveld J, Colpaert JV (2009) Cd-tolerant *Suillus luteus*: a fungal insurance for pines exposed to Cd. Environ Pollut 157 (5):1581–1588
Lerner IM (1954) Genetic homeostasis. Olivier and Boyd, Edinbourgh
Lin JX, Sampson DA, Ceulemans R (2001) The effect of crown position and tree age on resin-canal density in Scots pine (*Pinus sylvestris* L.) needles. Can J Bot 79:1257–1261
Linhart YB, Grant MC (1996) Evolutionary significance of local genetic differentiation in plants. Annu Rev Ecol System 27:237–277
Liu Q, Kasuga M, Sakuma Y, Abe H, Miura S, Goda H, Shimada Y, Yoshida S, Shinozaki K, Yamaguchi-Shinozaki K (1998) Two transcription factors, DREB1 and DREB2, with an EREBP/AP2 DNA binding domain separate two cellular signal transduction pathways in drought- and low-temperature-responsive gene expression, respectively, in *Arabidopsis*. Plant Cell 10:391–406
Longauer R, Gomory D, Paule L, Karnosky F, Mankovska B et al (2001) Selection effects of air pollution on gene pools of Norway spruce, European silver fir and European beech. Environ Poll 115:405–411
Macnair MR (1993) The genetics of metal tolerance in vascular plants. New Phytol 124:541–559
Macnair MR, Tilstone GH, Smith SE (2000) The genetics of metal tolerance and accumulation in higher plants. In: Banuelos G, Terry N (eds) Phytoremediation of contaminated soil and water. CRC, Boca Raton, pp 235–250

Manara A (2012) Plant responses to heavy metals toxicity. In: Furini A (ed) Plants and heavy metals, Briefs in biometals. Springer, New York, pp 27–52
Markert B, Wünschmann S, Diatta J, Chudzińska E (2012) Innovative observation of the environment: bioindicators and biomonitors: definitions, strategies and applications. Environ Protect Nat Resour 37(2):115–152
Marschner H (1995) Mineral nutrition of higher plants, 2nd edn. Academic, London
Mehes-Smith M, Nkongolo K, Cholewa E (2013) Coping mechanisms of plants to metal contaminated soil. In: Dr. Steven Silvern (ed) Environmental change and sustainability. InTech. ISBN:978-953-51-1094-1
Meyer CL, Kostecka AA, Saumitou-Laprade P, Creach A, Castric V, Pauwels M, Frerot H (2010) Variability of zinc tolerance among and within populations of the pseudometallophyte species *Arabidopsis halleri* and possible role of directional selection. New Phytol 185:130–142
Mičieta K, Murin G (1998) Tree species of genus *Pinus* suitable as bioindicators of polluted environment. Water Air Soil Poll 104:413–422
Mukhopadhyay A, Vij S, Tyagi AK (2004) Overexpression of a zinc-finger protein gene from rice confers tolerance to cold, dehydration, and salt stress in transgenic tobacco. Proc Natl Acad Sci U S A 101:6309–6314
Müller-Starck G (1985) Genetic differences between tolerant and sensitive beeches (*Fagus sylvatica* L.) in an environmentally stressed adult forest stand. Silv Genet 34:241–246
Nevo E, Beharav A, Meyer MC, Hackett CA, Forster BP, Russell JR, Handley LL, Peña L, Séguin A (2001) Recent advances in the genetic transformations of trees. Trends Biotechnol 19 (12):500–506
Nieminen T, Helmisaari H-S (1996) Nutrient translocation in the foliage of *Pinus sylvestris* L. growing along a heavy metal pollution gradient. Tree Physiol 16:825–831
Nkongolo KK, Deck A, Michael P (2001) Molecular and cytological analyses of *Deschampsia cespitosa* populations from Northern Ontario (Canada). Genome 44:818–825
Novikova TN, Milyutin LI (2006) Variation in certain characters and properties of Scotch pine needles in geographic cultures. Russ J Ecol 37(2):90–96
Nowakowska J (2006) Zastosowanie markerów DNA (RAPD, SSR, PCR-RFLP i STS) w genetyce drzew leśnych, entomologii, fitopatologii i łowiectwie. Leśne Prace Badawcze 1:73–101
Oficerov E, Igonina S (2008) Genetic consequences of irradiation in a Scots Pine (*Pinus sylvestris* L.) population. Russ J Genet 45(2):183–188
Ogawa K, Iwabuchi M (2001) A mechanism for promoting the germination of *Zinnia elegans* seeds by hydrogen peroxide. Plant Cell Physiol 42:286–291
Oleksyn J, Prus-Głowacki W, Giertych M, Reich PB (1994) Relation between genetic diversity and pollution impact in an experiment with East-European *Pinus sylvestris* provenances. Can J Forest Res 24:2390–2394
Palowski B (2000) Seed yield from polluted stands of *Pinus sylvestris* L. New Forests 20(1):15–22
Potters G, Pasternak TP, Guisez Y, Palme K, Jansen M (2007) Stress-induced morphogenic responses: growing out of trouble? Trends Plant Sci 12(3):98–102
Prasad MNV, Hagemeyer J (2004) Heavy metal stress in plants: from molecules to ecosystems. Springer, Berlin
Prus-Głowacki W, Chudzińska E, Wojnicka-Półtorak A, Kozacki L, Fagiewicz K (2006) Effects of heavy metal pollution on genetic variation and cytological disturbances in the *Pinus sylvestris* L. population. J Appl Genet 47(2):99–108
Prus-Głowacki W, Godzik S (1991) Changes induced by zinc smelter pollution in the genetic structure of pine (*Pinus sylvestris* L.) seedling population. Silv Genet 40(5/6):184–188
Rauser WE (1999) Structure and function of metal chelators produced by plants. The case for organic acids, amino acids, phytin and metallothioneins. Cell Biochem Biophys 31:19–48
Rice KJ, Emery NC (2003) Managing microevolution: restoration in the face of global change. Front Ecol Environ 1(9):469–478
Rocha EP, Matic I, Taddei F (2002) Over-representation of repeats in stress response genes: a strategy to increase versatility under stressful conditions? Nucleic Acids Res 30:1886–1894

Salt DE, Rauser WE (1995) MgATP-dependent transport of phytochelatins across the tonoplast of oat roots. Plant Physiol 107:1293–1301

Schaberg PG, DeHayes D, Hawley GJ, Nijensohn S (2008) Anthropogenic alterations of genetic diversity within tree populations: Implications for forest ecosystem resilience. Forest Ecol Manag 256:855–862

Sharma P, Dubey RS (2005) Lead toxicity in plants. Braz J Plant Physiol 17:135–152

Schat H, Llugany M, Bernhard R (2000) Metal-specific patterns of tolerance, uptake and transport of heavy metals in hyperaccumulating and nonhyperaccumulating metallophytes. In: Terry N, Banuelos G (eds) Phytoremediation of contaminated soils and water. CRC, Boca Raton, pp 171–188

Scholz F (1991) Population-level processes and their relevance to the evolution in plants under gaseous air pollutants. In: Taylor GE, Pitelka LF, Clegg MT (eds) Ecological genetics and air pollution. Springer-Verlag, New York, pp 167–176

Schützendübel A, Polle A (2002) Plant responses to abiotic stresses: heavy metal-induced oxidative stress and protection by mycorrhization. J Exp Bot 53:1351–1365

Schützendübel A, Schwanz P, Teichmann T, Gross K, Langenfeld-Heyser R, Godbold DL, Polle A (2001) Cadmium-induced changes in antioxidative systems, hydrogen peroxide content, and differentiation in Scots pine roots. Plant Physiol 127:887–898

Sedelnikova TS, Muratova EN (2001) Karyological study of *Pinus sylvestris* (Pinaceae) with Witches' Broom growing on bog. Bot Zurnau 86(12):50–60

Stockinger EJ, Gilmour SJ, Thomashow MF (1997) *Arabidopsis thaliana* CBF1 encodes an AP2 domain-containing transcriptional activator that binds to the C-repeat/DRE, a cis-acting DNA regulatory element that stimulates transcription in response to low temperature and water deficit. Proc Natl Acad Sci U S A 94:1035–1040

Tang W, Charles TC, Newton RJ (2005) (2005) Overexpression of the pepper transcription factor CaPF1 in transgenic Virginia pine (*Pinus virginiana* Mill.) confers multiple stress tolerance and enhances organ growth. Plant Mol Biol 59:603–617

Theodorakis CW (2001) Integration of genotoxic and population genetic endpoints in biomonitoring and risk assessment. Ecotoxicology 10:245–256

Tomsett AB, Thurman DA (1988) Molecular biology of metal tolerances of plants. Plant Cell Environ 11:383–394

Van Assche F, Cardinaels C, Clijsters H (1998) Induction of enzyme capacity in plants as a result of heavy metal toxicity: dose–response relations in *Phaseolus vulgaris* L., treated with zinc and cadmium. Environ Pollut 52(2):103–115

Van Straalen N, Timmermans MJTN (2002) Genetic variation in toxicant-stressed populations: an evaluation of the "genetic erosion" hypothesis. Hum Ecol Risk Assess 8(5):983–1002

Vitalis R, Dawson K, Boursot P (2001) Interpretation of variation across marker loci as evidence of selection. Genetics 158:1811–1823

Waalkes MP, Goering PL (1990) Metallothionein and other cadmium-binding proteins: recent developments. Chem Res Toxicol 3:281–288

Xue TT, Li XZ, Zhu W, Wu CA, Yang GD, Zheng CC (2009) Cotton metallothionein *GhMT3a*, a reactive oxygen species scavenger increased against abiotic stress in transgenic tobacco and yeast. J Exp Bot 60:339–349

Yadav SK (2010) Heavy metals toxicity in plants: an overview on the roe of glutathione and phytochelatins in heavy metal stress tolerance of plants. S Afr J Bot 76:167–179

Chapter 12
Symbioses of Plants with Rhizobia and Mycorrhizal Fungi in Heavy Metal-Contaminated Tropical Soils

Fatima Maria de Souza Moreira, Paulo Ademar Avelar Ferreira, Laíze Aparecida Ferreira Vilela, and Marco Aurélio Carbone Carneiro

12.1 Heavy Metal (HM) Contamination in Tropical Soils

Most HMs are found naturally in the earth's crust in sufficient concentrations to nourish living systems without causing toxicity in these systems. However, pollution by HMs has been increased substantially due to factors such as increased mining activities, the disposal of industrial effluents and the use of pesticides and fertilisers.

Mining activities are crucial for obtaining materials required for the manufacture of goods indispensable to humans. These activities provide many benefits to society and to the economic development. Mining and metal smelting have always played an important role in the development of most societies and civilisations. However, these activities can cause environmental impacts that lead to damage to the environment, which affects ecosystems, population health and the lives of future generations.

The Brazilian state of Minas Gerais (MG) garnered fame due to the gold cycle, which began in the sixteenth century, although the major mines were discovered in 1693. MG holds 92 % of the Brazilian zinc (Zn) reserves, accounting for almost 100 % of its production (Bigarelli and Alves 1998). The municipalities of Vazante and Paracatu in MG currently support mining activities for which the ore processing and tailing deposition are conducted in Três Marias (TM). In this unit for Zn processing, the deposition of tailings contaminated with various HMs spreads to

F.M.d.S. Moreira (✉) • L.A.F. Vilela • M.A.C. Carneiro
Department of Soil Science, Federal University of Lavras (Universidade Federal de Lavras - UFLA), Caixa Postal 3037 Lavras, Minas Gerais, Brazil
e-mail: fmoreira@dcs.ufla.br; laizevilela@gmail.com; marcocarbone@dcs.ufla.br

P.A.A. Ferreira
Department of Soil, Federal University of Santa Maria (Universidade Federal de Santa Maria - UFSM), Santa Maria, Rio Grande do Sul, Brazil
e-mail: ferreira.aap@gmail.com

I. Sherameti, A. Varma (eds.), *Heavy Metal Contamination of Soils*, Soil Biology 44, DOI 10.1007/978-3-319-14526-6_12

an 18-ha area. In areas adjacent to the Zn processing unit in Três Marias, contamination occurred via deposition of tailings on the soil, which was localised in the beginning but spread by physical processes such as the erosive action of rain and programmed soil movement. Although there is nonhomogeneous horizontal and vertical distribution, the soil has extremely high levels of various metals in potentially toxic forms, namely, Zn = 18,600 mg kg^{-1}, Cd = 135 mg kg^{-1}, Cu = 596 mg kg^{-1} and Pb = 600 mg kg^{-1} (aqua regia extracted). In areas of industrial waste disposal, there is a predominance of exchangeable forms of Zn and Cd, which pose a greater environmental risk, whereas Cu and Pb are preferentially found in the residual form. Therefore, the high levels of these elements in the soil are responsible for the advanced degradation state of the area and for the difficulty in spontaneous revegetation. These situations can lead to side effects such as the loss of soil structure and consequently increased soil erosion, which promotes the transport of contaminated soil to water sources near the area and compromises the quality of the soil, the water and human health (Soares and Borma 2002).

In the São Francisco River Basin in MG, specifically in the Paracatu River subbasin, mining activity raises concerns about environmental impacts due to its coverage in terms of the occupied area and its economic importance in the region. During gold mining activity, mineral waste, called sterile materials, the final product and tailings are produced. Sterile materials are derived from mine stripping and contain compounds from the sulphide family, such as arsenopyrite, pyrite, pyrrhotite and chalcopyrite (Soares and Borma 2002). During ore processing or treatment that involves physical and chemical separations, mineral wastes called tailings are produced. Tailings are disposed of in containment dams as a material in pulp form that basically consists of mineral and liquid wastes rich in arsenic, and the dissolution of arsenic minerals, such as arsenopyrite, that are disposed of in these dams is a permanent contamination source (Ladeira et al. 2002).

In addition to these regions, surface coal mining activities over the past decades in Santa Catarina and Rio Grande do Sul states (Brazil) were not developed with adequate planning, and the necessary, essential recovery standards to maintain the quality of the environment in the vicinity of mining areas were not observed. Many areas were simply abandoned, which caused several problems, such as acid mine drainage, visual impact and the erosion and release of gases into the atmosphere, which compromised the quality of the air, water and soil; the resulting impacts still persist in the coal mining region. The substrate of mined areas is characterised by a high degree of disintegration, acidic pH, HM toxicity (such as that of Cd, Al, Mn, Zn, Cu, Pb and Fe) and a very crumbly texture, which makes it very susceptible to erosion (Silva et al. 2013). Its physical properties and low reserves of essential nutrients for plant growth impose severe restrictions on its use for agriculture and forestry.

Currently, most of these areas are in an advanced stage of degradation, and the risks of environmental impact are increasing. Although there is still microbial activity in the soil, various biological processes essential to its functionality are compromised because of the excessive metal content (Dias-Junior et al. 1998;

Santos et al. 2013), which prevents spontaneous revegetation in most locations. Therefore, there is a strong need for human intervention through the use of appropriate techniques to stop the advance of degradation and to reverse the environmental hazard threat.

In tropical agricultural soils, the accumulation of HMs mainly occurs through the application of pesticides, inorganic fertilisers, biosolids and organic fertilisers, such as pig manure and poultry litter (Fontes and Gomes 2003; Andrade et al. 2014; Mishra et al. 2009; Khairiah et al. 2006; Uchimiya et al. 2012; Huang and Jin 2008). Among inorganic fertilisers, phosphate fertilisers are considered the major source of HMs in agricultural soils (Kabata-Pendias 2011; Atafar et al. 2010; Huang and Jin 2008; Thawornchaisit and Polprasert 2009). However, biosolids and organic fertilisers, such as pig manure and poultry litter, are also worrisome sources of HMs in soil (Cai et al. 2007; Parkpain et al. 2000; Dœlsch et al. 2006).

In general, the prolonged use of phosphate fertilisers, biosolids and pesticides in tropical soils has substantially increased the natural levels of Cd, Cr, Cu, Pb and Zn in soils and crops, whereas other metals, such as As, Ba, Hg, Mo, Ni, Mn and Fe, are less reported and studied, most likely due to less significant changes or more regional accumulations (dos Anjos and Mattiazzo 2000; Appel and Ma 2002; Cai et al. 2007; Atafar et al. 2010; Camilotti et al. 2012; Andrade et al. 2014; Araújo et al. 2014; Dœlsch et al. 2006; Fernandes et al. 2007a; Fontes and Gomes 2003; Gonçalves Júnior et al. 2000; Huang and Jin 2008; Khairiah et al. 2006; Maftoun et al. 2005; Mishra et al. 2009; Nogueira et al. 2013; Parkpain et al. 2000; Parkpian et al. 2003; Sharma et al. 2009; Soumare et al. 2003; Udom et al. 2004). In Brazil, the application of biosolids from a sewage treatment station on red Latosol (Oxisol) for 12 consecutive years resulted in a significant accumulation of HMs, such as As, Ba, Cd, Cr, Cu, Hg, Ni, Pb and Zn. Moreover, in only a little more than a decade, the values of Cd (1.98 mg dm^{-3}), Cr (162 mg dm^{-3}), Cu (147 mg dm^{-3}) and Ni (31 mg dm^{-3}) exceeded the prevention value established by Resolution 420 (CONAMA 2009), a law that establishes the prevention values for Brazilian soils (Andrade et al. 2014).

The availability of these metals in the soil depends on many soil processes, such as dissolution, sorption, complexation, precipitation, occlusion and diffusion in minerals, volatilisation and absorption and sorption by soil microbes (Pierangeli et al. 2004; Kabata-Pendias 2011). Some factors, such as the physico-chemical properties of the metal, redox potential, pH and competition among cations in the soil solution, affect these reactions and alter the mobility of metals in the soil, resulting in increased absorption by plants or leaching (McBride 1989; Udom et al. 2004; Kabata-Pendias 2011). In general, the properties of tropical soils differ significantly from those of soils in temperate regions. Tropical soils are predominantly weathered, with a low density of surface charges, clay minerals with variable charges (pH dependent) and a high point of zero charge (PZC), i.e. with a great predominance of positive charges in common pH values (4.5–5.5). Thus, due to this characteristic, the soils of tropical regions display a lower affinity for and, consequently, a lower adsorption of HMs. In this case, there are increased mobility in the soil and availability of these metals to plants, which indicates that tropical soils are more vulnerable to contamination than soils in temperate regions (McBride 1989; Appel and Ma 2002; Fontes and Gomes 2003).

HM soil contamination is currently a major environmental problem because when these metals accumulate in the soil, they are introduced into the food chain of organisms or are leached into the bodies of water, endangering the quality of ecosystems (Pierangeli et al. 2004). The literature reports that the presence of HMs in agricultural soils has caused the accumulation of these metals in different crops, such as peanut (Camilotti et al. 2012), lettuce, cabbage, potato, carrot, beet, tomato (Fernandes et al. 2007b), spinach, cauliflower, okra (Maftoun et al. 2005; Sharma et al. 2009), sugar cane (Nogueira et al. 2013), soy (de Silva et al. 2007; Gonçalves Júnior et al. 2000), beans (Santos et al. 2003) and corn (Rangel et al. 2006; dos Anjos and Mattiazzo 2000).

One of the techniques to recover areas contaminated with HMs is phytoremediation, in which plants that form root symbioses with plant growth-promoting organisms—rhizobia and mycorrhizal fungi—contribute to plant nutrition with the two most limiting elements in tropical soils: N and P. However, the benefits of mutualistic symbioses to plants are not restricted only to the nutritional aspect but are also widely observed to mitigate the stress caused by HMs.

12.2 Diversity of Mycorrhizal Fungi in HM-Contaminated Soils

Arbuscular mycorrhizal fungi (AMF) have widespread occurrence and colonise more than 80 % of vascular plants, including in disturbed and degraded soils and in those contaminated with HMs. Heavy metal contamination, however, changes the abundance and diversity of the mycorrhizal fungi population in the soil (Del Val et al. 1999). Nonetheless, studies on the occurrence and diversity of AMF in tropical systems contaminated with HMs are rare (Table 12.1).

In general, it has been noted that although HM contamination does not completely eliminate mycorrhizal symbiosis, it causes quantitative and qualitative changes in the mycorrhizal community of these areas, where there is a reduction in species richness and in diversity indices (Silva et al. 2005; Del Val et al. 1999; Klauberg-Filho et al. (2002).

Regarding taxonomic diversity in tropical soils contaminated by HMs, the AMF genera most commonly reported in the literature are *Glomus* and *Acaulospora* (Gonzalez-Chavez et al. 2009; Klauberg-Filho et al. 2002; Zarei et al. 2010), as illustrated in Table 12.1. However, the AMF species vary according to the vegetation, metal type and soil contamination level. It is known that AMF diversity is directly related to the vegetation, which is heavily influenced by HM contamination (Macnair 1993). The adaptation potential of AMF to excess HMs is not restricted to specific genera or species and appears to be quite widespread (Klauberg-Filho et al. 2002). These results suggest that AMF are more tolerant than vascular plants

Table 12.1 Arbuscular mycorrhizal fungi associated with plants occurring in areas contaminated with heavy metals

Site characterisation	AMF species	References
Tannery effluent-contaminated soils with 153–201 mg kg^{-1} Cr, 1.8–2.1 mg kg^{-1} Zn, 0.87–0.99 mg kg^{-1} Cu and 0.97–1.19 mg kg^{-1} Pb	*Claroideoglomus etunicatum*[4], *Funneliformis constrictum*[5], *Funneliformis geosporum*[6], *Gigaspora albida*, *Gigaspora gigantea*, *Glomus aggregatum*, *Glomus citricola*, *Glomus heterosporum*, *Glomus microcarpum*, *Glomus multicaule*, *Rhizophagus fasciculatus*[12], *Sclerocystis rubiformis*[14], *Scutellospora erythropus*[16]	Raman and Sambandan (1998)
Addition of sewage-amended sludge containing Zn (44–88 mg kg^{-1}), Cd (0.3–2.8 mg kg^{-1}), Cu (9.8–92 mg kg^{-1}), Ni (8.3–29 mg kg^{-1}) and Pb (29–111 mg kg^{-1})	*Claroideoglomus claroideum*[3], *Funneliformis mosseae*[7], *Gigaspora rosea*, *Glomus* sp. *1*, *Glomus* sp. *2*, *Glomus* sp. *3*, *Glomus* sp. *4*	Del Val et al. (1999)
Sites of Zn extraction and industrialisation contaminated soil containing Zn (4.45–16.08 mg kg^{-1}), Cu (265–812 mg kg^{-1}), Cd (18–90 mg kg^{-1}) and Pb (52–125 mg kg^{-1})	*Acaulospora colombiana*[1a], *Acaulospora foveata*, *Acaulospora mellea*, *Acaulospora morrowiae*, *Acaulospora scrobiculata*, *Acaulospora* sp. *1*, *Acaulospora* sp. *2*, *Ambispora leptoticha*[2], *Gigaspora albida*, *Gigaspora margarita*, *Glomus microaggregatum*, *Paraglomus occultum*[8], *Racocetra castanea*[9], *Racocetra fulgida*, *Racocetra persica*, *Rhizophagus clarus*[10], *Rhizophagus intraradices*[13], *Scutellospora heterogama*, *Scutellospora pellucida*, *Scutellospora* sp. *1*	Klauberg-Filho et al. (2002)
Copper mining area. Contaminated soil with Cu (2.7–606.7 mg kg^{-1}), Mn (33.4–53.4 mg kg^{-1}), Zn (1.6–5.7 mg kg^{-1}) and Fe (33.6–434.4 mg kg^{-1})	*Acaulospora scrobiculata*, *Acaulospora tuberculata*, *Ambispora leptoticha*[2], *Claroideoglomus etunicatum*[4], *Entrophospora infrequens*, *Funneliformis mosseae*[7], *Gigaspora margarita*, *Glomus albidum*, *Glomus invermaium*, *Glomus microaggregatum*, *Glomus macrocarpum*, *Glomus microcarpum*, *Glomus* sp. *1*, *Glomus* sp. *2*, *Glomus* sp. *3*, *Glomus* sp. *4*, *Glomus tortuosum*, *Paraglomus occultum*[8], *Rhizophagus clarus*[10], *Rhizophagus diaphanus*[11], *Sclerocystis sinuosa*[15], *Scutellospora gilmorei*	Silva et al. (2005)
Soil contaminated at a mine highly polluted with Pb (17.2–4,892 mg kg^{-1}) and Zn (221–5,606 mg kg^{-1})	*Funneliformis mosseae*[7], *Glomus* sp. *1*, *Glomus* sp. *2*, *Glomus* sp. *3*, *Rhizophagus clarus*[10], *Rhizophagus intraradices*[13], *Sclerocystis sinuosa*[15]	Zarei et al. (2008a)
Soil contaminated at a mine highly polluted with Pb (7.63–4,126.69 mg kg^{-1}) and Zn (106.28–5,688.89 mg kg^{-1})	*Claroideoglomus claroideum*[3], *Funneliformis mosseae*[7], *Glomus versiforme*, *Rhizophagus intraradices*[13]	Zarei et al. (2010)

[a]Superscript numbers describe the previous nomenclature used in the original paper. [1]*Entrophospora colombiana*; [2]*Glomus leptotichum*; [3]*Glomus claroideum*; [4]*Glomus etunicatum*; [5]*Glomus constrictum*; [6]*Glomus geosporum*; [7]*Glomus mosseae*; [8]*Glomus occultum*; [9]*Scutellospora castanea*; [10]*Glomus clarus*; [11]*Glomus diaphanum*; [12]*Glomus fasciculatum*; [13]*Glomus intraradices*; [14]*Sclerocystis pachycaulis*; [15]*Glomus sinuosum*; [16]*Scutellospora erythropa*

to these types of stress conditions; thus, this diversity may be the key function of mycorrhizae in contaminated sites (Meharg 2003).

Anthropogenic activities such as the long-term application of sewage sludge, for example, promote a significant increase in HMs in the soil, reducing the abundance and diversity of AMF in the soil. In soil treated with sewage sludge (100–300 m^3 ha^{-1} $year^{-1}$) for 10 consecutive years, Del Val et al. (1999) observed increased levels of Zn (43–295 mg dm^{-3}), Cd (0.3–2.8 mg dm^{-3}), Cu (9.8–92 mg dm^{-3}), Ni (8.3–29 mg dm^{-3}) and Pb (29–111 mg dm^{-3}) in the soil. Under these conditions, only the genus *Glomus* was identified, with the following six species: *Glomus claroideum*, *G. mosseae* and four unidentified *Glomus* species (*III*, *IV*, *V* and *VI*). However, only *G. claroideum* and *Glomus* sp. *V* were more abundant under the highest contamination levels, whereas *Glomus* sp. *III* were virtually non-existent under these conditions.

The practice of only surveying AMF species in samples directly from the field has limited the study of the mycorrhizal community of contaminated areas because although some species are present, they exhibit low sporulation, making it difficult to identify them in the field. In an area contaminated with Zn (4,146–16,080 mg dm^{-3}), Cu (265–812 mg dm^{-3}), Cd (18–90 mg dm^{-3}) and Pb (52–125 mg dm^{-3}), Klauberg-Filho et al. (2002) observed that while *G. clarum*, *G. occultum* and *A. mellea* exhibited high occurrence in the field (>50 %), some species were identified only in trap crops (*A. morrowiae*, *A. tuberculata*, *E. colombiana*, *G. intraradices*, *G. leptotichum*, *G. margarita*, *G. microaggregatum*, *Scutellospora heterogama*, *S. pellucida*, *S. persica*, *S. castanea* and *S.* sp. *#1*). Among the six *Scutellospora* species, for example, only *S. fulgida* occurred in the field samples. To date, the consequences of reduced diversity and changes in the composition of AMF populations caused by HM soil contamination on the revegetation process of these areas are still unknown.

In addition to the characteristics of the fungus, the plant species and environmental conditions must also be considered when selecting isolates tolerant to HMs (Nogueira 1996; Klauberg-Filho et al. 2002; Gonzalez-Chavez et al. 2009). The survey of areas contaminated by HMs is the best way of trying to understand the relationship between metals and AMF. However, few studies have investigated symbiotic associations under these conditions.

12.3 Tolerance of Tropical Rhizobia to HMs

HMs are a group of pollutants of great ecological and environmental importance because they can affect the populations of symbiotic nitrogen-fixing bacteria of legumes, which are commonly called rhizobia, and reduce their population density in the soil, thus interfering with the establishment of their symbiosis with legumes (Trannin et al. 2001b; Ferreira et al. 2013). However, there are no studies on the occurrence and diversity of rhizobia in tropical soils contaminated with HMs.

The selection and identification of nitrogen-fixing bacteria tolerant to and capable of the bioremoval of HMs in contaminated areas is important because it enables the use of the intrinsic characteristics of these bacteria combined with the phytoremediation technique, promoting plant growth and favouring the removal of HMs through different mechanisms in different stages of the remediation process (Germaine et al. 2013).

Sixty *Bradyrhizobium*, *Rhizobium*, *Sinorhizobium*, *Mesorhizobium* and *Azorhizobium* strains from different regions (Atlantic Forest, Amazon, agricultural crops and experiments with HM) and from host species belonging to the subfamilies Papilionoideae, Mimosoideae and Caesalpinioideae were evaluated regarding their tolerance to Zn, Cu and Cd in YMA (yeast mannitol agar) medium (Vincent 1970), also known as medium 79 (Fred and Waksman 1928), modified by the addition of biological buffers (HEPES, *N*-2-hydroxy-piperazine-*N*′-2-ethanesulfonic acid, and MES, 2-(*N*-morpholino)-ethanesulfonic acid) and supplemented with Cu (0, 2.5, 5, 10, 20, 40 and 60 mg L^{-1}), Cd (0, 5, 10, 20, 30, 40 and 60 mg L^{-1}) and Zn (0, 200, 400, 500, 600, 800 and 1,000 mg L^{-1}). Given the growth patterns attributed to cultures under different metal concentrations, the maximum tolerated concentrations and toxic doses of these metals to reduce growth by 25 % (DT_{25}) and 50 % (DT_{50}) were evaluated. The order of tolerance to HMs, considering the maximum concentrations tolerated, was *Bradyrhizobium* > *Rhizobium* = *Mesorhizobium* = *Sinorhizobium* > *Azorhizobium*. Some strains later identified as *Burkholderia* (INPA353B and BR3460) exhibited a similar tolerance to that of *Rhizobium*. The DT_{25} and DT_{50} were useful to differentiate strains/isolates of the same genus, which reached the same maximum tolerated concentration of Zn, Cu and Cd. The order of toxicity of the metals studied was Cu > Cd > Zn (Matsuda et al. 2002b). These strains showed higher tolerance to HMs than that of strains isolated from temperate climate soils (Angle et al. 1993).

In this group of 60 strains, the two most tolerant [BR-4406 (recommended strain for *Enterolobium* spp.) and UFLA-01-457 (isolated from contaminated soil), both belonging to the genus *Bradyrhizobium*] and the two most sensitive (UFLA-01-486 and UFLA-01-510, isolated from contaminated soil, belonging to the genus *Azorhizobium*) were tested regarding their tolerance to different HM concentrations in multi-contaminated soil. For this purpose, different mixtures of a dark-red Latosol (Oxisol) that contained 0, 15 and 45 % (v/v) of a red-yellow plinthic Latosol (Oxisol) contaminated with Zn, Cd, Pb and Cu were used. The contaminated soil mixtures were inoculated with 20-mL cultures of the mentioned strains in the log phase in YMA medium. The average number of cells that survived after 28 days of incubation was (in CFU g^{-1} soil) $10^{10,36}$, $10^{10,29}$ and $10^{9,70}$ for *Bradyrhizobium* and $10^{9,36}$, $10^{7,54}$ and 0 for *Azorhizobium* in 0, 15 and 45 % contaminated soil mixtures, respectively. Therefore, *Bradyrhizobium* survived better than *Azorhizobium* in the soil, which indicates the greater tolerance of the first genus to HMs. Because *Bradyrhizobium* was also more tolerant in vitro, the results indicate that there is a relationship between its behaviour in contaminated soil and that in culture medium with HMs (Matsuda et al. 2002a).

Ferreira et al. (2012) determined the tolerance of 35 *Cupriavidus necator* strains to Zn, Cd, Cu and Pb that were able to form effective nodules in species of the family Fabaceae and subfamilies Faboideae and Mimosoideae. The $ZnSO_4{\cdot}7H_2O$, $CdSO_4{\cdot}8H_2O$, $CuSO_4{\cdot}5H_2O$ and $PbCl_2$ concentrations applied to the culture medium were 2.5, 5, 7.5, 10, 12.5 and 15 mmol L^{-1}. Of the 35 strains evaluated, 31 strains grew in up to 2.5 mmol L^{-1} of all of the metals; only 5 strains were able to grow at a 10 mmol L^{-1} Zn and Cu concentration, and one strain grew in up to a 7.5 mmol L^{-1} Pb concentration. None of the isolates grew in Cd concentrations higher than 2.5 mmol L^{-1}. Therefore, the toxicity of the metals for the strains decreased in the following order: Cd, Pb, Cu and Zn. These strains were isolated from nodules of different legume species in uncontaminated environments (Florentino et al. 2009; Silva et al. 2012), which suggests that there is no influence of the location of origin on the metal concentrations tolerated by this genus. The addition of up to 10 mmol L^{-1} of Zn and Cu stimulated the growth of *Cupriavidus necator* strains. For essential metal ions (Zn and Cu), there is a range of concentrations that may limit the optimum growth of microorganisms. However, above certain limits, the HM concentration may become toxic (Chaudhary et al. 2004). Conversely, non-essential metal ions cause toxicity at any concentration, but there is a limit up to which the organism can tolerate the adverse effects.

Studies have shown the influence of HMs on microorganisms, which affect their growth, morphology and biochemical activities. Among soil microorganisms, bacteria from the genus *Cupriavidus* have received scientific, economic and ecologic interest due to their ability to fix nitrogen and to tolerate high metal concentrations. The evaluation of four *Cupriavidus necator* strains in a culture medium with different Zn + Cd concentrations or when inoculated in multi-contaminated soil solution indicated that the strains were highly tolerant to these elements. The strain UFLA 02–71 bioaccumulated 93.76 μmol g $cell^{-1}$ of Zn and 16.03 μmol g $cell^{-1}$ of Cd at the total concentration of 9,140 μmol L^{-1} (9,000 Zn + 140 Cd), and in soil solution, the bioaccumulation was 16.98 μmol g $cell^{-1}$ of Cd (Ferreira 2014).

Microorganisms have developed various mechanisms of tolerance/resistance to survive in sites contaminated by HMs. These mechanisms include the active transport of metal ions from the inside of the cell to the outside of the cell efflux systems (Rensing et al. 1998; Silver and Phung 1996), intracellular sequestration via bonding with specific peptides and organic acids (Cook et al. 1998; Purchase et al. 1997), extracellular sequestration (Scott and Palmer 1990) and the reduction of metal ions to a less toxic form by enzymatic action (Noctor and Foyer 1998). The tolerance/resistance mechanisms are very diverse, but they have only been found in a limited number of species, and their mode of action is not always well known.

Because bacteria are able to inhabit the most diverse ecological niches, including those with high HM concentrations, they have developed different resistance mechanisms to a wide variety of these metals (Gadd and White 1993). The extracellular sequestration of HMs was observed in *Cupriavidus metallidurans* CH34, in which the metal cations Zn and Cd were complexed with carbonates, bicarbonates and hydroxides. Exopolysaccharides (EPS) and outer membrane proteins also appear to be involved with the immobilisation of these elements (Collard et al. 1994).

Some nitrogen-fixing bacteria have the ability to remove HMs from the environment (Ferreira 2014). The biosorption capacity and the mechanisms of accumulation may vary widely according to the species or even between different isolates of the same species. Cells and secreted products (cell wall and membrane and polysaccharides, respectively) are effective bioaccumulators of soluble and insoluble forms of HMs (Lovley et al. 1993; Ledin et al. 1996; Silver and Phung 1996; Niu et al. 1993; Kazy et al. 2002; Comte et al. 2008).

Because bioremediation-based technologies are usually linked to the genetic and biochemical ability of microorganisms to interact with and survive the toxic effects of metals, studies related to the remediation ability of these microorganisms also represent a great potential for exploration.

12.4 Tolerance of Mycorrhizal Fungi to HMs

Currently, there is increasing interest in the ability of AMF to retain metals in their mycelium (Gonzalez-Chavez et al. 2002), thereby reducing the translocation of metals to the plant shoots and increasing its tolerance. However, similar to plants, AMF are also adversely affected by excess HMs (Grazziotti et al. 2001).

According to Meharg (2003), the tolerance of organisms to HMs may be adaptive, constitutive or induced. Constitutive tolerance is the tolerance observed in isolates present in contaminated soil whose vegetation and microbial diversity have been greatly reduced but in which the growth of this isolate is similar to that of non-contaminated sites. Adaptive tolerance is characterised by reduced sensitivity to metal toxicity through the natural selection of genes that confer this characteristic. Induced tolerance, in turn, refers to the activation of specific enzymes responsible for the increasing tolerance that occurs after HM exposure.

Soil contamination by HMs may delay, reduce or even suppress spore germination, germ tube growth, mycelial growth and the mycorrhizal colonisation rate (Siqueira et al. 1999a; Del Val et al. 1999; Cardoso et al. 2002; Silva et al. 2001; Klauberg-Filho et al. 2002; Gonzalez-Chavez et al. (2002); Andrade et al. 2003; Andrade et al. 2004; Soares et al. 2006; Mergulhão et al. 2007; Zarei et al. 2008b; Shahabivand et al. 2012). However, different AMF species may have different degrees of tolerance to HMs (Del Val et al. 1999), as observed in Table 12.2. Still, Meier et al. (2012) state that continuous exposure to high metal concentrations may induce tolerance and promote the development of some specialised populations.

Weissenhorn et al. (1994) suggested that after a year of exposure to Cd, AMF spores showed high tolerance to Cd and Zn, which indicates that there is phenotypic plasticity resulting from rapid adaptation to local adverse conditions.

Still, relatively high mycorrhizal colonisation rates can be found in plants from contaminated areas, as observed in areas of Pb and Zn deposition from mining

Table 12.2 Effects of heavy metal contamination on arbuscular mycorrhizal fungi in tropical soils

AMF species	Conditions	Results	References
–	Soil treated with sewage sludge (Pb, Cd, Cr, Cu, Ni, Zn and Hg) for 10 years	The highest dose of sewage sludge (300 m^3 ha^{-1} $year^{-1}$) decreased by 1,700 % the spore density, with only 30 spores 100 g^{-1}	Del Val et al. (1999)
P. occultum[1a] and *R. clarus*[2]	Ratios of contaminated soil (Zn, Cd, Pb and Cu) and formononetin	There was a 75 % reduction in mycorrhizal colonisation at the highest level of contamination even with the presence of formononetin (400 mg kg^{-1} de solo)	Siqueira et al. (1999a)
C. etunicatum[3], *G. margarita* and *A. scrobiculata*	Different contaminated/non-contaminated soil ratios	Mycorrhizal colonisation was almost been eliminated in proportion with 80 % of contaminated soil	Siqueira et al. (1999b)
–	Mining area contaminated with Cu, Mn, Zn and Fe	In dry season, the area with the highest contamination showed no infective propagules and sporulation; in rainy season, these were very low. There was a negative correlation between the number of spores and the contents of Cu and Fe in the soil	Silva et al. (2001)
A. morrowiae, *A. appendicula*[4], *S. heterogama*, *G. margarita*, *G. macrocarpum*, *C. etunicatum*[3]	In vitro evaluation of Mn doses	Mn reduced the germination of *G. margarita*, *A. morrowiae*, *G. macrocarpum* and *C. etunicatum*. *A. appendicula* was severely affected. *S. heterogama* germination was stimulated by 30 mg kg^{-1} Mn	Cardoso et al. (2002)
F. mosseae[5], *F. caledonium*[6] and *G. rosea*	Soil containing As (0, 50 and 220 mg dm^{-3})	As reduced the germination of *G. rosea* and *F. mosseae*. *F. caledonium* was stimulated by the addition of As, even at the highest dose	Gonzalez-Chavez et al. (2002)

(continued)

Table 12.2 (continued)

AMF species	Conditions	Results	References
F. caledonium[6], *C. claroideum*[7] and *F. mosseae*[5], all isolated from the contaminated soil with As and Cu	–	*F. caledonium* had lower adsorption capacity of Cu in the extraradical mycelium than *F. mosseae* and *C. claroideum*. X-ray spectrometry showed the accumulation of metals on the external wall of hyphae mucilage and within the cytoplasm	Gonzalez-Chavez et al. (2002)
G. macrocarpum	Pb levels (0, 7.5, 37.5, 150 and 300 mg dm^{-3})	40 % reduction in mycorrhizal colonisation	Andrade et al. (2003)
G. macrocarpum, *C. etunicatum*[3], *G. margarita*	Mn levels (0, 4, 8 and 12 mg dm^{-3})	Reduction in the absorption of Mn. There was no effect in the content in the roots, except at the highest dose, in which *G. macrocarpum* increased Mn content and other AMF reduced absorption in the roots	Cardoso et al. (2003)
G. macrocarpum	Pb levels (0, 150, 300 and 600 mg dm^{-3})	In higher levels of Pb, the colonisation and sporulation was decreased	Andrade et al. (2004)
G. rosea and *F. mosseae*[5] isolated from the contaminated soil with Cu	Cu levels (0, 5, 10 and 20 μM)	Glomalin reached up to 4.3 mg g^{-1} Cu, 0.08 mg g^{-1} Cd and 1.12 mg g^{-1} Pb In vitro, produced glomalin by *Gigaspora rosea* accumulated to 28 mg g^{-1} Cu	Gonzalez-Chavez et al. (2004)
C. etunicatum[3] and *G. macrocarpum*	P and Mn concentrations	Reduction of root colonisation (6 % of colonised roots) and the length of the extraradical mycelium, especially *C. etunicatum*	Nogueira et al. (2004)
C. etunicatum[3], *R. intraradices*[8] and *G. macrocarpum*	Cd addition (0 and 5 μmol L^{-1})	Cd addition did not affect mycorrhizal colonisation, but reduced the growth of extraradical mycelium	Andrade et al. (2005)

(continued)

Table 12.2 (continued)

AMF species	Conditions	Results	References
C. etunicatum[3]	Proportions of soil from the mining area contaminated with Cu (0, 25, 50, 75 and 100 %)	Higher sporulation was observed at 25 % mixture of the contaminated soil At 50 % proportion of contaminated soil, there was no inhibition of mycorrhizal colonisation	Lins et al. (2006)
C. etunicatum[3]	Zn concentrations (2, 75, 150 and 225 μmol L^{-1})	High doses (150 and 225 mmol L^{-1}) were highly inhibitory to growth and mycorrhizal colonisation. AMF colonisation was inhibited from 54 % by Zn at a dose of 2 mmol L^{-1} to 21 % at the highest dose of Zn	Soares et al. (2006)
–	Gypsum mining areas	Reduction in the most probable number of propagules and spore density in areas of mining, in both periods evaluated (dry and wet). Mycorrhizal colonisation was higher in areas with higher contamination (vicinity of the mine site and tailings), reaching 56.8 % colonisation	Mergulhão et al. (2007)
C. etunicatum[3] and *G. macrocarpum*	P concentrations (30 and 45 mg kg^{-1}) in soil containing 508 mg kg^{-1} Mn	Mycorrhizal plants had lower Fe and Mn concentrations in shoots The increase in P uptake in mycorrhizal plants alleviated the damage of Fe and Mn toxicity	Nogueira et al. (2007)
F. mosseae[5] and *Sclerocystis sinuosa*[9]	Irrigated areas for 5, 35, 65 and 95 years with sewage sludge	Spore abundance decreased with time irrigation. Mycorrhizal colonisation (intraradical hyphae, arbuscules) was affected after 35 years of application	Ortega-Larrocea et al. (2007)
G. macrocarpum	Cd addition(0 and 20 μmol L^{-1})	Cd addition reduced mycorrhizal colonisation and the growth of	Andrade and Silveira (2008)

(continued)

Table 12.2 (continued)

AMF species	Conditions	Results	References
		extraradical mycelium, especially at low concentrations of P	
R. intraradices[8]	Cd addition (20 μmol L^{-1})	Cd addition had no effect on mycorrhizal colonisation, but reduced the growth of extraradical mycelium	Andrade et al. (2008)
–	Area near copper smelting industry and deposits of metal-rich waste area, especially Cu and Zn	There was storage of Cu (5.6–28.5 mg g^{-1}) and Zn (3.3–4.8 mg g^{-1}) binding to glomalin, and there was high correlation between the content of metals and glomalin. The amount of metal binding to glomalin corresponds to 27 and 5.8 % of the Cu and Zn total amount, respectively	Cornejo et al. (2008)
–	Deposition area from Pb and Zn mining	*Glomus* spores were abundantly found in all assessed areas, while *Acaulospora* spores were found only in a few areas. Despite the high contamination, there was complete elimination of AMF propagules in the soil. The spore density was more affected than the mycorrhizal colonisation, which reached up to 85 %	Zarei et al. (2008b)
–	Deposition area from Pb and Zn mining	Increasing contamination caused a drastic reduction in the number of spores AMF colonisation was affected, ranging from 32 % (most contaminated area) to 77 %, and there was considerable increase in the number of arbuscules and vesicles in areas with less contamination	Zarei et al. (2008a)

(continued)

Table 12.2 (continued)

AMF species	Conditions	Results	References
FMAs indígenas	Natural vegetation area in contaminated wastewater canal with Cd and Zn	Contamination affected mycorrhizal colonisation, reducing the amount of vesicles, arbuscules and hyphae	Rashid et al. (2009)
–	Attenuation area in contaminated slag with high Cd availability	All plant species showed a low rate of mycorrhizal colonisation, the largest in *Sida rhombifolia*, with 46 % of colonised roots	Gonzalez-Chavez et al. (2009)
R. clarus[2], *G. margarita* and *Acaulospora* sp.	Cu (0, 50, 150 and 450 mg kg^{-1}) and Zn (0, 100, 300 and 900 mg kg^{-1}) concentrations	Zn and Cu doses decreased very little mycorrhizal colonisation of seedlings of coffee, ranging 44–37 % in the presence of Cu and 36–33 % in the presence of Zn	Andrade et al. (2010)
Glomus deserticola	In vitro Zn concentrations (0, 10, 100, 500 and 1,000 mg kg^{-1})	There was a significant reduction in the length of hyphae and mycorrhizal colonisation, which reached 10 %	Arriagada et al. (2010)
–	Ore area contaminated with Zn and Pb	Contamination reduced the density and spores of the mycorrhizal colonisation rate	Zarei et al. (2010)
R. clarus[2], *Glomus* sp. and *G. gigantea*	Kinetics and retention capacity of Cu, Zn, Cd and Pb	*R. clarus* showed higher retention of Cu, Cd and Pb in the mycelium than the other species, whereas for Zn, there was greater retention in the *G. gigantea* mycelium. AMF-independent isolates retaining the speed decrease in the order $Cu > Cd \gg Zn > Pb$	Cabral et al. (2010)
–	Ratios of soil contaminated with Pb (7.5, 15 and 30 %)	Contamination did not affect the sporulation of AMF, but reduced the number of infective propagules in the soil	Gattai et al. (2011)
F. mosseae[5]	Cd concentrations (0, 0.3, 0.6 and 0.9 mM)	Mycorrhizal colonisation decreased with increasing Cd	Shahabivand et al. (2012)

(continued)

Table 12.2 (continued)

AMF species	Conditions	Results	References
		concentrations (23–45 % at the highest concentration)	

[a]Superscript numbers describe the previous nomenclature used in the original paper. [1]*Glomus occultum*; [2]*Glomus clarum*; [3]*Glomus etunicatum*; [4]*Acaulospora appendicula*; [5]*Glomus mosseae*; [6]*Glomus caledonium*; [7]*Glomus claroideum*; [8]*Glomus intraradices*; [9]*Glomus sinuosum*

companies in Iran, where mycorrhizal colonisation rates of up to 85 % were observed in species native to the area (Zarei et al. 2010; Zarei et al. 2008a, b).

In moderate cases of HM contamination, an increase in AMF propagules in the soil has been observed and may be related to the response of the fungus to stress, making it more competitive and resulting in it colonising the roots and completing its cycle (Del Val et al. 1999). Cardoso et al. (2002) found that although up to 30 mg kg^{-1} Mn inhibited the germination of *Acaulospora morrowiae*, *Gigaspora margarita*, *Glomus macrocarpum* and *Claroideoglomus etunicatum*, there was a significant stimulation for the germination of *Scutellospora heterogama* spores. A similar pattern was observed by Gonzalez-Chavez et al. (2002), who found that *Funneliformis caledonium* isolated from a mine contaminated with As had its germination stimulated with the addition of 220 mg kg^{-1} As.

Another interesting fact is the ability of AMF to produce a recalcitrant glycoprotein through their hyphae that complexes with soil HMs, making them unavailable to plants (Rillig and Steinberg 2002). Recent studies have noted the great capacity of this glycoprotein to accumulate Cu, Cd, Pb and Zn (Gonzalez-Chavez et al. 2004; Cornejo et al. 2008; Gonzalez-Chavez et al. (2002)). In areas close to the deposits of Cu from the smelting industry in Chile, Cornejo et al. (2008) found that the amount of metals bound to glomalin corresponded to 27 and 5.8 % of the total Cu and Zn content in the soil, respectively. These data demonstrate the active role of glomalin in sequestering potentially toxic metals in the soil and their effectiveness in mitigating HM stress in plants growing in contaminated soils (Meier et al. 2012).

12.5 Effects of Rhizobia and/or Mycorrhizal Fungi on Plant Growth in HM-Contaminated Soils

The soil microbial community is functionally complex and contains key groups of microorganisms with an essential role in maintaining soil fertility that is directly related to the mineral nutrition of plants (Ghorbani et al. 2002; Obbard 2001). Improved soil fertility may occur through the direct action of microbial decomposers of organic matter (OM) that transform OM into mineral forms available to plants and also through symbiotic relationships between microorganisms and plants

(Giller et al. 1998). The symbiotic relationships between nitrogen-fixing bacteria and AMF in different plant species may be a promising alternative to phytoremediation programmes because when tolerant to metals, these bacteria and AMF stimulate plant growth in these environments.

Some tropical legume trees have been identified as promising for programmes to recover soils degraded by excess HMs, such as *Enterolobium contortisiliquum* and *Acacia mangium*, which are able to form an efficient symbiosis with *Bradyrhizobium* (Trannin et al. 2001b; Marques et al. 2000), *Mimosa caesalpiniaefolia* and *Leucaena leucocephala*, which form symbioses with *Cupriavidus necator* (Ferreira et al. 2012). Other symbioses, such as that between *Sesbania virgata* and *Azorhizobium doebereinerae*, are highly efficient in uncontaminated conditions (Moreira et al. 2006) but are not tolerant in contaminated soils (Trannin et al. 2001a).

Thus, the selection of symbiotic associations between nitrogen-fixing bacteria and legumes tolerant to HMs is of great importance for soil nitrogen accumulation and OM increase. This relationship is highly relevant to the growth of plants in contaminated soils because these soils tend to have low OM levels (Ferreira et al. 2013). However, high HM contamination levels in soils may inhibit the biological fixation of nitrogen, affecting the infectious process and nodulation, nitrogenase activity and leghaemoglobin production (Trannin et al. 2001a; Hasan et al. 2008; Ibekwe et al. 1995; Mårtensson and Witter 1990). Plants nourished with mineral N were less sensitive than those dependent on the symbiosis with rhizobia.

The process of the formation of effective nodules in contaminated soils is greatly affected by the presence of high HM concentrations (Ferreira et al. 2013). The evaluation of the symbiotic effectiveness of *Azorhizobium* and *Bradyrhizobium* strains in soils from a mining area contaminated with 750, 111, 22 and 65 mg kg^{-1} of Zn, Cu, Cd and Pb had a deleterious effect on the formation of nodules and decreased nitrogenase activity (Trannin et al. 2001a). The presence of high HM concentrations in the soil reduces the number of root hairs, which compromises the nodulation process and the fixed nitrogen rate (Trannin et al. 2001a; Ferreira et al. 2013). The fact that plants are continuously absorbing Zn and Cd results in phytotoxic effects caused by these metals, in addition to physiological disorders (Castaldi et al. 2009; Cunha et al. 2008). As a consequence, there is a reduction in the carbohydrate flow to the root system, which reduces the energy supply needed to maintain the functionality of the nodules.

Strategies that mitigate environmental contamination may be employed by applying mitigating substances to the soil to reduce HM toxicity to plants (Ferreira et al. 2013). The application of limestone and calcium silicate in soil contaminated with Zn and Cd provided significant increases in pH with reductions of up to 99 and 94 % in the availability of Zn and Cd in the soil solution, respectively, favouring better nodulation of *Mimosa caesalpiniaefolia*, *Leucaena leucocephala* and *Mimosa pudica* under the different contamination levels. The nodulation in the root system by the strain UFLA01-659 in *Leucaena leucocephala* and *Mimosa pudica* and by the strain UFLA02-71 in *Mimosa caesalpiniaefolia* effectively increased the nitrogen levels in the shoots to 36, 40 and 39 g kg^{-1}, respectively.

Table 12.3 Effects of mycorrhizal fungi on plant growth in soil contaminated by heavy metals from tropical areas

AMF species	Study conditions	Plant species	Results	References
P. occultum[1a], *R. clarus*[2]	Ratios of contaminated soil (Zn, Cd, Pb and Cu)	*Zea mays*	Formononetin application alleviated the toxicity of heavy metals by stimulating mycorrhizal colonisation	Siqueira et al. (1999a)
C. etunicatum[3], *G. margarita*, *A. scrobiculata*	Ratios of contaminated soil	*Senna multijuga*, *Luehea grandiflora*, *Enterolobium contortisiliquum*, *Albizia lebbeck*, *Senna macranthera*	Inoculation reduced Zn concentration in the shoot of all species, with greater effects in *Senna multijuga and Albizia lebbeck* P increased with the increasing proportion of contaminated soil, and inoculation promoted its increase only in mixtures without contaminated soil	Siqueira et al. (1999b)
A. scrobiculata, *P. occultum*[1], *C. etunicatum*[3], *G. margarita*	Ratios of contaminated soil with Zn, Cd, Pb and Cu	Grass and mustard	Inoculation increased 18 and 35 % of the dry matter of shoots and roots of grasses, respectively, and increased the concentration of Cd, Zn and Pb in the roots compared to shoots	Carneiro et al. (2001)
G. macrocarpum	Pb concentrations (0, 7.5, 37.5, 150 and 300 mg dm^{-3})	Soybean	Inoculation reduced by 30 % the Cd absorption in soybean plants	Andrade et al. (2003)
G. macrocarpum, *C. etunicatum*[3], *G. margarita*	Mn concentrations (0, 4, 8 and 12 mg dm^{-3})	Soybean	AMF reduced the absorption of Mn in shoots. In roots, *G. macrocarpum* increased the Mn content, and the other species reduced absorption in the roots	Cardoso et al. (2003)

(continued)

Table 12.3 (continued)

AMF species	Study conditions	Plant species	Results	References
G. macrocarpum, *C. etunicatum*[3], *R. intraradices*[4]	Substrate containing Mn	Soybean	In sandy substrate, inoculation, especially *R. intraradices* and *C. etunicatum*, reduced the concentration of Mn in shoots and roots In clayey substrate, *R. intraradices* and *C. etunicatum*, especially *C. etunicatum*, promoted increased levels of Mn in roots and shoots	Nogueira and Cardoso (2003)
G. macrocarpum	Pb concentrations (0, 150, 300 and 600 mg dm^{-3})	Soybean	Mycorrhization stimulated the nodulation of plants in the presence of Pb and reduced absorption of Pb in shoots at up to 30 % compared to non-inoculated plants	Andrade et al. (2004)
C. etunicatum[3], *G. macrocarpum*	P and Mn concentrations	Soybean	Mycorrhizal plants receiving additional P had more tolerance to Mn	Nogueira et al. (2004)
C. etunicatum[3], *R. intraradices*[4], *G. macrocarpum*	Cd addition (0 and 5 μmol L^{-1})	Jack bean	Mycorrhization increased the concentration of Cd in roots and shoots of the plant	Andrade et al. (2005)
A. morrowiae, *A. scrobiculata*, *A. spinosa*, *G. albida*, *G. gigantea*, *G. margarita*, *R. clarus*[2],	Influence of AMF isolates on the growth and extraction of Zn, Cd, Cu and Pb in contaminated soil	*Brachiaria decumbens*	Isolates promoted an increase on an average of 84 % in dry matter production, reducing 20, 28 and 63 % levels of Zn, Cd	Silva et al. (2006)

(continued)

Table 12.3 (continued)

AMF species	Study conditions	Plant species	Results	References
C. etunicatum[3], *P. occultum*, *R. gregaria*[5], *S. heterogama*, *S. pellucida*			and Cu in the shoot, respectively, and increased Cu concentration in more than 1,000 % roots AMF increased the amount of metals extracted from the soil 845, 142, 68 and 54 % for Cu, Pb, Zn and Cd, respectively *A. spinosa*, *A. morrowiae* and *G. gigantea* increased four simultaneous extractions of contaminating metals	
C. etunicatum[3], *G. macrocarpum*	P concentrations (30 and 45 mg kg^{-1}) in soil containing Mn	Soybean	Mycorrhizal plants had lower concentrations of Fe and Mn in shoots	Nogueira et al. (2007)
C. etunicatum[3]	Cd concentrations (0, 5, 10 and 15 μmol L^{-1})	*Trema micrantha*	Inoculation did not alleviate effects in plants against Cd contamination	Soares et al. (2007)
G. macrocarpum	Cd (0 and 20 μmol L^{-1}) and P (5 and 10 mg L^{-1})	Maize	There was no difference in Cd concentration between inoculated and non-inoculated plants; however, there was a greater accumulation of Cd in roots Mycorrhizal plants had higher P/Cd, N/S and Cd/Cd	Andrade and Silveira (2008)

(continued)

Table 12.3 (continued)

AMF species	Study conditions	Plant species	Results	References
R. intraradices[4]	Cd addition (20 μmol L^{-1})	Sunflower	Inoculation alleviated the phytotoxic effects of Cd, promoting better growth Inoculated plants showed higher Cd content in shoots (23 % higher)	Andrade et al. (2008)
A. morrowiae, *G. albida*, *R. clarus*[2]	Contaminated soil with Zn, Cu, Cd and Pb and P concentrations (25, 100 and 400 μg g^{-1})	*Brachiaria decumbens*	At low concentration of P, inoculated plants showed lower bioaccumulation factor of Zn, Cd and Cu than non-inoculated plants in shoots. In roots, there's an inversion behaviour In high concentration of P, the behaviour of the shoot was similar to that of low P concentration; however, the roots of inoculated plants had lower bioaccumulation factors than non-inoculated plants	Soares and Siqueira (2008)
F. mosseae[6], *Glomus* sp.	Rations contaminated soil with Cd, Pb and Zn	Maize	Concentrations of heavy metals were lower in shoots of mycorrhizal plants Mycorrhization reached 50 % reduction in the concentration of Pb, Zn and Cd in shoots and Pb in the roots	Liang et al. (2009)

(continued)

Table 12.3 (continued)

AMF species	Study conditions	Plant species	Results	References
R. clarus[2], *G. margarita*, *Acaulospora* sp.	Cu (0, 50, 150 and 450 mg kg^{-1}) and Zn (0, 100, 300 and 900 mg kg^{-1}) concentrations	*Coffea arabica*	Inoculated plants plus Cu had higher leaf Cu + K, P, S and Zn concentrations and reduced levels of Ca, Mg and Mn. In roots, inoculation increased the levels of P, Ca, Mg, S and Zn and reduced the levels of K Inoculated plants plus Zn showed increases in the + Zn foliar concentrations of K, P and Cu but reduced the levels of Ca, Mg and S. The roots have reduced levels of K, Ca, Mg, S and Cu	Andrade et al. (2010)
G. deserticola	Zn (0, 10, 100, 500 and 1,000 mg kg^{-1})	Barley	Inoculation alleviated excess Zn, with higher growth of shoots and roots in the inoculated plants, and increased the concentration of Zn in roots and shoots	Arriagada et al. (2010)
F. mosseae[6]	Cd concentrations (0, 0.3, 0.6 and 0.9 mM)	Wheat	Colonised plants had higher growth and higher chlorophyll content The presence of *F. mosseae* reduced the concentration of Cd in roots and shoots	Shahabivand et al. (2012)

(continued)

Table 12.3 (continued)

AMF species	Study conditions	Plant species	Results	References
C. etunicatum[3], *A. morrowiae*, *G. gigantea*, *Acaulospora* sp.	Gold mining soil contaminated with As	*Acacia mangium*, *Crotalaria juncea*, *Enterolobium contortisiliquum*, *Stizolobium aterrimum*	Phytoprotective effect and mechanisms involved varied according to plant and AMF species. *S. aterrimum*, regardless of AMF inoculation, and the symbiosis of *C. juncea* and *Acaulospora* sp. demonstrate the potential use in As phytostabilisation in As-contaminated soils. They were capable of immobilising high As concentration in roots and increasing P/As ratio in shoots, providing a greater production of shoot biomass	Rangel et al. (2014)

[a]Superscript numbers describe the previous nomenclature used in the original paper. [1]*Glomus occultum*; [2]*Glomus clarum*; [3]*Glomus etunicatum*; [4]*Glomus intraradices*; [5]*Scutellospora*

The mitigating effect of calcium silicate favoured greater dry matter production in the shoots, a greater number of nodules and higher rates of fixed nitrogen. The results of this previous study show that the combination between rhizobia-legumes-silicate may represent a key factor for the revegetation of areas contaminated by HMs.

In addition to these strategies, mycorrhizal symbiosis significantly changes the plant response to HM soil contamination. The effects of HMs on the interaction between plants and AMF differ among different mycorrhizal fungi and plants (Siqueira et al. 1999b), as shown in Table 12.3.

Similar to other soil microbial populations, AMF have developed mechanisms that confer greater resistance or tolerance to the toxic effects caused by HMs. In fungi, Meier et al. (2012) cite the following main mechanisms: (I) adsorption of metals to the cell wall surface (biosorption), (II) transport and intracellular incorporation (bioaccumulation) and (III) transformation of metals through reduction, oxidation or methylation reactions. In addition to the mechanisms mentioned, the ability of AMF to retain and accumulate metals in glomalin, a glycoprotein produced by AMF hyphae, has been reported (González-Chávez et al. 2004; Arriagada et al. 2010; Cornejo et al. 2008; Gonzalez-Chavez et al. 2002). Additionally, membrane transporters in AMF arbuscules may transport metals to the contact zone between the plasma membrane of the fungus and that of the plant cell, with subsequent incorporation of the metals into plant tissues (Meier et al. 2012); this could explain the accumulation of metals in the shoots of plants in some of the previously noted cases.

Several studies have demonstrated a reduction in metal toxicity as a result of mycorrhizal association (Schützendübel and Polle 2002). Siqueira et al. (1999b) emphasise, however, that AMF benefits are related to the degree of stress to which the host plant is subjected, i.e. under high degrees of stress, a dysfunction may occur in symbiosis, eliminating its mutualistic character.

AMF reduce HM absorption under conditions of high concentrations in the soil, temporarily immobilising them in the hyphae and, consequently, reducing the translocation of these elements to the shoots, they may even mitigate the nutritional imbalance caused by the excess metals (Carneiro et al. 2001; Siqueira et al. 1999a; Meharg 2003).

Recently, AMF were shown to improve the tolerance of hosting plants to As in contaminated soils (Rangel et al. 2014). In general, all leguminous species had a low As translocation index from the roots to shoots when inoculated with AMF species. The inoculation of *Crotalaria juncea* with *Acaulospora* sp. significantly improved As accumulation in roots and decreased the activity of ascorbate peroxidase (APX) and superoxide dismutase (SOD), highlighting its phytoprotective effect and the potential use of this symbiosis for the phytoremediation of As-contaminated soils. However, *Stizolobium aterrimum* has also shown the potential for phytoremediation irrespective of AMF inoculation. APX was a good indicator of the phytoprotective effect against As contamination in *C. juncea* and *Acacia mangium*. In general, the P/As ratio in shoots was the best indicator of the phytoprotective effect of all AMF species in all plant species.

12.6 Concluding Remarks

The great potential of rhizobia and arbuscular mycorrhizal fungi in mitigating the stress caused by HMs in plants is evident. The use of these microorganisms in contaminated areas is an important biotechnological tool for the revegetation process of these areas and may be integrated successfully into bioremediation programmes for contaminated areas.

The results mainly show the beneficial effects of mycorrhizal fungi and rhizobia for promoting the establishment and survival of plants in soils contaminated with HMs. However, although a wide range of studies has been conducted, few investigations have examined the specific mechanisms involved in the benefits of these organisms under these conditions. There is also a huge gap in information on topics related to these organisms such as the occurrence under different contamination conditions, the identity of species tolerant to certain HMs under specific conditions and, especially, the symbiotic efficiency of mycorrhizal fungi and rhizobia species in plants native to contaminated areas.

Thus, for the application of this wide range of positive results obtained in remediation programmes for contaminated soils, field and molecular studies should be conducted to elucidate the remaining issues, which will enable the understanding of the complex dynamics of mycorrhizal fungi and their effects on plants in contaminated areas.

References

Andrade SA, Silveira AP (2008) Mycorrhiza influence on maize development under Cd stress and P supply. Braz J Plant Physiol 20(1):39–50

Andrade MG, Lima AST, Melo WJ, Santos EJ, Herrmann AB (2014) Elementos-traço em dois latossolos após aplicações anuais de lodo de esgoto por treze anos. Semina Ciências Agrárias 35(1):135–148

Andrade S, Abreu C, Abreu M, Silveira A (2003) Interação de chumbo, da saturação por bases do solo e de micorriza arbuscular no crescimento e nutrição mineral da soja. Rev Bras Ciênc Solo 27(5):945–954

Andrade S, Abreu C, De Abreu M, Silveira A (2004) Influence of lead additions on arbuscular mycorrhiza and *Rhizobium* symbioses under soybean plants. Soil Ecol Appl 26(2):123–131

Andrade SAL, Jorge RA, Silveira APD (2005) Cadmium effect on the association of jackbean (Canavalia ensiformis) and arbuscular mycorrhizal fungi. Sci Agric 62(4):389–394

Andrade SAL, Silveira APD, Jorge RA, Abreu MF (2008) Cadmium accumulation in sunflower plants influenced by arbuscular mycorrhiza. Int J Phytoremediat 10(1):1–13

Andrade S, Silveira A, Mazzafera P (2010) Arbuscular mycorrhiza alters metal uptake and the physiological response of Coffea arabica seedlings to increasing Zn and Cu concentrations in soil. Sci Total Environ 408(22):5381–5391

Angle JS, McGrath SP, Chaudri AM, Chaney RL, Giller KE (1993) Inoculation effects on legumes grown in soil previously treated with sewage sludge. Soil Biol Biochem 25(5):575–580

dos Anjos ARM, Mattiazzo ME (2000) Metais pesados em plantas de milho cultivadas em Latossolos repetidamente tratados com biossólido. Sci Agric 57(4):769–776

Appel C, Ma L (2002) Concentration, pH, and surface charge effects on cadmium and lead sorption in three tropical soils. J Environ Qual 31(2):581–589

Araújo SR, Demattê JA, Vicente S (2014) Soil contaminated with chromium by tannery sludge and identified by vis-NIR-mid spectroscopy techniques. Int J Remote Sens 35(10):3579–3593

Arriagada C, Pereira G, García-Romera I, Ocampo J (2010) Improved zinc tolerance in *Eucalyptus globulus* inoculated with *Glomus deserticola* and *Trametes versicolor* or *Coriolopsis rigida*. Soil Biol Biochem 42(1):118–124

Atafar Z, Mesdaghinia A, Nouri J, Homaee M, Yunesian M, Ahmadimoghaddam M, Mahvi AH (2010) Effect of fertilizer application on soil heavy metal concentration. Environ Monit Assess 160(1–4):83–89

Bigarelli W, Alves F (1998) Minas Gerais: Mantendo-se como líder na mineração brasileira. Brasil Miner 162:14–18

Cabral L, Siqueira JO, Soares CRFS, Pinto JEBP (2010) Retenção de metais pesados em micélio de fungos micorrízicos arbusculares. Quim Nova 33(1):25–29

Cai Q-Y, Mo C-H, Wu Q-T, Zeng Q-Y, Katsoyiannis A (2007) Concentration and speciation of heavy metals in six different sewage sludge-composts. J Hazard Mater 147(3):1063–1072
Camilotti F, Silva ARB, Marques MO (2012) Biomass and yield of peanut grown on tropical soil amended with sewage sludge contaminated with lead. Appl Soil Ecol 2012:1–6
Cardoso E, Navarro R, Nogueira M (2002) Manganês e germinação de esporos de fungos micorrízicos arbusculares in vitro. Rev Bras Ciênc Solo 26(3):795–799
Cardoso E, Navarro R, Nogueira M (2003) Absorção e translocação de manganês por plantas de soja micorrizadas, sob doses crescentes deste nutriente. Rev Bras Ciênc Solo 27(3):415–423
Carneiro MAC, Siqueira JO, Moreira FMS (2001) Estabelecimento de plantas herbáceas em solo com contaminação de metais pesados e inoculação de fungos micorrízicos arbusculares. Pesquisa Agropecuária Brasileira 36(12):1443–1452
Castaldi P, Melis P, Silvetti M, Deiana P, Garau G (2009) Influence of pea and wheat growth on Pb, Cd, and Zn mobility and soil biological status in a polluted amended soil. Geoderma 151 (3):241–248
Chaudhary P, Dudeja S, Kapoor K (2004) Effectivity of host-*Rhizobium leguminosarum* symbiosis in soils receiving sewage water containing heavy metals. Microbiol Res 159(2):121–127
Collard J-M, Corbisier P, Diels L, Dong Q, Jeanthon C, Mergeay M, Taghavi S, Van Der Lelie D, Wilmotte A, Wuertz S (1994) Plasmids for heavy metal resistance in *Alcaligenes eutrophus* CH34: mechanisms and applications. FEMS Microbiol Rev 14(4):405–414
Comte S, Guibaud G, Baudu M (2008) Biosorption properties of extracellular polymeric substances (EPS) towards Cd, Cu and Pb for different pH values. J Hazard Mater 151(1):185–193
CONAMA (2009) Dispõe sobre critérios e valores orientadores de qualidade do solo quanto à presença de substâncias químicas e estabelece diretrizes para o gerenciamento ambiental de áreas contaminadas por essas substâncias em decorrência de atividades antrópicas. Resolução 420. Diário Oficial da União, Brasília
Cook WJ, Kar SR, Taylor KB, Hall LM (1998) Crystal structure of the cyanobacterial metallothionein repressor SmtB: a model for metalloregulatory proteins. J Mol Biol 275 (2):337–346
Cornejo P, Meier S, Borie G, Rillig MC, Borie F (2008) Glomalin-related soil protein in a Mediterranean ecosystem affected by a copper smelter and its contribution to Cu and Zn sequestration. Sci Tot Environ 406(1):154–160
Cunha KP, do Nascimento CW, Pimentel RM, Ferreira CP (2008) Cellular localization of cadmium and structural changes in maize plants grown on a cadmium contaminated soil with and without liming. J Hazard Mater 160(1):228–234
Del Val C, Barea J, Azcon-Aguilar C (1999) Diversity of arbuscular mycorrhizal fungus populations in heavy-metal-contaminated soils. Appl Environ Microbiol 65(2):718–723
Dias-Junior H, Moreira FMS, Siqueira J, Silva R (1998) Metais pesados, densidade e atividade microbiana em solo contaminado por rejeitos de indústria de zinco. Rev Bras Ciênc Solo 22(4)
Dœlsch E, Deroche B, Van de Kerchove V (2006) Impact of sewage sludge spreading on heavy metal speciation in tropical soils (Réunion, Indian Ocean). Chemosphere 65(2):286–293
Fernandes AR, de Paiva H, de Carvalho JG, de Miranda JRP (2007a) Crescimento e absorção de nutrientes por mudas de freijó (Cordia goeldiana Huber) em função de doses de fósforo e de zinco. Revista Árvore 31(4):599–608
Fernandes R, Luz W, Fontes M, Fontes L (2007b) Avaliação da concentração de metais pesados em áreas olerícolas no estado de Minas Gerais. Revista Brasileira de Engenharia Agricola e Ambiental 11(1):81–93
Ferreira PAA (2014) Tolerância de Cupriavidus necator a cádmio e zinco e sua eficiência simbiótica em leguminosas. Universidade Federal de Lavras, Lavras, Tese
Ferreira PAA, Bomfeti CA, Silva Júnior R, Soares BL, Soares CRFS, Souza Moreira FMS (2012) Eficiência simbiótica de estirpes de *Cupriavidus necator* tolerantes a zinco, cádmio, cobre e chumbo. Pesquisa Agropecuária Brasileira 47(1):85–95
Ferreira PAA, Lopes G, Bomfeti CA, de Oliveira Longatti SM, de Sousa Soares CRF, Guilherme LRG, de Souza Moreira FMS (2013) Leguminous plants nodulated by selected strains of

Cupriavidus necator grow in heavy metal contaminated soils amended with calcium silicate. World J Microbiol Biotechnol 29(11):2055–2066

Florentino LA, Guimarães AP, Rufini M, da Silva K, de Moreira FMS (2009) Sesbania virgata stimulates the occurrence of its microsymbiont in soils but does not inhibit microsymbionts of other species. Sci Agric 66(5):667–676

Fontes MPF, Gomes PC (2003) Simultaneous competitive adsorption of heavy metals by the mineral matrix of tropical soils. Appl Geochem 18(6):795–804

Fred EB, Waksman SA (1928) Laboratory manual of general microbiology. Book Company, New York

Gadd GM, White C (1993) Microbial treatment of metal pollution—a working biotechnology? Trends Biotechnol 11(8):353–359

Gattai G, Pereira S, Costa C, Lima C, Maia L (2011) Microbial activity, arbuscular mycorrhizal fungi and inoculation of wood plants in lead contaminated soil. Braz J Microbiol 42:859–867

Germaine KJ, McGuinness M, Dowling DN (2013) Improving phytoremediation through plant-associated bacteria. In: Brujin F (ed) Molecular microbial ecology of the rhizosphere: vol 1 and 2. Wiley, Hoboken, pp 961–973

Ghorbani N, Salehrastin N, Moeini A (2002) Heavy metals affect the microbial populations and their activities. Proc 17th World Congr. Soil Sci 2234:1–11

Giller KE, Witter E, Mcgrath SP (1998) Toxicity of heavy metals to microorganisms and microbial processes in agricultural soils: a review. Soil Biol Biochem 30(10):1389–1414

Gonçalves Júnior AC, Luchese EB, Lenzi E (2000) Avaliação da fitodisponibilidade de cádmio, chumbo e crômio, em soja cultivada em Latossolo Vermelho escuro tratado com fertilizantes comerciais. Quím Nova 23(2):173–177

Gonzalez-Chavez C, D'Haen J, Vangronsveld J, Dodd J (2002a) Copper sorption and accumulation by the extraradical mycelium of different Glomus spp. (arbuscular mycorrhizal fungi) isolated from the same polluted soil. Plant Soil 240(2):287–297

Gonzalez-Chavez M, Carrillo-Gonzalez R, Gutierrez-Castorena M (2009) Natural attenuation in a slag heap contaminated with cadmium: the role of plants and arbuscular mycorrhizal fungi. J Hazard Mater 161(2):1288–1298

Gonzalez-Chavez M, Carrillo-Gonzalez R, Wright S, Nichols K (2004) The role of glomalin, a protein produced by arbuscular mycorrhizal fungi, in sequestering potentially toxic elements. Environ Pollut 130(3):317–323

Gonzalez-Chavez C, Harris P, Dodd J, Meharg A (2002b) Arbuscular mycorrhizal fungi confer enhanced arsenate resistance on Holcus lanatus. New Phytol 155(1):163–171

Grazziotti P, Siqueira J, Moreira FMS, Carvalho D (2001) Efeito de Zn, Cd e Cu no comportamento de fungos ectomicorrízicos em meio de cultura. Rev Bras Ciênc Solo 25 (4):831–837

Hasan S, Hayat S, Ali B, Ahmad A (2008) 28-Homobrassinolide protects chickpea (*Cicer arietinum*) from cadmium toxicity by stimulating antioxidants. Environ Pollut 151(1):60–66

Huang S-W, Jin J-Y (2008) Status of heavy metals in agricultural soils as affected by different patterns of land use. Environ Monit Assess 139(1–3):317–327

Ibekwe A, Angle J, Chaney R, Van Berkum P (1995) Sewage sludge and heavy metal effects on nodulation and nitrogen fixation of legumes. J Environ Qual 24(6):1199–1204

Lins CEL, Cavalcante UMT, Sampaio EV, Messias AS, Maia LC (2006) Growth of mycorrhized seedlings of Leucaena leucocephala (Lam.) de Wit. in a copper contaminated soil. Appl Soil Ecol 31(3):181–185

Kabata-Pendias A (2011) Trace elements in soils and plants, 4th edn. CRC, Boca Raton

Kazy SK, Sar P, Singh S, Sen AK, D'Souza S (2002) Extracellular polysaccharides of a copper-sensitive and a copper-resistant *Pseudomonas aeruginosa* strain: synthesis, chemical nature and copper binding. World J Microbiol Biotechnol 18(6):583–588

Khairiah J, Lim K, Ahmad-Mahir R, Ismail B (2006) Heavy metals from agricultural soils from cameron highlands, pahang, and cheras, Kuala Lumpur, Malaysia. Bull Environ Contam Toxicol 77(4):608–615

Klauberg-Filho O, Siqueira J, Moreira FMS (2002) Fungos micorrízicos arbusculares em solos de área poluída com metais pesados. Rev Bras Ciênc Solo 26(1):125–134

Ladeira ACQ, Ciminelli VST, Nepomuceno AL (2002) Seleção de solos para a imobilização de arsênio. Rem Rev Esc Minas 55(3):215–221

Ledin M, Krantz-Rülcker C, Allard B (1996) Zn, Cd and Hg accumulation by microorganisms, organic and inorganic soil components in multi-compartment systems. Soil Biol Biochem 28 (6):791–799

Lovley DR, Widman PK, Woodward JC, Phillips E (1993) Reduction of uranium by cytochrome c3 of *Desulfovibrio vulgaris*. Appl Environ Microbiol 59(11):3572–3576

Macnair MR (1993) The genetics of metal tolerance in vascular plants. New Phytol 124 (4):541–559

Maftoun M, Moshiri F, Karimian N, Ronaghi A (2005) Effects of two organic wastes in combination with phosphorus on growth and chemical composition of spinach and soil properties. J Plant Nutr 27(9):1635–1651

Marques TS, Moreira FMS, Siqueira J (2000) Crescimento e teor de metais de muda de especies arboreas cultivadas em solo contaminado com metais pesados. Pesquisa Agropecuária Brasileira 35:121–132

Mårtensson A, Witter E (1990) Influence of various soil amendments on nitrogen-fixing soil microorganisms in a long-term field experiment, with special reference to sewage sludge. Soil Biol Biochem 22(7):977–982

Matsuda A, Moreira FMS, Siqueira J (2002a) Sobrevivência de *Bradyrhizobium* e *Azorhizobium* em misturas de solo contaminadas com metais pesados. Rev Bras Ciênc Solo 26(1):249–256

Matsuda A, Moreira FMS, Siqueira JO (2002b) Tolerância de rizóbios de diferentes procedências ao zinco, cobre e cádmio. Pesquisa Agropecuária Brasileira 37(3):343–355

McBride M (1989) Reactions controlling heavy metal solubility in soils. Advances in soil science. Springer, New York, pp 1–56

Meharg AA (2003) The mechanistic basis of interactions between mycorrhizal associations and toxic metal cations. Mycol Res 107(11):1253–1265

Meier S, Borie F, Bolan N, Cornejo P (2012) Phytoremediation of metal-polluted soils by arbuscular mycorrhizal fungi. Crit Rev Environ Sci Technol 42(7):741–775

Mergulhão A, Oliveira J, Burity H, Maia L (2007) Potencial de infectividade de fungos micorrízicos arbusculares em áreas nativas e impactadas por mineração gesseira no semiárido brasileiro. Hoehnea 34(3):341–348

Mishra VK, Upadhyay AR, Tripathi B (2009) Bioaccumulation of heavy metals and two organochlorine pesticides (DDT and BHC) in crops irrigated with secondary treated waste water. Environ Monit Assess 156(1–4):99–107

Moreira FMS, Cruz L, Miana de Faria S, Marsh T, Martínez-Romero E, Oliveira Pedrosa F, Maria Pitard R, Peter W, Young J (2006) *Azorhizobium doebereinerae* sp. Nov. Microsymbiont of *Sesbania virgata* (Caz.) Pers. Syst Appl Microbiol 29(3):197–206

Niu H, Xu XS, Wang JH, Volesky B (1993) Removal of lead from aqueous solutions by *Penicillium biomass*. Biotechnol Bioeng 42(6):785–787

Noctor G, Foyer CH (1998) Ascorbate and glutathione: keeping active oxygen under control. Annu Rev Plant Biol 49(1):249–279

Nogueira AV (1996) As micorrizas e o excesso de metais. In: Siqueira JO (ed) Avanços em fundamentos e aplicação de micorrizas. Universidade Federal de Lavras, Lavras, pp 135–174

Nogueira MA, Cardoso EJBN (2003) Mycorrhizal effectiveness and manganese toxicity in soybean as affected by soil type and endophyte. Sci Agric 60(2):329–335

Nogueira MA, Magalhães GC, Cardoso EJ (2004) Manganese toxicity in mycorrhizal and phosphorus-fertilized soybean plants. J Plant Nutr 27(1):141–156

Nogueira M, Nehls U, Hampp R, Poralla K, Cardoso E (2007) Mycorrhiza and soil bacteria influence extractable iron and manganese in soil and uptake by soybean. Plant Soil 298 (1–2):273–284

Nogueira TAR, Franco A, He Z, Braga VS, Firme LP, Abreu-Junior CH (2013) Short-term usage of sewage sludge as organic fertilizer to sugarcane in a tropical soil bears little threat of heavy metal contamination. J Environ Manage 114:168–177

Obbard J (2001) Ecotoxicological assessment of heavy metals in sewage sludge amended soils. Appl Geochem 16(11):1405–1411

Ortega-Larrocea M, Siebe C, Estrada A, Webster R (2007) Mycorrhizal inoculum potential of arbuscular mycorrhizal fungi in soils irrigated with wastewater for various lengths of time, as affected by heavy metals and available P. Appl Soil Ecol 37(1):129–138

Parkpain P, Sreesai S, Delaune R (2000) Bioavailability of heavy metals in sewage sludge-amended Thai soils. Water Air Soil Pollut 122(1–2):163–182

Parkpian P, Leong ST, Laortanakul P, Thunthaisong N (2003) Regional monitoring of lead and cadmium contamination in a tropical grazing land site, Thailand. Environ Monit Assess 85 (2):157–173

Pierangeli M, Guilherme L, Curi N, Anderson S, Lima J (2004) Adsorção e dessorção de cádmio, cobre e chumbo por amostras de Latossolos pré-tratadas com fósforo. Rev Bras Ciênc Solo 28 (2):377–384

Purchase D, Miles RJ, Young TW (1997) Cadmium uptake and nitrogen fixing ability in heavy-metal-resistant laboratory and field strains of *Rhizobium leguminosarum* biovar trifolii. FEMS Microbiol Ecol 22(1):85–93

Rangel OJP, Silva CA, Bettiol W, Dynia JF (2006) Efeito de aplicações de lodos de esgoto sobre os teores de metais pesados em folhas e grãos de milho. Rev Bras Ciênc Solo 30(3):583–594

de Rangel WM, Schneider J, de Costa ETS, Soares CRFS, Guilherme LRG, de Moreira FMS (2014) Phytoprotective effect of arbuscular mycorrhizal fungi species against arsenic toxicity in tropical leguminous species. Int J Phytoremediat 16:840–858

Raman N, Sambandan K (1998) Distribution of VAM fungi in tannery effluent polluted soils of Tamil Nadu, India. Bull Environ Contam Toxicol 60(1):42–150

Rashid A, Ayub N, Ahmad T, Gul J, Khan AG (2009) Phytoaccumulation prospects of cadmium and zinc by mycorrhizal plant species growing in industrially polluted soils. Environ Geochem Health 31(1):91–98

Rensing C, Sun Y, Mitra B, Rosen BP (1998) Pb (II)-translocating P-type ATPases. J Biol Chem 273(49):32614–32617

Rillig MC, Steinberg PD (2002) Glomalin production by an arbuscular mycorrhizal fungus: a mechanism of habitat modification? Soil Biol Biochem 34(9):1371–1374

Santos F, Amaral Sobrinho N, Mazur N (2003) Conseqüências do manejo do solo na distribuição de metais pesados em um agrossistema com feijão de vagem (Phaseolus vulgaris L.). Rev Bras Ciênc Solo 27(1):191–198

Santos JV, Melo Rangel W, Guimaraes AA, Jaramillo PMD, Rufini M, Marra LM, López MV, Silva MAP, Soares CRFS, Souza Moreira FMS (2013) Soil biological attributes in arsenic-contaminated gold mining sites after revegetation. Ecotoxicology 22(10):1526–1537

Schützendübel A, Polle A (2002) Plant responses to abiotic stresses: heavy metal-induced oxidative stress and protection by mycorrhization. J Exp Bot 53(372):1351–1365

Scott J, Palmer S (1990) Sites of cadmium uptake in bacteria used for biosorption. Appl Microbiol Biotechnol 33(2):221–225

Shahabivand S, Maivan HZ, Goltapeh EM, Sharifi M, Aliloo AA (2012) The effects of root endophyte and arbuscular mycorrhizal fungi on growth and cadmium accumulation in wheat under cadmium toxicity. Plant Physiol Biochem 60:53–58

Sharma RK, Agrawal M, Marshall FM (2009) Heavy metals in vegetables collected from production and market sites of a tropical urban area of India. Food Chem Toxicol 47(3):583–591

Silva GA, Trufem SFB, Júnior OJS, Maia LC (2005) Arbuscular mycorrhizal fungi in a semiarid copper mining area in Brazil. Mycorrhiza 15(1):47–53

da Silva G, Maia LC, da Silva F, Lima PCF (2001) Potencial de infectividade de fungos micorrízicos arbusculares oriundos de área de caatinga nativa e degradada por mineração, no Estado da Bahia, Brasil. Rev Bras Bot 24(2):135–143

Silva K, Florentino LA, Silva KB, Brandt E, Vandamme P, Souza Moreira FMS (2012) *Cupriavidus necator* isolates are able to fix nitrogen in symbiosis with different legume species. Syst Appl Microbiol 35(3):175–182

Silva LF, Fdez-Ortiz de Vallejuelo S, Martinez-Arkarazo I, Castro K, Oliveira ML, Sampaio CH, de Brum IA, de Leão FB, Taffarel SR, Madariaga JM (2013) Study of environmental pollution and mineralogical characterization of sediment rivers from Brazilian coal mining acid drainage. Sci Total Environ 447:169–178

de Silva MLS, Vitti GC, Trevizam AR (2007) Concentração de metais pesados em grãos de plantas cultivadas em solo com diferentes níveis de contaminação. Pesq agropec bras Brasília 42 (4):527–535

Silver S, Phung LT (1996) Bacterial heavy metal resistance: new surprises. Annu Rev Microbiol 50(1):753–789

Siqueira J, Pereira M, Simão J, Moreira FMS (1999a) Efeito da formononetina (7 Hidroxi, 4″ metoxi Isoflavona) na colonização micorrízica e crescimento do milho em solo contendo excesso de metais pesados. Rev Bras Ciênc Solo 23(3):561–567

Siqueira J, Pouyú E, Moreira FMS (1999b) Micorrizas arbusculares no crescimento póstransplantio de mudas de árvores em solo com excesso de metais pesados. Rev Bras Ciênc Solo 23:569–580

Soares C, Siqueira JO, de Carvalho JG, Guilherme LRG (2006) Micorriza arbuscular e nutrição fosfática na toxidez de zinco para a trema [Trema micrantha (L.) Blum]. Rev Bras Ciênc Solo 30(4):665–675

Soares PSM, Borma LDS (2002) Drenagem ácida e gestão de resíduos sólidos de mineração. In: Trindade R, Barbosa Filho O (eds) Extração de Ouro: Princípios, Tecnologia e Meio Ambiente. Centro de Tecnologia Mineral—CETEM, Ministério da Ciência e Tecnologia, Rio de Janeiro, pp 242–266

Soumare M, Tack F, Verloo M (2003) Distribution and availability of iron, manganese, zinc, and copper in four tropical agricultural soils. Commun Soil Sci Plant Anal 34(7–8):1023–1038

Thawornchaisit U, Polprasert C (2009) Evaluation of phosphate fertilizers for the stabilization of cadmium in highly contaminated soils. J Hazard Mater 165(1):1109–1113

Trannin I, FMS M, Siqueira J, Lima A (2001a) Tolerância de estirpes e isolados de *Bradyrhizobium* e de *Azorhizobium* a zinco, cádmio e cobre *in vitro*. Rev Bras Ciênc Solo 25 (2):305–316

Trannin I, Moreira FMS, Siqueira J (2001b) Crescimento e nodulação de *Acacia mangium*, *Enterolobium contortisiliquum* e *Sesbania virgata* em solo contaminado com metais pesados. Rev Bras Ciênc Solo 25(3):743–753

Uchimiya M, Wartelle LH, Boddu VM (2012) Sorption of triazine and organophosphorus pesticides on soil and biochar. J Agric Food Chem 60(12):2989–2997

Udom B, Mbagwu J, Adesodun J, Agbim N (2004) Distributions of zinc, copper, cadmium and lead in a tropical ultisol after long-term disposal of sewage sludge. Environ Int 30(4):467–470

Vincent JM (1970) A manual for the practical study of root-nodule bactéria, vol 15, International biological programme handbook. Blackwell Scientific Publications, Oxford

Weissenhorn I, Glashoff A, Leyval C, Berthelin J (1994) Differential tolerance to Cd and Zn of arbuscular mycorrhizal (AM) fungal spores isolated from heavy metal-polluted and unpolluted soils. Plant and Soil 167(2):189–196

Zarei M, Hempel S, Wubet T, Schäfer T, Savaghebi G, Jouzani GS, Nekouei MK, Buscot F (2010) Molecular diversity of arbuscular mycorrhizal fungi in relation to soil chemical properties and heavy metal contamination. Environ Pollut 158(8):2757–2765

Zarei M, König S, Hempel S, Nekouei MK, Savaghebi G, Buscot F (2008a) Community structure of arbuscular mycorrhizal fungi associated to *Veronica rechingeri* at the Anguran zinc and lead mining region. Environ Pollut 156(3):1277–1283

Zarei M, Saleh-Rastin N, Jouzani GS, Savaghebi G, Buscot F (2008b) Arbuscular mycorrhizal abundance in contaminated soils around a zinc and lead deposit. Eur J Soil Biol 44(4):381–391

Chapter 13
Functional Diversity of Microorganisms in Heavy Metal-Polluted Soils

Anelia Kenarova and Silvena Boteva

13.1 Introduction

Soil is a major component of ecosystems that contributes to the primary productivity of natural and agricultural ecosystems through nutrient cycling (Zak et al. 2003), decomposition of dead organic matter (Wardle et al. 2004), and energy flow (Preston et al. 2001). High microbial diversity insures ecosystems against declines in function that occur under environmental fluctuations, mainly extreme or rapidly changing environmental conditions, and under human activities causing different types and levels of pollution. Because soil microorganisms are the most sensitive component of the soil ecosystem, measurements of their community parameters have often been used for soil quality evaluation (Hofman et al. 2004) and for environmental monitoring at contaminated sites (Pennanen 2001). Microbial communities have the ability to adapt to changing environments in different ways, such as modification of individual activity, increased reproduction of species with favorable abilities, or by developing new capabilities via horizontal gene transfer, which makes them highly responsive to environmental changes and stress (Winding et al. 2005).

Among the widespread and persistent soil pollutants are heavy metals (HMs), which originate from agricultural (fertilizers and contaminated water for irrigation) or industrial activities (metal mining and smelting). Most of the analyzed effects of pollution are from long-term HM-polluted areas impacted by mining and tailing activities (Niklińska et al. 2006; Kenarova et al. 2014), deposition of industrial wastes (Ellis et al. 2002, 2003), and sewage application as manure (Brandt et al. 2010; Bérard et al. 2014).

A. Kenarova (✉) • S. Boteva
Department of Ecology and Environmental Protection, Faculty of Biology, Sofia University "St. Kl. Ohridski", 8 Dragan Tsankov Blvd., Sofia 1164, Bulgaria
e-mail: akenarova@abv.bg; silvenab@abv.bg

I. Sherameti, A. Varma (eds.), *Heavy Metal Contamination of Soils*, Soil Biology 44, DOI 10.1007/978-3-319-14526-6_13

Many studies showed that HMs, with different concentrations and speciation, have negative impacts on microbial abundance/biomass (Wang et al. 2007), enzyme activities (Kenarova and Rradeva 2010; Yang et al. 2007), and microbial diversity (Gremion et al. 2004; Lewis et al. 2010; Ellis et al. 2003), which can lead to serious threats to the normal function of the ecosystem and long-term soil productivity (Dai et al. 2004). The effects of HM pollution are complex and strongly dependent on local geographic and climatic factors, soil characteristics, and origin of the impacted microorganisms.

This chapter focuses on the end effects of HM pollution on soil community-level physiological profiles (CLPPs) as an indicator for microbial community function under HM stress.

13.2 Methods for Studying the Functional Diversity of Soil Microorganisms

Carbon is a key factor governing microbial growth in soil, and the functional aspects of diversity related to substrate use are extremely important for long-term stability of ecosystems. One of the functional approaches that has gained increasing importance in soil microbial ecology is community-level physiological profiling based on sole-carbon-source use by Biolog (Garland and Mills 1991; Insam 1997) or MicroResp (Campbell et al. 2003) assays. The difference between Biolog and MicroResp techniques is in the principle of carbon (C) source use rate measurement using soil cell extracts (Biolog) or intact soil samples (MicroResp), profiling the catabolic activity and functional diversity of bacteria (Biolog GN and Eco plates), fungi (Biolog SF plates), and the whole microbial community (MicroResp). The Biolog GN plate (95 C sources) was initially designed to identify medical bacterial isolates (Bochner 1989) and has since been adapted for environmental purposes (Garland and Mills 1991). The Biolog Ecoplate (31 C sources each with three replicates) was developed by Insam (1997) for soil analysis containing substrates that are known to be plant root exudates or that have previously been found to have a high discriminatory power among soil communities. Campbell et al. (2003) applied the MicroResp technique in an effort to overcome some drawbacks, which are inherent in the cultivation-dependent technique of Biolog (Haack et al. 1995; Campbell et al. 1997). MicroResp combines the advantages of substrate-induced respiration (Degens and Harris 1997) and Biolog (Garland and Mills 1991; Insam 1997) assays.

In general, Biolog plate design consists of multiple C substances, each contained in a separate well, to which a minimal growth medium and tetrazolium violet are added. After inoculation with soil extracts and incubation, microbial catabolic activity is detected spectrophotometrically based on the color development of tetrazolium violet, which turns purple in the presence of electron transfer. The MicroResp system consists of a microplate with deep wells where the soil samples and C sources are placed and a top microplate where CO_2 emitted from the catabolism of C sources is trapped by absorption in agar containing a pH indicator dye. The rate of microbial activity is detected by spectrophotometric reading to determine the amount of CO_2 trapped in the agar plate.

Catabolic activity and/or functional diversity of bacterial communities under different environmental conditions are calculated based on the number, types, and rates at which the C sources in the plates are used by the heterotrophic members of the microbial communities.

The limitations of the Biolog assay are similar to those for culturable methods, as the results are biased by the inoculum density (Garland and Mills 1991), the time (Garland and Mills 1991; Haack et al. 1995), and the temperature (Haack et al. 1995) of incubation, and are dominated by soil heterotrophic bacteria that are easily extractable (Balser 2000) and grow rapidly in a nutrient-rich environment (Heuer and Smalla 1997; Konopka et al. 1998). Soil and inoculum pretreatments, such as sieving and storage of soil samples (Shishido and Chanway 1998; Goberna et al. 2005) and filtering or sedimentation of inocula (Hollibaugh 1994), are also reflected in Biolog results. Finally, Biolog GN and Ecoplates, the most applicable for environmental studies, cannot determine the metabolic activity of fungi (Buyer et al. 2001; Nannipieri et al. 2003) and some bacteria, which cannot reduce the redox dye (Winding and Hendriksen 1997).

Several approaches have been used to avoid or minimize the abovementioned biases, such as the design of an optimal sampling scheme (Balser 2000), standardization of the Biolog procedure (including cell extraction), standardization of inoculum density (Garland and Mills 1991; Insam and Goberna 2004), standardization of the time and temperature of incubation, and data management (Haack et al. 1995; Heuer and Smalla 1997; Konopka et al. 1998; Insam and Goberna 2004). The MicroResp technique overcomes most of the abovementioned drawbacks, but its use is restricted mainly by cost and, in the case of radio-labeled substrate use, the difficulties associated with management and disposal of the radioactive waste.

13.3 Heavy Metal Pollution and Functional Diversity of Soil Microorganisms

Most studies focus on bacteria rather than fungi as indicators of HM stress, neglecting the latter because of their well-known resistance to different types of pollutants and the high versatility of their metabolic pathways, widening the capacity of fungi to metabolize different carbon sources (Nannipieri et al. 2003). According to Bååth and Díaz-Raviña (2005), the metabolic activity of bacteria in soil is dominated by the culturable component of the community, and hence they are more severely impacted by anthropogenic stresses than the whole microbial community (Ellis et al. 2003). Cultivable bacteria represent an ecologically relevant portion of the soil bacterial community and are the most appropriate component to measure the alterations in "soil health" (Pascaud et al. 2012).

13.3.1 Catabolic Activity of Microorganisms

CLPP (Biolog and MicroResp) is an indicator of microbial catabolic activity of the whole community and diversity in the use rates of individual carbon sources provided in the microplates. The rate of use of the set of C sources is measured using the average well color development (AWCD), the average value of the optical density of wells in each plate. The amount of each substrate contributing to the AWCD varies depending on the amount of substrate used, which is governed by their availability to the local microbial community. Thus, the value of AWCD may be determined by the use rates of few, some, or most of the C sources in the set.

The AWCD was used along with other microbial indicators, and relationships were found between the levels of HM pollution and the rates of bacterial catabolic activity (Epelde et al. 2008; Kong et al. 2006; Ellis et al. 2002; Kenarova et al. 2014; Gremion et al. 2004; Wang et al. 2012). Ellis et al. (2002) clustered the highest (>0.9) and lowest (<0.15) AWCDs into two different clusters associated with low and high rates of HM pollution in soils, respectively. Assessing the impacts of uranium mining activities on soil bacterial communities, Kenarova et al. (2014) distinguished the most polluted environment of a mine waste pile with an AWCD of 0.076 from the unpolluted one with an AWCD of 1.28.

While some studies have found a linear relationship between the level of HM pollution and the rate of bacterial catabolic activity, several other studies (Yuangen et al. 2006; Jusselme et al. 2012) did not find any difference in the AWCDs between the different levels of HM-polluted soil samples. These contrasting results might occur because some studies used soils with low levels of HM pollution (Yuangen et al. 2006). In other studies, the existence of compensation mechanisms is possible, including the metabolic shift of stressed microbial communities from a low and stable energy level in uncontaminated soils to an unstable and more energy-demanding state in heavily polluted soils (Yuangen et al. 2006; Wang et al. 2012; Masto et al. 2011).

13.3.2 Functional Diversity of Soil Microorganisms

An approach that is often used to assess HM toxicity is to calculate the functional diversity of impacted microbial communities using the Shannon-Weaver diversity index (H') based on CLPP. One of the underlying premises of the CLPP approach is that a functionally diverse microbial community is able to break down a broad range of organic molecules that vary in structural complexity at more similar rates compared with a less functionally diverse community (Degens et al. 2000). The diversity index of Shannon-Weaver consists of two components, richness (the number of the carbon sources used) and evenness (*E*, the use rates of substrates), having different weights in the formula in different cases. It was reported that different traits in bacterial metabolic diversity were dependent on the level and speciation of HM pollution. In general, there was a lack of adverse effects of HMs on bacterial functional diversity in the case of low to moderate soil pollution (Epelde et al. 2008; Wang et al. 2010, 2012; Kenarova et al. 2014) and strong reduction in the diversity/richness in the case of high HM pollution (Gremion et al. 2004; Dobler et al. 2000; Kong et al. 2006). In contrast, Avidano et al. (2005) and Jusselme et al. (2012) reported increases in the bacterial functional diversity/richness with increasing levels of HM pollution. For instance, the number of C sources used by bacteria (Biolog) increased with increasing soil concentrations of Cu, As and Pb (Avidano et al. 2005), and Pb (Jusselme et al. 2012) from 53 to 73 and from 72 to 82, respectively. This increase in functional diversity may be explained by the stimulatory effects of low concentrations of bioavailable forms of some HMs on bacterial growth (Zeng et al. 2007) or by an increase in the weight of *E* in the Shannon-Weaver formula (Kong et al. 2006; Kenarova et al. 2014). For instance, when analyzing a soil-Pb-rice system, Zeng et al. (2007) reported the stimulatory effects of low Pb concentrations on microbial growth expressed as microbial carbon content (C_{mic}). Additionally, Yan et al. (2000) found that functional diversity (Biolog) increased with increasing C_{mic} up to a threshold of 105.6 μg g^{-1} and remained constant above this value. Chodak et al. (2009) reported an even lower threshold value of C_{mic} (35 μg g^{-1}) for reclaimed mine soils afforested with pine. In laboratory experiments, Kong et al. (2006) recorded a dramatic decrease in functional diversity and evenness at 100 μM Cu, and stabilization of H' and an increase in *E* at 300 μM Cu pollution of soil samples. An increase in substrate evenness by increasing the level of HM pollution was also recorded by Kenarova et al. (2014). The substrate evenness results from shifts in the catabolism occurring in impacted microbial communities, where some C sources may show a reduced use rate and others will show an increased use rate at the same time.

13.3.3 Use of Biochemical Substrate Categories

C sources in the microplates were classified into several biochemical categories (guilds): carbohydrates, carboxylic acids, amino acids, amines/amides, polymers (Garland and Mills 1991; Insam 1997), phenolics (Insam 1997), and miscellaneous compounds (Garland and Mills 1991).

According to some studies, the catabolism of carbohydrates, polymers, and carboxylic acids were highly sensitive to HM toxicity (Tischner et al. 2008; Kong et al. 2006; Muñiz et al. 2014; Yuangen et al. 2006; Bérard et al. 2014; Campbell et al. 2003; Banning et al. 2012; Lalor et al. 2007). In most studies using Biolog, reductions in the use rates of amino acids and amines were reported, although they were smaller than the reductions observed for other substrate guilds (Boivin et al. 2006; Kong et al. 2006; Epelde et al. 2008; Chodak et al. 2009). In contrast, in studies using samples with low levels of HM pollution (Yuangen et al. 2006) or using the MicroResp technique (Bérard et al. 2014; Campbell et al. 2003; Banning et al. 2012; Lalor et al. 2007), shifts from carbohydrate use to preferential use of amino acids were observed. These contradictory results obtained using MicroResp are likely because of the contribution of soil fungi to the metabolic profile of microbial communities, especially in the case of high HM pollution where the proportion of fungi in soil microbial communities increases because of their HM resistance (Khan and Scullion 2000, 2002; Müller et al. 2001; Rajapaksha et al. 2004).

HMs stimulate the use of some C sources, such as amino acids (L-serine (Muñiz et al. 2014; Epelde et al. 2008), L-phenylalanine (Muñiz et al. 2014), and L-threonine (Epelde et al. 2008)), phenolic acids (4-hydroxybenzoic acid (Muñiz et al. 2014; Banning et al. 2012; Kenarova et al. 2014) and 2-hydroxybenzoic acid (Kenarova et al. 2014)), carboxylic acids (α-ketobutyric acid and D-malic acid (Banning et al. 2012; Epelde et al. 2008; Muñiz et al. 2014)), amines (*N*-acetyl-D-glucosamine and D-glucosaminic acid (Epelde et al. 2008; Muñiz et al. 2014)), sugar alcohols (glycerol (Tischner et al. 2008) and D-mannitol (Muñiz et al. 2014; Kenarova et al. 2014)), and carbohydrate (D-cellobiose (Epelde et al. 2008)). According to Banning et al. (2012) and Degens et al. (2000), the observed shift in microbial catabolic structure from the use of carbohydrates to amino acids/amines reflects differences in the profile and bioavailability of C sources in contaminated soils and/or differences in the microbial community composition of these soils. Metal stress induces several cellular mechanisms for detoxification in microorganisms, involving proteins and amino acids—macromolecules that can later be metabolized by the same microbiota (Hall 2002). Plants were also stressed by metal contamination and may exude, for example, protein metal chelators via their roots (Sharma and Dietz 2006; Hall 2002). Both phenomena could select microbial species that are more competent in the use of amino acids than other C sources under metal stress.

13.3.4 Pollution-Induced Community Tolerance

There are many local environmental factors at the site of pollution, which may modify the effects of HMs and the final response of microorganisms to the metal stress. To overcome the problem of confounding effects of factors other than toxicants, the measurement of pollution-induced community tolerance (PICT) was proposed (Blanck et al. 1988) and used in several studies (Niklińska et al. 2006; Stefanowicz et al. 2009; van Beelen et al. 2004; Boivin et al. 2006). The PICT concept is based on the assumption that at contaminated sites the proportion of species tolerant to the pollutant(s) increases, resulting in an increased average tolerance of the community (Blanck 2002). Thus, the agent/agents causing the disturbance can be identified because induced tolerance will only be observed for those pollutants that have exerted selection pressure on the community. PICTs based on CLPPs were analyzed for several metal pollutants, including Zn (van Beelen et al. 2004; Stefanowicz et al. 2009; Niklińska et al. 2006; Boivin et al. 2006), Cu (van Beelen et al. 2004; Stefanowicz et al. 2009; Brandt et al. 2010; Niklińska et al. 2006; Boivin et al. 2006), Cr (van Beelen et al. 2004), Ni (van Beelen et al. 2004; Stefanowicz et al. 2009), Cd (Niklińska et al. 2004), and Pb (Stefanowicz et al. 2009; Boivin et al. 2006). Although the appearance of PICT in bacterial communities due to metal pollution is well documented (Stefanowicz et al. 2009; van Beelen et al. 2004), particular metals may have no effect on the level of bacterial tolerance (Gong et al. 2002; Boivin et al. 2006; Stefanowicz et al. 2009). One reason for the lack of PICT might be low levels of soil pollution, where the metal tolerance of bacteria cannot be developed. Additionally, it was shown that even at high concentrations of Cu, PICT was not developed (van Beelen et al. 2004; Boivin et al. 2006), indicating a limited bioavailability of this metal in soils (Mebride et al. 1997). According to van Beelen et al. (2004), PICT can distinguish between the toxic effects of Cr^{6+} and Cr^{3+}, observing Cr^{6+} tolerance at 923 mg kg^{-1} and zero tolerance to Cr^{3+}, although it is the predominant form of the metal in chromium-polluted soils (Rudel et al. 2001). On the contrary, Gong et al. (2002) found that bacterial community tolerance altered and increased as soil Cr concentrations increased, regardless of the form of the metal. The most commonly recorded PICTs are for Zn and Pb, which are highly labile and soluble in soils (Izquierdo et al. 2013). Izquierdo et al. (2013) determined the labile pools of Zn, Pb, and Cd in contaminated alluvial soils, confirming the high solubility of Zn (25–594 mg kg^{-1}) and Pb (11–350 mg kg^{-1}), compared with Cd (0.5–14 mg kg^{-1}). In most of the cases when PICT was not found, the differences in metal concentrations between the polluted and unpolluted sites were not large enough to allow detection of a community tolerance in the polluted soils (Stefanowicz et al. 2009; Boivin et al. 2006).

Some studies used the Biolog microplate to determine the PICT for every C source from the microplate. Stefanowicz et al. (2009), van Beelen et al. (2004), and Niklińska et al. (2006) recorded an increased bacterial tolerance to HMs on D-mannitol (Cu, Zn, Cr^{6+}), i-erythritol (Pb, Zn, Cu), Tween 40/80, L-asparagine and L-serine (Zn, Ni), 4-hydroxybenzoic acid and itaconic acid (Zn, Cu), L-phenylalanine and *N*-acetyl-D-glucosamine (Pb, Zn, Cr^{6+}), putrescine (Ni, Cr^{6+}), D-cellobiose (Cu, Cr^{6+}), α-cyclodextrin (Pb, Zn), D-xylose (Zn, Cr^{6+}), D-galacturonic acid (Pb, Cu, Ni, Cr^{6+}), D-malic acid (Cu, Pb), and α-D-lactose (Pb, Cr^{6+}). There are metal-specific C sources that induce higher bacterial tolerance to HMs, such as glycogen, γ-hydroxybutyric acid and glycyl-L-glutamic acid (Zn), 2-hydroxybenzoic acid and L-threonine (Ni), pyruvic acid methyl ester (Cu), and β-methyl-D-glucoside, D-glucosaminic acid, and D-galacturonic acid γ-lactone (Cr^{6+}). Most of the C sources found to develop high PICT under HM stress were mentioned above as highly available substrates for metal-impacted bacteria (Muñiz et al. 2014; Epelde et al. 2008; Banning et al. 2012; Kenarova et al. 2014).

The metal stress imposed on soil microbial communities, resulting in an increased tolerance (PICT), does not necessarily lead to a decrease in metabolic diversity (Davis et al. 2004). Davis et al. (2004) indicated that accumulation of Zn up to 2,000 mg kg^{-1} did not decrease the metabolic diversity in the culturable fraction of the microbial community although the Zn-PICT was detected at a soil Zn concentration of approximately 300 mg kg^{-1}.

13.4 Environmental Effects on Microbial Functional Diversity in Metal-Polluted Soils

Soil properties (pH, texture, humidity, vegetation) were found to have more significant effects on microbial structural and functional diversity than HM contamination (Gremion et al. 2004; Li et al. 2011; Jusselme et al. 2012; Niklińska et al. 2005; Muñiz et al. 2014; Bérard et al. 2014; Yang et al. 2007).Soils could affect microbial communities directly, by providing habitat for selection of specific microorganisms, or indirectly, by affecting plant root exudation (Garbeva et al. 2004).

Soil pH is one of the most important environmental factors influencing bacterial growth and activity (Niklińska et al. 2005), availability of nutrients (Dakora and Phillips 2002), and solubility of HMs (Hornburg and Brümmer 1993). The toxicity of HMs is related to their water-soluble (bioavailable) concentrations (Zhang et al. 2001), although many studies have found relationships between HMs and bacterial diversity for the total concentrations, rather than the bioavailable concentrations of HMs (Davis et al. 2004; van Beelen et al. 2004; Li et al. 2011). This suggests a significant influence of habitat change on microbial community structure and function under the impact of HM pollution.

Vegetation at HM-polluted soils has a positive effect on substrate use. In comparison with unplanted soils, AWCD, functional diversity (Li et al. 2011), and the number of substrates used (57 vs. 19) were higher in planted soil samples (Gremion et al. 2004). Li et al. (2011) found that *Erigeron annuus*, *Lysimachia clethroides*, *Pueraria lobata*, *Imperata cylindrica*, and *Equisetum ramosissimum* differentially affected the structure and function of the rhizosphere bacterial communities at a mine tailings site, suggesting different composition of the root exudates and litter, which varied among plant species. Complex plant communities may produce higher diversity of root exudates and therefore select for complex microbial communities, with increased functional diversity when compared with monoculture plantings (Yang et al. 2007).

It is clear that different plants develop beneficial rhizobacteria by the release of specific sugars, carboxylic acids, and amino acids into the rhizosphere (Kowalchuk et al. 2002). The adaptation of soil microorganisms to plant exudates depends not only on their origin but also on the extent of microbial exposure (Banning et al. 2012; Kenarova et al. 2013). Different root exudates have different depth distribution patterns in soils (Bölter 1990; Buyer and Drinkwater 1997; Buyer et al. 2001), hence different availability for soil microorganisms. For instance, Bölter (1990) and Buyer and Drinkwater (1997) found low concentrations of carbohydrates in the top soil (7–10 times lower than that of free amino acids), and Buyer et al. (2001) reported that carbohydrates migrate easily to the deeper soil layers and that their content in the soil surface layer is very low to negligible. According to Buyer et al. (2001), the distribution of proteins through the soil layers depends on the type of environment; amino acids were located in the topsoil of vegetated environments and migrated into the deeper layers of bare soil, making them unavailable for bacteria from the topsoil. Van Hees et al. (2003) found that concentrations of carboxylic acids in forest soils were higher in the presence of ectomycorrhizal fungi, and the deforestation of mine areas decreased the fungal abundance and activity, followed by a decrease in bacterial exposure to carboxylic acids.

Similar to root exudation, the presence of earthworms increases the growth and activity of soil microbial communities. It was shown in a Pb-polluted soil pot experiment that the presence of earthworms significantly increased the biomass, metabolic activity (except carbohydrates), and functional diversity of bacterial communities but did not change the community structure (Jusselme et al. 2012). Muñiz et al. (2014) highlighted the modulating effects of earthworms in Cd-polluted soil (EwCd), reporting a shift in the bacterial CLPP compared with the Cd-polluted soil without earthworms (Cd). The EwCd soil pot was characterized by faster degradation of D-cellobiose, α-cyclodextrin, and β-methyl-D-glucoside, in contrast to the faster degradation of *N*-acetyl-D-glucosamine and D-xylose in the Cd soil pot. Earthworms were more important in explaining time course changes in bacterial functional diversity than Cd concentration (320 mg g^{-1}) in the experiment of Muñiz et al. (2014).

13.5 Conclusion

The commercially available assays based on substrate use patterns of microbial communities are fast, reliable, easy-to-use techniques that can provide insight into the physiological ecology of microbial communities in metal-polluted environments. The toxicity of HMs to microbial catabolic activity and functional diversity, especially in highly polluted soils, can be measured based on the changes in CLPPs and PICTs. Hence, various aspects of CLPPs and PICTs may be used in soil quality evaluation and environmental risk assessment.

References

Avidano L, Gamalero E, Cossa GP, Carraro E (2005) Characterization of soil health in an Italian polluted site by using microorganisms as bioindicators. Appl Soil Ecol 30:21–33

Bååth E, Díaz-Raviña BLR (2005) Microbial biomass, community structure and metal tolerance of a naturally Pb-enriched forest soil. Microb Ecol 50:496–505

Balser TC (2000) Linking soil microbial communities and ecosystem functioning. PhD dissertation. University of California-Berkeley, Berkeley, CA
Banning NC, Lalor BM, Cookson WR, Grigg AH, Murphy DV (2012) Analysis of soil microbial community level physiological profiles in native and post-mining rehabilitation forest: which substrates discriminate? Appl Soil Ecol 56:27–34
Bérard A, Mazzia C, Sappin-Didier V, Capowiez L (2014) Use of the MicroResp™ method to assess pollution-induced community tolerance in the context of metal soil contamination. Ecol Indicat 40:27–33
Blanck H (2002) A critical review of procedures and approaches used for assessing pollution-induced community tolerance (PICT) in biotic communities. Hum Ecol Risk Assess 8:1003–1034
Blanck H, Wängberg S-Å, Molander S (1988) Pollution-induced community tolerance—a new ecotoxicological tool. In: Cairns J, Pratt JR (eds) Functional Testing of Aquatic Biota for Estimating Hazards of Chemicals. American Society for Testing Materials, Philadelphia PA, pp 219–230
Bochner BR (1989) Sleuthing out bacterial identities. Nature 339:157–158
Boivin MEY, Greve GD, Kools SAE, van der Wurff AWG, Leeflang P, Smit E, Breure AM, Rutgers M, van Straalen NM (2006) Discriminating between effects of metals and natural variables in terrestrial bacterial communities. Appl Soil Ecol 34:103–113
Bölter M (1990) Microbial ecology of soil from Wilkes Land Antarctica: 1. The bacterial population ant its activity in relation to dissolved organic matter. Proc NIPR Symp Polar Biol 3:104–119
Brandt KK, Frandsen RJN, Holm PE, Nybroe O (2010) Development of pollution-induced community tolerance is linked to structural and functional resilience of a soil bacterial community following a five-year field exposure to copper. Soil Biol Biochem 42:748–757
Buyer JS, Drinkwater LE (1997) Comparison of substrate utilization assay and fatty acid analysis of soil microbial communities. J Microbiol Methods 30:3–11
Buyer JS, Roberts DP, Millner P, Russek-Cohen E (2001) Analysis of fungal communities by sole carbon source utilization profiles. J Microbiol Methods 45:53–60
Campbell CD, Chapman SJ, Cameron CM, Davidson MS, Potts JM (2003) A rapid microtiter plate method to measure carbon dioxide evolved from carbon amendments so as to determine the physiological profiles of soil microbial communities by using whole soil. Appl Environ Microbiol 69:3593–3599
Campbell CD, Grayston SJ, Hirst D (1997) Use of rhizosphere C sources in sole C sources to discriminate soil microbial communities. J Microbiol Methods 30:33–41
Chodak M, Pietrzykowski M, Niklińska M (2009) Development of microbial properties in a chronosequence of sandy mine soils. Appl Soil Ecol 41:259–268
Dai J, Becquer T, Rouiller JH, Reversat G, Bernhard-Reversat F, Lavelle P (2004) Influence of heavy metals on C and N mineralisation and microbial biomass in Zn-, Pb-, Cu-, and Cd-contaminated soils. Appl Soil Ecol 25:99–109
Dakora FD, Phillips DA (2002) Root exudates as mediators of mineral acquisition in low-nutrient environments. Plant Soil 245:35–47
Davis MRH, Zhao F-J, McGrath SP (2004) Pollution-induced community tolerance of soil microbes in response to a zinc gradient. Environ Toxicol Chem 23:2665–2672
Degens BP, Harris JA (1997) Developments of a physiological approach to measuring the catabolic diversity of soil microbial communities. Soil Biol Biochem 29:1309–1320
Degens BP, Schipper LA, Sparling GP, Vojvodic-Vukovic M (2000) Decrease in organic C reserves in soils can reduce the catabolic diversity of soil microbial communities. Soil Biol Biochem 32:189–196
Dobler R, Saner M, Bachofen R (2000) Population changes of soil microbial communities induced by hydrocarbon and heavy metal contamination. Biorem J 4:41–56

Ellis RJ, Best JG, Fry JC, Morgan P, Neish B, Trett MW, Weightman AJ (2002) Similarity of microbial and meiofaunal community analyses for mapping ecological effects of heavy-metal contamination in soil. FEMS Microbiol Ecol 20:113–122
Ellis RJ, Morgan P, Weightman AJ, Fry JC (2003) Cultivation-dependent and –independent approaches for determining bacterial diversity in heavy-metal-contaminated soil. Appl Environ Microbiol 69:3223–3230
Epelde L, Becerril JM, Hernández-Allisa J, Barrutia O, Garbisu C (2008) Functional diversity as indicator of the recovery of soil health derived from *Thlaspi caerulescens* growth and metal phytoextraction. Appl Soil Ecol 39:299–310
Garbeva P, van Veen JA, van Elsas JD (2004) Microbial diversity in soil: selection of microbial populations by plant and soil type and implications for disease suppressiveness. Annu Rev Phytopathol 42:243–270
Garland JL, Mills AL (1991) Classification and characterization of heterotrophic microbial communities on the basis of patterns of community-level sole-carbon-source utilization. Appl Environ Microbiol 57:2351–2359
Goberna M, Insam H, Klammer S, Pascual JA, Sanchez J (2005) Microbial community structure at different depths in disturbed and undisturbed semiarid Mediterranean forest soils. Microb Ecol 50:315–326
Gong P, Siciliano SD, Srivastava S, Greer CW, Sunahara GI (2002) Assessment of pollution-induced microbial community tolerance to heavy metals in soil using ammonia-oxidizing bacteria and Biolog assay. Hum Ecol Risk Assess 8:1067–1081
Gremion F, Chatzinotas A, Kaufmann K, von Sigler W, Harms H (2004) Impacts of heavy metal contamination and phytoremediation on a microbial community during a twelve-month microcosm experiment. FEMS Microbiol Ecol 48:273–283
Haack SK, Garchow H, Klug MJ, Forney LJ (1995) Analysis of factors affecting the accuracy, reproducibility, and interpretation of microbial community carbon source utilization patterns. Appl Environ Microbiol 61:1458–1468
Hall JL (2002) Cellular mechanisms for heavy metal detoxification and tolerance. J Exp Bot 53:1–11
Heuer H, Smalla K (1997) Evaluation of community-level catabolic profiling using BIOLOG GN microplates to study microbial community changes in potato phyllosphere. J Microbiol Methods 30:49–61
Hollibaugh JT (1994) Relationship between thymidine metabolism, bacterioplankton community metabolic capacities, and source of organic matter. Microb Ecol 28:117–131
Hofman J, Dušek K, Klánová J, Bezchlebová J, Holoubek I (2004) Monitoring microbial biomass and respiration in different soils from the Czech Republic—a summary of results. Environ Int 30:19–30
Hornburg V, Brümmer GW (1993) Verhalten von schwermetallen in böden. 1. Untersuchungen zur schwermetallmobilität. Z Pflanz Bodenkunde 156:467–477
Insam H (1997) A new set of substrates proposed for community characterization in environmental samples. In: Insam H, Ranger A (eds) Microbial communities. Functional versus structural approaches. Springer, Berlin, pp 259–260
Insam H, Goberna M (2004) Community level physiological profiles (Biolog substrate use tests) of environmental samples. In: Akkermans ADL, van Elsas JD, DeBruijn FJ (eds) Molecular Microbial Ecology Manual, 2nd edn. Kluwer Academic Publ, Dordrecht, pp 1–8
Izquierdo M, Tye AM, Chenery SR (2013) Lability, solubility and speciation of Cd, Pb and Zn in alluvial soils of the River Trent catchment UK. Environ Sci Process Impacts 15:1844–1858
Jusselme MD, Poly FP, Miambi E, Mora P, Blouin M, Pando A, Rouland-Lefévre C (2012) Effect of earthworms on plant *Lantana camara* Pb-uptake and on bacterial communities in root-adhering soil. Sci Total Environ 416:200–207
Kenarova A, Encheva M, Chipeva V, Chipev N, Hristova P, Moncheva P (2013) Physiological diversity of bacterial communities from different soil locations on Livingston Island, South Shetland Archipelago, Antarctica. Polar Biol 36:223–233

Kenarova A, Rradeva G (2010) Inhibitory effects of total and water soluble concentrations of heavy metals on microbial dehydrogenase activity. C R Acad Bulg Sci 63:1029–1033
Kenarova A, Radeva G, Traykov I, Boteva S (2014) Community level physiological profiles of bacterial communities inhabiting uranium mining impacted sites. Ecotoxicol Environ Saf 100:226–232
Khan M, Scullion J (2000) Effect of soil on microbial responses to metal contamination. Environ Pollut 110:115–125
Khan M, Scullion J (2002) Effects of metal (Cd, Cu, Ni, Pb or Zn) enrichment of sewage-sludge on soil micro-organisms and their activities. Appl Soil Ecol 20:145–155
Kong WD, Zhu YG, Fu BJ, Marschner P, He JZ (2006) The veterinary antibiotic oxytetracycline and Cu influence functional diversity of the soil microbial community. Environ Pollut 143:129–137
Konopka A, Oliver L, Turco R (1998) The use of carbon substrate utilization patterns in environmental and ecological microbiology. Microb Ecol 35:103–115
Kowalchuk GA, Buma DS, De Boer W, Klinkhamer PG, van Veen JA (2002) Effects of above-ground plant species composition and diversity on the diversity of soil-borne microorganisms. Antonie Van Leeuwenhoek 81:509–520
Lalor BM, Cookson WR, Murphy DV (2007) Comparison of two methods that assess soil community level physiological profiles in a forest ecosystem. Soil Biol Biochem 39:454–462
Lewis DE, White JR, Wafula D, Athar R, Dickerson T, Williams HN, Chauhan A (2010) Soil functional diversity analysis of a bauxite-mined restoration chronosequence. Microb Ecol 59:710–723
Li J, Jin Z, Gu Q (2011) Effect of plant species on the function and structure of the bacterial community in the rhizosphere of lead-zinc mine tailings in Zhejiang, China. Can J Microbiol 57:569–577
Masto RE, Ahirwar R, George J, Ram LC, Selvi VA (2011) Soil biological and biochemical response to Cd exposure. Open J Soil Sci 1:8–15
Mebride MB, Richards BK, Steenhuis T, Russo JJ, Sauve S (1997) Mobility and solubility of toxic metals and nutrients in soil fifteen years after sludge application. Soil Sci 162:487–500
Müller AK, Westergaard K, Christensen S, Sørensen SJ (2001) The effect of long-term mercury pollution on the soil microbial community. FEMS Microbiol Ecol 36:11–19
Muñiz S, Lacarta J, Pata M, Liménez JJ, Navarro E (2014) Analysis of the diversity of substrate utilization of soil bacteria exposed to Cd and earthworms activity using generalized additive models. PLoS One 9(1):e85057. doi:10.1371/journal.pone.0085057
Nannipieri P, Ascher J, Ceccherini MT, Landi L, Pietramellara G, Renella G (2003) Microbial diversity and soil functions. Eur J Soil Sci 54:655–670
Niklińska M, Chodak M, Laskowski R (2005) Characterization of the forest humus microbial community in a heavy metal polluted area. Soil Biol Biochem 37:2185–2194
Niklińska M, Chodak M, Laskowski R (2006) Pollution-induced community tolerance of micro-organisms from forest soil organic layers polluted with Zn or Cu. Appl Soil Ecol 32:265–272
Niklińska M, Chodak M, Stefanowicz A (2004) Community level physiological profiles of microbial communities from forest humus polluted with different amounts of Zn, Pb, and Cd—preliminary study with Biolog ecoplates. Soil Sci Plant Nutr 50:941–944
Pascaud A, Soulas M-L, Amellal S, Soulas G (2012) An integrated analytical approach for assessing the biological status of the soil microbial community. Eur J Soil Biol 49:98–106
Pennanen T (2001) Microbial communities in boreal coniferous forest humus exposed to heavy metals and changes in soil pH-a summary of the use of phospholipids fatty acids, Biolog and H-thymidine incorporation methods in field studies. Geoderma 100:91–126
Preston S, Wirth S, Ritz K, Griffiths BS, Young IM (2001) The role played by microorganisms in the biogenesis of soil cracks: importance of substrate quantity and quality. Soil Biol Biochem 33:1851–1858
Rajapaksha RMCP, Tobor-Lapłan MA, Bååth E (2004) Metal toxicity affects fungal and bacterial activities in soil differently. Appl Environ Microbiol 70:2966–2973

Rudel H, Wenzel A, Terytze K (2001) Quantification of soluble chromium(VI) in soils and evaluation of ecotoxicological effects. Environ Geochem Health 23:219–224
Sharma SS, Dietz K-J (2006) The significance of amino acids and amino acid-derived molecules in plant response to xeric and mesic conditions. Soil Biol Biochem 41:1882–1893
Shishido M, Chanway CP (1998) Storage effects on indigenous soil microbial communities and PGPR efficacy. Soil Biol Biochem 30:939–947
Stefanowicz AM, Niklińska M, Laskowski R (2009) Pollution-induced tolerance of soil bacterial communities in meadow and forest ecosystems polluted with heavy metals. Eur J Soil Biol 45:363–369
Tischner S, Tanneberg H, Guggenberger G (2008) Microbial parameters of soils contaminated with heavy metals: assessment for ecotoxicological monitoring. Pol J Ecol 56:471–479
Van Beelen P, Wouterse M, Posthuma L, Rutgers M (2004) Location-specific ecotoxicological risk assessment of metal-polluted soils. Environ Toxicol Chem 23:2769–2779
Van Hees PAW, Godbold DL, Jentschke G, Jones DL (2003) Impact of ectomycorrhizas on the concentrations and biodegradation of simple organic acids in a forest soil. Eur J Soil Sci 54:697–706
Wang Q, Dai J, Yu Y, Zhang Y, Shen T, Liu J, Wang R (2010) Efficiencies of different microbial parameters as indicator to assess slight metal pollutions in a farm field near a gold mining area. Environ Monit Assess 161:495–508
Wang Q, Wang R, Tian C, Yu Y, Zhang Y, Dai J (2012) Using microbial community functioning as the complementary environmental condition indicator: a case study of an iron deposit tailing area. Eur J Soil Biol 51:22–29
Wang Y, Shi J, Wang H, Lin Q, Chen X, Chen Y (2007) The influence of soil heavy metals pollution on soil microbial biomass, enzyme activity, and community composition near a copper smelter. Ecotoxicol Environ Saf 67:75–81
Wardle DA, Bardgett RD, Klironomos JN, Setälä H, van der Putten WH, Wall DH (2004) Ecological linkages between aboveground and belowground biota. Science 304:1629–1633
Winding A, Hendriksen NB (1997) Biolog substrate utilization assay for metabolic fingerprints of soil bacteria: incubation effects. In: Insam H, Rangger A (eds) Microbial communities. Functional versus structural approaches. Springer, Berlin, pp 195–205
Winding A, Hund-Rinke K, Rutgers M (2005) The use of microorganisms in ecological soil classification and assessment concept. Ecotoxicol Environ Saf 62:230–248
Yan F, McBratney AB, Copeland L (2000) Functional substrate diversity of cultivated and uncultivated A horizons of vertisols in NW New South Wales. Geoderma 96:321–343
Yang R, Tang J, Chen X, Hu S (2007) Effects of coexisting plant species on soil microbes and soil enzymes in metal lead contaminated soils. Appl Soil Ecol 37:240–246
Yuangen Y, Campbell CD, Clark L, Cameron CM, Paterson E (2006) Microbial indicators of heavy metal contamination in urban and rural soils. Chemosphere 63:1942–1952
Zak DR, Holmes WE, White DC, Peacock AD, Tilman D (2003) Plant diversity, soil microbial communities, and ecosystem function: are there any links? Ecology 84:2042–2050
Zeng LS, Liao M, Chen LC, Huang YC (2007) Effects of lead contamination on soil enzymatic activities, microbial biomass and rice physiological indices in soil-lead-rice (*Oryza sativa L.)* system. Ecotoxicol Environ Saf 67:67–74
Zhang H, Ahao FJ, Sun B, Davison W, McGrath SP (2001) A new method to measure effective soil solution concentration predicts copper availability to plants. Environ Sci Technol 35:2602–2607

Part IV
Approaches for the Remediation of Contaminated Soils

Chapter 14
Environmentally Sound Combustion of Ligneous Plants Grown in Heavy Metal-Contaminated Soil

Helga Kovacs, Katalin Szemmelveisz, and Arpad B. Palotas

14.1 Introduction

Heavy metals occur naturally in the environment and are taken up in minute amounts by living organisms. As trace elements, they provide essential nutrients for plants and herbivores, while at toxic levels, they cause harmful effects to the whole ecosystem.

Excessive heavy metal concentrations are regularly encountered at areas of extensive mining spoil and metal processing (mine tailings, plant sites). In any case, measures should be taken to prevent the spread of pollution and to reduce toxic levels of pollutant concentration—whether apparently detectable or not. For the remediation of misused industrial sites, various recovery techniques are known. Biological treatments include bioaccumulation, which is based on the capacity of plants and microorganisms to absorb and store chemical substances in concentrations higher than typical for the original environmental source (soil). The absorbed chemicals are accumulated in the plant cells and tissues.

By extracting large quantities of toxic contaminants (e.g. lead, cadmium, zinc) from soil and groundwater, certain plant species effectively contribute to the remediation of heavy metal-contaminated sites (Yaapar et al. 2008). The metals are soaked up by the roots and transported through conductive tissues to the above-ground plant parts (shoots), yielding easily harvestable biomass. This vegetal activity effectively improves soil quality and helps restoring the ecological balance (Tian et al. 2009).

The effect of heavy metal concentration on vegetal growth has been described by kinetic models in studies targeting plant-soil interactions in metalliferous systems. The actual bioaccumulative potential of plants is a function of several variables,

H. Kovacs • K. Szemmelveisz • A.B. Palotas (✉)
Department of Combustion Technology, University of Miskolc, Miskolc, Hungary
e-mail: kovacs.helga@uni-miskolc.hu; tuzszemt@uni-miskolc.hu; arpad.palotas@uni-miskolc.hu

I. Sherameti, A. Varma (eds.), *Heavy Metal Contamination of Soils*, Soil Biology 44, DOI 10.1007/978-3-319-14526-6_14

among them the biochemical properties of the given species, the metallic properties of the contaminants, soil type and other environmental factors (biotic and abiotic) (Gualab et al. 2010).

Besides the overall effect of heavy metals on seedling growth, the distribution of bioaccumulated heavy metals (i.e. the rate of metal bioaccumulation in the distinctive plant parts) has been studied in a number of papers. Regarding the chemical composition of different biomass samples (expressed as a percentage of dry weight), no substantial differences were found. The basic values for the elemental components tended to change within a limited range. In descending order of weight, the following components were detected: C, O, H, N, Ca, K, Si, Mg, Al, S, Fe, P, Cl, Na, Mn and Ti (Vassilev et al. 2010).

For seedlings grown on contaminated soils, Turner and Dickinson measured elevated levels of heavy metals in the leaves and the bark. The woody stem contained considerably less metallic compounds (Turner and Dickinson 1993). This follows from natural transport mechanisms. Metallic substances are taken up with other nutrients from the soil and get conveyed through the veins to remote storage tissues.

In woody plants, two types of conductive tissues can be distinguished: xylem and phloem. Numerous studies are devoted to the function of these in the plant's water and nutrient supply. Phloem belongs to the vascular system and is considered to play a central role in transporting essential nutrients from the shoots to the storage tissues or ground tissues (*vacuoles*) (Dinant and Lemoine 2010; Zhong et al. 1998; Harada 2010).

The intense accumulation of metals in the most vital plant parts (shoots, leaves) has been reported by several authors. Pulford and Watson measured high levels of Zn and Cd in the foliage of poplar species. In another study, excess concentrations of Cu, Pb and Cr were detected in the stem of willow coppices grown on treated silty soil, while Zn, Cd and Ni were found to accumulate in the crown. Metal distributions in the stems gave non-uniform data either (Pulford and Watson 2003). Therefore, it is of practical importance to determine the level of phytoextracted heavy metals with respect to all plant parts (including stems, branches, leaves).

From an environmental point of view, phytoremediation is an effective means to reduce high heavy metal concentrations in soils and thereby mitigate the ecological risks deriving from hazardous pollutant discharge. It should be remembered, however, that the toxic substances get translocated into the plant organs, giving way to further concerns. Soil recovery through phytoextraction yields large amounts of contaminated biomass, the proper disposal of which should be provided for. In a summary report of 2004 (Sas-Nowosielska et al. 2004), the following disposal methods were discussed:

- Composting: American scholars have shown the water solubility of bioaccumulated zinc and other metallic components (Zhao et al. 2000); therefore this method is only applicable with strict control requirements. Total heavy metal content should be minimized through mixing the metal-enriched biomass with high proportions of uncontaminated dry matter and other biodegradable substances. The technology requires the close and continuous monitoring of mixture composition. On the other hand, the phytoextracted metals are returned to the soil in small doses to provide for nutrient recovery. Ex situ composting is proposed as a postharvest treatment for biomass by which the disposal of hazardous plant material can be made more effective (Guangwei et al. 2009).

- Pyrolysis: Flash pyrolysis can be used for the disposal of heavy metal-contaminated biomass (Bridgwater et al. 1999; Al Chami et al. 2014; Stals et al. 2010b; Lievens et al. 2008). Low-temperature pyrolysis renders liquid fuel, with the metallic substances concentrated in the solid residues and the by-products (Stals et al. 2010a). Unless it is further processed as recyclables, the produced coke breeze should be treated as hazardous waste.
- Direct disposal: The direct disposal of the harvested biomass waste would cause environmental problems; therefore it is forbidden (Šyc et al. 2012). Disposal as hazardous waste is a feasible yet cost-demanding solution; thus, it is less preferred.
- Leaching: this disposal technology is based on the behaviour of soluble metals to percolate from the carrying medium—which is the very property that accounts for the categorization of metal-contaminated biomass as hazardous waste. After enrichment, toxic metals are leached with different solvents from the compacted biomass. The leached product (residual biomass matter) can be treated as nonhazardous material (Sas-Nowosielska et al. 2004). There are viable technologies to recover the metallic components from the leached solution at relatively high costs, though (Mulligan et al. 2001). The leachability (i.e. recovery rate) of toxic metals from biomass is generally determined as a function of time and pH value (Saeeda et al. 2005).
- Incineration (smelting): this process is based on the thermal degradation of contaminated woody biomass into manageable volumes of metal-containing ash. By substantially reducing the volume of hazardous waste material, both transport and disposal can be made more cost-effective. Countless descriptions of the technology are known (Garbisu and Alkorta 2001), but there is a paucity of data on the efficiency and technical feasibility of the process. Note that the thermal energy obtained from smelting can be utilized for energy purposes, which adds to the benefits of this method.

The incineration of heavy metal-contaminated biomass is a promising yet not fully mature technology. Various combustion systems are currently under development for field-scale application. In the near future, this method is likely to mean an environmentally sound and economically acceptable alternative (Sas-Nowosielska et al. 2004).

Disposal through incineration is an excellent option to replace the expensive treatment and costly transportation of the harvested biomass to hazardous waste disposal facilities. In this way, up to 99 % volume reduction of the contaminated material can be reached, with the pollutants concentrated in the solid combustion residues (ash, fly ash). The final product is easy to mobilize and handle in a controlled, environmentally acceptable manner.

14.2 Heavy Metals: Phase Behaviour and Emissions from Biomass Incineration

The incineration of phytoextracted biomass entails associated environmental hazards. On top of evident GHG emission problems, the discharge of pollusive heavy metals from the effluent gas and the solid combustion residues of the contaminated biomass needs to be considered.

From an environmental point of view, two basic sources of danger are encountered:

- On one hand, solid combustion products (bottom ashes) are enriched with toxins, which basically determine the viable options for safe disposal.
- On the other hand, the operating temperature of the combustion equipment is high enough to allow the transformation of volatile heavy metals into vapour-phase products. With decreasing flue gas temperature, these metal oxide species steadily condensate onto the surface of solid particles in the effluent gas flow (Pál Károlyné 1999). Depending on flue gas temperature, the volatilized fractions of heavy metals leave the exhaust system either in solid state (deposited onto the fly ash) or as metal vapours. The resulting emissions—particularly of Cr, Pb, Cd, Zn and Hg—are highly dangerous and pose a massive threat to ecological and human health.

We find a number of research devoted to the inspection of residues left from the combustion of contaminated and non-contaminated biomass wastes (Ryua et al. 2006; Simoneit et al. 1993). The distribution of metals in the different material streams (residual solids and gases) is commonly called the *partitioning* of metals—implying that the metallic components are partitioned between the emissive sources (Randall Seeker 1991).

The primary aim being the safe management and disposal of metal-enriched plant material, a reasonable subgoal of incineration is the capture of heavy metals in the bottom ashes. A more challenging task is the prevention of undesirable pollutant discharge deriving from the release and entrainment of volatilized heavy metals with the exit gas. This requires proper and reliable knowledge on the distribution of toxic metals in the material streams (by the end of the combustion process). Analysing the composition of the flue gas after cooling (to about 180–200 °C), one of the authors found that most heavy metals get deposited on the fly ash particles. This did not hold, however, to As, Co and Sb, of which 30–70 % tended to remain in the gas phase (Pál Károlyné 1999).

With fluidized bed combustion technology, pollusive heavy metals (Cd, Pb, Zn) generally get volatilized in the combustion chamber (Khan et al. 2009). The distinctive compounds show different condensation temperatures and various thermodynamic behaviours, which need to be considered (Verhulst et al. 1996).

Based on thermochemical calculations, the volatility of various metals (at 1 Pa operating pressure) is shown in Table 14.1—in the presence and absence of chlorides.

Table 14.1 Volatility temperatures and chemical formulae of metals at 1 Pa operating pressure (Randall Seeker 1991)

Heavy metal	No chlorides: Volatility temperature (°C)	No chlorides: Form	With 10 % of chlorides: Volatility temperature (°C)	With 10 % of chlorides: Form
Cr	1,613	CrO_2/ CrO_3	1,610	CrO_2/ CrO_3
Ni	1,210	$Ni(OH)_2$	693	$NiCl_2$
Pb	627	Pb	−15	$PbCl_4$
Cd	214	Cd	214	Cd

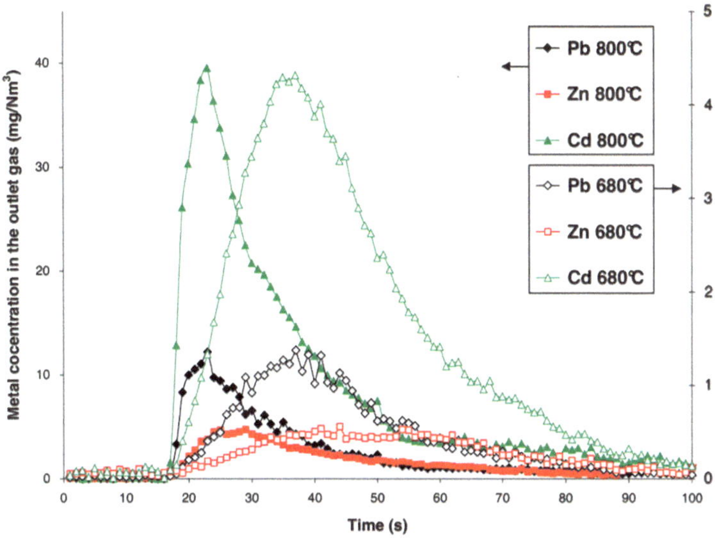

Fig. 14.1 Concentration curves of lead, zinc and cadmium in the flue gas (plotted against time) (Randall Seeker 1991)

The data presented in Table 14.1 clearly indicate the proneness of metals (e.g. Pb, Cd) to volatilize at normal operating temperature (i.e. of biomass combustion). It also implies the increased volatility of chloride forms, especially that of Pb (Randall Seeker 1991).

The above observations are affirmed by the results of a recent experiment, where excessive heavy metal contents were studied with fluidized bed combustion technology. At a temperature of 700–800 °C, the volatile compounds of lead, zinc and cadmium were found to rapidly leave the combustion chamber in gas phase. Evidence was obtained from the chemical analysis of particulate matter in the flue gas (Falcoz et al. 2010) (Fig. 14.1).

The release of heavy metal particles from the incinerated waste and their subsequent absorption onto the fly ash depend on several factors, among them volatility, operating conditions and ash/fly ash transport (Evans and Williams 2000).

The relative fraction of metals (Cd, Fe, Pb, Zn) in the emitted gas is closely associated with the formation of distinctive chlorine, sulphur, carbon, nitrogen, fluorine and other compounds during the thermal processes (burning and cooling). By performing X-ray diffraction analysis on fly ash samples from municipal waste incinerators, the following metallic compounds were found (at 420 °C fly ash temperature and 953 °C combustion temperature):

- Oxides: Al_2O_3, MnO, Pb_3SiO_5, $Pb_3Sb_2O_7$, $PbSiO_4$, Fe_3O_4, Fe_2O_3
- Other compounds: $Pb_3O_2SO_4$, $Cd_5(AsO_4)_3Cl$, $CdSO_4$, K_2ZnCl_4, $ZnCl_2$, $ZnSO_4$

meaning that in the absence of sulphur, mainly oxides and chlorides are formed (Evans and Williams 2000).

Given the above, three sources of emission are targets for research and comparison (with respect to the initial heavy metal fractions in the incinerated biomass): residual ashes, fly ash and emitted gas. Investigations based on this principle have revealed considerable differences in the partitioning of the distinctive metals in the combustion streams. While mercury, for example, mostly got released into the atmosphere, the bulk of copper remained in the solid residues. Cadmium and lead tended to be more evenly distributed among the sources (Randall Seeker 1991).

Particle formation during the transition from vapour-gas into solid phase (with decreasing flue gas temperature) is driven by two mechanisms: new particle formation via homogeneous nucleation and heterogeneous condensation to the surface of existing particles (Jöller et al. 2007; McNallan et al. 1981). Once metallic vapours become supersaturated, condensation reactions occur, facilitating the formation of ultra-fine particles (Jiao et al. 2013).

Various chemicals can be used to leach out metals from solid combustion residues (ashes, fly ash) left from burning. Leachability varies with the operating conditions and the combustion parameters (Randall Seeker 1991).

For the qualitative and quantitative analysis of fly ash, different solvents can be reacted with the flue gas. It has been shown that by using distilled water, the amount of heavy metals leachable from flue gas residues increases with time. A test solution was prepared by mixing 100 g of solid material in 1 l of distilled water. Leaching efficiency was demonstrated for various metallic elements as a function of time. The plot profiles suggest that on a longer time scale, a persistently growing fraction of pollusive substances are getting discharged—from rainwater-leached ash/fly ash deposits—into the environment (Shi and Kan 2009).

Conclusively, the two most pivotal tasks are to assess the ecological hazards of the combustion ashes and to estimate and reduce the emitted amount of toxins released into the atmosphere with the flue gas. For the latter, an obvious solution would be the separation of particulate matter (thus of the condensed heavy metal compounds) from the flue gas—supposing that flue gas temperature is reasonably low and the condensation of toxic vapours terminates by the time the stream leaves the exhaust system. Should these conditions not be met, the cleaning of the waste gas and the proper removal of the metallic pollutants should be provided.

14.3 Experiment, Methods

The incineration of heavy metal-contaminated plant material combines effective waste disposal with heat production. It is of economical reasonability to utilize the produced heat for energy purposes. Special attention should be paid, however, to the conditions of safe operation and the mitigation/elimination of environmental risks.

Since residential heating systems typically operate under different conditions than large industrial units, they generally fall out of the scope of technical standards and specifications pertinent to high-performance industrial systems. If contaminated biomass is fed into such small-scale units, there is no way to legally regulate and control the combustion process. Unless preventive measures and proper monitoring is adopted, collateral pollutant emissions cannot be excluded.

According to literary data, uncontrolled combustion entails the risk of heavy pollution due to the massive discharge of toxic metals into the environment. Considering this, a series of combustion experiments have been conducted on a woodchip-fuelled heating equipment under controlled conditions.

The tested biomass was sampled from a heavily contaminated site recorded as brownfield land near Gyöngyösoroszi (Hungary). It is, in fact, an abandoned mining area where industrial lead and zinc production commenced in 1926. The excavated ore had been crashed on-site and then enriched through a flotation process. Sphalerite (zinc sulphide) and galena (lead sulphide) were separated in powdered form and then smelted abroad. Subsequent to the processing of the four main products—lead, copper lead, zinc and pyrite in high-purity powdered form—the refuse ore was transported to the flotation spoil bank through a pipeline. Due to exchange rate fluctuations in the late 1970s, the operation of the mine was longer deemed profitable. The national mining authority decided to suspend mining operations and cease ore processing in 1986.

Due to massive heavy metal contamination and toxic levels of metallic components detected in the groundwater system of Toka, the Gyöngyösoroszi mining site came to the focus of attention in the late 1980s. By 1992, a complex land rehabilitation programme was worked out for the environmental cleanup of the delft and its vicinity (including the enrichment plant) by the Hungarian Institute of Environmental Management. Since then, several environmental and research centres have been conducting research on the recovery potential of the mine tailings.

Phytoremediation seems a viable option for Gyöngyösoroszi, supposing that proper measures are taken for the safe management of the contaminated wood waste. In order to assess the environmental impacts associated with the disposal of the phytoextracted biomass, a series of combustion experiments were conducted on oak and poplar samples from the respective area. Prior to burning, the samples were cut into woodchips instead of logs. With respect to the hyperaccumulation of metallic elements in smaller twigs and shoots—as described in the literature—woodchipping seemed practical for the effective degradation of the contaminated biomass. This assumption was affirmed by the comparative analysis of Zn, Cu, Cd,

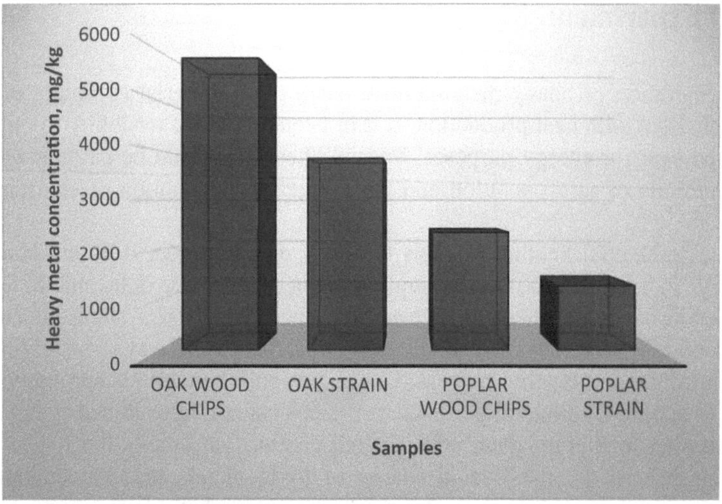

Fig. 14.2 Zn, Cu, Cd, Pb, Cr, Ni Co and Fe content of logged vs. chipped oak and poplar samples

Pb, Cr, Ni, Co and Fe levels measured preliminarily for the logged stems vs. the chipped parts of the sampled wood species. As seen in Fig. 14.2, considerable differences arose between the two sample types (logs vs. chips). Based on the results, most of the accumulated heavy metal content was assumed to be removable in solid form by chipping and subsequent combustion.

The combustion experiments were carried out in a Binder RRK 400–600 type woodchip-, briquette- and bark-fuelled heating system. The operational parameters were automatically set by the temperature control, so that the circulating water was kept at a constant temperature. The control unit was placed directly adjacent to the boiler. Prior to the actual combustion experiment, the system was fitted for the measurement purposes and for multiple samplings. For the stepwise detection of the process temperature, the number of sample collection points was increased.

A series of tests were performed to determine flue gas composition—including gaseous components and particulate matter content—at the outlet of the exhaust pipe (flue stack). Thereby, the applicability of the chosen semi-plant scale equipment for the risk-free disposal of hazardous biomass waste has been tested. Solid combustion residues (ash, fly ash) were analysed by taking samples from preselected points. The order of the sampling points followed the sequence of solid formation:

- Bottom ash
- Fly ash separated by the cyclone dust separator
- Fly ash emitted at the outlet of the exhaust pipe
- Residual solids (ash + separated fly ash) in the bottom ash hopper

Figure 14.3 shows the schematic illustration of the measurement apparatus. The sampling points are indicated with grey dots. Three ash samples were taken from

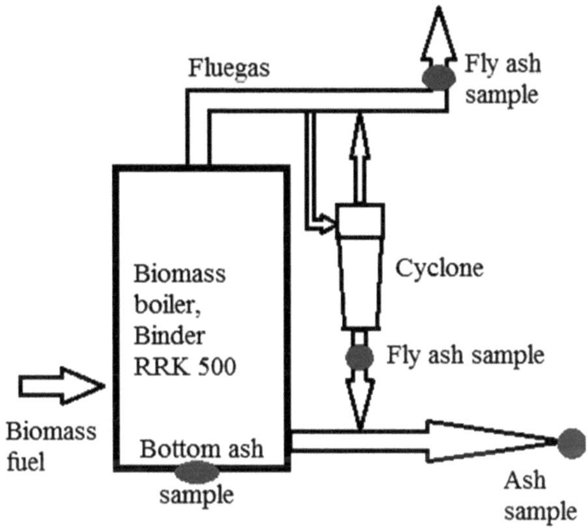

Fig. 14.3 Measurement apparatus with the sampling points

the combustion chamber for each wood type, respectively. The combustion residues (left in the ash hopper) were sampled at the end of the combustion process. Samples from the fly ash separated by the cyclone dust separator were obtained from the dust collection pipe under the cyclone. For the qualitative and quantitative analysis of particulate matter emission, the exhaust pipe was fitted with a standard measurement unit. Testing was based on gravimetric method, and the level of particulate pollutant emission was determined by using isokinetic S-20 type cyclone microprobe. Four samples were taken for oak and five samples for poplar.

The heavy metal content was determined with ICP spectrometry, using a Varian 720 ES type Inductively Coupled Plasma-Optical Emission Spectrometer. For the calibration of the measurement, a certified ICP multielement standard solution (by MERCK) was used. The samples were prepared by digestion and dissolution in high-purity concentrated nitric acid at 130 °C for 120 min in a Teflon acid digestion bomb. After cooling, the solution was filled up to a final volume of 50 cm^3.

14.4 Results and Discussion

The twofold purpose of the measurements was to determine the heavy metal fractions contained by the ashes and the exhaust gases and to examine the differences between samples taken from two distinctive points of the flue gas system. The latter provided data on the capture efficiency of the cyclone dust separator system as regards the safe removal of the metallic compounds absorbed onto fly ash particles.

The chemical composition of the solid combustion residues from the experimental process is detailed in Table 14.2. In some cases, the sampled amount

Table 14.2 Measured fractions of heavy metals in the solid combustion residues of oak and poplar samples (The values are expressed as a percentage of the dry weight of the samples)

Sample	Zn	Cu	Cd	Pb	Cr	Ni	Co	Fe
Oak	$mg/kg_{residual\ solids\ sample}$							
Ash hopper	*148.63*	*69.05*	*3.37*	*21.44*	*90.90*	*95.63*	*10.72*	*5989.89*
Combustion chamber, sample 1	34.41	68.11	0.00	12.24	78.71	76.43	4.79	5820.19
Combustion chamber, sample 2	29.04	61.67	0.00	9.14	54.44	64.33	3.98	3534.68
Combustion chamber, sample 3	26.34	45.59	0.00	25.91	53.47	55.47	4.02	4021.65
Combustion chamber average	*29.93*	*58.46*	*0.00*	*15.76*	*62.21*	*65.41*	*4.26*	*4458.84*
Cyclone, sample 1	459.37	327.25	7.77	35.52	50.21	95.97	0.05	2741.80
Cyclone, sample 2	376.07	168.99	6.03	31.08	48.50	44.87	1.45	3019.97
Cyclone, sample 3	430.11	241.00	7.05	28.21	32.66	41.89	1.70	2196.37
Particulate matter (cyclone) average	*421.85*	*245.75*	*6.95*	*31.60*	*43.79*	*60.91*	*1.07*	*2652.71*
Flue stack, sample 1	2156.60	283.08	15.46	326.39	150.66	87.85	4.87	9425.73
Flue stack, sample 2*	N.A.*	243.75	12.44	74.10	271.36	145.12	1.00	2129.94
Flue stack, sample 3*	N.A.*	247.98	13.62	218.21	95.50	76.21	2.10	6210.18
Flue stack, sample 4*	N.A.*	235.68	14.45	117.73	125.11	94.24	1.25	4111.58
Flue stack, sample 5*	N.A.*	402.50	19.58	105.03	81.94	69.37	1.00	2200.25
Particulate matter (flue stack) average	*2156.60*	*282.60*	*15.11*	*168.29*	*144.91*	*94.56*	*2.04*	*4815.54*
Poplar	mg/kg_{sample}							
Ash hopper	*985.29*	*70.41*	*22.88*	*18.75*	*65.01*	*50.95*	*5.17*	*6209.00*
Combustion chamber, sample 1	72.11	100.71	0.00	5.18	54.54	49.23	6.80	4302.17
Combustion chamber, sample 2	74.89	84.05	0.00	1.73	86.03	72.72	7.83	11171.14
Combustion chamber, sample 3	47.47	58.97	0.00	0.45	108.11	58.02	7.73	15306.99
Combustion chamber average	*64.82*	*81.24*	*0.00*	*2.45*	*82.89*	*59.99*	*7.45*	*10260.10*
Cyclone, sample 1	2456.11	116.29	49.23	43.59	36.59	37.72	2.88	3150.99
Cyclone, sample 2	3350.18	136.07	72.21	39.65	29.68	30.52	3.91	2216.57
Cyclone, sample 3	3095.61	88.35	75.54	34.87	24.61	27.63	3.62	1665.74
Particulate matter (cyclone) average	*2967.30*	*113.57*	*65.66*	*39.37*	*30.29*	*31.95*	*3.47*	*2344.44*
Flue stack, sample 1	2517.29	211.84	20.17	314.88	90.67	52.13	5.44	9903.35
Flue stack, sample 2*	N.A.*	195.98	39.66	75.88	42.26	38.97	0.67	2738.51
Flue stack, sample 3*	N.A.*	189.02	42.89	84.10	47.76	124.98	1.00	604.27
Flue stack, sample 4*	N.A.*	227.48	73.96	69.16	43.70	38.14	4.81	1900.76
Flue stack, sample 5	1514.82	221.84	27.89	75.71	68.09	54.09	1.97	3091.94
Flue stack, sample 6	5883.86	259.51	74.88	80.12	102.86	74.32	1.34	1831.59
Particulate matter (flue stack) average	*3305.32*	*217.61*	*46.58*	*116.64*	*65.89*	*63.77*	*2.54*	*3345.07*

*No data are available due to the high Zn content of the filtering paper used for sample preparation and evaluation

was so small that sample preparation (i.e. acid digestion) should necessarily involve the filtering paper. Consequently, no reliable data for Zn could be obtained due to the high Zn content of the filtering paper. Such cases are indicated with *.

Based on Table 14.2, the following conclusions can be drawn:

Cadmium is either completely absent from the solid combustion residues or remains below the level of detection. This suggests that the Cd content of the incinerated woodchips rapidly volatilizes and exits the combustion chamber in gaseous form. It is also apparent that the concentration levels of heavy metals are considerably higher in the exhaust gases gas than in the bottom ashes. For the determination of the exact quantities, see Tables 14.3 and 14.4.

In complete agreement with the literary data, the measurement results suggest that at a combustion temperature of 800 °C, most metallic compounds get released from the combustion chamber in vapour-gas phase. As the furnace off-gases cool on passing the flue gas system, the heavy metals undergo several condensation reactions. Still, even at a final flue gas temperature of 250 °C, some metal vapour species are suspected to remain uncondensed and be directly emitted from the flue gas stack.

14.4.1 Heavy Metals Trapped in the Solid Combustion Residues of Incinerated Wood Waste

Compositional analyses were conducted to determine the percentage of heavy metals in the residual solids (bottom ashes) at a temperature of about 800 °C. First, the ash content of the sample fuel was determined. Then, large quantities of woodchips were combusted at 815 °C under laboratory conditions and the ashes were sampled. Finally, chemical analysis was performed on the sample fuel and on the ash samples, respectively. The results are summarized in Table 14.3.

Based on the measured ash content (first dataset), the weight percent of the ash produced from 1 kg of incinerated woodchips was calculated (second dataset). Then, based on the measured heavy metal fractions (fourth dataset), the percentage of heavy metals in 1 kg of incinerated woodchip residuals has been calculated. Accordingly, the third and fifth datasets represent the percentage values of heavy metals per unit volume (1 kg) of raw and incinerated woodchips, respectively—calculated from the following correlation:

$$\pi = h \cdot \pounds \tag{14.1}$$

where:

π is the fraction of heavy metals per unit volume (1 kg) of incinerated woodchips, $mg_{heavy\ metals}/kg_{woodchips}$.

h is the measured ash content, $kg_{ash}/kg_{woodchips}$.

Table 14.3 Percentage values of heavy metals in the solid combustion residues of oak and poplar (relative to sample weight)

Dataset No.	Oak			Poplar		
1	Ash content, % m/m		1.02	Ash content, % m/m		0.86
2	Ash produced from 1 kg of incinerated woodchips (h), $kg_{ash}/kg_{incinerated\ woodchips}$		0.0102	Ash produced from 1 kg of incinerated woodchips (h), $kg_{ash}/kg_{incinerated\ woodchips}$		0.0086
3	Initial heavy metal content of oak woodchips (ε) $mg_{heavy\ metals}/kg_{incinerated\ woodchips}$	Zn	18.20	Initial heavy metal content of poplar woodchips (ε), $mg_{heavy\ metals}/kg_{incinerated\ woodchips}$	Zn	82.40
		Cu	43.70		Cu	45.50
		Cd	0.79		Cd	2.19
		Pb	2.06		Pb	79.40
		Cr	40.70		Cr	86.50
		Ni	19.40		Ni	80.10
		Co	2.20		Co	0.95
		Fe	440.00		Fe	212.00
	Total, $mg_{heavy\ metals}/kg_{incinerated\ woodchips}$		567.05	Total, $mg_{heavy\ metals}/kg_{incinerated\ woodchips}$		589.04
4	Heavy metal fractions in 1 kg of ash (£), $mg_{heavy\ metals}/kg_{ash}$	Zn	222.00	Heavy metal fractions in 1 kg of ash (£), $mg_{heavy\ metals}/kg_{ash}$	Zn	2297.00
		Cu	342.00		Cu	476.00
		Cd	5.93		Cd	19.10
		Pb	8.86		Pb	20.70
		Cr	51.60		Cr	564.00
		Ni	91.10		Ni	399.00
		Co	3.72		Co	11.90
		Fe	7404.00		Fe	7187.00
	Total, $mg_{heavy\ metals}/kg_{ash}$		8129.21	Total, $mg_{heavy\ metals}/kg_{ash}$		10974.70
5	Heavy metal fractions in the ash of 1 kg woodchips (π), $mg_{heavy\ metals}/kg_{incinerated\ woodchips\ (ash)}$	Zn	2.27	Heavy metal fractions in the ash of 1 kg woodchips (π), $mg_{heavy\ metals}/kg_{incinerated\ woodchips\ (ash)}$	Zn	19.67
		Cu	3.50		Cu	4.08
		Cd	0.06		Cd	0.16
		Pb	0.09		Pb	0.18
		Cr	0.53		Cr	4.83
		Ni	0.93		Ni	3.42
		Co	0.04		Co	0.10
		Fe	75.70		Fe	61.56
	Total, $mg_{heavy\ metals}/kg_{incinerated\ woodchips\ (ash)}$		83.11	Total, $mg_{heavy\ metals}/kg_{incinerated\ woodchips\ (ash)}$		94.00

£ is the fraction of heavy metals per unit volume (1 kg) of ash, $mg_{heavy\ metals}/kg_{ash}$.

Based on the data, the partitioning of metals between the incineration residues and the flue gas stream could be calculated—expressed as percentages of the initial metal content of the incinerated biomass. Thus:

Table 14.4 Heavy metal partitioning in the solid and gaseous combustion streams of oak and poplar samples

		Zn	Cu	Cd	Pb	Cr	Ni	Co	Fe	Total
Sample		%								
Oak	Bottom ash	12.47	8.00	7.67	4.40	1.30	4.80	1.73	17.20	14.66
	Gas stream	87.53	92	92.33	95.6	98.7	95.2	98.27	82.8	85.34
Poplar	Bottom ash	23.88	8.96	7.47	0.22	5.58	4.27	10.73	29.04	15.96
	Gas stream	76.12	91.04	92.53	99.78	94.42	95.73	89.27	70.96	84.04

$$\beta = \frac{\pi}{\varepsilon} \cdot 100 \qquad (14.2)$$

where:

β is the fraction of heavy metals trapped in the bottom ash, expressed as a percent weight of the incinerated biomass, %.

π is the fraction of heavy metals per unit volume (1 kg) of incinerated woodchips, $mg_{heavy\ metals}/kg_{woodchips}$.

ε is the initial heavy metal content of the biomass, $mg_{heavy\ metals}/kg_{incinerated\ woodchips}$.

And:

$$\tau = 100 - \beta \qquad (14.3)$$

where:

τ is the fraction of volatilized metals expressed as a percent weight of the incinerated biomass, %.

The results are summarized in Table 14.4.

As seen from Table 14.4, about 15 % of the initial metal content remained trapped in the bottom ash, and the rest entered the flue gas system in volatile form. The metal vapours partly got condensed onto the surface of solid particles and partly got deposed to the pipe walls. A small fraction was emitted from the system through the flue gas stack. Evaluating the results for each metal distinctively, we find high ratios of Fe and Zn to remain in the bottom ashes. Regarding other elements, 12–24 % of Zn, 8–9 % of Cu, 7–8 % of Cd, 0–5 % of Pb, 1–6 % of Cr, 4–5 % of Ni, 1–11 % of Co and 17–29 % of Fe are trapped in the solid residues, while over 70 % of the metallic compounds volatilize and leave the combustion chamber in vapour-gas form. A summary is given in Fig. 14.4.

The volatilized metal compounds are subject to condensation reactions along the entire length of the flue gas system. Whether residual fractions of metallic vapours are likely to be discharged through the flue gas stack basically depends on exit gas temperature.

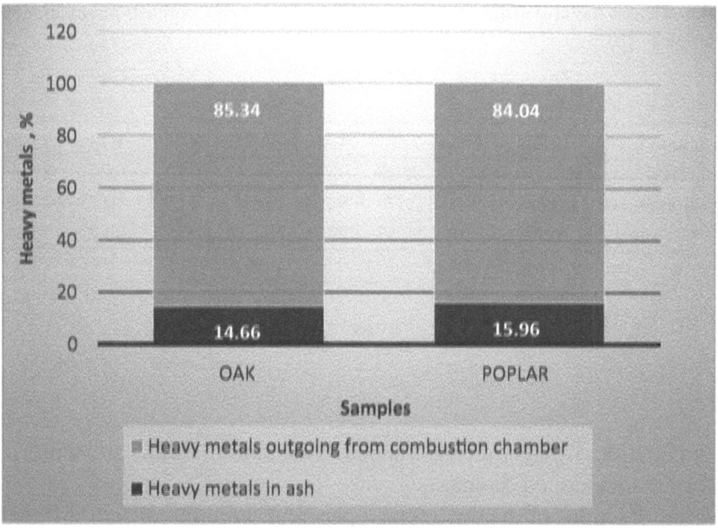

Fig. 14.4 The heavy metal content of the ash and the flue gas

14.4.2 Refitting the Experimental Boiler for the Safe Disposal of Contaminated Biomass

As demonstrated in the previous section, over 70 % of the metallic elements (Zn, Cu, Cd, Pb, C, Ni, Co, Fe) get released from the incinerated biomass and leave the combustion chamber in volatile form, transported by the flue gas stream until condensation occurs with decreasing flue gas temperature.

For the removal of the condensed matter, the cooling gas passes through a cyclone dust separator, where metal-containing solid particles are captured before the stream exits through the flue stack.

Due to the limited capacity of the cyclone dust collection system to reduce particulate load, fine and ultra-fine particles do remain in the flue gas and got released into the atmosphere with the outlet gas. Thus, the given experimental apparatus will inevitably entail a certain degree of discharge of the absorbed heavy metals at the final phase. The fully effective and safe disposal of the contaminated biomass would require the development of an advanced gas cleaning technology, including the application of (extra) dust filters. In a secondary filtering stage, the waste gas would preferably pass through a porous medium, where so far uncollected solid particles would be filtered. Depending on the filter type, the porous filtering layer might consist of tissues, fibres or granules. Filters are classified according to their material, which should be considered when flue gas temperature is high (250 °C or above). Natural fibres are applicable up to 100 °C and synthetic fibres up to 250 °C, while glass fibres tolerate even 350 °C.

Another risk is posed by the massive presence of metallic vapours in the hot flue gas. Ideally, the circulating gas stream is cooled down to a temperature below the condensation temperature of the volatilized metals; thus, condensation terminates before the final emission of the gas from the system. This can be promoted by air cooling the flue gas from 250 to 130 °C prior to the application of fabric filters. This technology would further reduce the risk of undesirable pollutant discharge, since it would retain the condensed heavy metal components within the system.

The combined application of flue gas cooling and fine particle filtering is therefore highly recommended.

14.5 Conclusions

Bioremediation is one of the most eco-friendly ways of removing heavy metals from contaminated soils. Nonetheless, particular attention should be paid to the safe and sound disposal of the phytoextracted biomass—e.g. through incineration. For this purpose, a series of combustion experiments have been conducted on heavy metal-containing wood samples from an abandoned Hungarian mining area.

In close correlation with the results presented in the literature, the preliminary testing of the wood material indicated the hyperaccumulation of metallic elements in external plant parts like smaller twigs, sprouts and bark. This lead to the conclusion that woodchips is a preferable form of fuel (over logging) for the effective thermal degradation of the accumulated heavy metal content of the biomass. Incineration has the double benefit of producing heat energy and drastically reducing the volume/mass of the toxic residues to be handled (i.e. wood material into ashes).

Multiple samples were taken from the solid combustion residues and the fly ash obtained by incinerating the biomass in a Binder RRK 500 type furnace/combustion equipment. The samples were analysed, and the presence of heavy metals was demonstrated for each sample set. The chemical analysis of the bottom ashes made it obvious that at a combustion/operating temperature of about 800 °C, cca. 76–88 % of Zn, 91–92 % of Cu, 92–93 % of Cd, 95–100 % of Pb, 94–99 % of Cr, 95–96 % of Ni, 89–99 % of Co and 71–83 % of Fe leave the combustion chamber in vapour-gas phase. Since most of the released heavy metal compounds get condensed onto the surface of flue gas particles (fly ash), proper dust separation technology is of crucial importance. With the air cooling of the flue gas to 105–130 °C and the capture of contaminated flue gas particles via the application of fabric filters, unwanted emissions of particulate pollutant matter can be considerably reduced or fully eliminated.

Acknowledgements This research was carried out in the framework of the Center of Excellence of Sustainable Resource Management at the University of Miskolc.

References

Al Chami Z et al (2014) Evaluation of flash and slow pyrolysis applied on heavy metal contaminated Sorghum bicolor shoots resulting from phytoremediation. Biomass Bioenergy 63:268–279

Bridgwater AV, Meier D, Radlein D (1999) An overview of fast pyrolysis of biomass. Org Geochem 30:1479–1493

Dinant S, Lemoine R (2010) The phloem pathway: new issues and old debates. C R Biol 333 (4):307–319

Evans J, Williams P (2000) Heavy metal adsorption onto flyash in waste incineration flue gases. Process Saf Environ Prot 78(1):40–46

Falcoz Q et al (2010) A general kinetic law for heavy metal vaporization during municipal solid waste incineration. Process Saf Environ Prot 88:125–130

Garbisu C, Alkorta I (2001) Phytoextractiona cost-effective plant-based technology for the removal of metals from the environment. Bioresour Technol 77:229–236

Gualab SD, Vega FA, Covelo EF (2010) The dynamics of heavy metals in plant–soil interactions. Ecol Model 221(8):1148–1152

Guangwei Y et al (2009) In-situ stabilisation followed by ex-situ composting for treatment and disposal. J Environ Sci 21:877–883

Harada T (2010) Grafting and RNA transport via phloem tissue in horticultural plants. Sci Hortic 125(4):545–550

Jiao F et al (2013) Effect of HCl, SO_2 and H_2O on the condensation of heavy metal vapors in flue gas cooling section. Fuel Process Technol 105:181–187

Jöller M, Brunner T, Obernberger I (2007) Modeling of aerosol formation during biomass combustion for various furnace and boiler types. Fuel Process Technol 88:1136–1147

Károlyné P (1999) Heavy metals in fly ash (Hungiarian: Nehézfémek a pernyében), 18th edn. Környezetvédelmi Füzetek, Hungary. ISBN 963 593 416 5

Khan AA, de Jong W, Jansens PJ, Spliethoff H (2009) Biomass combustion in fluidized bed boilers: potential problems and remedies. Fuel Process Technol 90(1):21–50

Lievens C, Yperman J, Vangronsveld J, Carleer R (2008) Study of the potential valorisation of heavy metal contaminated biomass via phytoremediation by fast pyrolysis: part I. Influence of temperature, biomass species and solid heat carrier on the behaviour of heavy metals. Fuel 87 (10–11):1894–1905

Mcnallan MJ, Yurek GJ, Elliot JF (1981) The formation of inorganic particulates by homogeneous nucleation in gases produced by the combustion of coal. Combust Flame 42:45–60

Mulligan CN, Yong RN, Gibbs BF (2001) Remediation technologies for metal-contaminated soils and groundwater: an evaluation. Eng Geol 60:193–207

Pulford ID, Watson C (2003) Phytoremediation of heavy metal-contaminated land by trees—a review. Environ Int 29:529–540

Randall Seeker WM (1991) Waste combustion. Symposium (International) on Combustion 23 (1):867–885

Ryua C et al (2006) Effect of fuel properties on biomass combustion: part I. Experiments—fuel type, equivalence ratio and particle size. Fuel 85(7–8):1039–1046

Saeeda A, Akhterb MW, Iqbal M (2005) Removal and recovery of heavy metals from aqueous solution using papaya wood as a new biosorbent. Sep Purif Technol 45(1):25–31

Sas-Nowosielska A et al (2004) Phytoextraction crop disposal—an unsolved problem. Environ Pollut 128(3):373–379

Shi H-S, Kan L-L (2009) Leaching behavior of heavy metals from municipal solid wastes incineration (MSWI) fly ash used in concrete. J Hazard Mater 164(2–3):750–754

Simoneit BRT et al (1993) Lignin pyrolysis products, lignans, and resin acids as specific tracers of plant classes in emissions from biomass combustion. Environ Sci Technol 27(12):2533–2541

Stals M et al (2010a) Flash pyrolysis of heavy metal contaminated hardwoods from phytoremediation: characterisation of biomass, pyrolysis oil and char/ash fraction. J Anal Appl Pyrolysis 89(1):22–29

Stals M et al (2010b) Flash pyrolysis of heavy metal contaminated biomass from phytoremediation: influence of temperature, entrained flow and wood/leaves blended pyrolysis on the behaviour of heavy metals. J Anal Appl Pyrolysis 87(1):1–7

Šyc M et al (2012) Willow trees from heavy metals phytoextraction as energy crops. Biomass Bioenergy 37:106–113

Tian D et al (2009) Heavy metal accumulation by panicled goldenrain tree (*Koelreuteria paniculata*) and common elaeocarpus (*Elaeocarpus decipiens*) in abandoned mine soils in southern China. J Environ Sci 21(3):340–345

Turner AP, Dickinson NM (1993) Survival of *Acer pseudoplatanus* L. (sycamore) seedlings on metalliferous soils. New Phytol 123:509–521

Vassilev SV, Baxter D, Andersen LK, Vassilev CG (2010) An overview of the chemical composition of biomass. Fuel 89(5):913–933

Verhulst D, Buekens A, Spencer PJ, Eriksson G (1996) Thermodynamic behavior of metal chlorides and sulfates under the conditions of incineration furnaces. Environ Sci Technol 30 (1):50–56

Yaapar et al (2008) Phytoextraction. [Online] Available at: http://www.biology-online.org/articles/phytoremediation-a-lecture/phytoextraction.html. Accessed 30 Jan 2013

Zhao FJ, Lombi E, Breedon T, McGrath SP (2000) Zinc hyperaccumulation and cellular distribution in *Arabidopsis halleri*. Plant Cell Environ 23:507–514

Zhong W et al (1998) Phloem loading of inorganic cations and anions by the seedling of *Ricinus communis* L. J Plant Physiol 152(2–3):328–335

Chapter 15
Role of Earthworms on Phytoremediation of Heavy Metal-Polluted Soils

My Dung Jusselme, Edouard Miambi, Thierry Lebeau, and Corinne Rouland-Lefevre

15.1 Introduction

Heavy metal contamination of soil is a major concern in all parts of the world, in particular in emerging countries where there is an increasing need for soil for food. Heavy metals such as cadmium, copper, lead, mercury, and zinc remain in the soil where they accumulate as a result of activities such as mining and the application of urban sewage sludge for agriculture. The accumulation of heavy metals in the environment can affect the health of humans and animals. At microscale, heavy metals also have an adverse effect on bacterial populations which in turn affects the global functioning of ecosystems. Microorganisms play a key role in biogeochemical processes. Changes in the microbial communities may reduce their ability to maintain soil fertility in the long term. This has led to the recent development of techniques for cleaning up polluted soils and sites. One such technique is phytoremediation, which exploits the ability of certain plants to accumulate large amounts of heavy metals (Chaney et al. 1997; Salt et al. 1998; Padmavathiamma and Li 2007). Phytoremediation has many advantages: (1) it is the only method available for in situ extraction of heavy metals from soils; (2) it is economically viable as, at least theoretically, the energy required for the process is free (from the

M.D. Jusselme • E. Miambi
IEES-Paris Université Paris Est Créteil, 62 avenue du Général De Gaulle, 94000 Creteil, France
e-mail: mydung.jusselme@gmail.com; miambi@u-pec.fr

T. Lebeau
UMR LPGN 6112 CNRS, LUNAM, Université de Nantes, 2, rue de la Houssinière, BP 92208, 44322 Nantes, France
e-mail: thierry.lebeau@univ-nantes.fr

C. Rouland-Lefevre (✉)
Centre IRD France-Nord, 32 avenue Henri Varagnat, 93140 Bondy, France
e-mail: corinne.rouland-lefevre@ird.fr

I. Sherameti, A. Varma (eds.), *Heavy Metal Contamination of Soils*, Soil Biology 44, DOI 10.1007/978-3-319-14526-6_15

sun), and the harvestable parts of the plants that accumulate heavy metals can be used for energy, biological catalysts, insulators, etc.; and (3) it has a low environmental impact consistent with environmental protection policies and allows the soils to be exploited for crops after or at the same time as the soil is being remediated (Losfeld et al. 2012). However, phytoremediation has limitations: (1) the slow growth and low biomass require a considerable investment in time and/or money, and (2) the heavy metals accumulate slowly in the plants as the pools of heavy metals available to the plants at a given time are small. Chelating agents such as EDTA, DTPA, and citric acids have been tested successfully (Luo et al. 2005; Luo et al. 2006a, b) but may have undesirable effects such as toxicity for plants (Evangelou et al. 2007), and (3) plant growth is reduced by the phytotoxicity of the heavy metals (Shah and Nongkynrih 2007; Salt et al. 1998; Singh et al. 2003).

To improve the performance of phytoextraction, hyperaccumulating plants with high biomass (e.g., *Brassica juncea* or Indian mustard) are used. Recent research has concentrated on the role of the rhizosphere with a view to associating microbial bioaugmentation of soils with phytoextraction (Lebeau et al. 2008; Sessitsch et al. 2013; Wenzel 2009; Khan 2005), but few studies have considered the drilosphere compartment, the part of the soil influenced by earthworm secretions and castings (Aghababaei et al. 2014a; Du et al. 2014; Jusselme et al. 2012, 2013). However, earthworms as ecological engineers play an important role in their environment (Derouard et al. 1997; Bohlen et al. 2002; Dechaine et al. 2005; Tapia-Coral et al. 2006). The positive effects of earthworms on plant production (Table 15.1) have been extensively documented (Blouin et al. 2007, 2013; Wang et al. 2006) as well as their effects on heavy metal solubility and availability (Sizmur et al. 2011b, c, d). The interactions between heavy metals and earthworms depend on the earthworm species, the metal, and the physical and chemical properties of the soil (Weltje 1998; Morgan and Morgan 1999; Sizmur and Hodson 2009). Earthworms have an effect on metal speciation in soils, changing the bioaccessibility and bioavailability of the metals for other organisms, such as plants (Sizmur et al. 2011a). This chapter summarizes the current understanding of the interactions between earthworms, plants, and microorganisms in heavy metal-contaminated soil. It covers basic research as well as practical phytoremediation.

15.2 Earthworms as Ecosystem Engineers

The term "ecosystem engineers" was used by (Lawton 1994) to designate organisms that directly or indirectly influence the availability of resources to other species by causing physical state changes in biotic and abiotic materials. Earthworms in tropical soils are recognized as key ecosystem engineers as they modify, maintain, and create habitats (Jones et al. 1994; Lavelle 1996).

Earthworms (annelids, oligochaetes) are the dominant biomass of soil macrofauna in most terrestrial ecosystems. About 7,000 species have been

Table 15.1 Earthworm effects on plant biomass in heavy metal-polluted soils

				Plant biomass		
Soil conditions	Plants	Earthworms		Shoot (g)	Root (g)	Reference
Pb pollution at 1,000 mg kg^{-1}	*Lantana camara*	*Pontoscolex corethrurus*	Without	21.52	19.11	Jusselme et al. (2012, 2013)
			With	31.77	24.21	
Pb/Zn mine tailings at 1,202 mg Pb kg^{-1}	*Leucaena leucocephala*	*Pheretima guillelmi*	Without	30.0		Ma et al. (2006)
			With	34.0		
Cu pollution 400 mg kg^{-1}	Ryegrass (*Lolium multiflorum*)	*Metaphire guillelmi*	Without	1.23	0.33	Dandan et al. (2007)
			With	2.67	0.45	
Zn pollution at 400 mg kg^{-1}	Ryegrass	*Pheretima* sp.	Without	1.79	0.67	Wang et al. (2006)
			With	2.66	1.11	
	Indian mustard	*Pheretima* sp.	Without	1.04	0.12	
			With	1.41	0.27	
Cd pollution at 20 mg kg^{-1}	Ryegrass (*L. multiflorum*)	*Pheretima* sp.	Without	1.78	0.53	Yu et al. (2005)
			With	1.96	0.46	

identified, divided into 15 families, most of which live in the tropics (Lavelle 1997). They play an important role in environmental functioning through various physical and biological mechanisms that preserve the structure of the soil and improve its fertility (Stork and Eggleton 1992; Lavelle 1997). By modifying the physical and chemical properties of the soil, they also change the habitats of microbial communities (Lavelle 1997).

15.2.1 Main Geographic Origins and Taxonomy

Earthworms are found in all tropical and temperate soils with a high level of diversity. Quaternary glaciers caused earthworms to become locally extinct so that they are found in greater abundance in the tropics. Earthworms fall into three broad ecological categories—epigeic, anecic, and endogeic—depending on morphology and behavior (Bouche 1977).

1. *Epigeic earthworms* (Bouche 1977) are small (10–30 mm) and generally live in litter and decomposed organic matter. They are also found in the feces of large herbivores or in damp woods during decomposition. They live on the soil surface and are, therefore, particularly susceptible to predation, climate variability, and anthropogenic activities such as surface plowing and the application of pesticides. Epigeic earthworms play an important role in recycling organic matter.
2. *Anecic earthworms* (Bouche 1977) are medium to giant worms (10–110 cm) living in vertical or subvertical burrows with varying degrees of branching that open onto the soil surface. By ingesting soil and burying organic matter, they mix the organic matter and mineral fraction from the different soil horizons. These species are found throughout the entire depth of the soil profile and have strong muscles enabling them to adapt to a relatively high soil compaction and withstand human pressures in cultivated soils.
3. *Endogeic earthworms* (Bouche 1977) vary in size (1–20 cm). They account for 20–50 % of the biomass of fertile land and live in the soil, burrowing in any orientation. They feed on the organic matter in the soil and in poorer soils may need to be very mobile to find all the food they need. Lavelle (1981) defined three subcategories of endogeic earthworms—polyhumics, mesohumics, and oligohumics—based on the richness of the soil organic matter they eat, oligohumic earthworms being those which ingest organic matter in the poorest soils.

15.2.2 Impact of Earthworms on Soil

The impact of earthworms on soil depends on their ecological category, endogeic and anecic having the greatest effect (Brown et al. 2000). The main physical activities of earthworms include (1) the creation of galleries in which they move and (2) excretion in the galleries (feces) or on the soil surface (casts). The drilosphere is the area where soil functioning is influenced by earthworm activities. It includes all dependent physical structures of earthworms such as the contents of the digestive tract, casts, and galleries as well as associated communities of invertebrates and microorganisms. The structure and relative importance of the drilosphere are determined by the climate, soil parameters, and quality of organic inputs (Lavelle 1997).

15.2.2.1 Galleries

The gallery network (size, orientation, etc.) depends on the ecological category of the earthworm. The number of galleries in the soil depends on the abundance of earthworms but can be up to several hundreds per m^2. In sites with large earthworm communities, the volume of the galleries contributes significantly to the pore size of the soil, providing passageways for air and water in the soil (Bouché and Al-Addan 1997). These galleries improve the porosity and aeration. Experiments have shown that, in microcosms, the galleries of *L. terrestris* earthworms significantly increased the water flow (Joschko et al. 1989). Field studies clearly support these findings by showing the transfer of water through the *L. terrestris* galleries (Edwards et al. 1992) and the strong correlation between the infiltration rate and the length, area, and volume of the galleries of the anecic earthworm *Scherotheca gigas mifuga* (Bouche 1977).

Moreover, galleries make it possible to transfer compounds from different soil horizons, both passively by percolation and infiltration and by the active role of anecic burrowing species. Earthworms line their galleries with mucus- and nutrient-rich droppings as they pass through the soil (Binet and Curmi 1992). This makes the walls of galleries richer in organic carbon and nitrogen than the surrounding soil. These relationships stimulate the development of a high density of bacteria throughout the gallery walls, increasing respiratory activity and enzymatic digestion. Tiunov and Scheu (1999) showed that microbial biomass was higher by a factor of 2.3–4.7 in the walls of *L. terrestris* galleries than in the surrounding soil. This microbial growth increased the soil respiration by a factor of 3.7–9.1 in forest ecosystems.

15.2.2.2 Casts

Earthworms ingest soil and excrete waste onto the soil surface or in the galleries. They produce casts on the surface amounting to between 200 and 250 t ha^{-1} $year^{-1}$ in temperate soil and 40–50 t ha^{-1} $year^{-1}$ in grassland, representing a soil thickness of 3–4 mm. Binet and Le Bayon (1998) evaluated the production of casts from 2.5 to 3.2 kg m^{-2} $year^{-1}$ (dry weight) in a temperate maize crop.

However, the production and abundance of earthworm casts depend on environmental conditions (climate, soil type), the earthworm species, and the vegetal cover.

For example, the disappearance of earthworm casts accounts for about 70 % and 20 % in rainy season and dry season, respectively. The casts are gradually incorporated into the soil matrix during the dry season (Binet and Le Bayon 1998).

The feeding behavior of earthworms leads to considerable variability in the composition of their casts. The physical structure of the casts provides microenvironmental conditions that differ from the initial soils. Some earthworms feed selectively on the parts of the soil that are rich in organic matter (Zhang and Schrader 1993; Doube et al. 1994). Doube et al. (1997) showed that *L. terrestris* and *A. caliginosa* preferentially consume a mixture of fine inorganic particles and organic material rather than organic material on its own. Fungi are also an important food source for many species of earthworms (Edwards and Fletcher 1988). Earthworms may also feed on protozoa, bacteria, and algae. During transit through the digestive tract of earthworms, these microorganism populations are modified by the physical and chemical conditions in the intestine. The surviving microorganisms (in particular fungal spores, protozoa, and resistant bacteria) are present in the inoculums that subsequently colonize the casts and are responsible for the microbial processes (Brown 1995). Parle et al. (1963a) showed that *L. terrestris* casts had higher concentrations of bacteria and actinomycetes after the soil had passed through the earthworm gut where conditions were favorable for their development. Fungal hyphae developed on the surface of casts (Parle et al. 1963b), a phenomenon that was also observed for the geophagous earthworm *Pontoscolex corethrurus* (Barois et al. 1987). Many studies have demonstrated that earthworms can stimulate soil microbial activity, although the density of bacterial and fungal populations may be reduced after transiting the gut of endogeic earthworms (Krišrtuek et al. 1992). Enzyme activities can provide information on the functional diversity of the microbial community. Tiwari et al. (1989) showed that phosphatase, dehydrogenase, and urease activities were more intense in casts. High phosphatase activity has also been found in fresh casts of the endogeic earthworm *A. caliginosa* (Aira et al. 2010). These enzyme activities can affect the bioavailability of mineral elements such as phosphorus (Satchell and Martin 1984).

15.3 Earthworms and Heavy Metals

Earthworms are more sensitive to heavy metals than other invertebrates living in soils (Bengtsson et al. 1992), and their ability to accumulate heavy metals is often greater than for other animal species (Beyer et al. 1982). However, the effects of heavy metals depend on the earthworm species, stage of development, lifestyle (where they live and what they eat), and their ability to adapt to contaminants. These effects also depend on the nature and chemical forms of the metal and the physical and chemical properties of the soil.

An increase in the heavy metal content in soil above acceptable levels reduces the density of earthworms (Pizl and Josens 1995) and also reduces weight gain (Spurgeon and Hopkin 1996), sexual development, and cocoon production

(Spurgeon and Hopkin 1999). Spurgeon and Hopkin (1999) reported a significant reduction in the survival rate of four species of earthworm with zinc levels ranging from 2,000 to 3,600 mg Zn kg^{-1} of soil. They also reported significant weight loss with zinc levels ranging from 1,200 to 2,000 mg Zn kg^{-1}. Lukkari and Haimi (2005) suggested that one of the potential mechanisms of adaptation to pollution is avoidance as earthworms placed in contaminated soil may be able to differentiate organic matter according to its level of contamination.

The ability of earthworms to accumulate heavy metals was recognized in the literature as early as the late nineteenth century (Hopkin 1989). Heavy metals accumulate in the digestive tissues of earthworms after ingestion or by dermal exposure as earthworms have no protective cuticle and are in continuous contact with the polluted soil. Many studies have determined the factors controlling bioconcentration: (1) the earthworm species and its ecological category, (2) the heavy metal species, (3) the physical and chemical properties of the soil, (4) the season, and (5) the distance from the source of contamination. Studying heavy metal accumulation in earthworm tissue might be a good bioindicator of heavy metal availability in soils (Lanno and Mccarty 1997; Conder and Lanno 2000; Paoletti 1999; Oste et al. 2001).

15.4 Earthworms and Phytoremediation

15.4.1 Evidence of the Effect of Earthworms on Phytoremediation

The effects of earthworms on phytoremediation performance were described for the first time by (Ma et al. 2003) who found that the presence of the anecic earthworm *Pheretima guillelmi* increased the amount of Pb (mostly in roots) extracted by the leguminous plant *Leucaena leucocephala*. These results are in line with those of (Wang et al. 2006) who reported an increase of Zn phytoextraction by ryegrass and Indian mustard when the soil was inoculated with the earthworm *Pheretima* sp. The additional accumulation of heavy metals by plants as the result of earthworms was confirmed by (Dandan et al. 2007) with Cu uptake by ryegrass in the presence of *Metaphire guillelmi*. In recent years (Ruiz et al. 2009), used soil microcosms to show that the epigeic earthworm, *Eisenia fetida*, significantly increased the growth of maize (*Zea mays*) and barley (*Hordeum vulgare*) and resulted in the accumulation of heavy metals (Cu, Cd, Pb, and Zn). The plant growth led to a threefold increase in Zn extraction. The presence of the anecic earthworm *L. terrestris* also significantly increased the phytoextraction of Pb and Zn by maize and barley, although to a lesser extent (Ruiz et al. 2011). More recently, (Jusselme et al. 2012) studied the interaction between *Lantana camara* which is a hyperaccumulating plant for lead and cadmium and the endogenous tropical earthworm *P. corethrurus* (Oligochaeta, Glossoscolecidae) commonly found in both polluted and unpolluted areas. In this study, which used Pb-spiked soil in microcosms (500 and 1,000 mg kg^{-1}), most of the earthworms introduced into the microcosms remained alive (>90 %) after 1 month, and all the soil was burrowed by earthworms. With *P. corethrurus*, an increase of shoot and root biomass was recorded as well as an increase of lead uptake by plants (Jusselme et al. 2012). Eventually, Du et al. (2014) showed that the influence of the earthworm *Eisenia fetida* on the accumulation of Cd in leaves or stems of corn resulted in a Cd concentration in the soil of more than 1,000 mg Cd kg^{-1}. Conversely, *Pheretima* sp. only improved the phytoextraction of Zn and Pb by ryegrass *Lolium multiflorum* in moderately contaminated soils after the third harvest.

All these studies clearly demonstrated that all types of earthworms have a clear effect on the phytoremediation by various plants of soils contaminated by heavy metals. Some of the mechanisms by which earthworms influence phytoremediation are described below.

15.4.2 Mechanisms by Which Earthworms Influence Phytoremediation

15.4.2.1 Interactions Between Earthworms and Plants

Most studies showed that earthworms affected the growth of hyperaccumulating plants, in particular the roots (Table 15.2). The overall health of the plant is often given as the main factor for the increase in heavy metal phytoextraction performance. For example, Wang et al. (2006) showed that soil bioaugmentation by earthworms increased the biomass of ryegrass and Indian mustard which resulted in greater uptake and accumulation of zinc. The positive effect of earthworms on growth and heavy metal accumulation by plants may be direct and/or indirect through a positive effect on soil microorganisms such as arbuscular mycorrhizal fungi (AMF) (Eisenhauer et al. 2009; Ortiz-Ceballos 2007; Ma et al. 2006; Gaur and Adholeya 2004) and almost all plant growth-promoting bacteria (PGPB) (Sinha 2010; Wu et al. 2012) that are themselves known to improve phytoextraction performance (Lebeau et al. 2008; Sessitsch et al. 2013). In return, the earthworms use root exudates as a nutrient source to survive in polluted conditions. Earthworms increase the dispersion rate of viable mycorrhizal propagules and actinomycetes such as *Frankia* and PGPB, some of which are nitrogen fixing (Wu et al. 2006). The effect of microbial stimulation on the amount of nitrogen fixed by the plants could be an important part of the positive effect of earthworms. The microorganisms increase the primary biomass by stimulating the plant growth in various ways. Firstly, PGPB increase the plant biomass and root surface as well as reduce the toxicity of heavy metals to the plant. The amount of ethylene produced by the plant during induced heavy metal stress can be reduced by the degradation of ACC (aminocyclopropane carboxylic acid), a precursor of ethylene, by ACC-deaminase produced by PGPB (Ma et al. 2009a, b, 2011; Braud et al. 2009;

Table 15.2 Earthworm effects on heavy metal phytoextraction

	Barley		Maize		*Lantana camara*		Indian mustard		Ryegrass		Tomato	
Heavy metals	Without E	With E	Without E	With E	Without E	With E	Without E	With E	Without E	With E	Without E	With mucus E
Cd												
Shoot	n.d	2.8 ± 0.4	n.d	n.d							161.6 ± 14.1	265 ± 13.1
Root	n.d	14.4 ± 3.8	n.d	n.d							117.4 ± 82.8	2474.5 ± 119.8
Cu												
Shoot	10.8 ± 2.1[a]	9.5 ± 1.4	7.65 ± 0.57	10.2 ± 2.0					52.98[b]	69.52		
Root	14.8 ± 1.3	18.5 ± 2.2	8.26 ± 0.58	17.6 ± 3.6					854.97	1312.5		
Shoot	18.4 ± 4.2	15.2 ± 3.0	6.89 ± 2.15	9.5 ± 4.3								
Root	16.8 ± 2.0	19.0 ± 8.0	7.16 ± 1.94	17.7 ± 4.7								
Pb												
Shoot	26.5 ± 5.3	32.0 ± 6.8	26.6 ± 3.8[c]	20.8 ± 6.8	225.97 ± 74.9[d]	522.77 ± 135.1			5.76	13.36		
Root	109.1 ± 26.8	165.0 ± 58.7	101.3 ± 14.25	263.1 ± 66.1	1645.9 ± 590.9[d]	3284.08 ± 328.1			185.9	228.4		
Shoot	17.0 ± 3.6	21.9 ± 4.5	13.8 ± 5.0	13.2 ± 3.7								
Root	27.9 ± 4.7	51.7 ± 6.5	15.3 ± 5.4	30.6 ± 16.8								
Zn												
Shoot	123.6 ± 25.8	267. 0 ± 24.9	246.9 ± 30.9[c]	351.2 ± 36.0					210.5	285.5		
Root	454.6 ± 80.2	1305.4 ± 268.0	271.5 ± 44.5	567.7 ± 139.1					742.5	909		
Shoot	76.9 ± 3.8	86.2 ± 7.3	77.1 ± 3.3	86.6 ± 11.8			503.8[c]	774.5	621.11[c]	996.04		
Root	144 ± 17	198 ± 8.7	60.3 ± 4.2	137 ± 14			101.9[c]	240.3	467.75	928.17		

[a]Ruiz et al. (2009)
[b]Wang et al. (2007)
[c]Ma et al. (2003)
[d]Jusselme et al. (2012)
[e]Wang et al. (2006)

Denton 2007; Dimkpa et al. 2009; Grandlic et al. 2009). Secondly, absorption of iron by the plant can be facilitated by bacterial siderophores (Crowley et al. 1988, 1992; Bar-Ness et al. 1992; Glick 2003). Thirdly, some authors (Tomati et al. 1988; Castellanos Suarez et al. 2014) have tested the hypothesis that bacterial phytohormones, in particular indoleacetic acid (IAA), that stimulate root growth and are activated in earthworm casts are responsible for the increase in the root biomass and the available heavy metal for plants. The exploration of the casts by roots facilitates the assimilation of large amounts of resources with a reduced investment in carbon. The resulting carbon gain reduces catabolism, increases chlorophyll synthesis, and improves the rate of CO_2 fixation, consequently accelerating plant growth. This excess energy can also enable the plant to respond to environmental stress such as heavy metal toxicity.

15.4.2.2 Interactions Between Earthworms and Heavy Metals

Many studies have addressed the impact of earthworms themselves and their activities on the dynamics of heavy metals in the soil in terms of solubility, extractability, and bioavailability (Abdul Rida 1996; Devliegher and Verstraete 1996; Wen et al. 2004; Udovic and Lestan 2007). The presence of the earthworm *Lumbricus terrestris* increased Cu availability by 6 % (Devliegher and Verstraete 1996). Some earthworms such as *L. terrestris*, *L. rubellus*, and *Aporrectodea caliginosa* can survive in heavy metal-polluted soils (Langdon et al. 1999; Morgan and Morgan 1999; Kızılkaya 2008). They influence the mobility and availability of metal through their burrowing and casting activity (Sizmur and Hodson 2009; Sizmur et al. 2011a, b, c, d). However, the effect of these organisms on the bioavailability of heavy metals for plants remains very modest in heavy metal-contaminated sites (Abdul Rida 1996).

It was shown that earthworms can influence the heavy metal availability in soils by mixing deep soils, humus, and biological material in the earthworm gut (Hobbelen et al. 2006; Cheng and Wong 2002). Ma et al. (2002) demonstrated that the concentration of the available Pb was increased by up to 48.2 % by earthworm inoculation, and (Cheng and Wong 2002) suggested that earthworm burrowing and feeding activities increased Zn availability.

At present, most results concerning the effects of earthworms on the availability of heavy metal have been obtained from artificially contaminated soils and/or microcosm experiments. (Smolders et al. 2009) stated that soils artificially contaminated with fully soluble metal sources do not represent conditions prevailing in naturally metal-rich contaminated soils, and (Spurgeon and Hopkin 1995) reported that heavy metals in artificially contaminated soils are likely to be more bioavailable than in "naturally" polluted soils. Therefore, further studies are required to determine the extent to which differences between the results under controlled conditions and field results can be explained by the differences between artificially contaminated and "naturally" polluted soils.

While the application of earthworms to soils to increase heavy metal availability has in some instances increased metal extraction from soils, it is also important to note that there are some drawbacks that may have practical implications. Although heavy metal migration through the soil is very slow (e.g., 0.01 cm $year^{-1}$ for Pb considered to be relatively immobile) (Kylander et al. 2008), if the available heavy metals exceed the capacity of metal absorption by plants, it is possible that, in some soils, the increase of metal availability by the earthworm activity may lead to faster diffusion.

15.4.2.3 Earthworms and Soil Enzyme Activities

The capacity of plants to absorb heavy metals depends on the plants' health. Plant health in turn depends on soil quality. Moreover, a phytoextraction process must aim not only to remove the heavy metal from the soil but, more importantly, to restore soil quality (Doran and Safley 1997). Therefore, soil quality indicators are needed to assess the overall performance of a phytoextraction process. Of the various biological indicators, soil enzyme activity has been suggested as a good indicator of soil quality (Alkorta et al. 2003) as this plays an important role in mineralization processes that convert organic compounds into inorganic compounds. The role of enzymes in soils is expressed as the quantity of nutrients released such as nitrate, phosphorus (P), and potassium (K) that are important for plant growth.

In heavy metal-contaminated soils, most enzymes are inactive as they are inhibited by the protein-binding capacity of metals (Alkorta et al. 2003). Enzymes such as xylanase, cellulase, alkaline phosphatase, and fluorescein diacetate (FDA) activities are affected by the amount of heavy metal in the soils although (Jusselme et al. 2013) concluded that lead pollution ranging between 500 and 1,000 mg kg^{-1} of soils does not inhibit soil enzyme activities. These discrepancies can be explained by the nature and degree of inhibition of heavy metals as the effects depend largely on soil type, heavy metal levels, and soil physical and chemical properties.

Earthworms play a major role in promoting soil health, in particular soil enzyme activities. A comparison of soil enzyme activities in the presence and absence of earthworms by (Jusselme et al. 2013) in a Pb-phytoextraction experiment with Pb ranging between 500 and 1,000 mg kg^{-1} of soil clearly showed that the presence of earthworms significantly increased most enzyme activities. This was particularly true in the root-adhering soil of *Lantana camara*, the hyperaccumulating plant used in this study. However, these authors also showed that the increased activities of *N*-acetyl-D-glucosaminidase and urease involved in the nitrogen cycle in the presence of earthworms resulted in lower nitrogen availability for plants. This may be explained by the complex interactions between plants, nutrient availability, and earthworms: (1) initially, the earthworm activities increase N availability, improving plant growth and health and so stimulating the metal phytoextraction process, but the uptake of available nitrogen by the plants increases too fast and the available N becomes too scarce by the end of phytoextraction process, and (2) the earthworm activities improve plant growth and heavy metal uptake by mechanisms not based on nitrogen. Blouin et al. (2006) showed that earthworms (*Millsonia anomala*) do not increase rice growth by improving nitrogen mineralization. Phosphorus (P) is also an important nutrient for plant growth and reproduction. Plants use phosphorus for root development, flower initiation, and seed and fruit development (Fuhrman et al. 2005). James (1991) showed that P availability in earthworm casts could contribute about 50 % of the plants' requirements. Satchell and Martin (1984) suggested that higher levels of P availability in earthworm casts were based mainly on increased phosphatase activity, involved in the hydrolysis of organic P compounds in the casts, although it has not been established whether the increase in the activity is due to earthworm-derived enzymes or to increased microbial activity. The increase in P availability as a result of an increase in enzyme activities by earthworms improves plant growth as well as Pb uptake (Jusselme et al. 2012). Unlike nitrogen and phosphorus, potassium does not form any vital organic compounds in the plant. However, the presence of K is vital for plant growth because K is known to be an enzyme activator that boosts the metabolism. Jusselme et al. (2013) demonstrated that earthworms increased FDA activity, which could lead to increased K availability for plant uptake.

Since enzyme responses depend on the type of enzyme, enzyme activity could be used as a biological indicator to assess heavy metal-contaminated soil functioning. This result can be explained by the interaction of plant/microorganisms/earthworms as shown in Sect. 15.4.2.3.

15.4.2.4 Interactions Between Earthworms and Soil Microorganisms

Trace metals are known to be toxic to soil microorganisms (Del Val et al. 1999; Giller et al. 1998). They reduce the microbial activity (Lorenz et al. 2006; Oliveira and Pampulha 2006), diversity (Hassan et al. 2011; Hu et al. 2007), and abundance (Liu et al. 2012; Pasqualetti et al. 2012). However soil functioning is a result of tight interactions between microorganisms, plants, and soil macrofauna. Soil microorganisms are in part influenced by soil macrofauna (Aira et al. 2002, 2007, 2010). It has been shown that earthworms have a significant effect on the composition, distribution, and activity of soil fungi, in particular by (1) ingesting fungal spores and even ingesting certain fungi, (2) creating microsites favorable to fungal development, (3) dispersing fungal species, and (4) transforming and redistributing soil organic matter (Brown 1995; Lavelle 2002; Scheu et al. 2002). Despite a large body of literature on the impact of earthworms and microorganisms on heavy metals in soils, only a few studies have addressed the question of earthworm-assisted heavy metal phytoextraction (Wang et al. 2006; Yu et al. 2005) although without considering the role of microorganisms (Aghababaei et al. 2014b).

The question arises whether the beneficial effect of earthworms on phytoextraction performance results from the stimulation of soil microorganisms. Although earthworms may be able to increase metal bioavailability in heavy metal-contaminated soil, the mechanism remains unclear. Sizmur and Hodson (2009) suggested four principal mechanisms by which earthworm activities may change heavy metal mobility and bioavailability: (1) modification of soil pH, (2) modification of soil dissolved organic carbon (DOC), (3) heavy metal speciation and sequestration within the earthworm tissue, and (4) stimulation of the soil microbial population.

Although heavy metals are toxic to soil microorganisms, Jusselme et al. (2012) showed that the total microbial activity and fungal richness index based on DGGE patterns increased with Pb pollution in the root-adhering soil of *L. camara*. This did not agree with previous results for other heavy metals (As, Cd, Hg) in long-term contaminated sites (Lorenz et al. 2006; Oliveira and Pampulha 2006). There may be various explanations for this discrepancy: (1) the duration of the exposition in this study was short (1 month vs. several years), and (2) the growth and activity of *L. camara* roots may have stimulated microbial activity as a result of a higher amount of rhizodeposits and counterbalanced the toxicity of Pb.

Microbial activity increases in the presence of earthworms as shown in many studies (Aira et al. 2008, 2010; Dempsey et al. 2013; Gómez-Brandón et al. 2012; Tao et al. 2009; Tiwari and Mishra 1993). As earthworms digest decaying substrates, they increase the pool size of nutrients available for microorganisms, promoting microbial growth. Microorganisms are largely dormant in the soil waiting for favorable conditions which are provided by earthworm burrowing and casting (Lavelle 2002). Earthworms significantly increase total fungal abundance and all fungal diversity indices, as has already been shown (Krišrtuek et al. 1992). Del Val et al. (1999) and Jusselme et al. (2013) showed that, in heavy metal-spiked soils bioaugmented with earthworms, the structure of the fungal community is modified: the appearance of new bands indicated that minor populations in uncontaminated soils became dominant in heavy metal-spiked soils as the result of their tolerance. The positive effect of earthworms on the abundance of cultivable fungi counteracted the negative effect of heavy metals in the polluted soils in spite of the higher bioavailable concentration of heavy metals in soils. The effects of earthworms on the activity, abundance, and structure of the fungal community may (1) increase fungal growth by means of decaying substrates (Brown 1995) and (2) select fungal populations more adapted to heavy metal contamination (Hui et al. 2012). Smith and Reed also showed that the activity of mycorrhizal fungi was increased by earthworms resulting in an increase in the exchange surface between plants and soil, increasing plant uptake and biomass. The effect of earthworms alone on Cd availability is greater than that of AMF in Cd-polluted soils, and interactions between these organisms have a much greater effect on soil microorganisms than on Cd availability. Thus, the presence of both earthworms and AMF could mitigate the effects of Cd on soil microbial life (Aghababaei et al. 2014b).

These results suggest a combined positive effect of earthworms and soil microorganisms on (1) the availability of heavy metals in soil and (2) the availability of nutrients leading to a higher plant biomass and increasing heavy metal absorption and accumulation by plants. Earthworms, as soil ecosystem engineers, are known to change the microbial composition and to stimulate its activity (Brown 1995; Binet and Le Bayon 1998), while soil microorganisms, as decomposers, improve nutrient mineralization and availability for plants (Berg and Laskowski 2005).

15.5 Conclusion: Mutualistic Interactions Between Plants, Microorganisms, and Earthworms

The bioaugmentation of earthworms in soils spiked by heavy metals modifies microbial functioning, counterbalances the effect of heavy metals on the fungal community (abundance, diversity, and structure) and promotes the phytoextraction of heavy metals by plants. Positive interactions between plants, microorganisms, and earthworms form a virtuous circle: improving any one of the interactions improves all the others (Fig. 15.1). Taking just the activation of microorganisms (PGPB and AMF), there is a benefit for both the plants and earthworms.

The activation by earthworms of microorganisms producing compounds which target heavy metals reduces the toxic effect of these metals and increases the heavy metal availability for the plants. The result is an indirect positive effect on plants exposed to heavy metal pollution (Fig. 15.2). Apart from the direct toxicity of heavy

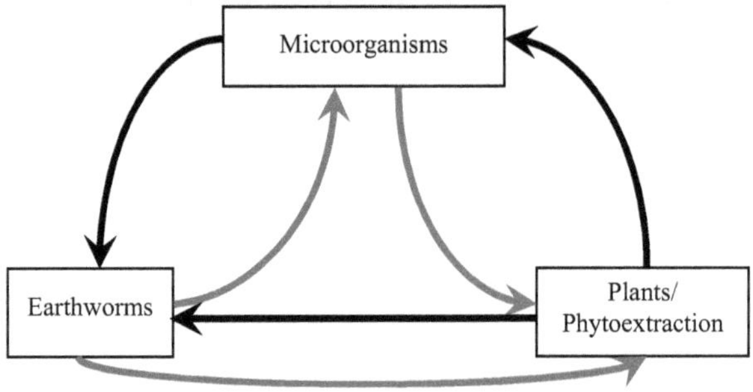

Fig. 15.1 Virtuous circle of interactions between plants, microorganisms, and earthworms

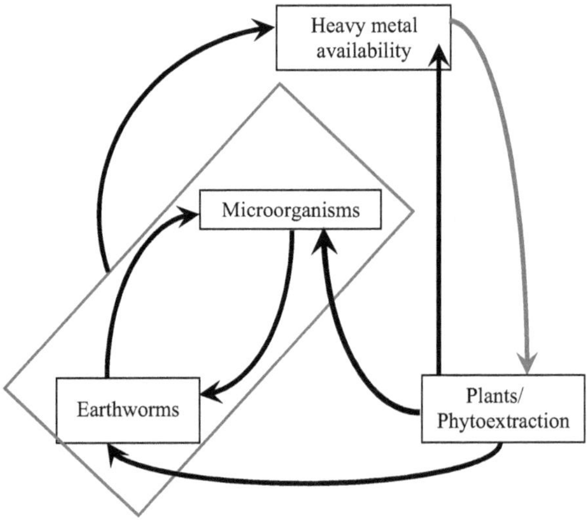

Fig. 15.2 Interactions between plants, microorganisms, and earthworms and Pb availability

metals to plants, improving any one of the interactions is beneficial for plants, microorganisms, and earthworms.

References

Abdul Rida AMM (1996) Concentrations et croissance de lombriciens et de plantes dans des sols contaminés ou non par Cd, Cu, Fe, Pb et Zn: interactions plante-sol-lombricien. Soil Biol Biochem 28:1037–1044

Aghababaei F, Raiesi F, Hosseinpur A (2014a) The combined effects of earthworms and arbuscular mycorrhizal fungi on microbial biomass and enzyme activities in a calcareous soil spiked with cadmium. Appl Soil Ecol 75:33–42

Aghababaei F, Raiesi F, Hosseinpur A (2014b) The influence of earthworm and mycorrhizal co-inoculation on Cd speciation in a contaminated soil. Soil Biol Biochem 78:21–29

Aira M et al (2002) How earthworm density affects microbial biomass and activity in pig manure. Eur J Soil Biol 38:7–10

Aira M, Monroy F, Domínguez J (2007) Earthworms strongly modify microbial biomass and activity triggering enzymatic activities during vermicomposting independently of the application rates of pig slurry. Sci Total Environ 385(1–3):252–261

Aira M et al (2008) Microbial communities of *Lumbricus terrestris* L. middens: structure, activity, and changes through time in relation to earthworm presence. J Soils Sediments 9(1):54–61

Aira M et al (2010) Ageing effects of casts of *Aporrectodea caliginosa* on soil microbial community structure and activity. Appl Soil Ecol 46:143–146

Alkorta I et al (2003) Soil enzyme activities as biological indicators of soil health. Rev Environ Health 18:65–73

Bar-Ness E et al (1992) Short-term effects of rhizosphere microorganisms on Fe uptake from microbial siderophores by maize and oat. Plant Physiol 100:451–456

Barois I et al (1987) Influence of the tropical earthworm Pontoscolex corethrurus (Glossoscolecidae) on the fixation and mineralization of nitrogen. In: Bonvicini AM, Omodeo P (eds) On earthworms. Mucchi, Bologna, pp 151–158

Bengtsson G, Ek H, Rundgren S (1992) Evolutionary response of earthworms to long-term metal exposure. Oikos 63:289–297

Berg B, Laskowski R (2005) Litter decomposition: a guide to carbon and nutrient turnover, vol 38. Academic, San Diego, p. 421

Beyer WN, Chaney RL, Mulhern BM (1982) Heavy metal concentrations in earthworms from soil amended with sewage sludge. J Environ Qual 11:381–385

Binet F, Curmi P (1992) Structural effects of *Lumbricus terrestris* (oligochaeta: lumbricidae) on the soil-organic matter system: micromorphological observations and autoradiographs. Soil Biol Biochem 24:1519–1523

Binet F, Le Bayon RC (1998) Space-time dynamics in situ of earthworm casts under temperate cultivated soils. Soil Biol Biochem 31:85–93

Blouin M, Barot S, Lavelle P (2006) Earthworms (Millsonia anomala, Megascolecidae) do not increase rice growth through enhanced nitrogen mineralization. Soil Biol Biochem 38:2063–2068

Blouin M, Lavelle P, Laffray D (2007) Drought stress in rice (*Oryza sativa* L.) is enhanced in the presence of the compacting earthworm Millsonia anomala. Env Exp Bot 60:352–359

Blouin M et al (2013) A review of earthworm impact on soil function and ecosystem services. Eur J Soil Sci 64:161–182

Bohlen PJ et al (2002) Indirect effects of earthworms on microbial assimilation of labile carbon. Appl Soil Ecol 20:255–261

Bouche M (1977) Strategies lombriciennes. Ecol Bull 25:122–132

Bouché MB, Al-Addan F (1997) Earthworms, water infiltration and soil stability: some new assessments. Soil Biol Biochem 29:441–452

Braud A et al (2009) Enhanced phytoextraction of an agricultural Cr- and Pb-contaminated soil by bioaugmentation with siderophore-producing bacteria. Chemosphere 74(2):280–286, Available at: http://dx.doi.org/10.1016/j.chemosphere.2008.09.013

Brown GG (1995) How do earthworms affect microfloral and faunal community diversity? Plant and Soil 170:209–231

Brown GG, Barois I, Lavelle P (2000) Regulation of soil organic matter dynamics and microbial activity in the drilosphere and the role of interactions with other edaphic functional domains. Eur J Soil Biol 36:177–198

Castellanos Suarez DE et al (2014) Combined effects of earthworms and IAA-producing rhizobacteria on plant growth and development. Appl Soil Ecol 80:100–107

Chaney RL et al (1997) Phytoremediation of soil metals. Curr Opin Biotechnol 8:279–284

Cheng J, Wong MH (2002) Effects of earthworms on Zn fractionation in soils. Biol Fertil Soils 36:72–78

Conder JM, Lanno RP (2000) Evaluation of surrogate measures of cadmium, lead, and zinc bioavailability to Eisenia fetida. Chemosphere 41:1659–1668

Crowley DE, Reid CP, Szaniszlo PJ (1988) Utilization of microbial siderophores in iron acquisition by oat. Plant Physiol 87:680–685

Crowley DE et al (1992) Root-microbial effects on plant iron uptake from siderophores and phytosiderophores. Plant and Soil 142:1–7

Dandan W et al (2007) Role of earthworm-straw interactions on phytoremediation of Cu contaminated soil by ryegrass. Acta Ecol Sinica 27:1292–1298

Dechaine J et al (2005) Correlation between earthworms and plant litter decomposition in a tropical wet forest of Puerto Rico. Pedobiologia 49:601–607

Del Val C, Barea JM, Azcón-Aguilar C (1999) Diversity of arbuscular mycorrhizal fungus populations in heavy-metal-contaminated soils. Appl Environ Microbiol 65:718–723

Dempsey MA et al (2013) Exotic earthworms alter soil microbial community composition and function. Soil Biol Biochem 67:263–270

Denton B (2007) Advances in phytoremediation of heavy metals using plant growth promoting bacteria and fungi. Biotechnology 3:1–5

Derouard L et al (1997) Effects of earthworm introduction on soil processes and plant growth. Soil Biol Biochem 29:541–545

Devliegher W, Verstraete W (1996) *Lumbricus terrestris* in a soil core experiment: effects of nutrient- enrichment processes (NEP) and gut-associated processes (GAP) on the availability of plant nutrients and heavy metals. Soil Biol Biochem 28:489–496

Dimkpa C, Merten D, Svatoš A (2009) Metal-induced oxidative stress impacting plant growth in contaminated soil is alleviated by microbial siderophores. Soil Biol Biochem 41:154–162

Doran JW, Safley M (1997) Defining and assessing soil health and sustainable productivity. In: Pankhurst C, Doube BM, Gupta VVSR (eds) Biological indicators of soil health. CAB International, Wallingford, pp 1–28

Doube BM, Buckerfield JC, Kirkegaard JA (1994) Short-term effects of tillage and stubble management on earthworm populations in cropping systems in southern New South Wales. Aust J Agr Res 45:1587–1600

Doube BM et al (1997) Influence of mineral soil on the palatability of organic matter for lumbricid earthworms: a simple food preference study. Soil Biol Biochem 29:569–575

Du Y et al (2014) Interactive effects between earthworms and maize plants on the accumulation and toxicity of soil cadmium. Soil Biol Biochem 72:193–202

Edwards CA, Fletcher KE (1988) Interactions between earthworms and microorganisms in organic-matter breakdown. Agr Ecosyst Environ 24:235–247

Edwards WM et al (1992) Role of *Lumbricus terrestris* (L.) burrows on quality of infiltrating water. Soil Biol Biochem 24:1555–1561

Eisenhauer N, König S, Sabais A (2009) Impacts of earthworms and arbuscular mycorrhizal fungi (*Glomus intraradices*) on plant performance are not interrelated. Soil Biol Biochem 41:561–567

Evangelou MWH, Ebel M, Schaeffer A (2007) Chelate assisted phytoextraction of heavy metals from soil: effect, mechanism, toxicity, and fate of chelating agents. Chemosphere 68:989–1003

Fuhrman JK et al (2005) Water-soluble phosphorus as affected by soil to extractant ratios, extraction times, and electrolyte. Commun Soil Sci Plant Anal 36:925–935

Gaur A, Adholeya A (2004) Prospects of arbuscular mycorrhizal fungi in phytoremediation of heavy metal contaminated soils. Curr Sci 86(4), Available at: http://www.currentscience.ac.in/Downloads/article_id_086_04_0528_0534_0.pdf. Accessed 11 Sep 2014

Giller KE, Witter E, Mcgrath SP (1998) Toxicity of heavy metals to microorganisms and microbial processes in agricultural soils: a review. Soil Biol Biochem 30:1389–1414

Glick B (2003) Phytoremediation: synergistic use of plants and bacteria to clean up the environment. Biotechnol Adv 21:383–393

Gómez-Brandón M, Lores M, Domínguez J (2012) Species-specific effects of epigeic earthworms on microbial community structure during first stages of decomposition of organic matter. PLoS One 7:e31895

Grandlic C, Palmer M, Maier R (2009) Optimization of plant growth-promoting bacteria-assisted phytostabilization of mine tailings. Soil Biol Biochem 41(8):1734–1740

Hassan SED et al (2011) Molecular biodiversity of arbuscular mycorrhizal fungi in trace metal-polluted soils. Mol Ecol 20(16):3469–3483, Available at: http://www.ncbi.nlm.nih.gov/pubmed/21668808. Accessed 11 Sep 2014

Hobbelen P, Koolhaas J, van Gestel C (2006) Effects of heavy metals on the litter consumption by the earthworm *Lumbricus rubellus* in field soils. Pedobiologia 50:51–60

Hopkin SP (1989) Ecophysiology of metals in terrestrial invertebrates. Elsevier Applied Science, London

Hu Q et al (2007) Bacterial diversity in soils around a lead and zinc mine. J Environ Sci 19 (1):74–79

Hui N, Liu XX, Kurola J, Mikola J, Romantschuk M (2012) Lead (Pb) contamination alters richness and diversity of the fungal, but not the bacterial community in pine forest soil. Boreal Env Res 17:46–58

James SW (1991) Soil, nitrogen, phosphorus, and organic matter processing by earthworms in tallgrass prairie. Ecology 72:2101–2109

Jones CG, Lawton JH, Shachak M (1994) Organisms as ecosystem engineers. Oikos 69:373–386

Joschko M, Diestel H, Larink O (1989) Assessment of earthworm burrowing efficiency in compacted soil with a combination of morphological and soil physical measurements. Biol Fertil Soils 8:191–196

Jusselme MD et al (2012) Effect of earthworms on plant *Lantana camara* Pb-uptake and on bacterial communities in root-adhering soil. Sci Total Environ 416:200–207

Jusselme MD et al (2013) Increased lead availability and enzyme activities in root-adhering soil of *Lantana camara* during phytoextraction in the presence of earthworms. Sci Total Environ 445–446:101–109

Khan AG (2005) Role of soil microbes in the rhizospheres of plants growing on trace metal contaminated soils in phytoremediation. J Trace Elem Med Biol 18:355–364

Kızılkaya R (2008) Dehydrogenase activity in *Lumbricus terrestris* casts and surrounding soil affected by addition of different organic wastes and Zn. Bioresour Technol 99:946–953

Krišrtuek V, Ravasz K, Pižl V (1992) Changes in densities of bacteria and microfungi during gut transit in *Lumbricus rubellus* and *Aporrectodea caliginosa* (Oligochaeta: Lumbricidae). Soil Biol Biochem 24:499–1500

Kylander ME et al (2008) Lead penetration and leaching in a complex temperate soil profile. Environ Sci Tech 42:3177–3184

Langdon CJ et al (1999) Resistance to arsenic-toxicity in a population of the earthworm *Lumbricus rubellus*. Soil Biol Biochem 31:1963–1967

Lanno RP, Mccarty LS (1997) Earthworm bioassays: adopting techniques from aquatic toxicity testing. Soil Biol Biochem 29:693–697

Lavelle P (1981) Stratégies de reproduction chez les vers de terre (in French, with English summary). Acta Oecol Oecol Gen 2:117–133

Lavelle P (1996) Diversity of soil fauna and ecosystem function. Biol Int 33:3–16

Lavelle P (1997) Biology and ecology of earthworms. Agr Ecosyst Environ 64:78–79
Lavelle P (2002) Functional domains in soils. Ecol Res (October 2001). Available at: http://onlinelibrary.wiley.com/doi/10.1046/j.1440-1703.2002.00509.x/full. Accessed 17 Dec 2013
Lawton JH (1994) What do species do in ecosystems? Oikos 71:367–374
Lebeau T, Braud A, Jézéquel K (2008) Performance of bioaugmentation-assisted phytoextraction applied to metal contaminated soils: a review. Environ Pollut (Barking, Essex: 1987) 153 (3):497–522, Available at: http://www.ncbi.nlm.nih.gov/pubmed/17981382. Accessed 4 Oct 2013
Liu Y et al (2012) Decline in topsoil microbial quotient, fungal abundance and C utilization efficiency of rice paddies under heavy metal pollution across South China. PLoS One 7(6):p. e38858, Available at: http://www.pubmedcentral.nih.gov/articlerender.fcgi?artid=3372496&tool=pmcentrez&rendertype=abstract. Accessed 11 Sep 2014
Lorenz N et al (2006) Response of microbial activity and microbial community composition in soils to long-term arsenic and cadmium exposure. Soil Biol Biochem 38(6):1430–1437, Available at: http://linkinghub.elsevier.com/retrieve/pii/S0038071705004104. Accessed 11 Sep 2014
Losfeld G et al (2012) The chemical exploitation of nickel phytoextraction: an environmental, ecologic and economic opportunity for New Caledonia. Chemosphere 89:907–910
Lukkari T, Haimi J (2005) Avoidance of Cu- and Zn-contaminated soil by three ecologically different earthworm species. Ecotoxicol Environ Saf 62:35–41
Luo C, Shen Z, Li X (2005) Enhanced phytoextraction of Cu, Pb, Zn and Cd with EDTA and EDDS. Chemosphere 59:1–11
Luo C, Shen Z, Lou L et al (2006a) EDDS and EDTA-enhanced phytoextraction of metals from artificially contaminated soil and residual effects of chelant compounds. Environ Pollut 144:862–871
Luo C, Shen Z, Li X et al (2006b) Enhanced phytoextraction of Pb and other metals from artificially contaminated soils through the combined application of EDTA and EDDS. Chemosphere 63:1773–1784
Ma Y, Dickinson MN, Wong M (2002) Toxicity of Pb/Zn mine tailings to the earthworm Pheretima and the effects of burrowing on metal availability. Biol Fertil Soils 36:79–86
Ma Y, Dickinson N, Wong M (2003) Interactions between earthworms, trees, soil nutrition and metal mobility in amended Pb/Zn mine tailings from Guangdong, China. Soil Biol Biochem 35:1369–1379
Ma Y, Dickinson N, Wong M (2006) Beneficial effects of earthworms and arbuscular mycorrhizal fungi on establishment of leguminous trees on Pb/Zn mine tailings. Soil Biol Biochem 38:1403–1412
Ma Y, Rajkumar M, Freitas H (2009a) Improvement of plant growth and nickel uptake by nickel resistant-plant-growth promoting bacteria. J Hazard Mater 166:1154–1161
Ma Y, Rajkumar M, Freitas H (2009b) Inoculation of plant growth promoting bacterium Achromobacter xylosoxidans strain Ax10 for the improvement of copper phytoextraction by Brassica juncea. J Environ Manage 90:831–837
Ma Y et al (2011) Plant growth promoting rhizobacteria and endophytes accelerate phytoremediation of metalliferous soils. Biotechnol Adv 29(2):248–258
Morgan JE, Morgan AJ (1999) The accumulation of metals (Cd, Cu, Pb, Zn and Ca) by two ecologically contrasting earthworm species (Lumbricus rubellus and Aporrectodea caliginosa): implications for ecotoxicological testing. Appl Soil Ecol 13:9–20
Oliveira A, Pampulha ME (2006) Effects of long-term heavy metal contamination on soil microbial characteristics. J Biosci Bioeng 102(3):157–161
Ortiz-Ceballos A (2007) Mycorrhizal colonization and nitrogen uptake by maize: combined effect of tropical earthworms and velvetbean mulch. Biol Fertil Soils 44:181–186
Oste LA et al (2001) Cadmium uptake by earthworms as related to the availability in the soil and the intestine. Environ Toxicol Chem 20:1785–1791
Padmavathiamma PK, Li LY (2007) Phytoremediation technology: hyper-accumulation metals in plants. Water Air Soil Pollut 184:105–126

Paoletti MG (1999) The role of earthworms for assessment of sustainability and as bioindicators. Agr Ecosyst Environ 74:137–155
Parle JN (1963a) Microorganisms in the intestines of earthworms. J Gen Microbiol 31:1–11
Parle JN (1963b) A microbiological study of earthworm casts. J Gen Microbiol 31:13–22
Pasqualetti M et al (2012) Effects of long-term heavy metal contamination on soil fungi in the Mediterranean area. Cryptogamie, Mycologie 33:43–57
Pizl V, Josens G (1995) Earthworm communities along a gradient of urbanization. Environ Pollut (Barking, Essex : 1987) 90:7–14
Ruiz E, Rodríguez L, Alonso-Azcárate J (2009) Effects of earthworms on metal uptake of heavy metals from polluted mine soils by different crop plants. Chemosphere 75(8):1035–1041
Ruiz E, Alonso-Azcárate J, Rodríguez L (2011) *Lumbricus terrestris* L. Activity increases the availability of metals and their accumulation in maize and barley. Environ Pollut 159:722–728
Salt DE, Smith RD, Raskin I (1998) Phytoremediation. Annu Rev Plant Biol 49:643–668
Satchell JE, Martin K (1984) Phosphatase activity in earthworm faeces. Soil Biol Biochem 16:191–194
Scheu S et al (2002) Effects of the presence and community composition of earthworms on microbial community functioning. Oecologia 133:254–260
Sessitsch A et al (2013) The role of plant-associated bacteria in the mobilization and phytoextraction of trace elements in contaminated soils. Soil Biol Biochem 60:182–194
Shah K, Nongkynrih JM (2007) Metal hyperaccumulation and bioremediation. Biol Plant 51:618–634
Singh OV et al (2003) Phytoremediation: an overview of metallic ion decontamination from soil. Appl Microbiol Biotechnol 61:405–412
Sinha RK (2010) Earthworms: Charles Darwin's "Unheralded Soldiers of Mankind": protective & productive for man & environment. J Environ Protect 01:251–260
Sizmur T, Hodson ME (2009) Do earthworms impact metal mobility and availability in soil?–a review. Environ Pollut (Barking, Essex: 1987) 157(7):1981–1989
Sizmur T, Watts MJ, Brown GD et al (2011a) Impact of gut passage and mucus secretion by the earthworm *Lumbricus terrestris* on mobility and speciation of arsenic in contaminated soil. J Hazard Mater 197:169–175
Sizmur T, Palumbo-Roe B, Watts MJ et al (2011b) Impact of the earthworm *Lumbricus terrestris* (L.) on As, Cu, Pb and Zn mobility and speciation in contaminated soils. Environ Pollut (Barking, Essex: 1987) 159(3):742–748, Available at: http://www.ncbi.nlm.nih.gov/pubmed/21185630. Accessed 4 Sep 2013
Sizmur T, Tilston EL, Charnock J et al (2011c) Impacts of epigeic, anecic and endogeic earthworms on metal and metalloid mobility and availability. J Environ Monit 13(2):266–273
Sizmur T, Palumbo-Roe B, Hodson ME (2011d) Impact of earthworms on trace element solubility in contaminated mine soils amended with green waste compost. Environ Pollut (Barking, Essex: 1987) 159(7):1852–1860
Smolders E et al (2009) Toxicity of trace metals in soil as affected by soil type and aging after contamination: using calibrated bioavailability models to set ecological soil standards. Environ Toxicol Chem 28:1633–1642
Spurgeon DJ, Hopkin SP (1995) Extrapolation of the laboratory-based OECD earthworm toxicity test to metal-contaminated field sites. Ecotoxicology 4:190–205
Spurgeon DJ, Hopkin SP (1996) The effects of metal contamination on earthworm populations around a smelting works: quantifying species effects. Appl Soil Ecol 4:147–160
Spurgeon DJ, Hopkin SP (1999) Tolerance to zinc in populations of the earthworm *Lumbricus rubellus* from uncontaminated and metal-contaminated ecosystems. Arch Environ Contam Toxicol 37:332–337
Stork NE, Eggleton P (1992) Invertebrates as determinants and indicators of soil quality. Am J Altern Agric 7:38
Tao J et al (2009) Effects of earthworms on soil enzyme activity in an organic residue amended rice–wheat rotation agro-ecosystem. Appl Soil Ecol 42:221–226

Tapia-Coral SC et al (2006) Effect of Pontoscolex corethrurus Muller, 1857 (Oligochaeta: Glossoscolecidae) inoculation on litter weight loss and soil nitrogen in Mesocosms in the Peruvian Amazon. Carib J Sci 42:410–418

Tiunov AV, Scheu S (1999) Microbial respiration, biomass, biovolume and nutrient status in burrow walls of *Lumbricus terrestris* L. (Lumbricidae). Soil Biol Biochem 31:2039–2048

Tiwari SC, Mishra RR (1993) Fungal abundance and diversity in earthworm casts and in uningested soil. Biol Fertil Soils 16:131–134

Tiwari SC, Tiwari BK, Mishra RR (1989) Microbial populations, enzyme activities and nitrogen-phosphorus-potassium enrichment in earthworm casts and in the surrounding soil of a pineapple plantation. Biol Fertil Soils 8:178–182

Tomati U, Grappelli A, Galli E (1988) The hormone-like effect of earthworm casts on plant growth. Biol Fertil Soils 5:288–294

Udovic M, Lestan D (2007) The effect of earthworms on the fractionation and bioavailability of heavy metals before and after soil remediation. Environ Pollut 148:663–668

Wang D et al (2006) Effect of earthworms on the phytoremediation of zinc-polluted soil by ryegrass and Indian mustard. Biol Fertil Soils 43:120–123

Wang Y et al (2007) The influence of soil heavy metals pollution on soil microbial biomass, enzyme activity, and community composition near a copper smelter. Ecotoxicol Environ Saf 67:75–81

Weltje L (1998) Mixture toxicity and tissue interactions of Cd, Cu, Pb and Zn in earthworms (Oligochaeta) in laboratory and field soils: a critical evaluation of data. Pergamon 36 (12):2643–2660

Wen B et al (2004) The role of earthworms (*Eisenia fetida*) in influencing bioavailability of heavy metals in soils. Biol Fertil Soils 40:181–187

Wenzel WW (2009) Rhizosphere processes and management in plant-assisted bioremediation (phytoremediation) of soils. Plant and Soil 321:385–408

Wu SC et al (2006) Effects of inoculation of plant growth-promoting rhizobacteria on metal uptake by Brassica juncea. Environ Pollut 140:124–135

Wu F et al (2012) Effects of earthworms and plant growth-promoting rhizobacteria (PGPR) on availability of nitrogen, phosphorus, and potassium in soil. J Plant Nutr Soil Sci 175:423–433

Yu X, Cheng J, Wong M (2005) Earthworm–mycorrhiza interaction on Cd uptake and growth of ryegrass. Soil Biol Biochem 37:195–201

Zhang H, Schrader S (1993) Earthworm effects on selected physical and chemical properties of soil aggregates. Biol Fertil Soils 15:229–234

Chapter 16
Phytochemical Removal of Heavy Metal-Contaminated Soils

Ali A. Abdel-Salam, Haythum M. Salem, Mohamed A. Abdel-Salam, and Mahmoud F. Seleiman

16.1 Phytochemical Remediation

Contamination of soils by heavy metals and metalloids has recently been a major concern particularly with the rapid expansion in industrial activities. It has become a global threat to plants, humans, animals, fish, and other living organisms. Such metals remain indefinitely in the soil (Bodar et al. 2006; Fotakis and Timbrell 2006). There are two methods for the remediation of the contaminated soils (i.e., chemo-remediation and phyto-remediation).

The cleanup of soils contaminated with heavy metals is one of the most difficult tasks for environmental engineering. A number of techniques have been developed to remove heavy metals from contaminated soil, including ex situ washing with physical-chemical methods (Anderson 1993) and in situ phyto-extraction (McGrath 1998 and Salt et al. 1998). In the ex situ washing methods, chelating agents or acids are used to enhance heavy metal removal.

16.1.1 Chemo-remediation

Chemo-remediation is remediating the soil through the use of some chemicals such as acids, chelators, and immobilizers of heavy metals. Treating contaminated soils

A.A. Abdel-Salam • H.M. Salem • M.A. Abdel-Salam
Department of soils and water, Faculty of Agriculture, Benha University, 13637, Benha, Egypt

M.F. Seleiman (✉)
Department of Agricultural Sciences, University of Helsinki, 27, 00014 Helsinki, Finland

Department of Crop Sciences, Faculty of Agriculture, Minufiya University, 32511 Shibin El-kom, Egypt
e-mail: mahmoud.seleiman@helsinki.fi

I. Sherameti, A. Varma (eds.), *Heavy Metal Contamination of Soils*, Soil Biology 44, DOI 10.1007/978-3-319-14526-6_16

with acids and chelators can increase the mobility of heavy metals either as soluble metal ions or as complexes of metal chelates, rendering the heavy metals easily removed from soil through leaching, electrokinetic extraction, or plant accumulation (Lestan et al. 2008). Chelants can either be mixed with the soil or added close to the plant roots (Kayser et al. 1999). Chelates could be added in one operation or in different successive applications. The latter one is more effective (Shen et al. 2002). Successive application by different combinations of chelates can be highly efficient in remediation, following three different combination operations. The first one is using a mixture which reduces soil pH, consequently increasing the solubility of heavy metals in the soil (Blaylock et al. 1997). The second is using a mixture which increases heavy metal mobility via reaction between the mixture components and heavy metals in soil (Tandy et al. 2004). The third is using a mixture which increases metal uptake by plants (Luo et al. 2006; Mathis and Kayser 2001). The method causes immobilization of heavy metals in soil by adding particular soil amendments to the contaminated soil to increase adsorption of heavy metals or decrease their mobility, hence decreasing their harmful effect. These amendments comprise inorganic materials (e.g., lime and fly ash), organic materials (e.g., organic residues), and synthetic materials (e.g., zeolites and Fe/Al hydrous oxides) (Bolan and Duraisamy 2003).

Mainly, there are four categories used in chemo-remediation: (a) synthetic aminopolycarboxylic acids, (b) natural biodegradable acids, (c) low-molecular-weight organic acids, and (d) humic substances including humic acids (Evangelou et al. 2007a). Synthetic aminopolycarboxylic acids are generally used to supply plants with micronutrients. They are used to increase the removed heavy metals either through uptake by plants grown on the soil or through leaching of the bare soil (Luo et al. 2005; Finzgar and Lestan 2007). They include ethylenediaminetetraacetic acid (EDTA), hydroxyethylene diamine tetraacetic acid (HEDTA), diethylene triamino pentaacetic acid (DTPA), trans-1,2-cyclohexylene dinitrilo tetraacetic acid (CDTA), *ethylene glycol tetraacetic acid* (EGTA), ethylenediamine-*N*,*N*′-bis (*o*-hydroxyphenyl)acetic acid (EDDHA), N-(2-hydroxyethyl)iminodia- cetic acid (HEIDA), and *N*,*N*′-bis(2-hydroxybenzyl) ethylenediamine-*N*,*N*′-diacetic acid (HBED) (Evangelou et al. 2007a). Although synthetic aminopolycarboxylic acids (EDTA in) are effective remediators, their low biodegradability makes them not very efficient (Nowack 2002). Natural biodegradable acids have been proposed as substitute for the synthetic aminopolycarboxylic acids due to their biodegradability, hence they are considered friendly to the environment. They include ethylene diamine disuccinate (EDDS) and nitrilotriacetic acid (NTA) (Lestan et al. 2008). Natural low-molecular-weight organic acids are created in plant roots. They affect the solubility of heavy metals directly and indirectly: directly, through reducing the soil pH, metal precipitation, metal chelating, and reduction-oxidation reactions, and indirectly, through the effect on growth and activity of microorganisms, on soil physical properties, and on plant root growth (Uren and Reisenauer 1988). Natural low-molecular-weight organic acids include citric, oxalic, and malic acids (Nigam et al. 2001). Humic substances are decomposed residues of plants, animals, and other organisms. Humic substances include humic acids, fulvic acid, and humin (McCarthy 2001).

Generally, the chemo-remediation method is costly; besides, leaching of different elements from contaminated soil can contaminate water of agricultural drains and consequently can contaminate the environment.

16.1.1.1 Soil Remediation by Extractants

In general, although a number of well-established extraction reagents have useful predictive power, many of them are specific to one element and are relevant to specific crops and may be restricted to particular soils. Perhaps the most useful extractants are 0.01 M or 0.05 M EDTA and 0.005 M DTPA.

AB-DTPA Extractant

Many soil testing laboratories use the 0.5 M sodium bicarbonate to extract available P and the 1.0 M ammonium acetate to extract available K. A universal extractant 1.0 M ammonium bicarbonate/0.005 M DTPA (i.e., AB-DTPA) is used to extract available P, K, Fe, Mn, Zn, Cu, and nitrate-N (Soltanpour 1991).

EDTA Extraction

Ethylenediaminetetraacetic acid (EDTA) is the most commonly used chelating extractant, because of its strong chelating ability for heavy metals (Norvell 1991). It is effective in removing Pb, Zn, Cu, and Cd from contaminated soils, although extraction efficiency depends on many factors such as the extent of heavy metals in soil, strength of the extractant, contents of electrolytes, soil pH, and soil matrix (Elliot and Brown 1989; Brown and Elliot 1992; Pichtel and Pichtel 1997; Elliot and Shastri 1999; Heil et al. 1999; Papassiopi et al. 1999). After its use in soil washing, the EDTA extractant may be recovered and reused through chemical reaction of the EDTA-heavy metal complexes with sodium sulfide (Na_2S), (Hong et al. 1999).

16.1.2 Phyto-remediation

Phyto-remediation is the method by which plants are used to sequester, remove, or accumulate different heavy metals from the environment (Cunningham and Ow 1996; Raskin et al. 1997). Phyto-remediation is defined as using plants to make soil contaminants nontoxic (Chaney et al. 1997). Plants used in phyto-remediation should be hyperaccumulators, which accumulate heavy metals in their tissues, most likely as a defense mechanism against the pathogens or to prevent predation (Boyd et al. 1994, Sagner et al. 1998). Plant species vary considerably in their ability of accumulating metals from contaminated soils. More than 400 plant species can accumulate heavy metals and metalloids (Baker and Walker 1989; McGrath and Zhao 2003); the majority of them are characterized by slow growth and low production of biomass. Extent of removal of heavy metals by plants depends mostly on the biomass production and the concentration of heavy metals in the plant biomass (Seleiman et al. 2012, 2013); the low level of either or both factors would lead to slow limited removal of heavy metals and metalloids (Cunningham et al. 1995; Ebbs et al. 1997).

Candidates for hyperaccumulator plants should be characterized by high growth rate, production of more aboveground parts than roots, well-distributed and well-branched root system, high accumulation of the targeted heavy metals, high tolerance to the adverse effects of the targeted heavy metals, translocation of the targeted heavy metals from roots to shoots, adaptation to negative environmental and climatic conditions, withstanding pathogens and pests, repelling feeding animals, and easy cropping and harvesting (Ali et al. 2013). There are different opinions, among researchers, regarding concentration of accumulated metals in hyperaccumulator plants. Some investigations were conducted in laboratories to screen metal hyperaccumulating plants of high biomass species (Kumar et al. 1995b; Blaylock et al. 1997; Huang and Cunningham 1996; Huang et al. 1997). Results from these studies indicate that metal hyperaccumulation can be achieved by using high biomass agronomic crop in conjunction with soil amendment application to the contaminated soils. Baker and Brooks (1989) reported that for a plant to be a hyperaccumulator grown in a polluted soil, the concentration of heavy metals in shoots (mg kg^{-1} dry weight) should exceed 100 for Cd; 1,000 for Ni, Cu, or Pb; and 10,000 for Zn or Mn. Van der Ent et al. (2013) suggested the following concentrations in dry foliage: 100 for Cd, Se, or Ti; 300 for Co, Cu, or Cr; 1,000 for Ni, Pb, or As; 3,000 for Zn; and 10,000 for Mn. Other researchers (Lasat 2002; Reeves 2006; Chaney et al. 2007) concluded that heavy metal concentrations in hyperaccumulators are 100- to 1,000-fold higher, especially in shoots, as compared with the non-accumulator plants. Bioavailability plays an important role in phyto-remediating soils contaminated with heavy metals. Soil properties affect the mobility and bioavailability of the heavy metals (Klos et al. 2012). According to Prasad (2003), there are three types of the heavy metal bioavailability: readily bioavailable (Cu, Ni, Cd, Ni, Zn, and As), moderately bioavailable (Mn, Fe, and Co), and least bioavailable (U, Cr, and Pb).

Phyto-remediation can be done using two methods: (1) natural methods and (2) induced one. In the natural method, remediation is done by continuously cropping hyperaccumulators. In the induced method, remediation is done by accelerating chelation of the chelating agents. Phyto-remediation is categorized into five techniques: (1) phyto-extraction, (2) phyto-filtration, (3) phyto-stabilization, (4) phyto-volatilization, and (5) phyto-organic degradation in the rhizosphere (Flathman and Lanza 1998; Alkorta et al. 2004). Phyto-extraction is the removal of the pollutants from soil through the uptake of plant species that has high biomass production, easily handled, genetically characterized, and with established cultivation practice. Many scientists are investigating different approaches to increase the phyto-extraction capacity of plants. Screening is used to select the more desirable plants, and efforts are directed at identifying soil amendments and metal-uptake-enhancing chemicals to increase the efficiency of phyto-extraction. Addition of chelators to contaminated soil during phyto-remediation increases Pb uptake, and translocation from roots to shoots of plants provided a major advance in phyto-extraction technology (Huang et al. 1997; Blaylock et al. 1997). Phyto-filtration is the absorption of the pollutants from the contaminated water. Phyto-stabilization is the hampering or decreasing the availability and mobility of pollutants, while phyto-volatilization is the conversion of a pollutant to a volatile substance, therefore getting rid of it to the atmosphere.

16.2 Hyperaccumulator Plants

Peterson (1971) defined metal accumulation in two concepts: (1) accumulation of an element within organisms to reach a concentration greater than found in the growth medium and (2) possession of quantities of an element in an organism greater than usually found in that organism. Such definitions create some problems in interpreting metal-accumulation patterns in plants. There are many definitions for hyperaccumulator plants. For instance, the term hyperaccumulator was introduced by Baker et al. (1994) as a plant species whose leaves contain higher than 100 mg kg^{-1} dry matter for elements of Cr or Cd; or 1,000 mg kg^{-1} dry matter for elements of Ni, or Cu; or more than 10,000 mg kg^{-1} dry matter for elements of Zn or Mn when grown on contaminated soil with high level of heavy metals and metalloids. The term hyperaccumulator describing plants with a highly abnormal level of metal accumulation appears to have been first applied by Jaffré et al. (1976) in the title of their paper on Ni concentration in the New Caledonian plant *Sebertia acuminata*. Brooks et al. (1977a) concluded it is possible that ultramafic rocks and soils are cases of high Ni concentration in leaves of some plant in herbarium specimens of certain genera. Hyperaccumulation was used to describe the accumulation of Ni to more than 1,000 mg kg^{-1} dry matter of leaves on the basis of such concentration is 100–1,000 times higher than the normal level found in plants on un-ultramafic origin soils and 10–100 times higher than found in most other plants grown on Ni-rich ultramafic soils. Although there was some arbitration in the choice of this criterion, it has been noted that in many ultramafic floras, Ni concentrations of 100–1,000 mg kg^{-1} are quite rare, and accumulating more than 1,000 mg kg^{-1} seems to represent a distinct plant response, implying some characteristics and unusual metabolic behavior. An attempt to give greater precision to the definition of hyperaccumulation was made by Reeves (1992) who stated that a Ni-hyperaccumulator plant has a Ni concentration of at least 1,000 mg kg^{-1}. Such a definition applies to the dry matter of any aboveground tissue, at least one specimen growing in its natural habitat. Definitions of hyperaccumulation have been extended to elements other than Ni. Malaisse et al. (1978) used the 1,000 mg kg^{-1} criterion for Cu accumulation. Reeves and Brooks (1983) used the same criterion for Pb but not for Mn and Zn which are normally present at higher and more widely varying concentrations (about 20–400 mg kg^{-1}). A threshold concentration of 10,000 mg kg^{-1} was suggested by Baker and Brooks (1989) for Mn, following the use of the term hyper-manganesophore for plants with this level of Mn reported by Jaffré (1980).

Ni, Zn, and Co hyperaccumulator plants accumulate such metals preferentially in their shoot, with lower concentration in the roots (Baker 1981; Gabbrielli et al. 1990; Homer et al. 1991; Baker et al. 1994; Brown et al. 1995; Lioyd-Thomas 1995; Krämer et al. 1996; Lasat et al. 1996; Köhl et al. 1997). This observation was made in field-collected as well as in hydroponically grown plants, irrespective of whether extracellularly bound metal was removed by desorption (Lasat et al. 1996). The shoot/root ratio of metal concentrations in all confirmed hyperaccumulators is therefore generally above one but invariably below one in non-accumulators. On this basis, it has been proposed (Baker 1981) that metal accumulator plants can be categorized on the basis of a leaf/root metal concentration ratio higher than one. This definition has a physiological basis and is probably more useful in studies designed to investigate the biological mechanisms underlying metal hyperaccumulation.

The success of using plants to extract metals from contaminated soils requires a better understanding of the mechanisms of metal uptake, translocation, and accumulation. The understanding of metal transport characteristics could help to design a more efficient cultivation management to enhance metal accumulation in plants. The efficiency of metal phyto-extraction is a function of a number of factors. The major factors are plant species, availability of metal to plant roots, uptake of metal by roots, translocation of metal from roots to shoots, and plant tolerance to the toxic metal (Blaylock and Huang 2000). The key of using hyperaccumulators (or any other plant species) in phyto-remediation relies on the rate of biomass production, as well as the concentration of the element absorbed by the plant (Reeves and Baker 2000). It has been suggested that a potential metal accumulator plant for remediating metal-contaminated sites should have the following traits stated by the United States Department of Energy, USDOE, (1994): (a) high accumulation rate, (b) ability to accumulate several heavy metals and metalloids, (c) high biomass production above ground, and (d) resistance to the different diseases and pests.

16.3 Sunflower, Sorghum, Hemp, Maize, and Elephant Grass as Example Species for Phyto-remediation

Roots of sunflower (*Helianthus annuus* L.) have been used to treat water contaminated with Pb, U, Co, Zn, and some other metals to bring the concentration of such metals below the accepted water standards (Dushenkov et al. 1995). Shahandeh and

Hossner (2000) suggested that high biomass production of sunflower and Indian mustard (*Brassica juncea*) are potential candidates for phyto-remediation of Cr, since they accumulate higher Cr than many other plant species. Sunflower was the superior indicator plant for the Cr accumulation. Sunflower growth and metal uptake were affected by the level of the Cr and form and rate of applied organic chelate. Hemp (*Cannabis sativa* L.) and maize (*Zea mays* L.) are suitable plant species for phyto-remediation, since they can accumulate in their tissues high concentration of heavy metals and metalloids (such as Pb, Ni, Cu, Cr, Cd, and As) and also give higher biomass production than oilseed rape (*Brassica napus* L. ssp. oleifera "Moench." Metzg.) when sludge was applied on cropland (Seleiman et al. 2013).

Khalil (1995) reported that the Co uptake by sorghum roots and the whole plant increased progressively with increasing the rate of Co application up to 50 mg Co kg^{-1} soil and that the rate of Co uptake by sorghum roots was higher than by shoots. For this reason, translocation of Co decreased with Co application. Such results could initiate a general statement that the rate and magnitude of Co accumulation in roots and its translocation to shoots is a phenomenon strongly dependent on soil properties. The Cr concentrations in sorghum shoots and roots and hence in the whole plants increased progressively with increasing the rate of Cr application up to 25 mg Cr kg^{-1} soil. Rabie (1984) found that addition of Co to soil at 1–5 mg kg^{-1} increased the uptake of Co in the shoots of faba beans grown on alluvial and calcareous soils and also with maize in both soils when Co was added at the rate of 4–20 mg kg^{-1}.

16.4 Efficiency and Translocation Factors of Plants

Plant efficiency in accumulating heavy metals in their tissues form the soil/media can be calculated through the following equation (Zhuang et al. 2007):

$$\text{Bio-concentration Factor (BCF)} = \frac{C \text{ in harvested tissue}}{C \text{ in soil}}$$

where C in the harvested tissue is the concentration of the targeted metal in tissues of harvested plant and C in soil is the concentration of the targeted metal in soil. Plant tendency to translocate accumulated metals from roots to shoots is calculated through the following equation (Padmavathiamma and Li 2007):

$$\text{Translocation Factor (TF)} = \frac{C \text{ in shoot}}{C \text{ in root}}$$

Where C in shoot is the concentration of the targeted metal in tissues of plant shoots, and C in root is the concentration of the targeted metal in plant roots.

Hyperaccumulator plants should have BCF and TF more than one (Cluis 2004, Badr et al. 2012).

Uptake of Cd and Pb in shoots by tobacco (*Nicotiana tabacum*) was investigated by Evangelou et al. (2007b). They found that addition of citric acid, oxalic acid, and EDTA enhanced Cd uptake, while tartaric acid caused no effect. Uptake of Pb by shoots was enhanced by EDTA, while citric, tartaric, and oxalic acids had no effect. According to Jankong et al. (2007), application of P fertilizer and rhizo-bacteria resulted in an increase in the uptake of As in silver fern (*Pityrogramma calomelanos*). They added that addition of such rhizo-fungi resulted in a decrease in As concentration in plant, but plant biomass was considerably increased, thus enhancing As phyto-stabilization.

References

Ali H, Khan E, Sajad MA (2013) Phytoremediation of heavy metals-concepts and applications. Chemosphere 91:869–881

Alkorta I, Hernández-Allica J, Becerril JM, Amezaga I, Albizu I, Garbisu C (2004) Recent findings on the phytoremediation of soils contaminated with environmentally toxic heavy metals and metalloids such as zinc, cadmium, lead, and arsenic. Rev Environ Sci Biote chnol 3:71–90

Anderson WC (ed) (1993) Soil washing flushing. American Academy of Environmental Engineers, Annaplois, MD

Badr N, Fawzy M, Al-Qahtani KM (2012) Phytoremediation: an economical solution to heavy-metal-polluted soil and evaluation of plant removal ability. World Appl Sci J 16:1292–1301

Baker AJM (1981) Accumulators and excluders-strategies in the response of plants to heavy metals. J Plant Nutr 3:643–654

Baker AJM, Brooks RR (1989) Terrestrial higher plants which hyperaccumulate metallic elements. A review of their distribution, ecology and phytochemistry. Biorecovery 1:81–126

Baker AJM, Walker PL (1989) Ecophysiology of metal uptake by tolerant plants. In: Shaw AJ (ed) Heavy metal tolerance in plants. Evolutionary aspects. CRC press, Boca Raton, FL, pp 155–176

Baker AJM, Reevs RD, Hajar ASM (1994) Heavy metal accumulation and tolerance in British populations of the metallophyte Thlaspi Caerulescens J&C presl (Brassicaceae). New Phytol 127:61–68

Blaylock MJ, Huang JW (2000) Phytoextraction of metals. In: Raskin I, Ensley B (eds) Phytoremediation of toxic metals using plants to clean up the environment. John Wiley & Sons Inc, New York

Blaylock MJ, Salt DE, Dushenkov S, Zakharova O, Gussman C, Kapulnik Y, Ensley BD, Raskin I (1997) Enhanced accumulation of Pb in Indian mustard by soil-applied chelating agents. Environ Sci Technol 31:860–865

Bodar CW, Pronk ME, Sijm DT (2006) The European Union risk assessment on zinc and zinc compounds: the process and the facts. Integr Environ Assess Manag 1:301–319

Bolan NS, Duraisamy VP (2003) Role of inorganic and organic soil amendments on immobilization and phytoavailability of heavy metals: a review involving specific case studies. Aust J Soil Res 41:533–555

Boyd RS, Shaw JJ, Martens SN (1994) Nickel hyperaccumulation defends Streptanthus polygaloides (Brassicaceae) against pathogens. Am J Bot 81:294–300

Brooks RR, Lee J, Reeves RD, Jaffré T (1977) Detection of nickeliferous rocks by analysis of herbarium specimens of indicator plants. J Geochem Explor 7:49–57
Brown GA, Elliot HA (1992) Influence of electrolytes on EDTA extraction of Pb from polluted soil. Water Air Soil Pollut 62:157–165
Brown SL, Chaney RL, Angle JS, Baker AJM (1995) Zinc and cadmium uptake by hyperaccumulator Thlaspi caerulescens grown in nutrient solution. Soil Sci Soc Am J 59:125–133
Chaney RL, Malik M, Li YM, Brown SL, Angle JS, Baker AJM (1997) Phytoremediation of soil metals. Curr Opin Biotechnol 8:279–284
Chaney RL, Angle JS, Broadhurst CL, Peters CA, Tappero RV, Sparks DL (2007) Improved understanding of hyperaccumulation yields commercial phytoextraction and phytomining technologies. J Environ Qual 36:1429–1443
Cluis C (2004) Junk-greedy Greens: phytoremediation as a new option for soil decontamination. BioTech J 2:61–67
Cunningham SD, Ow DW (1996) Promises and prospects of phytoremediation. Plant Physiol 110:715–719
Cunningham SD, Berti WR, Huang JW (1995) Phytoremediation of contaminated soils. Trends Biotechnol 13:393–397
Dushenkov V, Kumar PBAN, Motto H, Raskin I (1995) Rhizofiltration: the use of plants to remove heavy metals from aqueous streams. Environ Sci Technol 29:1239–1245
Ebbs SD, Lasat MM, Brady DJ, Cornish J, Gordon R, Kochian LV (1997) Phytoextraction of cadmium and zinc from a contaminated site. J Environ Qual 26:1424–1430
Elliot HA, Brown GA (1989) Comparative evaluation of NTA and EDTA for extractive decontamination of Pb-polluted soils. Water Air Soil Pollut 45:361–369
Elliot HA, Shastri NL (1999) Extractive decontamination of metal polluted soils using oxalate. Water Air Soil Pollut 110:335–346
Evangelou MWH, Bauer U, Ebel M, Schaeffer A (2007a) Evaluation of the effect of small organic acids on phytoextraction of Cu and Pb from soil with tobacco Nicotiana tabacum. Chemosphere 68:345–353
Evangelou MWH, Ebel M, Schaeffer A (2007b) Chelate assisted phytoextraction of heavy metals from soil. Effect, mechanism, toxicity, and fate of chelating agents. Chemosphere 68:989–1003
Finzgar N, Lestan D (2007) Multi-step leaching of Pb and Zn contaminated soils with EDTA. Chemosphere 66:824–832
Flathman PE, Lanza GR (1998) Phytoremediation: current view on an emerging green technology. J Soil Contam 7:415–432
Fotakis G, Timbrell JA (2006) Role of trace elements in cadmium chloride uptake in hepatoma cell lines. Toxicol Lett 164:97–103
Gabbrielli R, Pandolfini T, Verganano O, Palandri MR (1990) Comparison of two serpentine species with different nickel tolerance strategies. Plant Soil 122:271–277
Heil DM, Samani Z, Hanson AT, Rudd B (1999) Remediation of lead contaminated soil by EDTA. I-Batch and column studies. Water Air Soil Pollut 113:77–95
Homer FA, Morrison RS, Brooks RR, Clemens J, Reeves RD (1991) Comparative studies of nickel, cobalt and copper uptake by some nickel hyperaccumulators of the genus Alyssum. Plant Soil 138:195–205
Hong APK, Li C, Banerji SK, Regmi T (1999) Extraction, recovery, and biostability of EDTA for Remediation of heavy metal-contaminated soil. J Soil Contam 8:81–103
Huang JA, Cunningham SD (1996) Lead phytoextraction: species variation in lead uptake and translocation. New Phytol 134:75–84
Huang JW, Chen JJ, Berti WR, Cunningham SD (1997) Phytoremediation of lead-contaminated soils: role of synthetics chelates in lead phytoextraction. Environ Sci Technol 31:800–805
Jaffré T (1980) Etude Écologique du peuplement végétal des sols dérivés de roches ultrabasiques en Nouvelle Calédonie. vol 124, Travaux et Documents de l ORSTOM, (Paris)

Jaffré T, Brooks RR, Lee J, Reeves RD (1976) Sebertia acuminata: a hyperaccumulator of nickel from new caledonia. Science 193:579–580

Jankong P, Visoottiviseth P, Khokiattiwong S (2007) Enhanced phytoremediation of arsenic contaminated land. Chemosphere 68:1906–1912

Kayser A, Schulin R, Felix H (1999) Field trials for the phytoremediation of soils polluted with heavy metals. In: Umweltbundesamt (Ed.), Proc. Int. Workshop am Fraunhofer Institut fur Umweltchemic und Okotoxikologie,Schmallenberg, Germany, 1–2 Dec. 1997. Erich Schmidt Verlag, Berlin, pp. 170–182.

Khalil MEA (1995) Studies on some heavy metals in soil and plant. Ph.D. Thesis, Fac. of Agric., Moshtohor, Zagazig Uinv., Egypt

Kłos A, Czora M, Rajfur M, Wacławek M (2012) Mechanisms for translocation of heavy metals from soil to epigeal mosses. Water Air Soil Pollut 223:1829–1836

Köhl KI, Smith JAC, Baker AJM (1997) Defining a metal-hyperaccumulator plant: The relationship between metal uptake, allocation and tolerance. Abstract Plant Physiol 114:124

Krämer U, Cotter-howells JD, Charnock JM, Baker AJM, Smith JAC (1996) Free histidine as a metal chelator in plants that accumulate nickel. Nature 379:635–638

Kumar P, Dushenkov V, Motto H, Raskin I (1995b) Phytoextraction: a novel strategy for the removal of toxic metals from the environment using plants. Biotechnology 13:468–474

Lasat MM (2002) Phytoextraction of toxic metals: a review of biological mechanisms. J Environ Qual 31:109–120

Lasat MM, Baker AJM, Kochion LV (1996) Physiological characterization of root Zn^{2+} absorption and translocation to shoots in Zn hyperaccumulator and non accumulator species of Thlapi. Plant Physiol 112:1715–1722

Lestan D, Luo C, Li X (2008) The use of chelating agents in the remediation of metal-contaminated soils: A review. Environ Pollut 153:3–13

Lioyd-Thomas DH (1995) Heavy metal accumulation by Thlaspi Caerulescens. J. & C. PressI. Ph. D. Thesis, Univ. Sheffield, UK

Luo CL, Shen ZG, Li XD (2005) Enhanced phytoextraction of Cu, Pb, Zn and Cd with EDTA and EDDS. Chemosphere 59:1–11

Luo CL, Shen ZG, Li XD, Baker AJM (2006) The role of root damage in the EDTA-enhanced accumulation of lead by Indian mustard plants. Int J Phytoremediation 8:323–337

Malaisse F, Gregoire J, Brooks RR, Morrison RS, Reeves RD (1978) Aeolanthus Bioformifolius De wild: a hyperaccumulator of copper from Zaïre. Science 199:887–888

Mathis P, Kayser A (2001) Plant uptake of heavy metals following glyphosate treatment. In: International Society for Trace Element Biogeochemistry (ed.) Proceedings of the Sixth International Conference on the Geochemistry of Trace Elements (ICOBTE), Guelph, ON, Canada, 29 July—2 August 2001, p. 484

McCarthy P (2001) The principles of humic substances. Soil Sci 166:738–751

McGrath SP (1998) Phytoextraction for soil remediation. In: Brooks RR (ed) Plants that hyperaccumulate heavy metals. CAB International, Wallingford, pp 261–287

McGrath SP, Zhao FJ (2003) Phytoextraction of metals and metalloids from contaminated soil. Curr Opin Biotechnol 14:277–282

Nigam R, Srivastava S, Prakash S, Srivastava MM (2001) Cadmium mobilisation and plant availability – the impact of organic acids commonly exuded from roots. Plant Soil 230:107–113

Norvell WA (1991) Reaction of metal chelates in soils and nutrient solution. In: Mortvedt JJ, Cox FR, Shuman LM, Welch RM (eds) Micronutrient in agriculture, 2nd edn. Soil Science Society of America, Wiscosin, Madison, pp 87–227

Nowack B. (2002) Environmental chemistry of aminopolycarboxylate chelating agents. Environ Sci Technol 36:4009–4016

Padmavathiamma PK, Li LY (2007) Phytoremediation technology: hyperaccumulation metals in plants. Water Air Soil Pollut 184:105–126

Papassiopi N, Tambouris S, Kontopoulos A (1999) Removal of heavy metals from calcareous contaminated soil by EDTA leaching. Water Air Soil Pollut 109:1–15

Peterson PJ (1971) Unusual accumulation of elements by plants and animals. Sci Prog Oxford 59:505–526

Pichtel J, Pichtel TM (1997) Comparison of solvents for ex situ removal of chromium and lead from contaminated soil. Environ Eng Sci 14:97–104

Prasad MNV (2003) Phytoremediation of metal-polluted ecosystems: hype for commercialization. Russ J Plant Physiol 50:686–700

Rabie MH (1984) Studies on some heavy metals in soils of A.R.E. PhD Thesis, Fac. of Agric. Ain Shams Univ., Egypt

Raskin I, Smith RD, Salt DE (1997) Phytoremediation of metals: using plants to remove pollutants from the environment. Curr Opin Biotechnol 8:221–226

Reeves RD (1992) Hyperaccumulation of nickel by serpentine plants. In: Proctor J, Baker AJM, Reeves RD (eds) The vegetation of ultramafic (serpentine) soils. Intercept Ltd, Andover, pp 253–277

Reeves RD (2006) Hyperaccumulation of trace elements by plants. In: Morel JL, Echevarria G, Goncharova N (eds) Phytoremediation of metal-contaminated soils, NATO science series: IV: earth and environmental sciences. Springer, New York, pp 1–25

Reeves RD, Baker AJM (2000) Metal accumulating plants. In: Phytoremediation of toxic metals using plants to clean up the environment. Raskin I, Ensley B.D. (eds.) John Wiley & Sons Inc, New York

Reeves RD, Brooks RR (1983) Hyperaccumulation of lead and zinc by two metallophytes from a mining area in central Europe. Environ Pollut 31:277–287

Sagner S, Kneer R, Wanner G, Cosson J-P, Deus-Neumann B, Zenk MH (1998) Hyperaccumulation, complexation and distribution of nickel in Sebertia acuminata. Phytochemistry 47:339–347

Salt DE, Smith RD, Raskin I (1998) Phytoremediation. Annu Rev Plant Physiol Plant Mol Biol 49:643–668

Seleiman MF, Santanen A, Stoddard FL, Mäkelä PSA (2012) Feedstock quality and growth of bioenergy crops fertilized with sewage sludge. Chemosphere 89:1211–1217

Seleiman MF, Santanen A, Jaakkola S, Ekholm P, Hartikainen H, Stoddard FL, Mäkelä PSA (2013) Biomass yield and quality of bioenergy crops grown with synthetic and organic fertilizers. Biomass Bioenergy 59:477–485

Shahandeh H, Hossner LR (2000) Plant screening for Chromium phytoremediation. Int J Phytoremediation 2:34–54

Shen ZG, Li XD, Wang CC, Chen HM, Chua H (2002) Lead phytoextraction from contaminated soil with high-biomass plant species. J Environ Qual 31:1893–1900

Soltanpour PN (1991) Determination of nutrient availability element toxicity by ABDTPA. Soil Test and ICPS. Adv Soil Sci 16:165–190

Tandy S, Bossart K, Mueller R, Ritschel J, Hauser L, Schulin R, Nowack B (2004) Extraction of heavy metals from soils using biodegradable chelating agents. Environ Sci Technol 38:937–944

Uren NC, Reisenauer HM (1988) The role of root exudates in nutrient acquisition. Adv Plant Nutrition 3:79–114

van der Ent A, Baker AJM, Reeves RD, Pollard AJ, Schat H (2013) Hyperaccumulators of metal and metalloid trace elements: facts and fiction. Plant Soil 362:319–334

Zhuang P, Yang Q, Wang H, Shu W (2007) Phytoextraction of heavy metals by eight plant species in the field. Water Air Soil Pollut 184:235–242

Chapter 17
Sludge Amendment and Tree Vegetation Could Improve the Quality of Mine Soils

Verónica Asensio, Alfonso Rodríguez-Vila, and Emma F. Covelo

17.1 Introduction

Reclaiming areas where mining has previously carried out is necessary in order to restore mine soils. Mine tailings occupy huge areas, requiring substantial resources to improve the quality of their soils. Planting vegetation is a common practice to reclaim mine areas (Chodak and Niklińska 2010; De Varennes et al. 2010; Shrestha and Lal 2008), although in most cases this is not enough due to the severely degraded condition of these soils (extreme pH, nutrient deficiency, poorly aerated). Amending with organic residues has proven to be a cheap and effective treatment to increase soil quality (Bendfeldt et al. 2001; Brown et al. 2003; Hemmat et al. 2010), as well as being a good way to reuse wastes.

In 1988, work began on recovering the depleted copper mine at the village of Touro (Galicia, Northwest Spain, Fig. 17.1). The mine soils were partially reclaimed by planting trees at the end of the copper extraction, although another company has been carrying out work to restore the whole area since 2000. This reclamation process involves planting pine trees (*Pinus pinaster* Aiton) and eucalypts (*Eucalyptus globulus* Labill) and amending with organic wastes (mainly sewage sludge and paper mill residue).

There are three main categories of soil quality indicators: chemical, physical and biological (Carter et al. 1997). We propose a method for developing soil quality

V. Asensio
Department of Plant Biology and Soil Science, Faculty of Biology, University of Vigo, As Lagoas-Marcosende, 36310 Vigo, Pontevedra, Spain

Department of Plant Nutrition, CENA, Universidade de São Paulo (CENA-USP), Av. Centenário 303, 13400-970 Piracicaba, SP, Brazil

A. Rodríguez-Vila • E.F. Covelo (✉)
Department of Plant Biology and Soil Science, Faculty of Biology, University of Vigo, As Lagoas-Marcosende, 36310 Vigo, Pontevedra, Spain
e-mail: emmaf@uvigo.es

I. Sherameti, A. Varma (eds.), *Heavy Metal Contamination of Soils*, Soil Biology 44, DOI 10.1007/978-3-319-14526-6_17

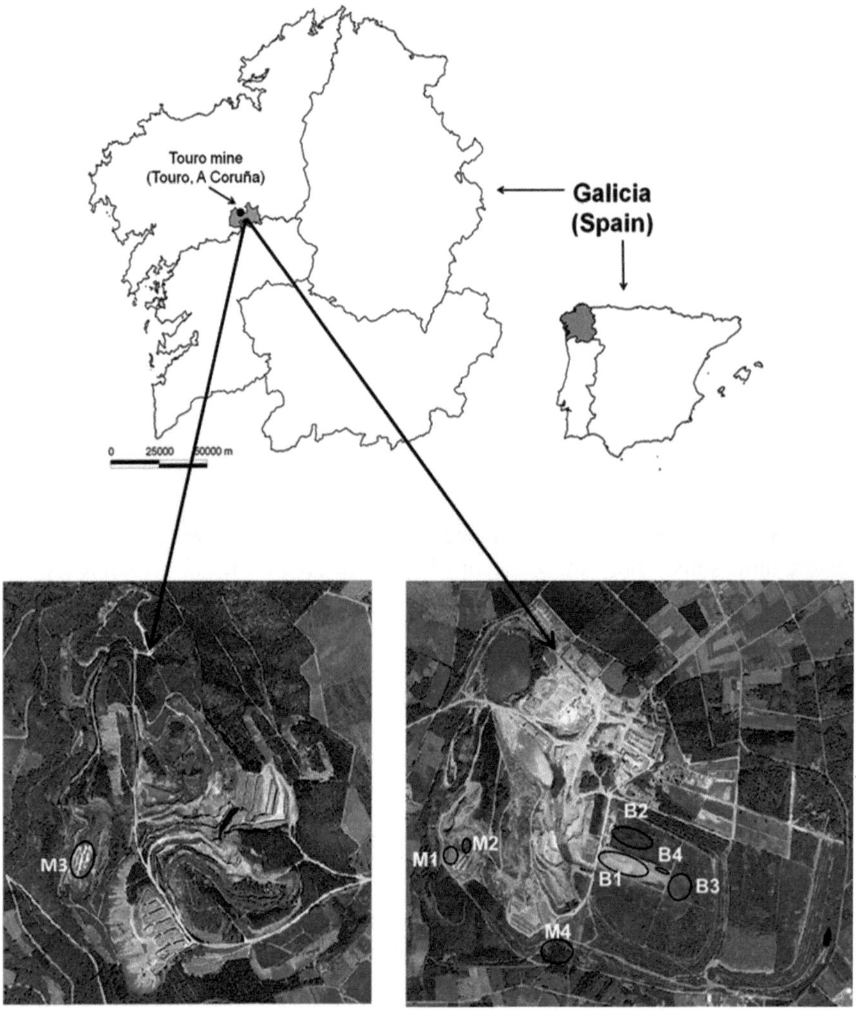

Fig. 17.1 Location of sampled areas in Touro mine. Source: ©Instituto Geográfico Nacional de España

indexes (SQI) for two types of mine soils (settling pond and mine tailing) reclaimed by planting trees and/or amending with wastes and to assess the quality of these soils under field conditions.

We hypothesised that vegetating with *E. globulus* Labill and *P. pinaster* Aiton and amending with sludges at the same time may increase the soil quality in mine soils polluted by metals.

17.2 Materials and Methods

17.2.1 Soil Sampling

The sampling area is located at the mine in Touro (Galicia, Northwest Spain) (Lat/Lon (Datum ETRS89): 8° 20′ 12.06″ W 42° 52′ 46.18″ N) (Fig. 17.1). The climate of the experimental site is Atlantic (oceanic), with precipitation reaching 1,886 mm per year (with an average of 157 mm per month) and a mean daily temperature of 12.6 °C. The average relative humidity is 77 % (AEMET 2012).

Copper was extracted from the Touro mine between 1973 and 1988 (15 years). This activity left behind a settling pond (71 ha) and a huge mine tailing (larger than 700 ha). Nowadays, apart from the extraction of material for road construction, the former mining areas are being reclaimed. Settling ponds are created by the accumulation of waste after the process of metal concentration in flotation banks. Mine tailings are the accumulation of thick material after mining activity. The settling pond in the Touro mine was created with waste from the copper flotation process, but today it is completely emerged and dry. There is an active oxidation zone near the surface where vegetation does not grow. The unvegetated or amended areas in the mine tailing were also completely bare. In order to evaluate the effectiveness of treatments in the settling pond and mine tailing, four areas were selected in each zone (Table 17.1, Fig. 17.1). The soil samples are described in greater detail below. Five soil samples were randomly collected in points sufficiently spaced to be representative of each site on 9 March 2010. Samples were stored in polyethylene bags, dried at room temperature and sieved to <2 mm prior to being analysed. All soils had only one horizon, except two of them, where both horizons were sampled (described below). Soils were classified according to the latest version of the FAO (2006).

Table 17.1 Location, depth and reclamation of soil samples

Zone	Name	Horizon	Treatment	Time of treatment
Settling pond	B1	AC	None	–
	B2v	AC	Tree vegetation	21 years
	B3v	AC	Tree vegetation	6 years
	B4Aw	AC	Waste amendment	5 months
	B4Bw	C	None	–
Mine tailing	M1	AC	None	–
	M2Av	AC	Tree vegetation	21 years
	M2Bv	C	None	–
	M3w	AC	Waste amendment	6 months
	M4vw	AC	Trees + wastes	10 years

17.2.2 Settling Pond Samples

The control sample from the settling pond (B1) was in an untreated area. B1 is a Spolic Technosol located 336 m above sea level, covering an area of 1.9 ha with an AC horizon 40 cm deep.

The second sampled area (B2v) was chosen for its old vegetation. Pine trees (*P. pinaster* Aiton) were planted here in 1989 (21 years ago). B2v is also a Spolic Technosol, with spontaneous vegetation: eucalyptus (*E. globulus* Labill), gorse (*Ulex* sp.), heather (*Erica* sp.), *Agrostis* sp. and bryophytes. The B2v soil covers 6200 m^2 and is 340 m above sea level, with an AC horizon 20 cm deep.

The third sample (B3v) was chosen because of its young vegetation, in order to compare it with B2v and to observe the effect of trees over time. The B3v area was vegetated with eucalyptus in 2004 and is also a Spolic Technosol. This area also has spontaneous vegetation: pine trees (*P. pinaster* Aiton), *Agrostis* sp., broom (*Cytisus* sp.), *Acacia* sp. and bryophytes. The B3v soil covers 1.15 ha and is 335 m above sea level, with an AC horizon 20 cm deep.

The fourth sample (B4w) was located in a recently amended area. Sewage sludge mixed with paper mill residue had been added 5 months before the sampling date. This waste was directly added with trucks and then spread on the soil surface, without being mixed in with the mine soil. The final depth of this new layer was 20 cm, with a volume of 280 t per ha. The properties of this amendment are similar to those reported by Camps et al. (2008). The general characteristics of the added waste were pH 8–10, more than 150 g kg^{-1} of total organic C, more than 100 mg kg^{-1} of total Cu and more than 300 mg kg^{-1} of total Zn. Only spontaneous herbaceous vegetation grew on B4w, measuring 100 m^2 and at a height of 339 m above sea level. B4w was an Urbic Technosol with two horizons: AC (B4Aw) and C (B4Bw). B4Bw was also sampled to compare it with the untreated sample (B1) as it was considered to be analogous of the control soil. B4Aw was 20 cm deep, and B4Bw was 40 cm deep.

As the use of waste amendments in this area began only 5 months before the sampling date, there are no samples that are representative of the long-term effects of amendment or both treatments at the same time (vegetation and amendment).

17.2.3 Mine Tailing Samples

The control sample in the mine tailing (M1) was in an untreated area and is classified as Spolic Technosol. M1 was 336 m above sea level and covered an area of 1.20 ha, with an AC horizon 20 cm deep.

The second sampled area (M2v) was chosen because of its old vegetation, as it was vegetated with pine trees in 1989. It also had spontaneous vegetation: gorse (*Ulex* sp.), heather (*Erica* sp.), *Agrostis* sp. and bryophytes. This soil is also a Spolic Technosol, covering an area of 0.60 ha and 340 m above sea level, with an AC horizon 20 cm deep. The C horizon of this soil was also sampled to be compared with the untreated one (M1), as it is similar to the original bare soil but with the amendment on top of it. The AC horizon was named M2Av and the C horizon M2Bv. M2Av was 4 cm deep and M2Bv was 20 cm deep.

The M3w soil is an Urbic Technosol created with sewage sludge and paper mill residue. These sludges were added 6 months before the sampling date by trucks, and then spread on the soil surface. The final depth of this new layer was around 3 m, covering an area of 0.8 ha. The amount of wastes added was around 158 t per ha. These wastes were the same as those used in the settling pond. This soil was 178 m above sea level and only had natural herbaceous vegetation.

The fourth area (M4vw) was selected because it was vegetated and amended at the same time. This area was vegetated with eucalyptus and amended in 2000 with the same type of wastes as M3w. The amount added in M4vw was 297 t per ha and the final depth of this new layer was around 70 cm. This area also had spontaneous vegetation (gorses, brambles, pine trees and bryophytes). The M4vw soil is also an Urbic Technosol, measuring 1.5 ha and 336 m above sea level.

17.2.4 *Analytical Methodology*

17.2.4.1 Physical Properties

Samples were air-dried, passed through a 2 mm sieve and homogenised in a vibratory homogeniser for solid samples (Fritsch Laborette 27 rotary sampler divider) prior to analysis. A portion of each sample was sieved at 10 mm to determine aggregate size distribution.

Standard procedures were used to estimate both soil stoniness (Eriksson and Holmgren 1996) and particle size distribution (Day 1965). Particle and bulk density (BD) were determined following standard procedures (Hao et al. 2008). Porosity was calculated as the difference between both densities as a percentage.

Aggregate size distribution was determined by sieving 50 g of >10 mm aggregates in a set of five sieves (5, 2, 1, 0.25 and 0.05 mm). The mean weight diameter (MWD) obtained of each sample was related to the stability of the aggregates (Le Bissonnais 1996).

The determination of water-stable aggregates (WSA) was carried out with standard equipment (Eijkelkamp Wet Sieving Apparatus) (Asensio et al. 2013).

17.2.4.2 Chemical Properties

Soil reaction was determined with a pH electrode in 1:2.5 water to soil extracts, while EC was determined according to Porta (1986). Exchangeable cations (Ca^{2+}, K^{+}, Mg^{2+}, Na^{+}, Al^{3+}, Fe^{2+} and Mn^{2+}) were extracted with 0.1 M $BaCl_2$ (Hendershot and Duquette 1986) and their concentrations determined by ICP-AES (Perkin-Elmer Optima 4300 DV). Effective cation exchange capacity (CECe) was calculated by adding up the total cation concentrations. The method developed by Mehra and Jackson (1960) was used to determine the free oxide concentrations. Aluminium, iron and manganese were determined in the extract by ICP-AES.

Both total and inorganic carbon (TC and IC) were determined in a module for solid analysis (SSM-5000) coupled with a TOC analyser (Shimadzu TNM-1, Japan). Soil organic carbon (SOC) contents were calculated from the difference TC–IC. Dissolved organic carbon (DOC) was extracted with bidistilled H_2O according to (Sanchez-Monedero et al. 1996) and the OC in the supernatant was determined with a TOC analyser (Shimadzu TNM-1, Japan).

The different chemical organic matter fractions labelled as humin C, humic acids (C_{HA}) and fulvic acids (C_{FA}) were separated following the method described in De Blas et al. (2010). Microbial biomass carbon (C_{mic}) was determined by the fumigation-extraction method according to (Joergensen 1996). Organic C in the fumigated and non-fumigated extracts was determined in a TOC analyser. The microbial biomass C was calculated as $C_{mic} = E_C/k_{EC}$, where E_C is the difference between organic C extracted from fumigated soils and organic C extracted from non-fumigated soils and k_{EC} is a constant. For soils with $pH > 4.5$, k_{EC} is 0.45 and when $pH < 4.5$, $k_{EC} = 0.30$ (Joergensen 1996; Vance et al. 1987).

Pseudototal concentrations of Cr, Cu, Ni, Pb and Zn were extracted with aqua regia by acid digestion in a microwave oven (Milestone ETHOS 1). The certified reference material CRM026-050 was analysed in parallel with samples to check the effectiveness and precision of the extraction analysis. Chromium, copper, nickel, lead and zinc were extracted with acidified 0.01 M $CaCl_2$ (Houba et al. 2000). The chemical form of heavy metals extracted with 0.01 M $CaCl_2$ is operationally defined as phytoavailable (Houba et al. 2000). Pseudototal and $CaCl_2$-extractable heavy metal concentrations were compared with the Generic Reference Level (GRL) established for Galician soils (Macías and Calvo de Anta 2009).

The total bioavailable phosphorus content was determined by the Olsen method (Olsen et al. 1954). The total Kjeldahl-N (TN) was determined according to Bremner (1996). The three inorganic N (IN) fractions, which are NH_4^+—N (ammonium), NO_2^-—N (nitrites) and NO_3^--N (nitrates), were extracted with 2 M KCl and distilled with their respective reagents: ammonium with MgO and nitrates + nitrites with Devarda's alloy (Bremner and Keeney 1965). To determine total and inorganic N contents, an aliquot of each extract was analysed by potentiometric titration with a titrator equipment (702 SM Titrino, Metrohm).

17.2.4.3 Statistical Analyses

All analytical determinations were performed in triplicate and the data obtained were statistically treated using the program SPSS 15.0 for Windows. Statistical analyses were performed by separating the data from the settling pond soils from the mine tailing, as the characteristics of the control soils at each zone are significantly different. Analyses of variance (ANOVAs) and test of homogeneity of variance were carried out. In case of homogeneity, a post hoc least significant difference (LSD) test was carried out. Otherwise Dunnett's T3 test was performed. The Mann-Whitney test was done when data were not parametric. An independent *t*-test was done to compare each control soil (B1 or M1) with their respective analogous (B4Bw or M2Bv). A correlated bivariate analysis was carried out from all soil samples in each zone.

Soil quality index proposed by Asensio et al. (2013) by totalling the scores of the minimum data set (MDS) selected through a principal component analysis (PCA) proved to be a valid and useful tool. These results revealed that the quality of these reclaimed soils should be determined by analysing both physical and chemical characteristics. In the case of settling pond sites, stoniness, pH, Al^{3+}, CECe, MWD, SOC, humin C, C_{HA}, C_{FA}, TN, clay, K $CaCl_2$ (extracted with $CaCl_2$), Mn $CaCl_2$, Ni $CaCl_2$, Zn $CaCl_2$ and pseudototal both Ni and Pb should be analysed. In the case of mine tailing sites, it is necessary to analyse stoniness, pH, BD, WSA, humin C, C_{HA}, C_{FA}, TN, C_{mic}, Ca $CaCl_2$, Mn $CaCl_2$, Zn $CaCl_2$ and pseudototal both Pb and Zn. The establishment of this set of key parameters for assessing the quality of reclaimed mine soils will help to save time and money by analysing a large number of soil characteristics.

To eliminate the effect of different units of variables, the PCA was carried out using the correlation matrix on the standardised values of the measured soil properties (Shukla et al. 2006). The principal components (PC) obtained had eigenvalues >1 and were subjected to varimax rotation with Kaiser. The variables selected for the minimum data set (MDS) to elaborate the soil quality index (SQI) were the highest weighted from each PC. We defined highly weighted as being within 10 % of the highest factor loading (using absolute values) (Andrews et al. 2002). When more than one variable was retained within a PC, we calculated their correlations to determine whether the variables could be considered redundant and, therefore, eliminated from the MDS (Andrews et al. 2002). If the highly weighted variables were not significantly ($p < 0.05$) correlated or the correlation was considered spurious (significantly affected by the reclamation treatments), then each was considered important and was retained in the MDS. The scores for values of each selected variable were created by assigning negative scores to values unfavourable for soil living organisms or hazardous for the environment, neutral value (zero) for values that do not affect positive or negative and finally positive scores for values that affect positive. The consideration of "negative" or "positive" for values were decided according to the references cited for each soil characteristic in Table 17.1. The scores of each selected MDS variable were then summed for each soil to obtain the soil quality index (SQI), based on the equation proposed by (Amacher et al. 2007) as follows:

$$SQI = \frac{\sum \text{individual soil characteristic score value}}{\text{maximum possible total SQI for characteristics measured}} \times 100 \quad (17.1)$$

Thus, missing characteristics do not contribute to the index. The highest score of the index is 100 %, which indicates very high soil quality. Scores around 0 or negative indicates very low quality.

Factor time is included in the SQI proposed due to sites where the same treatment applied in different years is used.

17.3 Results and Discussion

The untreated areas in the mine in Touro (Tables 17.2 and 17.3) were physically degraded: low porosity, high density and less than 50 % of water-stable aggregates. Both planted trees and added waste amendments significantly changed these physical properties in the mine soils, although each treatment did so a different way. The planted vegetation increased soil porosity and, consequently, decreased the bulk density. On the other hand, the added amendments increased the percentage of MWD and WSA probably due to the addition of organic matter from the wastes added to the soil. It was seen that the simultaneous use of both treatments improved the physical condition of the mine soils better than the use of only one treatment. Waste supplies organic matter and roots increase porosity and the percentage of small aggregates. The best treatment for improving the physical quality of mine soils is to plant trees and to add organic wastes at the same time.

The control soils (B1 and M1) had an extremely acidic pH (according to the USDA 1998) (Tables 17.4 and 17.5). The soil pH was still acidic 21 years after tree planting, but in amended sites the pH rose to 6–8, probably due to their high organic

Table 17.2 Physical characteristics of the settling pond soils

	Settling pond soils				
	B1	B2v	B3v	B4Aw	B4Bw
Stoniness (%)	20.1 ± 5.7d	34.9 ± 8.2b	30.9 ± 7.2c	81.2 ± 11.2a	11.3 ± 1.9e
Sand (%)	69.3 ± 13.5b	69.1 ± 0.7c	65.1 ± 17.5d	45.8 ± 10.9e	71.1 ± 17.1a
Silt (%)	21.1 ± 0.1c	19.9 ± 0.5d	24.6 ± 2.2b	27.9 ± 1.9a	19.7 ± 2.2e
Clay (%)	9.5 ± 1.5d	11.1 ± 0.3b	10.3 ± 2.5c	26.1 ± 6.1a	9.2 ± 3.1e
Texture	Sandy loam	Sandy loam	Sandy loam	Sandy clay loam	Sandy loam
Particle density ($g\ cm^{-3}$)	2.8 ± <0.1b	2.3 ± <0.1d	2.5 ± <0.1c	1.9 ± <0.1e	3.1 ± <0.1a
BD ($g\ cm^{-3}$)	1.7 ± 0.1a	0.8 ± 0.07c	1.5 ± 0.02b	0.3 ± 0.1e	0.7 ± 0.07d
Porosity (%)	38.4e	65.7c	40.6d	83.2a	77.6b
MWD (mm)	1.2 ± 0.2d	2.1 ± 0.2b	2 ± 0.2c	4.8 ± 0.3a	0.9 ± 0.2e
WSA (%)	43.3 ± 0.3c	82.1 ± 0.3b	97.6 ± 0.1a	93.2 ± 4.2a	44.4 ± 4.3c

Mean ± confidence interval (CI) of the mean of three replications. Values followed by different letters in each row differ significantly with $p < 0.05$

u.l. undetectable level, *WSA* water-stable aggregates, *MWD* mean weight diameter, *BD* bulk density

Table 17.3 Physical characteristics of the mine tailing soils

	Mine tailing soils				
	M1	M2Av	M2Bv	M3w	M4vw
Stoniness (%)	53.1 ± 8.7b	22.9 ± 6.4e	41.1 ± 3.5d	56.2 ± 6.9a	46.8 ± 7.7c
Sand (%)	69.1 ± 30.4a	51.1 ± 4.7e	59.3 ± 6.1d	66.2 ± 23.8b	60.8 ± 2.9c
Silt (%)	20.1 ± 4.9e	32.2 ± 6.2a	25.4 ± 9.1c	23.5 ± 4.5d	26.1 ± 1.9b
Clay (%)	10.9 ± 3.9d	16.7 ± 1.4a	15.3 ± 1.7b	10.3 ± 1.9e	13.1 ± 0.9c
Texture	Sandy loam	Sandy loam	Sandy loam	Sandy loam	Sandy loam
Particle density ($g\ cm^{-3}$)	2.8 ± <0.1a	2.3 ± <0.1d	2.7 ± <0.1b	2 ± <0.1e	2.4 ± <0.1c
BD ($g\ cm^{-3}$)	1.1 ± 0.05b	1.9 ± 0.03a	1.9 ± 0.02a	0.9 ± 0.05c	0.8 ± 0.1d
Porosity (%)	58.3b	19.1e	31d	53.1c	64.7a
MWD (mm)	2.8 ± 0.3c	1.3 ± 0.1e	3.5 ± 0.8b	3.8 ± 0.1a	2.3 ± 0.4d
WSA (%)	49.1 ± 5.9b	58.4 ± 6.3a	76.9 ± 5.7a	83.6 ± 6.0a	83.8 ± 9.5a

Mean ± confidence interval (CI) of the mean of three replications. Values followed by different letters in each row differ significantly with $p < 0.05$

u.l. undetectable level, *WSA* water-stable aggregates, *MWD* mean weight diameter, *BD* bulk density

carbon concentrations. The effective cation exchange capacity (CECe) significantly increased in the settling pond after amending with wastes but in the mine tailing only after the use of both treatments. The concentrations of aluminium (Al_2O_3) and manganese (MnO) oxides are higher in the treated soils than in the controls, whereas the opposite happens with iron oxide (Fe_2O_3). The soil total nitrogen concentration significantly increased with the use of both treatments, especially with waste amending.

The control sites (B1 and M1) were poor in both soil organic and inorganic carbon (SOC and IC) (Tables 17.4 and 17.5), since they had no carbon sources i.e., vegetation, animals, microorganisms or organic amendments. The SOC concentration was extremely low in both B1 and M1 in comparison to undisturbed sites with similar conditions (formed over schist and vegetated with pines and eucalyptus) in the surrounding area (92–126 g kg^{-1}) (De Blas et al. 2010). In the control soils, most of the SOC was still in the free organic matter (FOM) fraction and therefore was non-humified (Tables 17.4 and 17.5). Moreover, the microbial biomass C (C_{mic}) was very low in comparison with microbially active soils (1–4 %) (Anderson and Domsch 1990; Kandeler et al. 1996; Sparling 1992) showing that the organic matter in the untreated sites has not been biologically degraded. Planting trees significantly increased the total C concentration, even 6 years after starting this reclamation treatment (Tables 17.4 and 17.5). The inorganic C concentrations in the vegetated soils did not change significantly in comparison to control soils, and the increase in soil C was due to the organic C inputs. Most of the SOC in the vegetated sites was still in the FOM (Tables 17.4 and 17.5).

Both the organic and inorganic C concentration significantly increased in the mine soils after being amended with wastes (Tables 17.4 and 17.5). The carbon concentration shifted between the amended sites due to the heterogeneity of the

Table 17.4 Chemical characteristics of the settling pond soils

	Settling pond soils				
	B1	B2v	B3v	B4Aw	B4Bw
pH H_2O	3.8 ± 0.04d	4.8 ± 0.05b	4.7 ± 0.03b	6.9 ± 0.3a	4.3 ± 0.3c
TC (g kg^{-1})	1.7 ± 0.9e	10.7 ± 1.2c	11.4 ± 1.3b	115 ± 10a	1.8 ± 0.8d
SOC (g kg^{-1})	1.4d	10.4b	11.1b	112a	1.5c
IC (g kg^{-1})	0.3 ± 1.3b	0.3 ± 0.01b	0.3 ± 0.05b	3.7 ± 1.9a	0.3 ± <0.01b
DOC (g kg^{-1})	0.01 ± 0.007d	0.09 ± 0.01bc	0.1 ± 0.03b	1.2 ± 0.2a	0.03 ± 0.01 cd
Humin C (g kg^{-1})	0.51 ± 0.01d	4.23 ± 0.03c	4.64 ± 0.02b	59.55 ± 0.01a	0.3 ± 0.02e
C_{FA} (g kg^{-1})	0.65 ± 0.07c	2.03 ± 0.17b	2 ± 0.01b	12.3 ± 0.69a	0.48 ± 0.01c
C_{HA} (g kg^{-1})	0.89 ± 0.17c	2.16 ± 0.19b	1.89 ± 0.29b	36.60 ± 1.45a	0.68 ± 0.18c
C_{mic} (mg kg^{-1})	8.8 ± 11.6e	181 ± 57b	130 ± 26c	247 ± 38a	19.1 ± 10d
TN (g kg^{-1})	0.14 ± 0.03d	0.58 ± <0.06b	0.58 ± 0.04b	3.71 ± <0.73a	0.32 ± <0.06c
NH_4^+	100 ± 25b	120 ± 25b	110 ± 25b	1,500 ± 27a	120 ± 26b
$NO_2^- + NO_3^-$	20 ± 2b	340 ± 50ab	450 ± 25a	140 ± 26b	120 ± 26b
C/N	11.9c	18.4b	11.3d	115a	1.8e
EC (ms cm^{-1})	0.1 ± <0.1c	0.03 ± <0.1d	0.03 ± <0.1d	0.4 ± <0.1a	0.1 ± <0.1b
P (Olsen)	u.l.	u.l.	u.l.	424 ± 25.8a	18.98 ± 0.8b
P ($CaCl_2$-ext.)	u.l.	u.l.	u.l.	68.28 ± 1.38	u.l.
K ($CaCl_2$-ext.)	21.7 ± 1.61e	91.25 ± 3.85b	64.6 ± 0.65c	333 ± 7.99a	33.5 ± 5.36d
Ca ($CaCl_2$-ext.)	u.l.	u.l.	u.l.	704 ± 125a	135 ± 60b
Mg ($CaCl_2$-ext.)	13.38 ± 0.62e	41.40 ± 1.45c	34.97 ± 0.97d	777 ± 3.97a	60.5 ± 8.74b
Al ($CaCl_2$-ext.)	171 ± 5.7b	223 ± 10.7a	252 ± 6.8a	0.68 ± 0.3d	170 ± 5.6c
Mn ($CaCl_2$-ext.)	2.15 ± 0.05d	22.10 ± 0.65b	61.1 ± 1.42a	4.04 ± 0.13c	3.72 ± 0.94c
CECe (cmol kg^{-1})	2.1 ± <0.1b	2.5 ± <0.1b	2.4 ± <0.1b	58.4 ± 2.9a	2.2 ± 0.1b

Al_2O_3 (g kg^{-1})	2.54 ± 0.1d	7.41 ± 0.1b	5.06 ± <0.1c	15.08 ± 0.3a	2.38 ± <0.1d
Fe_2O_3 (g kg^{-1})	88.3 ± 4.1a	38.9 ± 0.6b	28.6 ± 0.1c	23.1 ± <0.1d	27.9 ± 0.5c
MnO (g kg^{-1})	0.02 ± <0.1c	0.6 ± <0.1b	0.8 ± <0.1a	0.8 ± <0.1a	0.06 ± <0.1c

Mean ± confidence interval (CI) of the mean of three replications. Values followed by different letters in each row differ significantly with $p < 0.05$

Abbreviations: *u.l.* undetectable level, *CECe* effective cation exchange capacity, *IC* inorganic carbon, *SOC* soil organic carbon, *DOC* dissolved organic carbon, *Humin C* humin carbon, C_{FA} fulvic acid carbon, C_{HA} humic acid carbon, C_{mic} microbial biomass carbon, *TN* total nitrogen, NH_4^+ ammonium, $NO_2^- + NO_3^-$ nitrates and nitrites, *EC* electric conductivity, *$CaCl_2$-ext.* extractable with $CaCl_2$

Table 17.5 Chemical characteristics of the mine tailing soils

	Mine tailing soils				
	M1	M2Av	M2Bv	M3w	M4vw
pH H_2O	3.6 ± 0.05e	6.6 ± 0.4b	5.2 ± 0.4d	5.8 ± 0.1c	8.2 ± 0.1a
TC (g kg^{-1})	3.5 ± 1.0e	35.3 ± 1.5c	6.9 ± 1.2d	56.6 ± 3.0b	88.3 ± 6.1a
SOC (g kg^{-1})	3.2c	35.1a	6.6b	56.3a	54.8a
IC (g kg^{-1})	0.3 ± 0.01b	0.3 ± 0.5b	0.3 ± <0.01b	0.3 ± 0.01b	33.5 ± 2.7a
DOC (g kg^{-1})	u.l.	u.l.	0.02 ± <0.01c	1.2 ± 0.04a	0.3 ± 0.1b
Humin C (g kg^{-1})	3.92 ± 0.02d	17.95 ± 0.01b	3.93 ± 0.02d	16.1 ± 0.03c	37.94 ± 0.04a
C_{FA} (g kg^{-1})	0.6 ± 0.17e	6.14 ± 0.99b	0.89 ± 0.25d	13.97 ± 0.11a	2.66 ± 0.1c
C_{HA} (g kg^{-1})	0.87 ± 0.08d	6.67 ± 0.21b	1.99 ± 0.17d	21.11 ± 1.79a	4.71 ± 0.32c
C_{mic} (mg kg^{-1})	7.6 ± 1.4e	64.2 ± 23.3b	12.5 ± 2.4d	669 ± 155a	53.1 ± 2.2c
TN (g kg^{-1})	0.25 ± 0.04d	0.74 ± 0.28b	0.28 ± 0.04d	3.7 ± 0.34a	0.6 ± 0.28c
NH_4^+	40 ± 26b	80 ± 27b	30 ± 25b	391 ± 12a	40 ± 5b
$NO_2^- + NO_3^-$	60 ± 27b	460 ± 25a	130 ± 25ab	220 ± 25ab	100 ± 25ab
C/N	3.5e	35.3c	6.9d	56.6b	88.3a
EC (ms cm^{-1})	0.1 ± <0.1c	0.1 ± <0.1c	0.1 ± <0.1c	1.9 ± 0.2a	0.2 ± <0.1b
P (Olsen)	u.l.	14.1 ± 6.1c	3.7 ± 1.5d	265 ± 16.4a	125 ± 13.9b
P ($CaCl_2$-ext.)	29.61 ± 0.30e	154 ± 7.41c	59.65 ± 5.00d	230 ± 2.85a	184 ± 9.20b
K ($CaCl_2$-ext.)	159 ± 43.2d	1,106 ± 43.9b	761 ± 20.3c	1,210 ± 39.1b	1,929 ± 91.3a
Ca ($CaCl_2$-ext.)	53.12 ± 3.1e	158 ± 3.61d	208 ± 6.68c	813 ± 27.18a	300 ± 9.97b
Mg ($CaCl_2$-ext.)	184 ± 11.1a	155 ± 53.4b	81.57 ± 10.0c	6.8 ± 0.4d	0.15 ± 0.2d
Al ($CaCl_2$-ext.)	17.2 ± 2.46b	93.12 ± 17.93a	20.98 ± 3.58b	4.31 ± 0.34c	u.l.
Mn ($CaCl_2$-ext.)	0.24 ± 0.03d	2.4 ± 0.02a	1.43 ± 0.1c	1.89 ± 0.16b	u.l.
CECe (cmol kg^{-1})	3.8 ± 0.1e	12.0 ± 0.4c	9.4 ± 0.5d	20.8 ± 1.1b	26.7 ± 0.7a

Al_2O_3 (g kg^{-1})	3.22 ± <0.1e	11.65 ± 0.2b	4.67 ± <0.1d	15.53 ± <0.1a	7.82 ± 0.1c
Fe_2O_3 (g kg^{-1})	132 ± 2.4a	74.8 ± 0.9c	111 ± 0.8b	22.1 ± 0.1d	7.4 ± 0.1e
MnO (g kg^{-1})	0.02 ± <0.1e	1.8 ± <0.1b	0.5 ± <0.1d	1.2 ± <0.1c	4.5 ± <0.1a

Mean ± confidence interval (CI) of the mean of three replications. Values followed by different letters in each row differ significantly with $p < 0.05$

Abbreviations: *u.l.* undetectable level, *CECe* effective cation exchange capacity, *IC* inorganic carbon, *SOC* soil organic carbon, *DOC* dissolved organic carbon, *Humin C* humin carbon, C_{FA} fulvic acid carbon, C_{HA} humic acid carbon, C_{mic} microbial biomass carbon, *TN* total nitrogen, NH_4^+ ammonium, $NO_2^- + NO_3^-$ nitrates and nitrites, *EC* electric conductivity, $CaCl_2$-*ext*. extractable with $CaCl_2$

wastes used, which had TOC concentrations that vary between 150 and 230 g kg^{-1} (Camps Arbestain et al. 2008). In fact, the organic C concentration was only within the range observed in undisturbed soils in the surroundings in one of the two amended sites (Tables 17.4 and 17.5). Therefore, if it is necessary to add organic C to soils, it is necessary to analyse the wastes for C prior to their use as amendments in order to achieve the optimal concentration for soils.

The settling pond and the mine tailing soils at the sites without reclamation treatment contain optimum levels of total N and K for the growth of *Eucalyptus globulus* Labill and *Pinus pinaster* Aiton but not for other plant species. The phytoavailable concentration of Ca and Mg was only optimum for eucalypts and pines and only in the mine tailing. Moreover, both the phytoavailable and Olsen P was undetectable in either site (Tables 17.4 and 17.5). The reclamation treatment consisting of planting these tree species increased the concentration of Mg in the settling pond up to adequate levels for eucalypts and pines but not for other plants. Nevertheless, the treatment consisting of amending with wastes (sewage sludges and paper mill residues) was able to increase the phytoavailable concentration of all macro- and micronutrients (N, P, K, Ca, Mg, Cu, Mn and Zn) up to adequate levels for most plant species (Tables 17.4 and 17.5). However, the site amended and vegetated for 10 years had a deficient N concentration and lower ammonium concentration that the recently amended sites.

Settling pond and mine tailing soils in the untreated mine sites (B1 and M1, respectively) were polluted by Cr and Cu (Tables 17.6 and 17.7). Pollution by these heavy metals should not be an environmental problem in the Touro mine soils at the present time, as the phytoavailable concentration was below the Generic Reference Level (Tables 17.6 and 17.7). The GRL for Galicia (Macías and Calvo de Anta 2009) established the following threshold limit values for metals in soils: 40 mg kg^{-1} for cobalt, 80 mg kg^{-1} for chromium, 50 mg kg^{-1} for copper,

Table 17.6 Metal content of the settling pond soils

		Settling pond soils				
		B1	B2v	B3v	B4Aw	B4Bw
Cr	$CaCl_2$-extractable (mg kg^{-1})	u.l.	u.l.	u.l.	u.l.	u.l.
Cu		12.6 ± 0.1a	1.6 ± <0.1c	6.7 ± 0.1b	0.7 ± <0.1d	13.4 ± 0.3a
Ni		u.l.	0.2 ± <0.1b	0.4 ± <0.1a	u.l.	u.l.
Pb		u.l.	0.4 ± <0.1a	0.4 ± <0.1a	u.l.	u.l.
Zn		u.l.	0.2 ± <0.1 cd	0.4 ± <0.1c	0.9 ± <0.1b	1.6 ± 0.3a
Cr	Pseudototal (mg kg^{-1})	**194 ± 2.8a**	**86.3 ± 7.1 cd**	71.8 ± 3.1d	**94.1 ± 5.6c**	**136 ± 18.1b**
Cu		**310 ± 8.1b**	**63 ± 5.8c**	**75.3 ± 7.5c**	**467 ± 21.1a**	**338 ± 16.1b**
Ni		16.2 ± 1.1c	66.7 ± 1.1a	55.6 ± 9.5b	50.5 ± 1.2b	15.6 ± 0.7c
Pb		19.1 ± 1.4c	31.9 ± 3.9bc	32.1 ± 1.7b	**97.1 ± 6.2a**	16.9 ± 0.6c
Zn		41.4 ± 0.7c	90.2 ± 8.3b	89.5 ± 10.4b	**538 ± 25.1a**	45.9 ± 1.9c

Mean ± confidence interval (CI) of the mean of three replications. Values followed by different letters in each row differ significantly with $p < 0.05$

u.l. undetectable level, *$CaCl_2$-ext.* extractable with $CaCl_2$. Values in bold are over the GRL—Generic Reference Level—established for Galician soils (Macías and Calvo de Anta 2009)

75 mg kg^{-1} for nickel and 200 mg kg^{-1} for zinc. However, in order to prevent the future bioavailability of Cr and Cu, it is important to observe whether planting trees or amending with wastes decreased the concentration of these heavy metals in the soils. The effect of the treatments will not be the same in each mine area due to the different soil characteristics of the settling pond and mine tailing (Tables 17.4 and 17.5).

Planting eucalyptus or pines decreased pseudototal Cr and Cu concentrations in the settling pond and mine tailing soils (Tables 17.6 and 17.7). The accumulation of Cr and Cu in M2Av can be explained because this horizon comprises the first 4 cm of soil, and heavy metal concentrations in soils usually decrease with depth. In fact, the subsuperficial horizon in this soil (M2Bv) had significantly less Cu than the untreated mine tailing and a similar Cr concentration (Tables 17.6 and 17.7). Nickel, lead and zinc concentrations in the vegetated soils behaved in the opposite way to chromium and copper. The sites with tree vegetation (B2v, B3v and M2Av) had more pseudototal Ni, Pb and Zn than the untreated sites.

Amending with wastes decreased the pseudototal Cr concentration in the mine soils, even to values below the GRL for Galicia in the mine tailing area (Table 17.7). Although the amended sites were still polluted by copper, a marked decrease in its concentration was observed in the mine tailing soil (M3w). However, the amended settling pond site (B4Aw) had a higher Cu concentration than its control (Table 17.6). Copper extracted from amended soil samples is attributable to the added wastes, since it has previously been reported that the sewage sludges and paper mill residues used contain more than 100 mg kg^{-1} of total Cu (Camps Arbestain et al. 2008). The results indicate that heavy metal concentrations in the wastes used could vary between batches, due to the heterogeneity caused by different production times. In spite of the high pseudototal Cu concentration in amended soils, this heavy metal was strongly retained and not available in the short term in amended sites, since very low concentrations were observed in the $CaCl_2$-extractable form (Tables 17.6 and 17.7). Pseudototal Ni, Pb and Zn concentrations were also higher in the soils treated with wastes than in the controls. The amended soils had Pb and Zn concentrations that were higher than the soil pollution limits for Galicia (Tables 17.6 and 17.7).

The use of both treatments at the same time (tree vegetation and waste amendments) decreased Cr and Cu concentrations in the mine soils more than the use of a single treatment (Tables 17.6 and 17.7) . Pseudototal Ni, Pb and Zn concentrations also increased with the use of both treatments, especially nickel, as they were in higher concentration than in M3w, even exceeding the GRL for Galicia (Table 17.7). It was clearly shown that Ni, Pb and Zn were added by the wastes used.

Table 17.7 Metal content of the mine tailing soils

		Mine tailing soils				
		M1	M2Av	M2Bv	M3w	M4vw
Cr	$CaCl_2$-extractable(mg kg^{-1})	u.l.	u.l.	u.l.	u.l.	u.l.
Cu		17.2 ± 0.9b	**93.1 ± 7.2a**	20.9 ± 1.4b	4.3 ± 0.1c	u.l.
Ni		0.2 ± <0.1d	2.4 ± <0.1a	1.4 ± <0.1c	1.9 ± <0.1b	u.l.
Pb		u.l.	u.l.	u.l.	u.l.	u.l.
Zn		0.8 ± <0.1d	9.2 ± 0.1b	1.7 ± <0.1c	26.8 ± 0.3a	u.l.
Cr	Pseudototal (mg kg^{-1})	**118 ± 18.6b**	**138 ± 9.9a**	**118 ± 3.1b**	68.9 ± 3.4c	31.7 ± 2.2d
Cu		**911 ± 63.7b**	**1,921 ± 126a**	**537 ± 31.8c**	**350 ± 20.8d**	**51.1 ± 3.5e**
Ni		15.3 ± 1.3d	**121 ± 1.9a**	31.4 ± 4.3c	55.6 ± 3.2b	**128 ± 10.1a**
Pb		19.3 ± 0.7d	24.7 ± 0.5c	16 ± 1.3d	71.1 ± 3.6a	34.4 ± 0.9b
Zn		78.2 ± 10c	176 ± 10.3b	59.7 ± 3.5c	**307 ± 20.5a**	135 ± 3.5b

Mean ± confidence interval (CI) of the mean of three replications. Values followed by different letters in each row differ significantly with $p < 0.05$
u.l. = undetectable level; $CaCl_2$-ext. = extractable with $CaCl_2$. Values in bold are over the GRL—Generic Reference Level—established for Galician soils (Macías and Calvo de Anta 2009)

17.3.1 Soil Quality in the Settling Pond Sites

The percentage of soil quality index (% SQI) was very low in the untreated settling pond site (B1), since most of the soil quality indicators were zero or had a negative score. The application of both types of reclamation treatments (planting pines and eucalyptus or amending with wastes) significantly increased the SQI at the settling pond (Fig. 17.2). The treatment that most increased the quality was amending with wastes, as it increased soil pH, CECe, MWD, concentrations of all chemical forms of soil organic C, total N concentration, clay percentage and phytoavailable K concentration (Tables 17.2,17.3,17.4,17.5,17.6 and 17.7). The increase of all these variables was due to the characteristics of the wastes used (sewage sludges and paper mill residues). The treatment consisting of planting trees also increased the

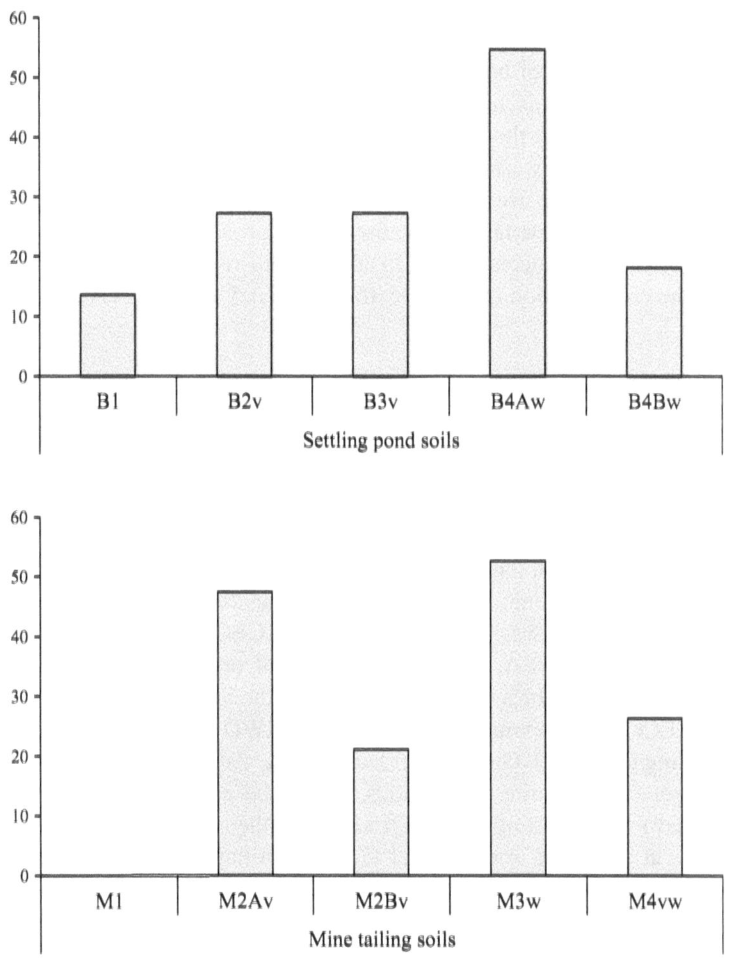

Fig. 17.2 Soil quality index (SQI) of the settling pond and mine tailing soils

quality of the settling pond soil, although less than by amending with wastes. The SQI increased by vegetating with trees basically due to the increase in soil pH, MWD and $CaCl_2$-extractable Mn. There was no difference in the % SQI between the two vegetated sites, despite the fact that one had been vegetated for 6 years and another for 21. It is likely that by planting pines or eucalyptuses alone, it is not possible to increase the quality of settling pond soils as much as by amending with wastes (according to the proposed index), although more studies under field conditions should support this hypothesis.

17.3.2 Soil Quality in the Mine Tailing Sites

The soil quality index percentage was also very low in the untreated mine site at the mine tailing zone (M1), as most soil quality indicators were zero or had negative values. The application of both types of reclamation treatments also significantly increased the SQI at this mine zone, and once again the treatment that most increased this index was the recent addition of wastes (M3w site). However, the site with both trees and wastes for 10 years showed lower SQI than the sites with only one treatment. The use of reclamation treatments (alone or in combination) increased the SQI of mine tailing because they increased pH, WSA, humin C and $CaCl_2$-extractable Ca concentrations. In addition, the recently amended site (M3w) increased the concentration of $CaCl_2$-extractable Zn, C_{FA}, C_{HA}, TN and C_{mic}.

17.3.3 Comparison of Reclamation Treatment Effects Between Both Mine Zones

The application of waste amendments to mine soils was the treatment that most increased the SQI in both the settling pond and the mine tailing, although its effect was different in each zone. The soil characteristics that contributed the most to increasing the SQI after waste amending were pH, CECe, $CaCl_2$-extractable K, Ca and Zn, clay, MWD, WSA, all chemical forms of organic C, total N and C_{mic}. However, all these variables were not selected for the SQI of both mine zones. The variables CECe, $CaCl_2$-extractable K, clay and MWD were only selected for the SQI of settling pond soils. On the other hand, WSA, $CaCl_2$-extractable Ca and C_{mic} were only selected for mine tailing soils. This is because the variables that were only selected for the settling pond SQI increased their values after waste amendments more at this mine zone than at the mine tailing. At the same time, those selected only for the mine tailing increased their values more in this mine zone. The high concentration of pseudototal Pb decreased the SQI in the amended settling pond site and the pseudototal Zn in the amended mine tailing. This indicates that the wastes used must be analysed prior to being added to soils. The differences in the soil characteristics of the amended sites (B4Aw and M3w) are due to the heterogeneity caused by different production times of the used wastes.

Planting trees also increased the SQI at both mine zones and, like the other reclamation treatment, had a different influence in each zone. The scores of pH and $CaCl_2$-extractable Mn at the vegetated sites contributed towards increasing the SQI in both mine zones. Humin C concentration of the vegetated sites, although it was selected for both SQI, only contributed to increase the index in the mine tailing. The MWD was only selected for the settling pond SQI because its value increased more at the vegetated site in this mine zone, while WSA and $CaCl_2$-extractable Ca were selected in the vegetated mine tailing site.

17.4 Conclusions

In order to assess the quality of any type of soil, a soil quality index (SQI) created by totalling the scores of the minimum data set (MDS) selected through a principal component analysis (PCA) proved to be a valid and useful tool. This index was tested in a settling pond and a mine tailing soil reclaimed by planting trees and amending with sewage sludges and paper mill residues. Our results revealed that the quality of these reclaimed soils should be determined by analysing both physical and chemical characteristics. In the case of settling pond sites, stoniness, pH, Al^{3+}, CECe, MWD, SOC, humin C, C_{HA}, C_{FA}, TN, clay, K $CaCl_2$ (extracted with $CaCl_2$), Mn $CaCl_2$, Ni $CaCl_2$, Zn $CaCl_2$ and pseudototal both Ni and Pb should be analysed . In the case of mine tailing sites, it is necessary to analyse stoniness, pH, BD, WSA, humin C, C_{HA}, C_{FA}, TN, C_{mic}, Ca $CaCl_2$, Mn $CaCl_2$, Zn $CaCl_2$ and pseudototal both Pb and Zn. The establishment of this set of key parameters for assessing the quality of reclaimed mine soils will help to save time and money by analysing a large number of soil characteristics.

In applying the proposed SQIs, we observed that the untreated mine sites had a very low quality and that the treated sites had a significantly higher index. The reclamation treatment that most increased the SQI of the mine soils was the recent addition of wastes. However, we recommend the regular addition of sewage sludges and paper mill residues to degraded sites because their favourable effect on soils decreases over time.

References

AEMET (Meteorological State Agency) (2012) Valores Climatológicos Normales. Santiago Compostela Aeropuerto. Ministerio de Agricultura, Alimentación y Medio Ambiente, Spain. http://www.aemet.es/es/serviciosclimaticos/datosclimatologicos/valoresclimatologicos?l=1428&k=gal. Accessed 12 Aug 2012

Amacher MC, Neill KPO, Perry CH, Service F (2007) Soil vital signs: a New Soil Quality Index (SQI) for assessing forest soil health. USDA Forest Research, Fort Collins, CO

Anderson TH, Domsch KH (1990) Application of eco-physiological quotients (qCO2 and qD) on microbial biomasses from soils of different cropping histories. Soil Biol Biochem 22:251–255

Andrews SS, Mitchell JP, Mancinelli R, Karlen DL, Hartz TK, Horwath WR, Pettygrove GS, Scow KM, Munk DS (2002) On-farm assessment of soil quality in California's central valley. Agron J 94:12

Asensio V, Guala SD, Vega FA, Covelo EF (2013) A soil quality index for reclaimed mine soils. Environ Toxicol Chem 32:2240–2248

Bendfeldt ES, Burger JA, Daniels WL (2001) Quality of amended mine soils after sixteen years. Soil Sci Soc Am J 65:1736–1744

Bremner JM (1996) Nitrogen-total. In: Sparks DL (ed) Methods of soil analysis, vol III. SSSA, Madison, WI, pp 1085–1121

Bremner JM, Keeney DR (1965) Steam distillation methods for determination of ammonium, nitrate and nitrite. Anal Chim Acta 32:485–495

Brown SL, Henry CL, Chaney R, Compton H, De Volder PS (2003) Using municipal biosolids in combination with other residuals to restore metal-contaminated mining areas. Plant Soil 249:203–215

Camps Arbestain M, Madinabeitia Z, Anza Hortalà M, Macías-García F, Virgel S, Macías F (2008) Extractability and leachability of heavy metals in Technosols prepared from mixtures of unconsolidated wastes. Waste Manag 28:2653–2666

Carter MR, Gregorich EG, Anderson DW, Doran JW, Janzen HH, Pierce FJ (1997) Concepts of soil quality and their significance. In: Gregorich EG, Carter MR (eds) Soil quality for crop production and ecosystem health. Elsevier Science, The Netherlands, pp 1–20

Chodak M, Niklińska M (2010) Effect of texture and tree species on microbial properties of mine soils. Appl Soil Ecol 46:268–275

Day PR (1965) Particle size analysis. In: Black CA (ed) Methods of soil analysis. American Society of Agronomy, Madison, WI, pp 545–566

De Blas E, Rodríguez-Alleres M, Almendros G (2010) Speciation of lipid and humic fractions in soils under pine and eucalyptus forest in northwest Spain and its effect on water repellency. Geoderma 155:242–248

De Varennes A, Abreu MM, Qu G, Cunha-Queda C (2010) Enzymatic activity of a mine soil varies according to vegetation cover and level of compost applied. Int J Phytoremediation 12:371–383

Eriksson CP, Holmgren P (1996) Estimating stone and boulder content in forest soils – evaluating the potential of surface penetration methods. Catena 28:121–134

FAO (2006) World reference base for soil resources. IUSS, ISRIC, FAO, Rome

Hao X, Ball BC, Culley JLB, Carter MR, Parkin GW (2008) Soil density and porosity. In: Carter MR, Gregorich EG (eds) Soil sampling and methods of analysis. Canadian Society of Soil Science, Boca Raton, FL, pp 743–759

Hemmat A, Aghilinategh N, Rezainejad Y, Sadeghi M (2010) Long-term impacts of municipal solid waste compost, sewage sludge and farmyard manure application on organic carbon, bulk density and consistency limits of a calcareous soil in central Iran. Soil Tillage Res 108:43–50

Hendershot WH, Duquette M (1986) A simple barium chloride method for determining cation exchange capacity and exchangeable cations. Soil Sci Soc Am J 50:605–608

Houba VJG, Temminghoff EJM, Gaikhorst GA, Van Vark W (2000) Soil analysis procedures using 0.01 M calcium chloride as extraction reagent. Commun Soil Sci Plant Anal 31:1299–1396

Joergensen RG (1996) The fumigation-extraction method to estimate soil microbial biomass: calibration of the kEC value. Soil Biol Biochem 28:25–31

Kandeler E, Kampichler C, Horak O (1996) Influence of heavy metals on the functional diversity of soil microbial communities. Biol Fertil Soils 23:299–306

Le Bissonnais Y (1996) Aggregate stability and assessment of soil crustability and erodibility: I. Theory and methodology. Eur J Soil Sci 47:425–437

Macías F, Calvo de Anta R (2009) Niveles genéricos de referencia de metales pesados y otros elementos traza en los suelos de Galicia. Xunta de Galicia, Xunta de Galicia, Spain
Mehra OP, Jackson ML (1960) Iron oxide removal from soils and clays by a dithionite-citrate system buffered with sodium bicarbonate. Clays and clay minerals 7th conference. pp 317–327
Olsen SR, Cole CV, Watanabe FS, Dean LA (1954) Estimation of available phosphorus in soils by extraction with sodium bicarbonate. United States Department of Agriculture. Circular 9
Porta J (1986) Técnicas y experimentos en Edafología. Collegi Oficial D'Enginyers Agronoms de Catalunya, Barcelona, Spain
Sanchez-Monedero MA, Roig A, Martínez-Pardo C, Cegarra J, Paredes C (1996) A microanalysis method for determining total organic carbon in extracts of humic substances. Relationships between total organic carbon and oxidable carbon. Bioresour Technol 57:291–295
Shrestha RK, Lal R (2008) Land use impacts on physical properties of 28 years old reclaimed mine soils in Ohio. Plant Soil 306:249–260
Shukla MK, Lal R, Ebinger M (2006) Determining soil quality indicators by factor analysis. Soil Tillage Res 87:194–204
Sparling GP (1992) Ratio of microbial biomass carbon to soil organic carbon as a sensitive indicator of changes in soil organic matter. Aust J Soil Res 30:195–207
USDA (1998) Soil quality indicators: pH.
Vance ED, Brookes PC, Jenkinson DS (1987) An extraction method for measuring soil microbial biomass C. Soil Biol Biochem 19:703–707

Chapter 18
Phytoextraction of Heavy Metals: The Potential Efficiency of Conifers

Gaëlle Saladin

18.1 Introduction

Heavy metals or metal trace elements (MTE) contaminate many soils worldwide. Their introduction in large quantities in various ecosystems can be of natural origin (erosion, volcanic activity) but is most often due to human activities (mining, industrial waste, pesticides). A high level of MTE in soils can be toxic and even lethal for microfauna and plant communities. Thereafter, the upper links in the food chain (including humans) may be affected. Indeed, many studies show that people exposed to MTE can develop diseases such as cancers or disturbance of the central nervous system, blood cells, and kidney (De Burbure et al. 2006).

The most common technique to clean soils contaminated with MTE is soil excavation and removal. This process is not only expensive but also environmentally unfriendly because it destroys microfauna and unattractive because the ecosystem is altered for a long period (Dermont et al. 2008). Phytoextraction (or phytoaccumulation) of MTE is an alternative method studied for 20–30 years. It consists of growing plants directly on contaminated soils in order to extract pollutants. Thereafter, aboveground organs are harvested and incinerated to produce gas and to concentrate minerals for additional fertilization and MTE for reuse in various sectors (Pan and Eberhard 2011; Nzihou and Stanmore 2013). The valorization of these by-products is of interest because it reduces the cost of the process. This technology has the advantage to be less expensive than the conventional process. Indeed, several authors indicate that the price can be reduced by 50–80 % and even by a factor of 10–100, depending on the type of pollutant and soil (Pulford and Watson 2003; Ghosh and Singh 2005; Vangronsveld et al. 2009). In

G. Saladin (✉)
Laboratoire de Chimie des Substances Naturelles (EA 1069), Faculté des Sciences et Techniques, Université de Limoges, 123 avenue Albert Thomas, 87060 Limoges Cedex, France
e-mail: gaelle.saladin@unilim.fr

I. Sherameti, A. Varma (eds.), *Heavy Metal Contamination of Soils*, Soil Biology 44, DOI 10.1007/978-3-319-14526-6_18

addition, phytoextraction preserves a part of biodiversity and maintains vegetation on the site, which is more aesthetic than soil deeply dug. However, the main drawback of phytoextraction is its duration. Indeed, this method requires several decades, whereas it usually takes 1 or 2 years with the conventional method (but a longer time to fertilize the decontaminated site and recreate a viable ecosystem) (Wang and Jia 2010). Phytoextraction can therefore be used on areas that are not intended to be used quickly, such as abandoned mine sites.

18.2 Why Use Conifers for Phytoextraction?

18.2.1 The Scarcity of Hyperaccumulating Trees

There are 550–600 hyperaccumulating plant species which are plants that are able to store in their aerial parts much more MTE than the average plants without lethal effects (Baker and Brooks, 1989; Verbruggen et al., 2009; van der Ent et al. 2012). The minimum level of MTE in the aerial parts for hyperaccumulators was set at 0.1 % of the dry matter (0.1 g kg^{-1} DW) for most MTE, but there are exceptions, as shown in Table 18.1 (Baker and Brooks 1989; Huang et al. 1998; Ma et al. 2001; Liang et al. 2009; van der Ent et al. 2012). This natural storage capacity is considered as a defense against herbivores and pathogens (Rascio and Navari-Izzo 2011; Sarma 2011). The main disadvantage of hyperaccumulators is that they are mainly herbaceous plants with a reduced root system, thus not optimal to clean soils deeply. In addition, biomass is quite low, which requires very regular plantations and harvests.

Trees thus appear as potentially interesting candidates since they have a more developed root system. In addition, they can stay on the site for several years without new plantings, which can reduce the cost of this process (Pulford and Watson 2003). However, the number of woody species among hyperaccumulators is currently between 130 and 140, corresponding to approximately 20 % of hyperaccumulators (Table 18.2). Moreover, among these woody species, 75 % are only Ni hyperaccumulators; thus, the choice for the extraction of other MTE is very limited. Furthermore, woody species known as hyperaccumulators are mainly localized in tropical areas. As a consequence, their use under temperate or cold climates for MTE extraction is not optimal.

Table 18.1 Minimal content of MTE required to consider plants as hyperaccumulators

MTE	Minimal content in aerial organs (mg kg^{-1} DW)
Fe, Mn, U, Zn	10,000
Al, As, Co, Cr, Cu, Ni, Pb, Se	1,000
Cd	100
Hg	1

Table 18.2 Woody species known as hyperaccumulators of at least one MTE

MTE	Hyperaccumulators	Families and woody species	References
Ni	≈100	Fabaceae (*Cassia siamea*), Sapotaceae (*Pycnandra acuminata*), Violaceae (*Hybanthus floribundus* subsp. *floribundus*, *Rinorea niccolifera*) Around 30 taxons from Cuba were reported by Reeves et al. (1999) among Acanthaceae, Clusiaceae, Myrtaceae, Oleaceae, Rubiaceae, and Tiliaceae More than 60 taxons from New Caledonia were reported by Jaffré et al. (2013) among Argophyllaceae, Capparaceae, Celastraceae, Cunoniaceae, Euphorbiaceae, Myrtaceae, Oncotheaceae, Phyllantaceae, Rubiaceae, Salicaceae, Sapotaceae, and Violaceae	Jaffré et al. (1976, 2013), Callahan et al. (2008), Fernando et al. (2014), Jambhulkar and Juwarkar (2009), Reeves et al. (1999)
Al	10	Theaceae (*Stewartia monadelpha*, *S. pseudocamellia*, *Camellia sinensis*, *C. sasanqua*, *C. japonica*, *Cleyera japonica*, *Eurya japonica*), Vochysiaceae (*Qualea grandiflora*, *Callisthene major*, *Vochysia pyramidalis*)	De Andrade et al. (2011), Osawa et al. (2013)
Mn	9	Araliaceae (*Chengiopanax sciadophylloides*), Celastraceae (*Maytenus cunninghamii*), Myrtaceae (*Gossia bamagensis*, *G. bidwillii*, *G. fragrantissima*, *G. sankowsiorum*, *G. gonoclada*), Phytolaccaceae (*Phytolacca acinosa*), Proteaceae (*Grevillea exul* var. *exul*), Theaceae (*Schima superba*)	Fernando et al. (2006), Rabier et al. (2007), Mizuno et al. (2008), Yang et al. (2008b), Xue et al. (2010)
Cd	7	Aquifoliaceae (*Ilex polyneura*), Araliaceae (*Evodiopanax innovans*), Ericaceae (*Rhododendron annae*), Oxalidaceae (*Averrhoa carambola*), Salicaceae (*Salix cathayana*, *S. dasyclados*, *Populus x canescens*)	Zu et al. (2004), Takenaka et al. (2009), Li et al. (2011), Dai et al. (2012), Fischerova et al. (2006)

(continued)

Table 18.2 (continued)

MTE	Hyperaccumulators	Families and woody species	References
Zn	3	Ericaceae (*Rhododendron annae*), Fabaceae (*Cassia siamea*), Salicaceae (*Salix dasyclados*)	Fischerova et al. (2006), Jambhulkar and Juwarkar (2009), Zu et al. (2004)
Pb	3	Aquifoliaceae (*Ilex polyneura*), Ericaceae (*Rhododendron annae*), Fabaceae (*Sesbania drummondii*)	Zu et al. (2004), Sahi et al. (2002), Venkatachalam et al. (2009)
Fe	1	Fabaceae (*Cassia siamea*)	Jambhulkar and Juwarkar (2009)
Hg	1	Fabaceae (*Sesbania drummondii*)	Sahi et al. (2002), Venkatachalam et al. (2009)

18.2.2 Phytoextraction by Non-hyperaccumulating Trees

Several studies have nevertheless shown that some non-hyperaccumulating tree species are able to store as much or more MTE than herbaceous hyperaccumulators. Until now, the trees studied for phytoextraction are angiosperms, especially fast-growing trees such as poplars or willows (Di Lonardo et al. 2011; Zacchini et al. 2011; He et al. 2013). For example, phytoaccumulation efficiency was compared between poplars, willows, and *Noccaea caerulescens* (formerly *Thlaspi caerulescens*), an herbaceous Cd and Zn hyperaccumulator. Several authors indicated that *N. caerulescens* can store 5 kg of Cd and Zn per year and per ha, whereas poplars or willows can accumulate up to 10 kg Cd and 20 kg Zn under similar conditions (Robinson et al. 2000; Vandecasteele et al. 2005; Yanai et al. 2006). Fischerova et al. (2006) compared the efficiency of remediation between two herbaceous hyperaccumulators, *Arabidopsis halleri* and *N. caerulescens*, and woody species (several willow and poplar species): they showed that Cd content was higher in *A. halleri* (80 mg Cd kg^{-1} DW in aboveground biomass) than *Populus trichocarpa* (30 mg) and *Salix caprea* (20 mg). However, the aerial biomass of trees was higher, and the authors obtained a better annual remediation factor for trees than herbaceous species. Moreover, they showed that *P. trichocarpa* and *P. nigra* were better for Pb remediation than both herbaceous species.

A major cause of soil contamination is mining activities. Trees naturally growing close to these MTE-enriched areas may be good candidates for phytoextraction. For example, Raju et al. (2008) studied three tree species (*Cassia siamea*, *Azadirachta indica*, and *Holoptelea integrifolia*) from a Mn mine tailing dump containing 1.3 g Mn kg^{-1} soil. Although they did not indicate the age of the trees, they concluded that *H. integrifolia* was the best species to use for phytoremediation with Mn contents of 1.2 and 1.7 g Mn kg^{-1} DW in stems and leaves, respectively. These concentrations are below the threshold for hyperaccumulators, but the high soil pH (8.84) was not optimal for plant growth. Another study highlighted the efficiency of several woody species (*Terminalia arjuna*, *Prosopis juliflora*, *Populus alba*, *Eucalyptus tereticornis*, and *Dendrocalamus strictus*) for the phytoextraction of many MTE in tannery sludges (Shukla et al. 2011). The authors indicated that after 1 year, trees were able to remove 70 % Cr, 50–60 % Ni/Cd/Fe, 20–30 % Mn/Zn/Cu, and 14 % Pb.

18.2.3 Conifers May Have Their Place in Phytoextraction Process

Conifers are often used as biomarkers of industrial pollutions and more particularly of atmospheric contaminations including MTE and radioelements (Čeburnis and Steinnes 2000; Monaci et al. 2000; Orlandi et al. 2002; Samecka-Cymerman et al. 2008; Gandoit and Probst 2012; Kuroda et al. 2013; Przybysz et al. 2014). MTE content in needles can be monitored and allowed, for example, to create pollution maps by using *Pinus sylvestris* (Dmuchowski and Bytnerowicz 1995). However, conifers are poorly studied for MTE phytoextraction despite several species exhibiting a fast growth rate and could be therefore good candidates.

Furthermore, the efficiency of phytoextraction depends on a good choice of plant species, not only for the type of soil and MTE to remove but also for climatic conditions. If poplars and willows can adapt to many soils, they are however more sensitive than conifers to low temperatures and freeze-thaw embolism (Carnicer et al. 2013). Conifers may therefore appear as models of interest for pollutions localized in cold regions. For example, many soils of boreal area contain non-negligible contents of Al and Fe and were submitted for decades to atmospheric pollution with a subsequent accumulation of other MTE in soils such as Pb (Bindler et al. 1999; Kiikkilä 2003; Steinnes and Friedland 2006). Orlandi et al. (2002) studied the evolution of MTE (Cd, Cr, Cu, Ni, Pb) in rings of *Larix decidua* in Western Italian Alps and highlighted the regular increase of MTE pollution since 1930.

In another way, dry soils may be another constraint for MTE phytoaccumulation. Indeed, the absorption of water and minerals and their transport from roots to leaves depend on the gradient of water potential between the soil and the atmosphere. Carnicer et al. (2013) compared angiosperms and gymnosperms in the Iberian Peninsula and reported that conifers exhibited a higher resistance to cavitation caused by drought and thus a lower risk of hydraulic failure. Moreover, the authors explained that conifers have a growth peak in autumn, whereas angiosperms may undergo an increase of embolism leading to a premature leaf abscission. However, the authors indicated that conifers have an earlier stomatal closure during summer, whereas angiosperms can maintain a non-negligible rate of transpiration during this period. This means that the growth of conifers in Mediterranean-like areas is lower than angiosperms in summer but more important in autumn. The drought tolerance of conifers was also mentioned in the review of Hamanishi and Campbell (2011) who reported that pine species were less sensitive than aspen to water deficit. Nevertheless, the difference between conifers and angiosperms is not so easy to establish because of intraspecific genetic variation. Indeed, the authors explained, for example, that Douglas firs (*Pseudotsuga menziesii*) or loblolly pines (*Pinus taeda*) have different levels of drought tolerance as a function of their geographical origin. Therefore, the efficiency of a conifer species (as well as angiosperm trees) for MTE phytoextraction may give variable results as a function of the environment.

18.3 Physiological Responses of Conifers to MTE

Few studies are available on the effect of MTE on the physiology of conifers. The majority of works is carried out under laboratory conditions, i.e., in vitro cultures and germinations or young trees grown in greenhouse, on contaminated soils, or hydroponic conditions.

18.3.1 Growth and Development

Conifers react differently as a function of MTE. For example, *Pinus sylvestris* is considered as highly sensitive to relatively low zinc concentrations according to Ivanov et al. (2011). Seeds were grown on medium contaminated with 150 μM Zn, and plantlets were harvested 6 weeks later. The authors observed that this Zn concentration reduced seed germination, lowered root growth (main and secondary roots), and reduced biomass of underground and aboveground organs. Moreover, photosynthetic pigment content was lower than control plantlets. Looking at other MTE, *P. sylvestris* appears to be sensitive to Ni but less to Cu: 4-year-old trees exposed for a season to Ni and/or Cu exhibited only 20 % of injury on fine roots for the treatment with 50 mg Cu kg^{-1}, whereas more than 60 % of fine roots were altered with only 5 mg Ni kg^{-1} (Kukkola et al. 2000). Moreover, needles were healthier after exposure to Cu than Ni. Another work with very young pines showed different level of tolerance as a function of pine species (Arduini et al. 1995): seedlings of *P. pinea* and *P. pinaster* were exposed for 4 weeks to 1 or 5 μM Cu. The results showed a similar inhibition of root elongation with 5 μM Cu and a reduction of taproot elongation with 1 μM (but recovery occurred after 7 days, suggesting a tolerance mechanism in roots). However, the 1 μM treatment increased the number of lateral roots in *P. pinaster* by contrast to *P. pinea*. This phenotype, known as stress-induced morphogenic response, is interpreted as an acclimation: plant growth, and particularly root growth, is redirected to reduce stress exposure (Potters et al. 2007). However, this phenotype was not observed when *P. pinaster* (as well as *P. halepensis*) plantlets were exposed for 3 weeks to 90 μM Zn (Disante et al. 2010). Nevertheless, in this study, the authors showed that Zn treatment reduced root biomass of *P. pinea* and *P. halepensis* but not *P. pinaster*, confirming that this pine species may be more tolerant to different MTE.

Ćurguz et al. (2012) compared the effect of several MTE (Cd, Pb, Cu, and Zn) on *Picea abies*. They observed that Cd was the most toxic metal for seed germination with a decrease by 77 % compared to Pb (58 %), Cu (52 %), and Zn (46 %). Moreover, they showed that Pb and Cu caused the highest reduction of root length and primary needle length, respectively. However, the concentrations of MTE in the medium were high for seed germination and the first stages of growth (33 ppm), suggesting that spruce may be quite tolerant to these MTE. The authors also concluded that *P. abies* could be selected for reforestation on sites contaminated with these MTE. Another work showed the tolerance of conifers to MTE: Moussavou Moudouma et al. (2013) exposed 4-week-old hybrid larches (*Larix* x *eurolepis*) grown in vitro to 1.5 mM Cd for 1 week. Despite this high Cd concentration, plantlet growth was not significantly altered: photosynthetic pigment content remained unchanged as well as biomass (aerial and root) and primary root length. Conifers selected for the phytoremediation of contaminated soils are not suspension cells or young plantlets, but these two studies highlight the tolerance of these conifer species to high MTE concentrations, which is promising for the future.

18.3.2 Antioxidative Pathway

It is known that MTE generate an oxidative stress and the accumulation of reactive oxygen species (ROS), but plants can stimulate antioxidative pathways to counteract the increase of ROS (Demidchik 2014). For example, Radotić et al. (2000) showed that 2-year-old spruce (*Picea abies*) exposed to 21 mg Cd kg^{-1} dry soil in greenhouse had a peroxidase activity in needles which was increased by a factor of 3 after 15 days of treatment. This activity was related to de novo synthesis of new isoenzymes, indicating that spruce was able to efficiently stimulate this defense pathway against cell oxidation generated by Cd. Moreover, when Cd exposure was extended to 2 months, the authors observed a decrease of activity in soluble fraction but a stimulation of peroxidase activity in cell wall. This suggests an additional strategy of protection, i.e., a stimulation of lignin synthesis. The stimulation of peroxidase activity was confirmed by Markkola et al. (2002) on *Pinus sylvestris*: they registered an increase by 20–30 % of root peroxidase activity in plantlets exposed for at least 7 months to urban polluted forest soils containing Cu, Cr, Pb, and Zn.

Another study on *P. sylvestris* showed that exposure to Zn led to an increase in activity of other antioxidative enzymes (Ivanov et al. 2012). Indeed, the authors treated plantlets for 6 weeks with 150 μM Zn and recorded a stimulation of superoxide dismutase (SOD) in whole organs (roots, stem, cotyledons, and needles). The H_2O_2 generated by SOD was then efficiently catalyzed by catalase in whole organs and by peroxidases in roots and stems but not in cotyledons and needle (but the authors showed that Zn accumulation was low in these organs). As a consequence, the level of malondialdehyde (a product of lipid peroxidation) remained close to control plants. Moreover, the authors highlighted that proline (but not phenolic compounds) accumulated as a response to Zn exposure. Data of literature often indicate that proline content increases as a response to drought and salinity, but it can be considered as a ROS scavenger too and thus can play a role in the defense against MTE (Szabados and Savouré 2010).

The family of glutathione-S-transferases (GST) is known to detoxify many organic xenobiotics by catalyzing GSH conjugation with electrophilic molecules but is also involved in antioxidative pathway to limit membrane lipid peroxidation (Edwards et al. 2005). It was shown that GST can be stimulated as a response to MTE (Schröder et al. 2003). The authors incubated spruce cells (*Picea abies*) for 16 h with different concentrations of MTE such as As (1.5–80 μM), Pb (10–150 μM), and Cd (50–500 μM) and observed in all cases an accumulation of H_2O_2. The authors indicated that GST activity was not significantly stimulated after Pb treatment by contrast to As and Cd exposures which enhanced enzyme activity up to 50 %.

18.3.3 MTE Scavenging

Oligopeptides containing cysteine residues can play a role in the tolerance to MTE. The main compounds are reduced glutathione (GSH) and phytochelatins (PC). GSH is a tripeptide with the formula γglutamate-cysteine-glycine (γGlu-Cys-Gly), the cysteine residue containing a sulfhydryl (or thiol) group which can potentially scavenge several MTE. GSH is the substrate of the phytochelatin synthase for PC synthesis, PC being usually more efficient MTE scavengers (Grill et al. 1985). The general formula of PC is $(\gamma\text{Glu-Cys})_n\text{Gly}$ with $n = 2$–11, but generally, two to four for most plants. GSH is involved in antioxidative pathway to maintain the redox level and can directly scavenge several MTE or indirectly via PC synthesis. The level of cysteine-rich oligopeptides can thus be a good biomarker to evaluate the tolerance of plants to MTE.

Works of Kukkola et al. (2000) on *Pinus sylvestris* are a good example: 4-year-old trees exposed for a season to Ni or Cu exhibit different levels of GSH in needles, thus indicating a different level of sensitivity as a function of MTE. Indeed, Ni exposures (5–25 mg kg^{-1} soil) decreased total glutathione content by a factor of 4 (and GSH level corresponded to less than 70 % of total glutathione), whereas total glutathione and GSH concentrations were not significantly modified by Cu exposures (25–50 mg kg^{-1} soil). Thangavel et al. (2007) used cell suspensions of red spruce (*Picea rubens*) to study the synthesis of GSH and PC as a response to Cd and Zn treatments. The highest exposures with Cd (200 μM) and Zn (800 μM) showed that the concentration of the GSH precursor (γglutamylcysteine or γEC) increased by a factor of 3.5 and 1.7 for Cd and Zn, respectively. Since GSH content was not modified, this increase allows maintaining a constant pool of GSH and an accumulation of PC. Indeed, the authors observed an increase of PC2 (i.e., 2 γGlu-Cys motifs) by 3.6 after Cd treatment and 1.7 after Zn treatment. Moreover, they detected a similar increase of 4 peaks of "long chain polythiols" (the authors did not identify them more precisely) by a factor of 3–9 for Cd and 2–3 for Zn as a function of the peak. Works of Moussavou Moudouma et al. (2013) on in vitro grown hybrid larches (4-week-old) exposed to 1.5 mM Cd for 1 week showed a stimulation by a factor of 1.5–2.4 of enzymes involved in GSH and PC synthesis (γglutamylcysteine synthetase, glutathione synthetase, and phytochelatin synthase) in both shoots and roots. In parallel, the concentration of cysteine-rich oligopeptides increased by a factor of 1.6 in shoots and 1.5 in roots. This confirms that these molecules can play a role of defense against MTE by a mechanism of scavenging (which limits the production of ROS).

However, these compounds are not the only ones to be involved in intracellular MTE scavenging. Indeed, it was reported on few tree species such as *Salix alba* or *Populus nigra* x *maximowiczii* that polyamines or organic acids could play a similar role (Mohapatra et al. 2010; Zacchini et al. 2011). The evolution of polyamines as a response to MTE in conifers is poorly documented. The works of Thangavel et al. (2007) previously described only mentioned that polyamine content was not modified after exposure of *Picea rubens* (suspension cells) to Cd or Zn. However, another study with *P. rubens* suspension cells showed that Al exposure increased intracellular putrescine and succinate contents as well as succinate and oxalate exudation (despite succinate accumulation and secretion was indicated by the authors as a by-product of putrescine degradation) (Minocha and Long 2004). Thus, the role of polyamine and/or organic acids in MTE scavenging could be not only species-dependent but also TE-dependent. In another way, metallothioneins could be other potential chelating agents as shown on hybrid aspen (*Populus tremula* x *tremuloides*) exposed to Cd and Zn (Hassinen et al. 2009). As for polyamines, data related to metallothioneins and conifers exposed to MTE are still lacking.

In addition to intracellular MTE scavenging, the cell wall is another important compartment since it can store at least half of MTE accumulated in plants (Peng et al. 2005; Sousa et al. 2008). When studying root tips of slash pine (*Pinus elliottii*) and loblolly pines (*Pinus taeda*) exposed to Al, Nowak and Friend (2005) showed that apoplasmic Al corresponded to 85–90 % of total Al with 30–40 % bound to the cell wall and 50–55 % as a labile fraction. Astier et al. (2014) observed that 3-year-old Douglas firs (*Pseudotsuga menziesii*) exposed for 9 months to Cd had lower lignin concentration and higher pectin content, more particularly low-methylesterified pectins. Although they did not measure Cd in the cell wall, the authors explained that these pectins could be binding sites for metallic cations. Indeed, most metallic cations are divalent and thus could replace Ca^{2+} in pectic structure called the "egg box" (Krzesłowska 2011). It was reported by several authors that 70–90 % of MTE (such as Al, Cd, Zn, or Cu) present in cell walls are bound to pectins, particularly to homogalacturonans (Peng et al. 2005; Sousa et al. 2008; Yang et al. 2008a, b).

18.3.4 Distribution of MTE

Whatever the plant species, MTE concentration is not similar in all organs. In seedlings, young organs are often protected in order to maintain growth and development. It was confirmed by Ivanov et al. (2011) on 6-week-old *Pinus sylvestris* plantlets treated with Zn: the authors explained that stems and cotyledons correspond to a kind of "barrier zone" to limit Zn transport to needles and thus to protect photosynthetic apparatus. In this study, Zn content was twice lower in needles than in the two other organs. Astier et al. (2014) showed something quite similar with older trees (3 years before Cd exposure): when separating aerial organs of Douglas firs, they observed higher Cd concentrations in the bark and wood than in needles and buds, probably to protect photosynthetic organs and future branches against oxidative stress caused by Cd. However, this phenomenon is not always observed: Mingorance et al. (2007) registered more Fe, Mn, and Pb in needles than in the bark and wood of *Pinus pinea*, whereas Cr, Cu, and Zn were preferentially stored in the bark. Furthermore, the distribution of MTE can depend on the initial concentration of each MTE and other minerals in soil: Scots pines (*P. sylvestris*) from three different sites (Nagu, Ostrobothnia, and Harjavalta) in Finland were compared (Saarela et al. 2005). The authors showed that Ni, Cu, Zn, Cd, and Pb were differently distributed between the bark and the wood as a function of the MTE and the site. For example, Ni preferentially accumulated in the wood only at the site of Nagu, whereas Cu preferentially accumulated in the wood only at the site of Ostrobothnia.

Otherwise, MTE are mostly stored in the root system for non-hyperaccumulating species, whereas MTE translocation to aboveground organs is more important for hyperaccumulators (McGrath et al. 2002). For example, previously described works of Ivanov et al. (2011) indicated that Zn content in roots of *P. sylvestris* plantlets was ten times the concentration in stems and cotyledons and 20 times the concentration in needles. The preferential distribution of MTE in roots was confirmed on *Pinus halepensis* exposed to Zn, Ni, or Cu for 3 or 8 months (Fuentes et al. 2007a, b). Another study with plantlets confirmed the preferential accumulation of MTE in roots: 4-week-old hybrid larch (*L.* x *eurolepis*) exposed for 1 week to Cd had twice more Cd in roots than in shoots (Moussavou Moudouma et al. 2013). Using older trees (3 years before MTE exposure), Astier et al. (2014) observed a similar distribution, i.e., more MTE in roots than in aerial organs (five times more). Moreover, when separating aerial organs, they observed higher Cd concentrations in the bark and wood than in needles and buds, probably to protect photosynthetic organs and future branches against oxidative stress caused by Cd. Some exceptions exist in conifers compared to the general trend of a greater accumulation of MTE in the roots: Bergqvist and Greger (2012) compared As distribution of many plant species and observed that *Picea abies* accumulated more As in shoots, whereas the other conifer tested (*Pinus sylvestris*) accumulated more As in roots.

However, several MTE such as Cd and Pb tend to accumulate preferentially in roots even for hyperaccumulating trees (Sahi et al. 2002; Dai et al. 2012).

Nevertheless, the advantage of trees is their high aerial biomass and the final quantity of MTE stored in aboveground organs. Indeed, this final quantity can be more important in aerial organs even if the concentration is lower than in the root. This was highlighted in hybrid larch grown in vitro and exposed to Cd (Moussavou Moudouma et al. 2013): Cd concentration was twice lower in shoots than in roots, but the final quantity of Cd stored in shoots was at least twice higher than in roots. However, this was not confirmed on Douglas firs (3-year-old trees exposed for 9 months to Cd) since the final quantity of Cd per tree was approximately 1.5 more important in root system than in aerial organs.

18.4 Efficiency of Phytoextraction by Conifers

Up to now, no conifer species is considered as hyperaccumulator, but little work has been done on this subject, and only few species have been tested. The question is to assess whether conifers can accumulate acceptable amounts of MTE to be used for phytoextraction. The limited amount of data available is mainly related to pines and particularly Scots pine (*Pinus sylvestris*). To estimate the potential efficiency of conifers, it could be interesting to compare conifer and angiosperm trees in similar experimental conditions, which is actually difficult to find in literature.

When comparing the efficiency of *Populus* x *canadensis* and *Larix olgensis* for the phytoextraction of Cd, Cu, and Zn in northeastern China, the authors concluded that poplar had higher MTE concentrations in aerial organs and higher biomass than larch (Wang and Jia 2010). According to their results, this larch species is not suitable for phytoaccumulation. However, the authors used a hybrid poplar (hybrids are generally more vigorous than wild species), and larches were three times smaller than poplars at the beginning of the experiments (in terms of tree height and total biomass) even if they had the same age. Moreover, at the end of the experiment, results showed that poplars exposed to MTE had a biomass reduced by 25 % compared to control poplars, whereas biomass of larches decreased only by 5 %. This suggests that the growth rate is less affected in this larch species than in poplar, which is important to take into account since the experiment was conducted during only one growing season (and trees have to be left for several years on polluted sites for phytoremediation).

An interesting study was conducted by Reimann et al. (2001) in northern Europe: the authors compared the accumulation of various MTE in the leaves of birch, willow, pine (*P. sylvestris*), and spruce (*Picea abies*) on several polluted sites. They observed that birch and willow stored more MTE except for Al which was more accumulated by pine (approximately ten times). *Pinus radiata* was compared with *Populus hybridus* and *Eucalyptus rostrata* growing close to a smelter in Argentina (Rodriguez et al. 2012): among MTE present in soil (Cd, Cu, Fe, Mn, Ni, and Zn), *P. radiata* was more efficient only for Fe storage (the authors only analyzed needles and leaves of the trees) with 40 and 80 % more Fe than poplar and eucalyptus, respectively. This better accumulation of Fe in needles by pines was already observed by Sawidis et al. (2001) who compared *Pinus nigra* and angiosperm trees (including poplar and willow) on areas contaminated with various MTE in Greece.

Conifers are not only present in cold areas but also in warmer climates such as the Mediterranean basin. However, few studies were conducted with pines native to this area. Pratas et al. (2005) compared *Pinus pinaster* and two *Quercus* species growing on abandoned mines of Portugal which were highly contaminated with arsenic, antimony, and tungsten. The most important difference was the strongest accumulation of As in needles of *P. pinaster* compared to the leaves of both oaks: young and old needles of pines contained 30–60 and 100–200 times more As than oak leaves, respectively. Mingorance et al. (2007) compared two Mediterranean species (*Pinus pinea* and *Nerium oleander*) growing in an industrial area in Spain: except Cu, stone pines often stored more MTE than *N. oleander* in all aerial organs (Al), in needles and bark (Fe, Pb), in needles and wood (Mn, Zn), or only in needles (Cr). Thus, in particular climates, local conifer species may be interesting for phytoextraction.

18.5 How to Improve Phytoextraction by Conifers?

Soil characteristics and pH affect solubilization and bioavailability of minerals absorbed by plants. These minerals include not only those that are essential for plant growth but also MTE: some are also essential for plant growth but in small quantities (Cu, Zn, Fe), but others are not required (Pb, Cd, Hg). To increase MTE extraction by plants, it is important that MTE are more easily absorbed. At least two strategies are possible. The first is to directly modify MTE bioavailability by suitable amendment, i.e., by addition of specific molecules on soils. The second strategy is to promote the development of mycorrhiza that could act on MTE solubilization in soil and thus promote absorption by roots.

18.5.1 Soil Amendment

Hodson and Sangster (1999) reported that acidic precipitations mobilize Al in soils and that this MTE is involved in the dieback of trees in Western Europe and North America. The authors studied needles of several conifers (*Picea glauca*, *Pinus strobus*, *Larix laricina*, and *Abies balsamea*) and found that Al was co-deposited with silicon. They indicated that this phenomenon could be a defense mechanism and thus an interesting tool to increase Al phytoaccumulation. This was confirmed by Prabagar et al. (2011) on Norway spruce (*Picea abies*). Indeed, the authors showed that Al absorption increased when Si was added and that free Al was reduced in the cell wall. The authors suggested that the formation of complexes between Al and Si observed in the cell wall and apoplasm avoids Al penetration in cells and thus protect intracellular compartment from Al toxicity.

Chelating agents such as ethylenediaminetetraacetic acid (EDTA) are often tested to increase MTE absorption. For example, Jarvis and Leung (2002) used H-EDTA and EDTA to determine the effect on Pb absorption by *Pinus radiata*. Although experiments were performed in hydroponic culture, their results clearly showed that phytoextraction process can be strongly improved by amendment. Indeed, chelating agents decreased Pb concentration in roots by a factor of 3 and increased Pb content in needles by a factor of 8 and 10 for H-EDTA and EDTA, respectively. However, synthetic chelating agents such as EDTA are toxic and rather persistent in the environment (Leštan et al. 2008). Therefore, it could be more attractive to replace these compounds by natural and/or cheaper compounds with a better biodegradability. For example, several organic acids (citric, oxalic, vanillic, and gallic acids) can solubilize non-negligible amounts of Zn, Ni, and Cd from the soil (do Nascimento 2006; do Nascimento et al. 2006). Unfortunately, it has not yet been tested if these compounds improve phytoextraction by conifers.

Another type of amendment is the use of fertilizers or sludge since the addition of essential minerals and organic matter can potentially have two advantages. Indeed, it can stimulate the growth and/or the defenses of plants and increase consequently the accumulation of MTE. The use of sludge has two interesting features: it can fertilize soil for trees and it recycles waste, reducing in this case the cost of amendment compared to commercial fertilizers. This method gave encouraging results on *Pinus sylvestris* according to Vaitkutė et al. (2010). Seedlings were planted on an experimental site amended or not with industrial sewage sludge and were analyzed 8 years later. The authors did not register a stimulating effect of sludge on tree growth (height or biomass) but showed that Cu and Pb contents in aboveground organs increased by a factor of 1.5 and 10, respectively. However, the amendment was not efficient for all MTE since it had no significant effect on Cd accumulation in aerial organs.

18.5.2 Mycorrhization

Mycorrhizal fungi can interact with more than 80 % of terrestrial plants as reported by Leung et al. (2013). They have a positive effect on plant nutrition by enhancing the absorption of minerals. Indeed, they cannot only transform organic compounds from the soil into available elements for roots but also increase the area in contact with soil and thus with nutrients. Mycorrhization could be an interesting tool to improve MTE phytoextraction because fungi could increase the absorption of MTE by the root system. Several works showed the improvement of phytoremediation (including conifer species such as pines and spruces) when mycorrhiza is present as reported by Leung et al. (2013). For example, Babu et al. (2014) isolated the endophytic fungus *Trichoderma* sp. PDR1-7 from a Pb-contaminated mine tailing soil. This fungus was able to increase MTE solubilization and nutrient availability in the soil and to remove Pb efficiently . The authors showed that this fungus was a benefit for *Pinus sylvestris* plantlets since they grew better with higher chlorophyll content, an enhanced activity of antioxidative enzymes and a decrease of malondialdehyde (a bioindicator of oxidative stress). Moreover, MTE content was higher in pine roots (As, Cd, Cu, Ni, Pb, Zn) when this fungus was present, indicating the improvement of phytoextraction with this strategy. Nevertheless, results showed that in shoots, only Pb and Zn accumulation was stimulated, suggesting that the combination *Trichoderma*/*Pinus* for phytoaccumulation is not efficient for all MTE. However, Pb and Zn accumulation in shoots was 30 and 45 % higher, respectively, which is promising for the depollution of soil contaminated by these MTE. Another fungus was tested on *Pinus pinaster*: Sousa et al. (2012) compared Cd accumulation in pines exposed for 6 months to 30 mg kg^{-1} Cd (plants were 11 months old at the end of the experiment) with or without inoculation with fungi (*Rhizopogon roseolus* and *Suillus bovinus*). They observed that *R. roseolus* was the most interesting since it was able to reduce Cd in roots by 60 % and increase Cd in shoots by 35 % without modifying pine growth. The final Cd concentration registered in shoots was 15 mg kg^{-1} when plants were inoculated with *R. roseolus*.

However, the efficiency of mycorrhiza to improve phytoextraction by conifers is not always proven, depending on the species, the strain of mycorrhiza, and the MTE (Jones et al. 1986; Galli et al. 1993; Godbold et al. 1998; Jentschke et al. 1998, 1999; Kozdrój et al. 2007; Leung et al. 2013). Indeed, when soil nutrients and MTE are solubilized, fungi can store MTE and then limit the absorption by roots and the translocation to shoots (Krznaric et al. 2010). The works of Sousa et al. (2012) described in the previous paragraph indicated a benefic effect of the fungus *R. roseolus* to reduce Cd in pine (*P. pinaster*) roots and increase the storage of this MTE in shoots. However, they published another work (Sousa et al. 2014) showing that this positive effect depended on pine genotype. Indeed, *R. roseolus* did not significantly increase Cd content in shoots of wild genotype and reduced Cd root content by 23 % instead of 60 % for the selected genotype.

Moreover, exudates secreted by fungi in soils are not always compounds allowing MTE solubilization. Indeed, exudates can sometimes form complexes with MTE and avoid their absorption. In this case, trees could be interesting for soil phytostabilization (meaning a reforestation to avoid MTE leaching) but not for phytoextraction. Krznaric et al. (2009) studied *Suillus luteus*, an ectomycorrhizal fungus of pine which is a good example illustrating this difference. The authors isolated Cd-tolerant and Cd-sensitive strains of *S. luteus* and analyzed their effect on *Pinus sylvestris* plantlets exposed to Cd. They showed that Cd-tolerant isolate increased Cd content in pine roots but not in needles and reduced root biomass by contrast to Cd-sensitive isolate. Thus, the Cd-sensitive isolate could be more interesting for phytoaccumulation (the authors registered a Cd content of 15 mg kg^{-1} in needles after 2 months of exposure), whereas the Cd-tolerant isolate would be more suitable for phytostabilization. In the same way, Ahonen-Jonnarth and Finlay (2001) showed that *Pinus sylvestris* inoculated with *Laccaria bicolor* accumulated more Ni and Cd in roots but less in shoots.

18.6 Conclusion

If MTE phytoextraction by trees is more and more studied, this technology is mainly focused on poplars and willows. Conifer species with a fast growth rate could nevertheless be an interesting tool in cold areas or even in dry zones (e.g., *Pinus pinaster* is located at the origin in the Mediterranean basin). In the northern hemisphere, pine species are quite widespread with more or less diversity as a function of geography. Indeed, *Pinus sylvestris* is well represented in Europe, but Asia and North America exhibit a higher biodiversity with approximately 100 species (against a dozen species for Europe and Mediterranean). Thus, the selection of pine species for phytoextraction should be more taken into account. Other conifers could be tested such as junipers (*Juniperus* sp. are often pioneer species) or cedars (*Cedrus* sp. can support more alkaline soils than other conifers). Moreover, other conifers are present in the southern hemisphere (*Araucaria*, *Podocarpus*) but were not studied for their potential use in phytoextraction. Furthermore, the selection of hybrid conifers can be of interest since they are often more vigorous than their parents and thus could grow better on contaminated soils. It was shown, for example, that hybrid larch (*Larix* x *eurolepis*) is more resistant to various biotic and abiotic stresses than its parents *L. decidua* and *L. kaempferi* (Bastien and Keller 1980).

Otherwise, other parameters may be refined to improve phytoextraction by conifers such as appropriate amendments to stimulate nutrient and MTE absorption or mycorrhization with fungi able to increase their own biomass and MTE availability in soils for higher subsequent absorption by conifers. Another alternative may be developed to enhance phytoremediation: overexpression of genes involved in metal absorption, transport, and/or vacuolar sequestration could increase MTE storage as reported by Eapen and D'Souza (2005). This strategy has already been successfully tested on poplars, for example, to enhance metal transport or glutathione synthesis or to reduce ROS production (Bittsánszky et al. 2005; Turchi et al. 2012; Shim et al. 2013). However, transgenic plants are not allowed on fields in all countries, particularly in Europe. This perspective is still difficult to generalize thus more "natural" selections of tolerant species and hybrids are probably a better approach.

References

Ahonen-Jonnarth U, Finlay RD (2001) Effects of elevated nickel and cadmium concentrations on growth and nutrient uptake of mycorrhizal and non-mycorrhizal *Pinus sylvestris* seedlings. Plant Soil 236:129–138

Arduini I, Godbold DL, Onnis A (1995) Influence of copper on root growth and morphology of *Pinus pinea* L. and *Pinus pinaster* Ait. Seedlings. Tree Physiol 15:411–415

Astier C, Gloaguen V, Faugeron C (2014) Phytoremediation of cadmium-contaminated soils by young Douglas fir trees: effects of cadmium exposure on cell wall composition. Int J Phytoremediation 16:790–803

Babu AG, Shea PJ, Oh BT (2014) *Trichoderma* sp. PDR1-7 promotes *Pinus sylvestris* reforestation of lead-contaminated mine tailing sites. Sci Total Environ 476–477:561–567

Baker AJM, Brooks RR (1989) Terrestrial higher plants which hyperaccumulate metallic elements: a review of their distribution, ecology and phytochemistry. Biorecovery 1:81–126

Bastien JC, Keller R (1980) Intérêts comparés du mélèze hybride (*Larix* x *eurolepis* Henry) avec les deux espèces parents. Revue Forestière Française 6:521–530

Bergqvist C, Greger M (2012) Arsenic accumulation and speciation in plants from different habitats. Appl Geochem 27:615–622

Bindler R, Brännvall ML, Renberg I (1999) Natural lead concentrations in Pristine boreal forest soils and past pollution trends: a reference for critical load models. Environ Sci Technol 33:3362–3367

Bittsánszky A, Kömives T, Gullner G, Gyulai G, Kiss J, Heszky L, Radimszky L, Rennenberg H (2005) Ability of transgenic poplars with elevated glutathione content to tolerate zinc(2+) stress. Environ Int 31:251–254

Callahan DL, Roessner U, Dumontet V, Perrier V, Wedd AG, O'Hair RAJ, Baker AJM, Kolev SD (2008) LC–MS and GC–MS metabolite profiling of nickel(II) complexes in the latex of the nickel-hyperaccumulating tree *Sebertia acuminata* and identification of methylated aldaric acid as a new nickel(II) ligand. Phytochemistry 69:240–251

Carnicer J, Barbeta A, Sperlich D, Coll M, Peñuelas J (2013) Contrasting trait syndromes in angiosperms and conifers are associated with different responses of tree growth to temperature on a large scale. Front Plant Sci 4(409):1–19

Čeburnis D, Steinnes E (2000) Conifer needles as biomonitors of atmospheric heavy metal deposition: comparison with mosses and precipitation, role of the canopy. Atmos Environ 34:4265–4271

Ćurguz VG, Raičević V, Veselinović M, Tabakovic-Tošić M, Vilotić D (2012) Influence of heavy metals on seed germination and growth of *Picea abies* L. Karst Pol J Environ Stud 21 (2):355–361

Dai HP, Wei Y, Zhang YZ, Wei AZ, Yang TX (2012) Subcellular localization of cadmium in hyperaccumulator *Populus* × *canescens*. Afr J Biotechnol 11(16):3779–3787

De Andrade LRM, Barros LMG, Echevarria GF, Velho do Amaral LI, Cotta MG, Rossatto DR, Haridasan M, Franco AC (2011) Al-hyperaccumulator Vochysiaceae from the Brazilian Cerrado store aluminum in their chloroplasts without apparent damage. Environ Exp Bot 70:37–42

De Burbure C, Buchet JP, Leroyer A, Nisse C, Haguenoer JM, Mutti A, Smerhovsky Z, Cikrt M, Trzcinka-Ochocka M, Razniewska G, Jakubowski M, Bernard A (2006) Renal and neurologic effects of cadmium, lead, mercury, and arsenic in children: evidence of early effects and multiple interactions at environmental exposure levels. Environ Health Perspect 144:584–590

Demidchik V (2014) Mechanisms of oxidative stress in plants: from classical chemistry to cell biology. Environ Exp Bot http://dx.doi.org/10.1016/j.envexpbot.2014.06.021

Dermont G, Bergeron M, Mercier G, Richer-Laflèche M (2008) Soil washing for metal removal: a review of physical/chemical technologies and field applications. J Hazard Mater 152:1–31

Di Lonardo S, Capuana M, Arnetoli M, Gabbrielli R, Gonnelli C (2011) Exploring the metal phytoremediation potential of three *Populus alba* L. clones using an *in vitro* screening. Environ Sci Pollut Res 18:82–90

Disante K, Fuentes D, Cortina J (2010) Sensitivity to zinc of Mediterranean woody species important for restoration. Sci Total Environ 408:2216–2225

Dmuchowski W, Bytnerowicz A (1995) Monitoring environmental pollution in Poland by chemical analysis of Scots pine (*Pinus sylvestris* L.) needles. Environ Pollut 87:87–104

Do Nascimento CWA (2006) Organic acids effects on desorption of heavy metals from a contaminated soil. Sci Agric 63(3):276–280

Do Nascimento CWA, Amarasiriwardena D, Xing B (2006) Comparison of natural organic acids and synthetic chelates at enhancing phytoextraction of metals from a multi-metal contaminated soil. Environ Pollut 140:114–123

Eapen S, D'Souza SF (2005) Prospects of genetic engineering of plants for phytoremediation of toxic metals. Biotechnol Adv 23:97–114

Edwards R, Dixon DP, Walbot V (2005) Plant glutathione S-transferases: enzymes with multiple functions in sickness and in health. Trends Plant Sci 5(5):193–198

Fernando DR, Batianoff GN, Baker AJ, Woodrow IE (2006) *In vivo* localization of manganese in the hyperaccumulator *Gossia bidwillii* (Benth.) N. Snow & Guymer (Myrtaceae) by cryo-SEM/EDAX. Plant Cell Environ 29:1012–1020

Fernando E, Quimado M, Doronila A (2014) *Rinorea niccolifera* (Violaceae), a new, nickel-hyperaccumulating species from Luzon Island, Philippines. PhytoKeys 37:1–13

Fischerova Z, Tlustos P, Szakova J, Sichorova K (2006) A comparison of phytoremediation capability of selected plant species for given trace elements. Environ Pollut 144:93–100

Fuentes D, Disante KB, Valdecantos A, Cortina J, Vallejo VR (2007a) Sensitivity of Mediterranean woody seedlings to copper, nickel and zinc. Chemosphere 66:412–420

Fuentes D, Disante KB, Valdecantos A, Cortina J, Vallejo VR (2007b) Response of *Pinus halepensis* Mill. seedlings to biosolids enriched with Cu, Ni and Zn in three Mediterranean forest soils. Environ Pollut 145:316–323

Galli U, Meier M, Brunold C (1993) Effects of cadmium on non-mycorrhizal and mycorrhizal Norway spruce seedlings *Picea abies* (L.) Karst.] and its ectomycorrhizal fungus *Laccaria laccata* (Scop, ex Fr.) Bk. & Br.: sulphate reduction, thiols and distribution of the heavy metal. New Phytol 125:837–843

Gandoit L, Probst A (2012) Localisation and mobility of trace metal in silver fir needles. Chemosphere 87:204–210

Ghosh M, Singh SP (2005) A review on phytoremediation of heavy metals and utilization of its byproducts. Appl Ecol Environ Res 3:1–18

Godbold DL, Jentschke G, Winter S, Marschner P (1998) Ectomycorrhizas and amelioration of metal stress in forest trees. Chemosphere 36(4–5):757–762

Grill E, Winnacker EL, Zenk MH (1985) Phytochelatins: the principal heavy-metal complexing peptides of plants. Science 230:674–676

Hamanishi ET, Campbell MM (2011) Genome-wide responses to drought in forest trees. Forestry 84(3):273–283

Hassinen V, Vallinkoski VM, Issakainen S, Tervahauta A, Kärenlampi S, Servomaa K (2009) Correlation of foliar MT2b expression with Cd and Zn concentrations in hybrid aspen (*Populus tremula* x *tremuloides*) grown in contaminated soil. Environ Pollut 157:922–930

He J, Ma C, Ma Y, Li H, Kang J, Liu T, Polle A, Peng C, Luo ZB (2013) Cadmium tolerance in six poplar species. Environ Sci Pollut Res Int 20(1):165–174

Hodson MJ, Sangster AG (1999) Aluminium/silicon interactions in conifers. J Inorg Biochem 76:89–98

Huang JW, Blaylock MJ, Kapulnik Y, Ensley BD (1998) Phytoremediation of uranium-contaminated soils: role of organic acids in triggering uranium hyperaccumulation in plants. Environ Sci Technol 32:2004–2008

Ivanov YV, Savochkin YV, Kuznetsov VIV (2011) Scots pine as a model plant for studying the mechanisms of conifers adaptation to heavy metal action: 1. Effects of continuous zinc presence on morphometric and physiological characteristics of developing pine seedlings. Russ J Plant Physiol 58(5):871–878

Ivanov YV, Savochkin YV, Kuznetsov VIV (2012) Scots pine as a model plant for studying the mechanisms of conifers adaptation to heavy metal action: 2. Functioning of antioxidant enzymes in pine seedlings under chronic zinc action. Russ J Plant Physiol 59(1):50–58

Jaffré T, Brooks RR, Lee J, Reeves RD (1976) *Sebertia acuminata*: a hyperaccumulator of nickel from New Caledonia. Science 193:579–580

Jaffré T, Pillon Y, Thomine S, Merlot S (2013) The metal hyperaccumulators from New Caledonia can broaden our understanding of nickel accumulation in plants. Front Plant Sci 4(279):1–7
Jambhulkar HP, Juwarkar AA (2009) Assessment of bioaccumulation of heavy metals by different plant species grown on fly ash dump. Ecotoxicol Environ Saf 72:1122–1128
Jarvis MD, Leung DWM (2002) Chelated lead transport in *Pinus radiata*: an ultrastructural study. Environ Exp Bot 48:21–32
Jentschke G, Marschner P, Vodnik D, Marth C, Bredemeier M, Rapp C, Fritz E, Gogala N, Godbold DL (1998) Lead uptake by *Picea abies* seedlings: effects of nitrogen source and mycorrhizas. J Plant Physiol 153:97–10
Jentschke G, Winter S, Godbold DL (1999) Ectomycorrhizas and cadmium toxicity in Norway spruce seedlings. Tree Physiol 19:23–30
Jones MD, Browning MHR, Hutchinson TC (1986) The influence of mycorrhizal associations on paper birch and Jack pine seedlings when exposed to elevated copper, nickel or aluminum. Water Air Soil Pollut 31:441–448
Kiikkilä O (2003) Heavy-metal pollution and remediation of forest soil around the Harja valta Cu-Ni smelter, in SW Finland. Silva Fennica 37(3):399–415
Kozdrój J, Piotrowska-Seget Z, Krupa P (2007) Mycorrhizal fungi and ectomycorrhiza associated bacteria isolated from an industrial desert soil protect pine seedlings against Cd(II) impact. Ecotoxicology 16:449–456
Krzesłowska M (2011) The cell wall in plant cell response to trace metals: polysaccharide remodeling and its role in defense strategy. Acta Physiol Plant 33(1):35–51
Krznaric E, Verbruggen N, Wevers JHL, Carleer R, Vangronsveld J, Colpaert JV (2009) Cd-tolerant *Suillus luteus*: a fungal insurance for pines exposed to Cd. Environ Pollut 157:1581–1588
Krznaric E, Wevers JHL, Cloquet C, Vangronsveld J, Vanhaecke F, Colpaert JV (2010) Zn pollution counteracts Cd toxicity in metal-tolerant ectomycorrhizal fungi and their host plant, *Pinus sylvestris*. Environ Microbiol 12(8):2133–2141
Kukkola E, Rautio P, Huttunen S (2000) Stress indications in copper- and nickel-exposed Scots pine seedlings. Environ Exp Bot 43:197–210
Kuroda K, Kagawa A, Tonosaki M (2013) Radiocesium concentrations in the bark, sapwood and heartwood of three tree species collected at Fukushima forests half a year after the Fukushima Dai-ichi nuclear accident. J Environ Radioact 122:37–42
Leštan D, Luo CL, Li XD (2008) The use of chelating agents in the remediation of metal contaminated soils:a review. Environ Pollut 153:3–13
Leung HM, Wang ZW, Ye ZH, Yung KL, Peng XL, Cheung KC (2013) Interactions between arbuscular mycorrhizae and plants in phytoremediation of metal-contaminated soils: a review. Pedosphere 23(5):549–563
Li JT, Liao B, Zhu R, Dai ZY, Lan CY, Shu WS (2011) Characteristics of Cd uptake, translocation and accumulation in a novel Cd-accumulating tree, star fruit (*Averrhoa carambola* L., Oxalidaceae). Environ Exp Bot 71:352–358
Liang HM, Lin TH, Chiou JM, Yeh KC (2009) Model evaluation of the phytoextraction potential of heavy metal hyperaccumulators and non-hyperaccumulators. Environ Pollut 157:1945–1952
Ma JF, Ryan PR, Delhaize E (2001) Aluminium tolerance in plants and the complexing role of organic acids. Trends Plant Sci 6:273–278
Markkola AM, Tarvainen O, Ahonen-Jonnarth U, Strömmer R (2002) Urban polluted forest soils induce elevated root peroxidase activity in Scots pine (*Pinus sylvestris* L.) seedlings. Environ Pollut 116:273–278
McGrath SP, Zhao FJ, Lombi E (2002) Phytoremediation of metals, metalloids and radionuclides. Adv Agron 75:1–56
Mingorance MD, Valdés B, Oliva SR (2007) Strategies of heavy metal uptake by plants growing under industrial emissions. Environ Int 33:514–520

Minocha R, Long S (2004) Effects of aluminum on organic acid metabolism and secretion by red spruce cell suspension cultures and the reversal of Al effects on growth and polyamine metabolism by exogenous organic acids. Tree Physiol 24:55–64

Mizuno T, Hirano K, Kato S, Obata H (2008) Cloning of ZIP family metal transporter genes from the manganese hyperaccumulator plant *Chengiopanax sciadophylloides*, and its metal transport and resistance abilities in yeast. Soil Sci Plant Nutr 54:86–94

Mohapatra S, Cherry S, Minocha R, Majumdar R, Thangavel P, Long S, Minocha SC (2010) The response of high and low polyamine-producing cell lines to aluminum and calcium stress. Plant Physiol Biochem 48:612–620

Monaci F, Moni F, Lanciotti E, Grechi D, Bargagli R (2000) Biomonitoring of airborne metals in urban environments: new tracers of vehicle emission, in place of lead. Environ Pollut 107:321–327

Moussavou Moudouma CF, Riou C, Gloaguen V, Saladin G (2013) Hybrid larch (*Larix* x *eurolepis* Henry): a good candidate for cadmium phytoremediation? Environ Sci Pollut Res 20(3):1889–1894

Nowak J, Friend AL (2005) Aluminum fractions in root tips of slash pine and loblolly pine families differing in Al resistance. Tree Physiol 25:245–250

Nzihou A, Stanmore B (2013) The fate of heavy metals during combustion and gasification of contaminated biomass—a brief review. J Hazard Mater 256–257:56–66

Orlandi M, Pelfini M, Pavan M, Santilli M, Colombini MP (2002) Heavy metals variations in some conifers in Valle d'Aosta (Western Italian Alps) from 1930 to 2000. Microchem J 73:237–244

Osawa H, Ikeda S, Tange T (2013) The rapid accumulation of aluminum is ubiquitous in both the evergreen and deciduous leaves of Theaceae and Ternstroemiaceae plants over a wide pH range in acidic soils. Plant Soil 363:49–59

Pan H, Eberhard TL (2011) Characterization of fly ash from the gasification of wood and assessment for its application as a soil amendment. BioResources 6(4):3987–4004

Peng HY, Yang X, Tian SK (2005) Accumulation and ultrastructural distribution of copper in *Elsholtzia splendens*. J Zhejiang Univ (Sci) 6B(5):311–318

Potters G, Pasternak TP, Guisez Y, Palme KJ, Jansen MAK (2007) Stress-induced morphogenic responses: growing out of trouble? TRENDS Plant Sci 12(3):98–105

Prabagar S, Hodson MJ, Evans DE (2011) Silicon amelioration of aluminium toxicity and cell death in suspension cultures of Norway spruce (*Picea abies* (L.) Karst.). Environ Exp Bot 70:266–276

Pratas J, Prasad MNV, Freitas H, Conde L (2005) Plants growing in abandoned mines of Portugal are useful for biogeochemical exploration of arsenic, antimony, tungsten and mine reclamation. J Geochem Explor 85:99–107

Przybysz A, Sæbø A, Hanslin HM, Gawroński SW (2014) Accumulation of particulate matter and trace elements on vegetation as affected by pollution level, rainfall and the passage of time. Sci Total Environ 481:360–369

Pulford ID, Watson C (2003) Phytoremediation of heavy metal-contaminated land by trees—a review. Environ Int 29:529–540

Rabier J, Laffont-Schwob I, Bouraïma-Madjebi S, Leon V, Prudent P, Viano J, Nabors MW, Pilon-Smits EAH (2007) Characterization of metal tolerance and accumulation in *Grevillea exul* var *exul*. Int J Phytoremediation 9:419–435

Radotić K, Dučić T, Mutavdić D (2000) Changes in peroxidase activity and isoenzymes in spruce needles after exposure to different concentrations of cadmium. Environ Exp Bot 44:105–113

Raju D, Kumar S, Mehta UJ, Hazra S (2008) Differential accumulation of manganese in three mature tree species (Holoptelia, Cassia, Neem) growing on a mine dump. Curr Sci 94 (5):639–643

Rascio N, Navari-Izzo F (2011) Heavy metal hyperaccumulating plants: how and why do they do it? And what makes them so interesting? Plant Sci 180:169–181

Reeves RD, Baker AJM, Borhidi A, Berazaín R (1999) Nickel hyperaccumulation in the serpentine flora of Cuba. Ann Bot 83:29–38

Reimann C, Koller F, Kashulina G, Niskavaara H, Englmaier P (2001) Influence of extreme pollution on the inorganic chemical composition of some plants. Environ Pollut 115:239–252

Robinson BH, Mills TM, Petit D, Fung LE, Green SR, Clothier BE (2000) Natural and induced cadmium-accumulation in poplar and willow: implications for phytoremediation. Plant Soil 227:301–306

Rodriguez JH, Wannaz ED, Salazar MJ, Pignata ML, Fangmeier A, Franzaring J (2012) Accumulation of polycyclic aromatic hydrocarbons and heavy metals in the tree foliage of *Eucalyptus rostrata*, *Pinus radiata* and *Populus hybridus* in the vicinity of a large aluminium smelter in Argentina. Atmos Environ 55:35–42

Saarela KE, Harju L, Rajander J, Lill JO, Heselius SJ, Lindroos A, Mattsson K (2005) Elemental analyses of pine bark and wood in an environmental study. Sci Total Environ 343:231–241

Sahi SV, Bryant NL, Sharma NC, Singh SR (2002) Characterization of a Lead hyperaccumulator shrub, *Sesbania drummondii*. Environ Sci Technol 36:4676–4680

Samecka-Cymerman A, Kolon K, Kempers AJ (2008) A preliminary investigation in using *Pohlia nutans* and *Larix decidua* as biomonitors of air pollution by the coke industry in Wałbrzych (SW Poland). Polish J Environ Stud 17(1):121–128

Sarma H (2011) Metal hyperaccumulation in plants: a review focusing on phytoremediation technology. J Environ Sci Technol 4:118–138

Sawidis T, Chettri MK, Papaioannou A, Zachariadis G, Stratis J (2001) Study of metal distribution from lignite fuels using trees as biological monitors. Ecotoxicol Environ Saf 48:27–35

Schröder P, Fisher C, Debus R, Wenzel A (2003) Reaction of detoxification mechanisms in suspension cultured spruce cells (*Picea abies* L. Karst.) to heavy metals in pure mixture and in soil eluates. Environ Sci Pollut Res 10(4):225–234

Shim D, Kim S, Choi YI, Song WY, Park J, Youk ES, Jeong SC, Martinoia E, Noh EW, Lee Y (2013) Transgenic poplar trees expressing yeast cadmium factor 1 exhibit the characteristics necessary for the phytoremediation of mine tailing soil. Chemosphere 90:1478–1486

Shukla OP, Juwarkar AA, Singh SK, Khan S, Rai UN (2011) Growth responses and metal accumulation capabilities of woody plants during the phytoremediation of tannery sludge. Waste Manag 31:115–123

Sousa AI, Caçador I, Lillebo AI, Pardal MA (2008) Heavy metal accumulation in *Halimione portulacoides*: intra- and extra-cellular metal binding sites. Chemosphere 70:850–857

Sousa NR, Ramos MA, Marques APGC, Castro PML (2012) The effect of ectomycorrhizal fungi forming symbiosis with *Pinus pinaster* seedlings exposed to cadmium. Sci Total Environ 414:63–67

Sousa NR, Ramos MA, Marques APGC, Castro PML (2014) A genotype dependent-response to cadmium contamination in soil is displayed by *Pinus pinaster* in symbiosis with different mycorrhizal fungi. Appl Soil Ecol 76:7–13

Steinnes E, Friedland AJ (2006) Metal contamination of natural surface soils from long-range atmospheric transport: existing and missing knowledge. Environ Rev 14:169–186

Szabados L, Savouré A (2010) Proline: a multifunctional amino acid. Trends Plant Sci 15 (2):89–97

Takenaka C, Kobayashi M, Kanaya S (2009) Accumulation of cadmium and zinc in *Evodiopanax innovans*. Environ Geochem Health 31:609–615

Thangavel P, Long S, Minocha R (2007) Changes in phytochelatins and their biosynthetic intermediates in red spruce (*Picea rubens* Sarg.) cell suspension cultures under cadmium and zinc stress. Plant Cell Tissue Organ Cult 88:201–216

Turchi A, Tamantini I, Camussi AM, Racchi ML (2012) Expression of a metallothionein A1 gene of *Pisum sativum* in white poplar enhances tolerance and accumulation of zinc and copper. Plant Sci 183:50–56

Vaitkuté D, Baltrénaité E, Booth CA, Fullen MA (2010) Does sewage sludge amendment to soil enhances the development of Silver birch and Scots pine? Hung Geogr Bull 59(4):393–410

Van der Ent A, Baker AJM, Reeves RD, Pollard AJ, Schat H (2012) Hyperaccumulators of metal and metalloid trace elements: facts and fiction. Plant Soil 362:319–334

Vandecasteele B, Meers E, Vervaeke P, De Vos B, Quataert P, Tack FMG (2005) Growth and trace metal accumulation of two Salix clones on sediment-derived soils with increasing contamination levels. Chemosphere 58:995–1002

Vangronsveld J, Herzig R, Weyens N, Boulet J, Adriaensen K, Ruttens A, Thewys T, Vassilev A, Meers E, Nehnevajova E, van der Lelie D, Mench M (2009) Phytoremediation of contaminated soils and groundwater: lessons from the field. Environ Sci Pollut Res 16:765–794

Venkatachalam P, Srivastava AK, Raghothama KG, Sahi SV (2009) Genes Induced in Response to Mercury-Ion-Exposure in Heavy Metal Hyperaccumulator *Sesbania drummondii*. Environ Sci Technol 43(3):843–850

Verbruggen N, Hermans C, Schat H (2009) Molecular mechanisms of metal hyperaccumulation in plants. New Phytol 181:759–776

Wang X, Jia Y (2010) Study on adsorption and remediation of heavy metals by poplar and larch in contaminated soil. Environ Sci Pollut Res 17:1331–1338

Xue S, Wang J, Zhou X, Liu H, Chen Y (2010) A critical reappraisal of *Phytolacca acinosa* Roxb. (Phytolaccaceae)—a manganese-hyperaccumulating plant. Acta Ecol Sinica 30:335–338

Yanai J, Zhao FJ, McGrath SP, Kosaki T (2006) Effect of soil characteristics on Cd uptake by the hyperaccumulator *Thlaspi caerulescens*. Environ Pollut 139:167–175

Yang JL, Li YY, Zhang YJ, Zhang SS, Wu YR, Wu P, Zheng SJ (2008a) Cell wall polysaccharides are specifically involved in the exclusion of aluminum from the rice root apex. Plant Physiol 146:602–661

Yang SX, Deng H, Li MS (2008b) Manganese uptake and accumulation in a woody hyperaccumulator, *Schima superba*. Plant Soil Environ 54(10):441–446

Zacchini M, Iori V, Scarascia Mugnozza G, Pietrini F, Massacci A (2011) Cadmium accumulation and tolerance in *Populus nigra* and *Salix alba*. Biol Plantarum 55:383–386

Zu YQ, Yuan L, Schvartz C, Langlade L, Fan L (2004) Accumulation of Pb, Cd, Cu and Zn in plants and hyperaccumulator choice in Lanping lead–zinc mine area, China. Environ Int 30:567–576

Chapter 19
Remediation of Heavy Metal-Contaminated Soils and Enhancement of Their Fertility with Actinorhizal Plants

Nathalie Diagne, Mariama Ngom, Pape Ibrahima Djighaly, Daouda Ngom, Babou Ndour, Maimouna Cissokho, Mathieu Ndigue Faye, Alioune Sarr, Mame Ourèye SY, Laurent Laplaze, and Antony Champion

19.1 Introduction

Land degradation is becoming a huge issue for agriculture development. This degradation is accelerated by environmental stresses that adversely affect ecosystem biodiversity and productivity. Among them, heavy metals are the most alarming and the most toxic inorganic substances which have contaminated a large area of land (Upadhyaya et al. 2010). Heavy metal pollution can be defined as an undesirable change in the physical, chemical, or biological characteristics of land and water that may or will harmfully affect animals and plants (Odum 1971). The commonly heavy metal found at contaminated sites are arsenic (As), cadmium

N. Diagne (✉)
Centre National de Recherches Agronomiques (CNRA/ISRA), Bambey, Sénégal

Laboratoire Commun de Microbiologie (IRD/UCAD/ISRA), Bel-Air, Dakar, Sénégal

Laboratoire mixte international Adaptation des Plantes et microorganismes associés aux Stress Environnementaux (LAPSE), Centre de Recherche de Bel Air, Dakar, Sénégal
e-mail: nathaliediagne@gmail.com

M. Ngom
Laboratoire Commun de Microbiologie (IRD/UCAD/ISRA), Bel-Air, Dakar, Sénégal

Laboratoire mixte international Adaptation des Plantes et microorganismes associés aux Stress Environnementaux (LAPSE), Centre de Recherche de Bel Air, Dakar, Sénégal

Laboratoire Campus de Biotechnologies Végétales (LCBV), Dakar, Sénégal

Département de Biologie Végétale, Université Cheikh Anta Diop de Dakar, Dakar, Sénégal

P.I. Djighaly
Laboratoire Commun de Microbiologie (IRD/UCAD/ISRA), Bel-Air, Dakar, Sénégal

Laboratoire mixte international Adaptation des Plantes et microorganismes associés aux Stress Environnementaux (LAPSE), Centre de Recherche de Bel Air, Dakar, Sénégal

Département d'Agroforesterie, Université Assane Seck de Ziguinchor, Ziguinchor, Sénégal

I. Sherameti, A. Varma (eds.), *Heavy Metal Contamination of Soils*, Soil Biology 44, DOI 10.1007/978-3-319-14526-6_19

(Cd), copper (Cu), chromium (Cr), lead (Pb), mercury (Hg), nickel (Ni), and zinc (Zn) (GWRTAC 1997; Tahar and Keltoum 2011; Wang and Björn 2014). Some of them like Cu, Fe, Mn, Zn, and Ni are necessary to plant and animal life, while others like Cd, Cr, and Pb are nonessential and/or toxic (Punz and Sieghardt 1993).

Increasing contamination of heavy metal in soils has several sources such as the rapid urbanization due to the growing population, industrialization, intensive agriculture, and animal manures (Upadhyaya et al. 2010; Nacke et al. 2013). Besides the deterioration of environmental quality, heavy metals are hazardous to humans and animals and affect agricultural productivity of crops (Gratão et al. 2005; Singh et al. 2011; Espína et al. 2014). It has been demonstrated that heavy metal-induced oxidative stress in plants disturbs the balance between oxidants and antioxidants in the cells (Upadhyaya et al. 2010; Juknys et al. 2012). These pollutants can be transferred to the food chain from the soil and affect human health (An et al. 2011). Heavy metals in soils are difficult to remediate, and without treatment the pollutants cannot be degraded. These environmental pollutants can remain in soil for long periods. Many physicochemical and biological strategies are used for the restoration of lands contaminated by pollutants (Wuana and Okieimen 2011). However, phytoremediation remains one of the most "eco-friendly," sustainable, and low-cost method (Raskin et al. 1997). Phytoremediation can be defined as an in situ

D. Ngom
Département d'Agroforesterie, Université Assane Seck de Ziguinchor, Ziguinchor, Sénégal

B. Ndour • A. Sarr
Centre National de Recherches Agronomiques (CNRA/ISRA), Bambey, Sénégal

M. Cissokho • M.N. Faye
Laboratoire Commun de Microbiologie (IRD/UCAD/ISRA), Bel-Air, Dakar, Sénégal

Laboratoire mixte international Adaptation des Plantes et microorganismes associés aux Stress Environnementaux (LAPSE), Centre de Recherche de Bel Air, Dakar, Sénégal

M.O. SY
Laboratoire mixte international Adaptation des Plantes et microorganismes associés aux Stress Environnementaux (LAPSE), Centre de Recherche de Bel Air, Dakar, Sénégal

Laboratoire Campus de Biotechnologies Végétales (LCBV), Dakar, Sénégal

Département de Biologie Végétale, Université Cheikh Anta Diop de Dakar, Dakar, Sénégal

L. Laplaze
Laboratoire mixte international Adaptation des Plantes et microorganismes associés aux Stress Environnementaux (LAPSE), Centre de Recherche de Bel Air, Dakar, Sénégal

Equipe Rhizogenèse, UMR DIADE, IRD, 911 avenue Agropolis, 34394, Montpellier Cedex 5, France

A. Champion
Laboratoire Commun de Microbiologie (IRD/UCAD/ISRA), Bel-Air, Dakar, Sénégal

Laboratoire mixte international Adaptation des Plantes et microorganismes associés aux Stress Environnementaux (LAPSE), Centre de Recherche de Bel Air, Dakar, Sénégal

Equipe Rhizogenèse, UMR DIADE, IRD, 911 avenue Agropolis, 34394, Montpellier Cedex 5, France

remediation strategy that uses vegetation and associated microbiota, soil amendments, and agronomic techniques to remove and contain environmental contaminants or render them harmless (Cunningham and Ow 1996; Helmisaari et al. 2007). Many species are used in phytoremediation to clean up toxic metal-contaminated sites (Fig. 19.1) (Kumar et al. 1995; Raskin et al. 1997). Among them, the actinorhizal plants which are nitrogen-fixing trees composed by eight families Betulaceae, Casuarinaceae, Coriariaceae, Datiscaceae, Elaeagnaceae, Myricaceae, Rhamnaceae, and Rosaceae are used in phytoremediation (Wall 2000). Plants of Casuarinaceae and Betulaceae families are the most widely planted around the world for the rehabilitation of degraded lands (Schwencke and Caru 2001; Diagne et al. 2013a, b). These plants are pioneer species well adapted to disturbed soils and abiotic stresses such as heavy metal (Sayed 2011; Lakhdari and Benabdeli 2012). They are able to be associated with different symbiotic soil microorganisms such as symbiotic mycorrhizal fungi and the nitrogen-fixing bacteria *Frankia* (Fig. 19.2)

Fig. 19.1 *C. equisetifolia* plantation in magnesite-mined outlands, India (It's Karthikeyan who took the photo)

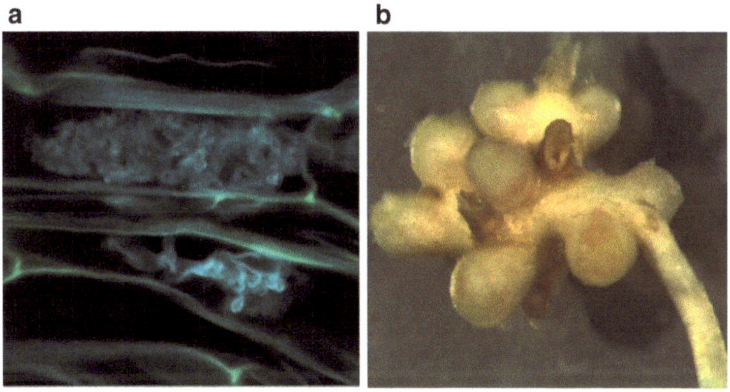

Fig. 19.2 (**a**) *C. equisetifolia* mycorrhizal roots; (**b**) actinorhizal nodule

that mitigate the adverse effect of heavy metal and improve plant performance in stressed environments (Sayed 2011; Parao 2012). In this study, we will discuss (1) the use of actinorhizal plants in phytoremediation to clean up toxic metal, (2) the impact of mycorrhizal fungi and/or the nitrogen-fixing bacteria on plant adaptation and performance in heavy metal-contaminated sites, and (3) the impact of actinorhizal plantations on soil fertility and plant productivity.

19.2 Remediation of Heavy Metal-Contaminated Soils with Actinorhizal Plants

Experimentations using actinorhizal trees and particularly plants of Casuarinaceae and Betulaceae families were conducted worldwide for the remediation of soils contaminated by heavy metal (Rosselli et al. 2003; Mertens et al. 2004; Lee et al. 2009; Sayed 2011) (Fig. 19.1). However, plants' responses to heavy metal vary considerably, and some species such as *Casuarina equisetifolia* and *Alnus maritima* tolerate high concentrations of heavy metal (Parao 2012). It has been reported that *C. equisetifolia* can tolerate 19 μg/g of Cr, 663 μg/g of Fe, and 13 μg/g of Ni (Claveria 2012). Some experiments in high level of fuel have also showed that *C. equisetifolia* can tolerate up to 10 g/kg of fuel in soil (Sun et al. 2004).

Studies carried out by Lakharadi and Benabdeli (2012) in Mascara in Algeria have showed that *C. equisetifolia* can be successfully applied in biomonitoring of air pollution and heavy metal soil remediation (Díaz-Martínez et al. 2013). These authors have found a higher heavy metal concentration in *C. equisetifolia* needles collected in sites with a high traffic density and frequency of car stoppage. These chemical analyses revealed the presence of Pb, Zn, Cu, and Ni in *C. equisetifolia* needles.

In addition, studies carried out by Parao (2012) showed that *Alnus* plays a major role in Fe and Mn remediation in soil. Samples from *Alnus* showed that the concentration of Fe varies from 3,982.12 ppm for the plantation of 8 months old to 93.33 ppm for *Alnus* trees years after plantation. For Mn, the concentration varies from 1,065.56 ppm after 8 months to 756.50 ppm 3 years after plantation. High concentrations of Cu have been also reported in *Betula* plants (Maurice and Lagerkvist 2000). Other ions like Zn are found accumulated at higher level in *Alnus glutinosa* leaves (Pulford et al. 2001). Black alders or *Alnus glutinosa* (L.) Gaertn grows naturally in the Taruskos forestry in Panevėžys region where 600 t of industrial sewage sludge was spread on the soil and accumulates some ions such as Ni, Cu, Pb, Zn, and Mn (Butkus and Baltrenaite 2007). These authors showed that *Alnus glutinosa* tolerates the presence of Zn up to 214 mg/kg. Compared to the black alder control, alders grown on sludge have accumulated about two times a large amount of Zn in leave (Butkus and Baltrenaite 2007). In naturally regenerated site in Yunnan province (Southwestern China), Nepalese alder (*Alnus nepalensis*) showed a great role in phytostabilization of Zn and Pb and phytoextraction of Cd (Jing et al. 2014).

A pilot experiment to reclaim dredging area with *A. firma* and *A. hursita* has shown after 4 years of plantation that *Alnus* plant took metals up from the soil in the following order: Pb greater than Zn greater than Cu greater than Cr greater than As greater than Cd (Lee et al. 2009). However, these results have shown that *A. firma* is more tolerant to these pollutants than *A. hursita*. Regarding the plant density, their authors found that low plant density resulted in higher heavy metal uptake per plant, but the total heavy metal concentration was not different for plants planted at low and high density, suggesting that the plant density effect might not be important with regard to total uptake by plants.

19.3 Impact of Mycorrhizal Fungi or the Nitrogen-Fixing Bacteria on Actinorhizal Plant Adaptation and Performance in Heavy Metal-Contaminated Sites

19.3.1 Impact of Frankia *on Actinorhizal Plant Adaptation and Performance in Heavy Metal-Contaminated Sites*

Species belonging to the Casuarinaceae family are able to form nitrogen-fixing symbioses with soil filamentous actinobacteria called *Frankia* (Wall 2000). They are collectively termed actinorhizal plants, and about 194 species and 24 genera have been identified (Benson and Silvester 1993). These symbiotic associations occur on the roots of host plant, and they lead to the formation of new root organs called actinorhizal root nodules (Fig. 19.2) where the bacteria are hosted and fix atmospheric nitrogen (Duhoux et al. 1996). *Frankia* formed extensive hyphae and multilocular sporangia (Fig. 19.3) located either terminally or in an intercalary position on the septate hyphae (Newcomb et al. 1979). Therefore, they provide to the host plant an unlimited source of nitrogen for its nutrition (Perrine-Walker

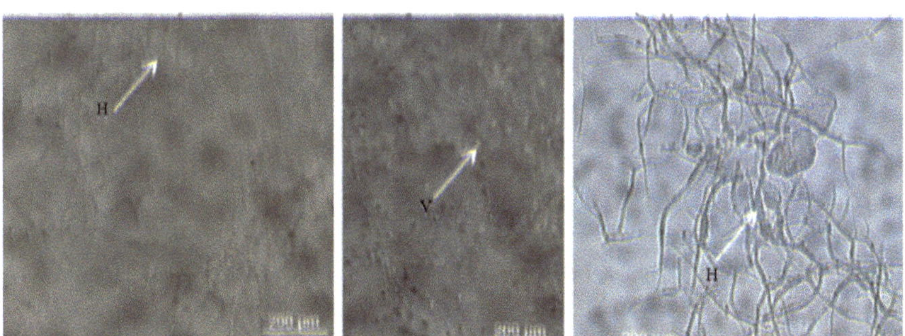

Fig. 19.3 Structures of the nitrogen-fixing actinobacteria *Frankia*. *H* hyphae, *V* vesicle, and *S* sporangia

et al. 2011), and in exchange, they benefit from plant carbon sources (Pawlowski and Demchenko 2012).

Resistance of *Frankia* strains and their impact on actinorhizal plant performance in heavy metal-contaminated soil are well documented. For instance, the sensitivity of 12 *Frankia* strains to heavy metals was determined by addition to the growth media at several concentrations of different metal ions including Cu^{2+}, Pb^{2+}, $AgNO_3$, Na2HAsO$_4$, $NaAsO_2$, $CdCl_2$, $CoCl_2$, K2CrO$_4$, $CuCl_2$, and $NiCl_2$ (Richards et al. 2002). Results showed that most of the *Frankia* strains were resistant to elevated levels of several heavy metals as AsO_4^{3-}, Pb^{2+}, SeO_2^{2-}, Cu^{2-}, and CrO_4^{2-}. This study suggested that the mechanism of SeO_2^{2-} resistance for *Frankia* strains seems to involve the reduction of the selenite oxyanion to insoluble elemental selenium which is much less toxic than selenite, whereas Pb^{2-} resistance and Cu^{2-} resistance may involve sequestration or binding mechanisms (Richards et al. 2002). Wheeler et al. (2001) showed that the yield of three *Frankia* strains was not affected significantly by 2.25 mM of nickel when cultured in vitro in the presence of propionate as carbon source and hydrolyzed casein as nitrogen source. It has been described that nickel is required for nitrogen-fixing bacteria including *Frankia* (Sellstedt and Smith 1990). Nickel is required for the synthesis of the uptake hydrogenases which catalyze the oxidation of hydrogen liberated by nitrogenase during the reduction of dinitrogen (Klucas et al. 1983). Oliveira et al. (2005) found that inoculation of *A. glutinosa* plants with *Frankia spp.* significantly increased the dry weight biomass and leaf N content by 197 % compared with the uninoculated controls when cultivated in an alkaline anthropogenic sediment. In the poor nutriment bauxite mine soils, in India, inoculated seedlings of *C. equisetifolia* by some suitable beneficial microbes including *Frankia* showed 90–100 % survival over the control seedlings (uninoculated plants). Plants inoculated with *Frankia* had also a significant growth as well as nutrient uptake (N, P, K) higher than the control after 3 months and also 2 years after planting (Karthikeyan et al. 2009). Inoculation of Alder (*Alnus cordata Loisel*) with selected *Frankia* strains in mine spoil afforestation plots showed a remarkable positive effect on alder aboveground biomass after 1 year planting (Lumini et al. 1994).

19.3.2 Impact of Arbuscular and Ectomycorrhizal Fungi on Actinorhizal Plant Adaptation to Heavy Metal-Contaminated Sites

A mycorrhiza is a symbiotic association between a fungus and the roots of vascular plants (Sadhana 2014). Several actinorhizal species like *Casuarina* and *Alnus* can also be colonized by both ectomycorrhizal (ECM) and arbuscular mycorrhizal (AMF) fungi (Gardner 1986). The mycorrhizal fungi act as extensions of the root system (Fig. 19.2) and improve host nutrition by their ability to take up nutrients and water more efficiently than roots alone (Roy et al. 2007; Smith and Read 2008).

They improve plant nutrient uptake and profit from plant carbon source. In addition, mycorrhizal fungi enhance plant performance and resistance to abiotic stresses (Sadhana 2014). Plants in symbiosis with AMF and ECM have the potential to be used for enhancing phytoremediation of heavy metal in contaminated soil thanks to their ability to take up heavy metal from an enlarged soil volume (Upadhyaya et al. 2010; Malekzadeh et al. 2012).

Some mycorrhizal fungi can grow in the presence of diverse environmental stresses including heavy metal. For example, the tolerance to heavy metal of five ECM strains was tested by adding to the medium several concentrations (0.1–400 μg/mL) of Al, Cu, Zn, Fe, Ni, Cd, Cr, Pb, and Hg (Tam 1995). Results showed that all ECM strains were tolerant to high concentrations of Fe and Pb and sensitive to low concentrations of Ni, Cd, and Hg. However, only *Pisolithus tinctorius* and *Scleroderma* sp. exhibited greater metal tolerance at high concentrations of Cu, while *Hymenogaster* sp. exhibited a great tolerance at high concentration of Al (Tam 1995).

Even if some metals are micronutrients necessary for plant and microorganism growth, such as Zn, Cu, Mn, Ni, and Co, high concentrations of these metals in soil are toxic to plant, bacteria, and fungi (Leyval et al. 1997). However, it has been reported that mycorrhizal fungi can considerably reduce the uptake of heavy metals into plant cells and therefore allow them to thrive on heavy metal-polluted sites (Hildebrandt et al. 2007). The protection provided by ECM is due to the mycelia affording a physical barrier or mantle and may include metabolic processes such as intracellular metal accumulation and the extracellular precipitation of metals by metabolites from exudates (Khan et al. 2000). In their study, Karthikeyan et al. (2009) found that inoculation of *C. equisetifolia* plants with *Glomus aggregatum* alone on bauxite mine spoils increases their growth both in nursery and field experiments. Inoculated seedlings showed also 90–95 % survival on bauxite mine spoils after 2 years compared to uninoculated plants (35 %). A significant increase of leaf biomass and N, P, and K content was equally showed. Inoculation of alders (*A. cordata* Loisel.) with spores of *G. fasciculatum* or *G. mosseae* showed a positive effect on plant biomass after 1 year of planting in mine spoil (Lumini et al. 1994).

19.3.3 Impact of Frankia*/AMF and/or EMF on Actinorhizal Plant Adaptation and Performance in Heavy Metal-Contaminated Sites*

Thanks to their ability to form symbiosis with nitrogen-fixing bacteria, *Frankia*, ECM, and/or AMF, actinorhizal plants are pioneer species which are able to grow well in a range of climates and disturbed soils. It is known that plant growth is often limited by the amount of available water, nitrogen, and phosphore in soil especially in degraded areas. Nitrogen is the most abundant component in the atmosphere, but only some prokaryotes as *Frankia* are able to use this form of nitrogen. These symbiotic relationships with *Frankia* and mycorrhizal fungi increase actinorhizal plant nutrition and water absorption unlike nonsymbiotic plants.

Several studies showed that dual inoculation with *Frankia* and mycorrhizal fungi enhances actinorhizal plant growth and fitness more than single inoculation. In bauxite mine spoils in India, inoculation of *C. equisetifolia* plants with both selected *Frankia* and AM fungi (*G. aggregatum*) increased significantly their growth 2 years after planting compared to single inoculation and uninoculated plants (Karthikeyan et al. 2009). The triple microbial inoculants (*Frankia* + AMF + phosphobacterium) showed significantly increased growth and biomass, nodule number, percentage of AMF fungal infection, and collar diameter than control and other treatments. N, P, and K contents were also higher in AMF + *Frankia* and AMF + *Frankia* + phosphobacterium-inoculated trees. A percentage of survival of 100 % was observed in trees inoculated with *Frankia* + AMF + phosphobacterium against more than 95 % for the combination *Frankia* + AMF, 90–95 % for the individual microbial inoculants, and 35 % for the controls. Similar results were observed by Lumini et al. (1994). In mine spoils, inoculation of alders by both *Frankia*/*G. fasciculatum* and *Frankia*/*G. mosseae* produced a high degree of nodulation and mycorrhizal infection. Plant aboveground biomass was significantly increased by inoculation with these combinations after 1 year of outplanting in mine spoils compared to single inoculation treatments and uninoculated plants. About 500 D^2H (cm^3) of the biomass for *Frankia*/*G. mosseae*-treated trees, more than 400 D^2H (cm^3) for *Frankia*/*G. fasciculatum*-treated trees against less than 100 D^2H (cm^3) for the controls and plants inoculated separately with each fungal (Lumini et al. 1994). Synergetic inoculation of *A. glutinosa* in a highly alkaline anthropogenic sediment by *G. intraradices* and *Frankia* spp. increased significantly the number of spores in dry soil, plant nodule number, and dry weight (Oliveira et al. 2005). The total leaf area, shoot height, root collar diameter, and total plant dry weight were significantly greater when compared to those inoculated with one symbiont or the uninoculated plants. There was a significant increase of 277 % in leaf N content, 240 % in leaf P content, and 129 % in leaf K content in plants inoculated with both symbionts simultaneously when compared to the uninoculated plant. The dual inoculation showed also an increase of 531 % in chlorophyll a + b content compared to the controls (Oliveira et al. 2005).

19.4 Actinorhizal Plants Improve Fertility of Heavy Metal-Contaminated Soil

As microorganisms are the key of organic matter decomposition, assessment of soil microbial community structure and function may be useful to study the effect of potential plant remediation on soil quality. It is well known that heavy metals decrease plant growth and ground cover and reduce soil biodiversity (Pal et al. 2010). Contamination with heavy metal can modify soil physicochemical properties, and it has been reported by Zloch et al. (2014) that the metabolic activity of endophytic populations was strongly influenced by soil physicochemical properties and was significantly lower at a more contaminated site.

In site planted with *C. equisetifolia*, Pérez et al. (2012) have reported that phytoremediation with *C. equisetifolia* improves soil quality and fertility by increasing the total N and P content in soil. On bauxite mine spoil planted with *C. equisetifolia* Frost, Karthikeyan et al. (2009) have found that inoculation with *Frankia*, AMF, and/or PSB increases *C. equisetifolia* N and P content. Given that the soil collected from bauxite mine spoils is acidic and poor in major nutrients (Karthikeyan et al. 2009), planting *C. equisetifolia* tree inoculated with *Frankia*, AMF, and/or PSB will be very useful because this species has an important biomass widely used in composting, and an increasing of N and P in *C. equisetifolia* needle will improve soil fertility of the site.

The positive role on the improvement of soil fertility with actinorhizal trees used in phytoremediation has been also observed with *Alnus* species. After 10 years of revegetation with *A. nepalensis* in a phosphate mine in Kunyang, Yunnan Province, SW China, the assessment of soil quality showed an improvement of N and K available in soil and a decrease of total P concentration. Their results showed that *A. nepalensis* cleans up heavy metal, improves nutrient content and stability of the soil, and promotes the growth of microorganisms as a bioindicator for soil health after contamination remediation (He et al. 2013).

19.5 Conclusion

To satisfy the high demand for food in Africa, some high-efficiency and low-cost options for heavy metal remediation in soil need to be promoting to make lands available for agricultural activities. Phytoremediation, is one of the best technologies for cleaning up contaminated environments. However, this practice needs to be optimized by better understanding the complex interactions between soil, contaminants, microbes, and plants which depend in pedo-climatic parameters.

As the main source of toxic heavy metal comes from inappropriate and illegal human actions, the population should be more aware to limit environmental heavy metal spreading to reduce food chain contamination by these pollutants for their well-being.

References

An LY, Pan YH, Wang ZB, Zhu C (2011) Heavy metal absorption status of five plant species in monoculture and intercropping. Plant Soil 345:237–245

Benson DR, Silvester WB (1993) Biology of *Frankia* strains, actinomycete symbionts of actinorhizal plants. Microbiol Rev 57:293–319

Butkus D, Baltrenaite E (2007) Transport of heavy metals from soil to *Pinus sylvestris* L. and *Betula pendula* wood. Ekologija 53:29–36

Claveria RJR (2012). Phytoremediation: an environmental option to mine rehabilitation. National Institute of Geological Sciences, Department of Environmental. Ateneo de Manila University. Oral Presentation

Cunningham SD, Ow DW (1996) Promises and prospects of phytoremediation. Plant Physiol 110:715–719

Diagne N, Arumugam K, Ngom M, Nambiar-Veetil M, Franche C, Narayanan KK, Laplaze L (2013a). Use of *Frankia* and actinorhizal plants for degraded lands reclamation. BioMed Res Int. http://dx.doi.org/10.1155/2013/948258.

Diagne N, Diouf D, Svistoonoff S, Kane A, Noba K, Franche C, Bogusz D, Duponnois R (2013b) Casuarina in Africa: distribution, role and importance of arbuscular mycorrhizal, ectomycorrhizal fungi and Frankia on plant development. J Environ Manag 128:204–209

Díaz-Martínez ME, Alarcón A, Ferrera-Cerrato R, Almaraz-Suarez JJ, Garcia-Barradas O (2013) Casuarina equisetifolia (Casuarinaceae) growth in soil with diesel and application of biostimulation and bioaugmentation. Rev Biol Trop 61:1039–1052

Duhoux E, Diouf D, Gherbi H, Franche C, Ahée J, Bogusz D (1996) Le nodule actinorhizien: les plantes actinorhiziennes. Acta Botanica Gallica 143:593–608

Espína S, Martínez-López EM, Jiménez P, María-Mojica P, García-Fernández AJ (2014) Effects of heavy metals on biomarkers for oxidative stress in Griffon vulture (*Gyps fulvus*). Environ Res 129:59–68

Gardner IC (1986) Mycorrhizae of actinorhizal plants. MIRCEN J Appl Microbiol Biotechnol 2:147–160

Gratão PL, Polle A, Lea PJ, Azevedo RA (2005) Making the life of heavy metal-stressed plants a little easier. Funct Plant Biol 32:481–494

GWRTAC (1997). Remediation of metals-contaminated soils and groundwater. Tech Rep TE-97-01, GWRTAC, Pittsburgh, PA

He L, Fang X, Meng G, Li G, Shao J, Chai Y, Kong J (2013) Effect of *Alnus nepalensis* cultivation on soil biological and physicochemical properties during restoration near a phosphate smelter in Kunyang, Yunnan Province, SW China. J Soil Sci Plant Nutr 13:355–366

Helmisaari HS, Salemaa M, Derome J, Kiikkilo O, Uhlig C, Nieminen TM (2007) Remediation of heavy metal contaminated forest soil using recycled organic matter and native woody plants. J Environ Qual 36:1145–1153

Hildebrandt U, Regvar M, Bothe H (2007) Arbuscular mycorrhiza and heavy metal tolerance. Phytochemistry 68:139–146

Jing Y, Cui H, Li T, Zhao Z (2014) Heavy metal accumulation characteristics of Nepalese alder (*Alnus nepalensis*) growing in a lead-zinc spoil heap, Yunnan, south-western China. iForest 7:204–208

Juknys R, Vitkauskaite G, Raĉaité M, Vencloviebe J (2012) The impact of heavy metal on oxidative stress and growth of spring barley. Cent Eur J Biol 7:299–306

Karthikeyan A, Deeparaj B, Nepolean P (2009) Reforestation in bauxite mine spoils with Casuarina equisetifolia frost. and beneficial microbes. Forest Tree Livelihood 19:153–165

Khan A, Kuek C, Chaudhry T, Khoo C, Hayes W (2000) Role of plants, mycorrhizae and phytochelators in heavy metal contaminated land remediation. Chemosphere 41:197–207

Klucas RV, Hanus FJ, Russell SA, Evans HJ (1983) Nickel: a micronutrient element for hydrogen-dependent growth of Rhizobium japonicum and for expression of urease activity in soybean leaves. Proc Natl Acad Sci 80:2253–2257

Kumar PBAN, Dushenkov V, Motto H, Raskin I (1995) Phytoextraction: the use of plants to remove heavy metals from soils. Environ Sci Technol 29:1232–1238

Lakhdari A, Benabdeli K (2012) Determining the heavy metal pollution in Mascara (Algeria) by using *Casuarina equisetifolia*. Ecologia Balkanica 4:1–7

Lee DB, Nam W, Kwak YS, Cho NH, Lee SS (2009) Phytoremediation of heavy-metal-contaminated soil in a reclaimed dredging area using *Alnus* species. J Ecol Field Biol 32:267–275

Leyval C, Turnau K, Haselwandter K (1997) Effect of heavy metal pollution on mycorrhizal colonization and function: physiological, ecological and applied aspects. Mycorrhiza 7:139–153

Lumini E, Bosco M, Puppi G, Isopi R, Frattegiani M, Buresti E, Favilli F (1994) Field performance of Alnus cordata loisel (Italian alder) inoculated with Frankia and VA-mycorrhizal strains in mine-spoil afforestation plots. Soil Biol Biochem 26:659–661

Malekzadeh P, Farshian SH, Ordubadi B (2012) Interaction of arbuscular mycorrhizal fungus (*Glomus intraradices* and *Glomus etunicatum*) with tomato plants grown under copper toxicity. Afr J Biotechnol 11:10555–10567

Maurice C, Lagerkvist A (2000) Using *Betula pendula* and *Thelephora caryophyllea* for soil pollution assessment. J Soil Contam 9:31–50

Mertens J, Vervaeke P, De Schrijver A, Luyssaert S (2004) Metal uptake by young trees from dredged brackish sediment: limitations and possibilities for phytoextraction and phytostabilisation. Sci Total Environ 326:209–215

Nacke H, Gonçalves AC Jr, Schwantes D, Nava IA, Strey L, Coelho GF (2013) Availability of heavy metals (Cd, Pb, and Cr) in agriculture from commercial fertilizers. Arch Environ Contam Toxicol 64:537–544. doi:10.1007/s00244-012-9867

Newcomb W, Callaham D, Torrey JG, Peterson R (1979) Morphogenesis and fine structure of the actinomycetous endophyte of nitrogen-fixing root nodules of Comptonia peregrina. Bot Gaz 140(Suppl):S22–S34

Odum EP (1971) Fundamentals of ecology, 3rd edn. Saunders, Philadelphia, p 574

Oliveira R, Castro P, Dodd J, Vosátka M (2005) Synergistic effect of *Glomus intraradices* and *Frankia* spp. on the growth and stress recovery of *Alnus glutinosa* in an alkaline anthropogenic sediment. Chemosphere 60:1462–1470

Pal S, Patra AK, Reza SK, Wildi W, Poté J (2010) Use of bio-resources for remediation of soil pollution. Natural Resources 1:110–125

Parao M (2012). Managing mine spoil through bioremediation efforts. Southeast Asian Regional Center for graduate study and research in agriculture. ISSN 2304–5035

Pawlowski K, Demchenko KN (2012) The diversity of actinorhizal symbiosis. Protoplasma 249:967–979

Perrine-Walker F, Gherbi H, Imanishi L, Hocher V, Ghodhbane-Gtari F, Lavenus J, Benabdoun FM, Nambiar-Veeti M, Svistoonoff S, Laplaze L (2011) Symbiotic signaling in actinorhizal symbioses. Curr Protein Pept Sci 12:156–164

Pérez A, Céspedes C, Almonte I, Ramírez DS, Cruz CE, Núñez PA (2012) Soil quality assessment of mined land after remediation management. Terra Latinoamericana 30:201–211

Pulford ID, Watson C, McGregor SD (2001) Uptake of chromium by trees: prospects for phytoremediation. Environ Geochem Health 23:307–311

Punz WF, Sieghardt H (1993) The response of roots of herbaceous plant to heavy metals. Environ Exp Bot 33:85–98

Raskin I, Smith RD, Salt DE (1997) Phytoremediation of metals: using plants to remove pollutants from the environment. Curr Opin Biotechnol 18:221–226

Richards JW, Krumholz GD, Chval MS, Tisa LS (2002) Heavy metal resistance patterns of *Frankia* strains. Appl Environ Microbiol 68:923–927

Rosselli W, Keller C, Boschi K (2003) Phytoextraction capacity of trees growing on a metal contaminated soil. Plant Soil 256:265–272

Roy S, Khasa DP, Greer CW (2007) Combining alders, frankiae, and mycorrhizae for the revegetation and remediation of contaminated ecosystems. Botany 85:237–251

Sadhana B (2014) Arbuscular Mycorrhizal Fungi (AMF) as a biofertilizer. A review. Int J Curr Microbiol Appl Sci 3:384–400

Sayed WF (2011) Improving Casuarina growth and symbiosis with Frankia under different soil environmental conditions-review. Folia Microbiol 56:19

Schwencke J, Caru M (2001) Advances in actinorhizal symbiosis: host plant-*Frankia* interactions, biology, and applications in arid land reclamation. A review. Arid Land Res Manage 15:285–327

Sellstedt A, Smith GD (1990) Nickel is essential for active hydrogenase in free-living *Frankia* isolated from Casuarina. FEMS Microbiol Lett 70:137–140

Singh R, Gautam N, Mishra A, Gupta R (2011) Heavy metals and living systems: an overview. Ind J Pharmacol 43(3):246–253

Smith SE, Read DJ (2008) Mycorrhizal symbiosis, 3rd edn. Academic, London

Sun WH, Lo JB, Robert FM, Ray C, Tang CS (2004) Phytoremediation of petroleum hydrocarbons in tropical coastal soils. I Selection of promising woody plants. Environ Sci Pollut Res Int 11:260–266

Tam PC (1995) Heavy metal tolerance by ectomycorrhizal fungi and metal amelioration by Pisolithus tinctorius. Mycorrhiza 5:181–187

Tahar K, Keltoum B (2011) Effects of heavy metals pollution in soil and plant in the industrial area, West ALGERIA. J Kor Chem Soc 55(6):1018

Upadhyaya H, Panda SK, Bhattacharjee MK and Dutta S (2010). Role of Arbuscular Mycorrhiza in heavy metal tolerance in plants: prospects for phytoremediation. J Phytol 2:16–27

Wall LG (2000) The actinorhizal symbiosis. J Plant Growth Regulat 19:167–182

Wang Y, Björn LO (2014) Heavy metal pollution in Guangdong Province, China and the strategies to manage the situation. Environ Sci. doi:10.3389/fenvs.2014.00009

Wheeler C, Hughes L, Oldroyd J, Pulford I (2001) Effects of nickel on *Frankia* and its symbiosis with *Alnus glutinosa* (L.) Gaertn. Plant Soil 231(1):81–90

Wuana RA, Okieimen FE (2011) Heavy metals in contaminated soils: a review of sources, chemistry, risks and best available strategies for remediation. ISRN Ecology 2011:1–20

Zloch M, Kowalkowski T, Szymanska S, Katarzyna Hrynkiewicz K (2014) Response of Birch and Alder root endophytes as well as rhizosphere and bulk soil microorganisms to heavy metal pollution. Pol J Ecol 62:37–53

Chapter 20
Chelating Agents and Heavy Metal Phytoextraction

Ali Akbar Safari Sinegani, Iman Tahmasbian, and Mahboobe Safari Sinegani

20.1 Introduction

Elements with density higher than 5 g cm^{-3} have been called heavy metals. From the metallic elements that naturally exist, 53 elements belong to the heavy metal group. Some of these heavy metals, such as iron, manganese, copper, zinc, and molybdenum, are known as micronutrients in agriculture, since their amount in soil is low and the little quantity of them is necessary for plant growth. Another group of these elements, such as cobalt and vanadium, has been shown to be useful for plant growth. The other category, such as nickel, chromium, selenium, lead, and cadmium, has toxic effects on plant and other organisms even in low level. The mobility and uptake of these elements and the appearing of their negative effect in the environment depend on soil, water, and plant relations.

The term "chelating agents" refers to ligands occupying multiple positions in the inner coordination of a metal ion. Heavy metals combine with one or more reactive groups or ligands like oxygen (–OH, –COO, –POO), nitrogen (–NH_2, –NH), or sulfur (–SH, –S–S). The result of this phenomenon is the formation of soluble multidentate metal–chelant complexes called "chelates" (Fig. 20.1) which came from a Greek word "chele" meaning the claw of a crab or scorpion (Tsang et al. 2012).

To know the increase of metal solubility in soil, it is important to figure out the metal ion–ligand interactions. The complex formation can be considered as equilibrium reaction between the ligand and the metal ion:

A.A.S. Sinegani (✉) • M.S. Sinegani
Department of soil science, College of Agriculture, Bu-Ali Sina University, Hamedan, Iran
e-mail: aa-safari@basu.ac.ir

I. Tahmasbian
Environmental Future Research Institute, School of Biomolecular and Physical Science, Griffith University, Brisbane, Australia

I. Sherameti, A. Varma (eds.), *Heavy Metal Contamination of Soils*, Soil Biology 44, DOI 10.1007/978-3-319-14526-6_20

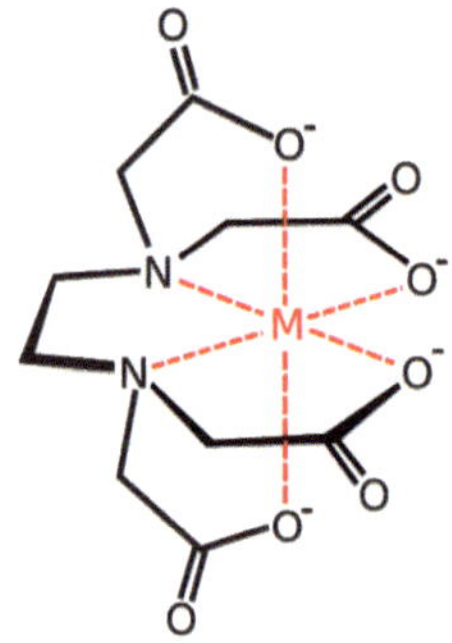

Fig. 20.1 Ethylene diamine tetra-acetic acid (EDTA) as a hexadentate ("six-toothed") ligand (Adopted from Wikimedia Commons)

$$M^{a+} + L^{b-} \leftrightarrow ML^{(a-b)}$$

where M is the metal ion (electron pair acceptor), a is the charge of M, L is the ligand (electron pair donor), and b is the charge of ligand.

According to the principle of mass action, the activities of M, L, and ML are as follows:

$$K_{ML} = \frac{\left[ML^{(a-b)}\right]}{\left[M^{a+}\right]\left[L^{b-}\right]}$$

where K_{ML} is the equilibrium constant (also denoted as the stability constant). In the case when pH should be taken into account, the conditional stability constant K_{cond} can be estimated according to the coefficient of ligand protonation (K_{HL}) and the coefficient of side reactions competing with the ligand for the metal ions (i.e., formation of metal hydroxides $(M(OH)_x)^{a-x}$ and of MLH or MLOH species). The amount of free L^{b-} increases with the increasing pH value.

Irrespective of soil fertility which can be increased by chelating agents, these materials have a special role in solubilizing heavy metals with low mobility such as Pb and Cr as well as enhancing the diffusion of them to plant roots. Therefore, in order to improve the uptake of these kinds of metals by plant, the phytoremediation enhanced by chelating agents was suggested (Huang et al. 1997; Blaylock et al. 1997a, b; Blaylock 2000; Cunningham et al. 1995).

Phytoremediation is a term which is used to describe the application of plant in order to ameliorate damaged or polluted substrates. The history of utilizing plants for controlling the degradation of soil and water goes back to the time when many early agriculturalists developed plant-based systems to reduce soil erosion, restore disturbed environments, and cleanse water. The potential of plants as extractor, however, began to be investigated in the twentieth century. Since then there has been rapid development of scientific methods regarding phytoremediation. Phytoremediation involves various processes, e.g., rhizodegradation, phytostabilization, phytovolatilization, and specially phytoextraction. Each process involves different mechanisms for the decontamination of pollutants from the environment. Such processes either decontaminate the soil or stabilize the pollutant within it.

The amount of metals being removed from the contaminated soil through plant uptake is the factor by which the successful implementation of phytoremediation in the field is assessed. For an effective phytoextraction, several conditions must be met. Firstly, the plant chosen for phytoextraction should be responsive to agricultural practices and be adapted to the environmental conditions that exist in the contaminated sites. Likewise, producing sufficient biomass coupled with high rates of metal uptake must be considered. The other point which should be taken into consideration is the availability of metals or other pollutants for plants in soil, a critical factor affecting the phytoremediation success. In other words, when pollutant metals in soil cannot be solubilized or made available for plant uptake, the uptake will be limited, which decreases the efficiency of phytoextraction. A number of soil characteristics are involved in metal solubility including soil pH type, and density of the charge on soil colloids (CEC and AEC), the reactive surface area, and complexation of metals with soluble ligands. Managing these factors during the phytoextraction will lead to more desirable results at the end of the process.

Therefore, chelating agents are utilized to increase the availability of heavy metals as well as nutrients in soil; both result in higher efficiency of phytoextraction by increasing the uptake of heavy metals by plant and increasing the produced biomass as a result of improving soil fertility, respectively. Besides, using the chelating agents can enhance the translocation of metals from roots to shoots. In this chapter the role of chelating agents in phytoextraction of heavy metals is discussed in detail. The objective of this chapter is to detail major interactions of chelating agents with metals with particular reference to effects on speciation, availability, and plant uptake. In addition, attention is drawn to their significance in plant growth and in environmental pollution and remediation.

20.2 Categories of Chelating Agents

Chelating agents can be categorized into natural and synthetic groups both used to increase the availability of soil nutrients and toxic metals for improving plant biomass production and metal uptake, respectively.

20.2.1 Synthetic Chelating Agents

The solubility of chelated metals depends on the stability constant of the complex in aqueous solution determined in a 1:1 (metal/chelant) ratio (Martell and Calvin 1958). Different rankings were reported to show the capacity of chelants to desorb heavy metal from the soil.

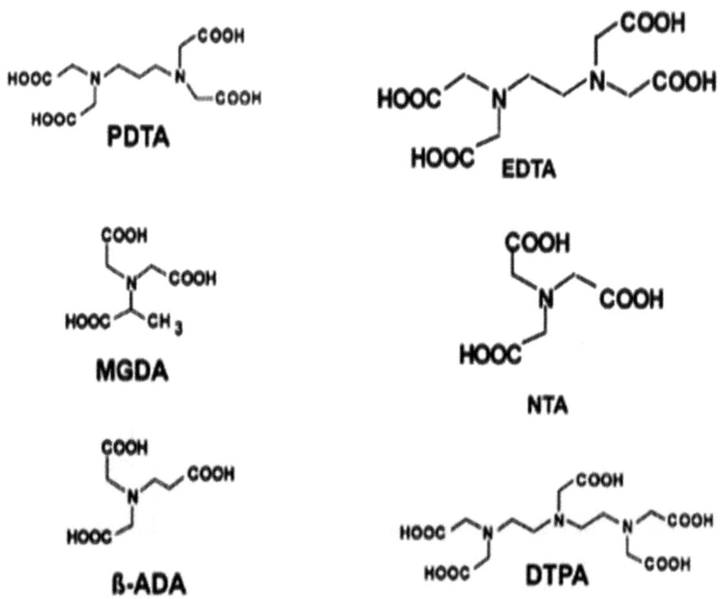

Fig. 20.2 Structural formula of the amino polycarboxylate chelating agents: *PDTA* 1,3-propylenedinitilo tetra-acetic acid, *EDTA* ethylene diamine tetra-acetic acid, *MGDA* methyl glycine diacetic acid, *NTA* nitrilo-triacetic acid, *β-ADA* ß-alanine diacetic acid, *DTPA* diethyl triamine penta-acetic acid (Adopted from Knepper 2003)

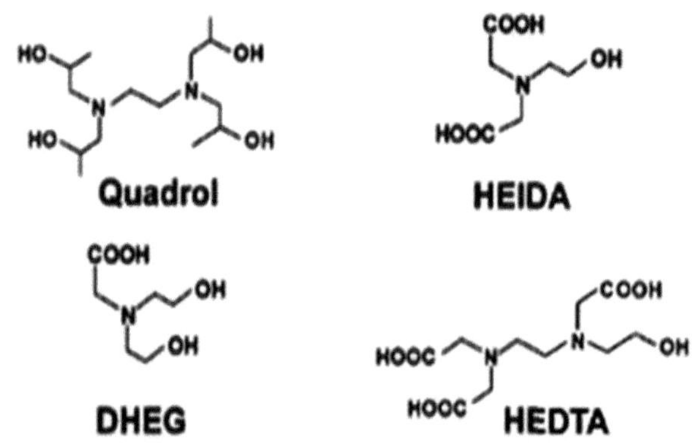

Fig. 20.3 Structural formula of the amino hydroxy carboxylate chelating agents: Quadrol (*N,N,N′, N′*-tetrakis-2-hydroxyisopropyl-ethylendiamine), HEIDA (*N*-(2-hydroxyethyl) iminodiacetic acid), DHEG (*N,N*-bis(2-hydroxyethyl) glycine), HEDTA (hydroxy ethyl-ethylene diamine triacetic acid) (Adopted from Knepper 2003)

Amino polycarboxylates (Fig. 20.2) and hydroxy carboxylates (Fig. 20.3) have a broad area of application as complexing agents because of their ability to complex metal ions in any kind of chemical process. Amino polycarboxylates are utilized due to their great stability as complexing agents in different application areas, mostly, together with the hydroxy carboxylates for the complexation of Ca^{2+} and Mg^{2+} ions (Sacher et al. 1998). These types of chelating agents not only can be used in remediation of soils and sludges but also in agriculture as fertilizer auxiliary materials and as sources of nitrogen (Knepper 2003).

Cooper and his colleagues (1999) by comparing some of these chelating agents in their capacity to desorb Pb from the soil reported the following ranking of chelants effectiveness for Pb:

HEDTA > CDTA (trans-1,2-cyclohexylene dinitrilotetraacetic acid) > DTPA > EGTA (ethylene bis[oxyethylenetrinitrilo] tetra-acetic acid) > HEIDA (N-(2-hydroxyethyl) iminodiacetic acid) > EDDHA(ethylene diamine, *N,N*-bis (o-hydroxyphenyl) acetic acid) ~ NTA.

In another study, Shen et al. (2002) showed that the effectiveness of chelating agents for solubilization of Pb is based on the following order: EDTA > HEDTA > DTPA > NTA > citric acid.

Each chelant usually has selective function for some metals. For instance, Wu et al. (1999) indicates that EDTA tends to complex Pb and Cd more than the other heavy metals in non-contaminated soils while NTA has a propensity to complex Cd and Cu. Higher concentration of these chelating agents reduces the selective function of them.

Among these synthetic chelating agents, EDTA is the most used chelant in phytoextraction of polluted soil. Huang et al. (1997) and Blaylock et al. (1997a, b) initially introduced the application of EDTA to enhance phytoextraction. EDTA is an aminopolycarboxylic acid and a colorless, water-soluble solid. Its usefulness arises because of its role as a hexadentate ("six-toothed") ligand and chelating agent, i.e., its ability to "sequester" metal ions. $EDTA^{4-}$ usually binds to a metal cation through its four carboxylates and two amines (Kirchner 1957). Since the carboxylates are protonated in very low pH, the solubility of EDTA and the stability of metal–EDTA complexes increase by increasing pH (Schmidt 2003).

It is widely reported that the application of EDTA in soil increased the soluble (water-extractable) and/or available (DTPA-extractable) form of heavy metals (Tahmasbian and Safari Sinegani 2013; Safari Sinegani and Khalilikhah 2008; Van Engelen et al. 2007), the ability which can be used to improve phytoextraction efficiency.

20.2.2 *Natural Chelating Agents*

Dissolved organic matters in soil significantly increase heavy metals' leaching (Ward et al. 2005). Soil humus includes different fractions, e.g., humic acid, fulvic acid, and humin. Each fraction has different reactions in soil based on their complicated structures. The reaction between organic matter and cations in soil results in soluble complexes playing the role of reducing agents and helping the movement of some elements.

Fulvic acid is made during the decomposition of organic matters and decreases with time, whereas humic acid increases with time (Aoyama 1996; Chefetz et al. 1996). Fulvic acid (simple organic acids with chelating ability) can affect the mobility of metals (Linehan and Shephard 1979; Schnitzer and Desjardins

1969; Sposito et al. 1982) and plays important roles in processes of podzolization in spodosols. Aluminum and iron cations are chelated and solubilized markedly by fulvic acids. They are leached and precipitated in lower layer of soil making the dark subsurface diagnosis horizon called spodic horizon.

Fulvic acid makes a stronger complex with Pb in comparison with the other elements (Hornick 1983); therefore, higher concentration of dissolved organic carbon enhances the mobility of lead in soil (Bergkvist 1986).

Heavy metals can be precipitated in oxy-hydroxide, phosphate, and carbonate forms in high Eh-pHs. There is a significant relationship between the amounts of soil organic carbon and heavy metal bioavailability, which more relates to organic matter–metal complexes than exchanging with exchangeable cations on organic matters (Basta and Grand Whoil 2000; Safari Sinegani and Mirahamdi Araki 2009). Moreover, increasing in organic ligand leads to blocking sorbing sites on mineral surfaces resulting in more solubility metals in soil (Sauvé et al. 2000). Bolt and Bruggenvert (1976) reported that the soluble complexes of Zn–organic matters come from amino acids and fulvic acids, while insoluble complexes relate to humic acids. Martin and Camazano (1993) by analyzing Cd in some polluted soils claimed that the absorption of Cd in soil more relates to the type of organic matters than the amount of them.

Of the organic matters used to improve soil phytoextraction efficiency and soil fertility, manures and their extracts such as cow, sheep, and poultry manures and extracts are more common and used by many scientists (Safari Sinegani and Khalilikhah 2008, 2010, 2011; Safari Sinegani and Ahmadi 2012; Tahmasbian Ghahfarokhi and Safari Sinegani 2013; Tahmasbian and Safari Sinegani 2014). Fresh organic matters such as manures contain soluble organic fractions being able to increase the solubility of metals shortly after adding to soil through the formation of organo-mineral complexes (Almas et al. 1999). Del-castilho and his colleagues (1993) stated that 30–70 % of soluble Cu and total Cd in treated soils were found to be in the form of organo-mineral complexes with the liquid part of cow manure. High concentration of inorganic anions, e.g., SO_4^{-2} and Cl^- in manure products, causes the metals to form metal–mineral complexes with high movement and bioavailability (Bril and Saalamons 1990; Japenga and Harmsen 1990).

Poultry manure, cow manure, and sewage sludge increase the soluble form of organic matters as a result of increasing the soluble organic carbon (SOC) and pH (Jackson and miller 2000). The increased SOC in turn increases the organo-metal complexes (Hesterberg et al. 1993; Safari Sinegani and Ahmadi 2012). Tahmasbian and Safari Sinegani (2013) by analyzing the active fractions of heavy metals in a soil treated with chelating agents reported that the concentration of soluble–exchangeable Pb and Zn reduced after the application of cow and poultry manure extract, while the concentration of organic-bound fraction of the metals increased. The results of Hashimoto et al. (2008) also indicated that there was a decrease in the concentration of Pb in soluble phase in poultry manure-treated soils, while, in residual phase, the amount of Pb had increased.

The other type of natural chelating agents is siderophores which are compounds produced by fungi, bacteria, and graminaceous plants for scavenging iron from the environment. Simply put, siderophores are categorized in microbial siderophores and phytosiderophores; both are low-molecular-weight compounds (500–1,500 Da) which possess a high affinity for iron (III) ($Kf > 10^{30}$). Their function is to supply iron to cells and the biosynthesis of which is regulated by iron levels (Hider and Kong 2010).

Without exception, siderophores possess a higher affinity for iron (III) than iron (II), considering that iron (III) is not readily bioavailable in aerobic environments due to the low solubility of iron (III). There is a strong logic for this selectivity. It is difficult to design ligands being selective for iron (II) over the biologically important cations with two positive charges such as zinc (II), manganese (II), nickel (II), and copper (II). This problem is less difficult for iron (III), since there are not many biologically important cations with three positive charges, just the kinetically inert Co (III). Therefore, a ligand which is selective for trivalent metals will effectively be selective for iron in biological matrices (Hider and Kong 2010). Although Al (III) can in principle compete with iron (III) for siderophore uptake processes, the radius of the aluminum (III) cation (0.54 Å) is much smaller than that of iron(III) (0.65 Å), and thus most hexadentate siderophores possess a much higher affinity for iron (III) than for aluminum (III). The other reason why microorganisms do not accumulate aluminum via siderophore transport systems is that in many such systems, iron is released from the siderophore by a redox process while this mechanism is not possible for aluminum (Hider and Kong 2010).

Phytosiderophores are nonproteinogenic amino acids belonging to the mugineic acid (MA) family of compounds (Mori 1999). Figure 20.4 illustrates the biosynthetic pathway of mugineic acid family of phytosiderophores (MAs) in graminaceous plants. Phytosiderophores are synthesized and secreted into rhizosphere to bind Fe (III) with high affinity. Fe (III)–phytosiderophores is transported across the cytoplasmic membrane using a proton symport mechanism (Mori 1998). Subsequently, the iron (III) complex is reduced to iron (II), and the iron is transferred to nicotianamine which translocates iron in phloem of all plants. Although nicotianamine is very similar to the phytosiderophores, it is selective for iron (II) and is not secreted by the roots (Corie et al. 2001).

The roots of most plants also develop a strong mutualistic association with fungi called mycorrhiza. The siderophore produced by *Wilcoxina rehmii* and *Cenococcum geophilum* is ferricrocin (Prabhu et al. 1996). In fact, the ferricrocin is a widespread siderophore in most forest environments (Winkelmann 2007). The iron associated with the ferrichrome and ferricrocin is made available to plant roots via a reductive mechanism (Winkelmann 2007; Bar-Ness et al. 1991).

Like graminaceous, bacteria and fungi, in response to low iron availability in their environment, synthesize siderophores called microbial siderophores. The soil is considered as a rich source of bacterial and fungal genera. From the gram-positive species of bacteria, those belonging to the Actinomycetales and species of the genera *Bacillus*, *Arthrobacter*, and *Nocardia* are common. Most of these organisms produce and secrete ferrioxamines that result in growth promotion of not

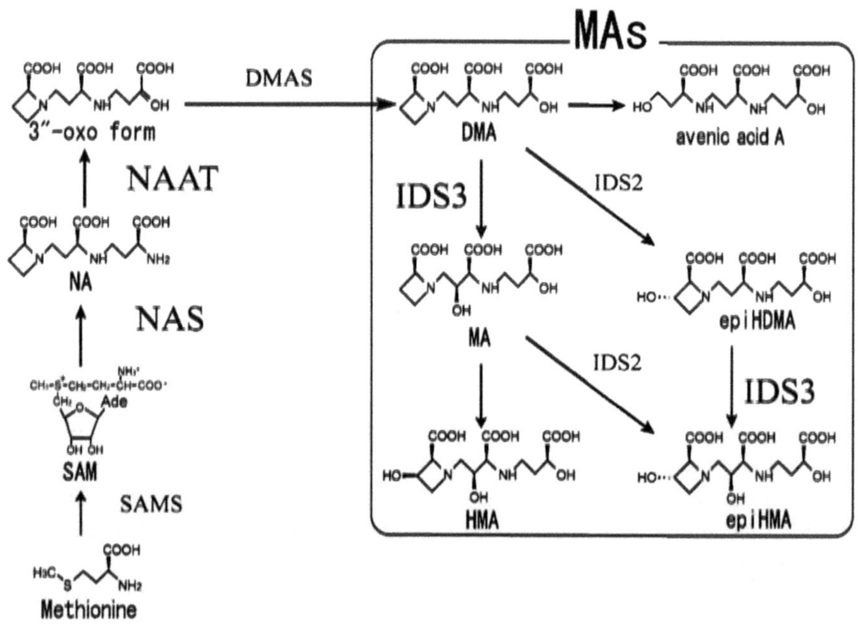

Fig. 20.4 The biosynthetic pathway of mugineic acid family of phytosiderophores (MAs) in graminaceous plants. *SAM* S-adenosyl-methionine, *NA* nicotianamine, *DMA* 2′deoxymugineic acid, *MA* mugineic acid, *HMA* 3-hydroxymugineic acid, *epi HDMA* 3-epihydroxy-2-′-deoxymugineic acid, *epi HMA* 3-epihydroxymugineic acid, *SAMS* S-adenosyl-methionine synthetase, *NAS* nicotianamine synthase, *NAAT* nicotianamine aminotransferase, *DMAS* 2′-deoxymugineic acid synthase, IDS2 dioxygenase catalyzing the hydroxylation of three positions of DMA and MA, IDS3 dioxygenase catalyzing the hydroxylation of 2′-positions of DMA and epiHDMA (Adopted from Masuda et al. 2008)

only the producing organisms but other microbial populations being able to utilize exogenous siderophores (Hider and Kong 2010).

Soil fungi including *Aspergillus* and *Penicillium* mostly produce ferrichromes which consist of cyclic hexapeptides. They are extremely resistant to environmental degradation associated with the wide range of hydrolytic enzymes present in humic soil (Winkelmann 2007). In the soils containing decaying material, pH values reach as low as 3–4 which is the best condition for those organisms that produce hydroxamate siderophores because of the extreme acid stability of these molecules in acidic conditions. Some soil bacteria, e.g., *Azotobacter* and *Agrobacterium*, are only able to synthesize catecholate siderophores and therefore are limited to scavenging Fe in neutral and alkaline calcareous soils (Hider and Kong 2010).

Now, two questions should be answered: whether microbial population can utilize phytosiderophores as sole source of carbon and iron and whether strategy II plants can directly utilize microbial siderophores. For the first question, in at least one case, the utilization of phytosiderophores as an iron source appears to be indirect, with ligand exchange occurring between the microbial siderophore and the phytosiderophores (Jurkevitch et al. 1993). Romheld (1991) suggested that physical separation on the root between zones of active release and uptake of phytosiderophores and also heavy colonization by microorganisms protect phytosiderophores not only from degradation but also from competition for iron with microbial siderophores.

To answer the second question, despite confliction data obtained due to the use of nonaxenic conditions, experiments suggest that microbial siderophores do not serve directly as sources of iron for strategy II plants (Bar-Ness et al. 1992). Nevertheless, they may serve indirectly, after degradation by microbes and release of iron (Cuerinot and Yi 1994).

20.3 Heavy Metals in Soil Environment

Heavy metals may be dissolved in soil solution, associated with organic and inorganic soil constituents in exchangeable and non-exchangeable forms, and precipitated as pure or mixed compounds, including primary or secondary minerals. The solubility of heavy metal in soil is governed by many chemical and biological processes including in organic and inorganic complexation, oxidation–reduction reactions, precipitation–dissolution, and adsorption–desorption. Heavy metals in soil solution can exist as free cations (e.g., Pb^{2+}, Cd^{2+}), soluble complexes with inorganic or organic ligands (e.g., $CdCl^{+}$, metal–citrates or metal–malates) and suspended connected with colloidal material. Common anionic ligands that can complex heavy metals in soil solution include chloride (Cl^{-}), sulfate (SO_4^{-2}), bicarbonate (HCO_3^{-}), hydroxyl (OH^{-}), carbonate (CO_3^{-2}), phosphate (PO_4^{-3}), and nitrate (NO_3^{-}). Organic complexing agents include humic and fulvic acids, aromatic and aliphatic acids, amines and amides, and carboxylic acids (McLean and Bledsoe 1992). The organic complexing agents generate in soil by biological activities.

In soil environment, the bioavailability of heavy metals is primarily controlled by adsorption−desorption reactions at the particle–solution interface (Backes et al. 1995). Although a great amount of research has been conducted examining these reactions, the accurate prediction of heavy metal potential bioavailability in soil environment has remained elusive (Glover et al. 2002). Soils are composed of a heterogeneous mixture of mineral and organic solid phases as well as a myriad of organic and inorganic solutes that compete with metals for adsorption sites on solids. This heterogeneity makes modeling the potential bioavailability of heavy metals problematic.

Colloidal materials affecting heavy metal bioavailability and transport include iron and manganese oxides, clay minerals, and organic matter. Soil microorganisms are agents of metal immobilization/mobilization and transport. Metal cation adsorption in soil environment is correlated with pH, redox potential, clay minerals, organic and inorganic matter, Fe and Mn oxides, and calcium carbonate content, and alterations in these may affect metal bioavailability.

The decreased kinetic of Pb bioavailability in the $Pb(NO_3)_2$-treated soils (400 μg Pb g^{-1}) has been studied in solid-state sterile and unsterile conditions at 15, 27, and 37 °C (Safari Sinegani and Mirahamdi Araki 2010a). Temperate soil samples with higher clay content, cation exchange capacity (CEC), organic carbon (OC), biological activity, and amorphous and crystalline Fe and Al had higher affinity surface sites for Pb sorption compared to semiarid soils with higher sand content, pH, and equivalent carbonate calcium (EEC). The added Pb in semiarid soils was more available, and temperate soils had a lower available Pb in each time of extraction. Metal sorption to soil organic matter (including organisms) is also highly important. Anyway, a greater quantity of Pb was sorbed to the unsterile soil surface compared to sterile soils. Sterilization and soil incubation at lower temperature (15 °C) reduced the rate of Pb sorption/precipitation processes (Safari Sinegani and Mirahamdi Araki 2010a).

Heavy metals can be sorbed onto minerals. Clay minerals, e.g., montmorillonite and kaolinite, and iron and manganese oxides are major inorganic determinants of metal bioavailability in soil. Metals associated with exchange sites (exchangeable fraction) are a significant reserve of potentially mobile metals, depending on environmental changes and biological activity. In addition, metals can precipitate as solids in soils, e.g., $CdCO_3$, $Pb(OH)_2$, as well as mixed compounds. Precipitation may be a significant process in highly contaminated calcareous soils.

Sequential extraction has been used as a suitable technique for identification of chemical forms of trace elements and their plant availability. In an experiment, the chemical forms of Pb in $Pb(NO_3)_2$-treated soils (400 μg Pb g^{-1}) have been studied in solid-state incubation at 27 °C in sterile and unsterile conditions (Safari Sinegani and Mirahamdi Araki 2010b). A sequential extraction scheme was used to fractionate Pb of the incubated samples into soluble–exchangeable (Sol-Exch), carbonate-associated (ACar), organic matter-associated (AOM), Mn oxide-associated (AMnOx), Fe oxide-associated (AFeOx), and residual (Res) forms after 20 min and 3,600 h. Soil Pb fractionation revealed that, in both groups of soils, Pb largely changed to exchangeable, carbonate-associated, and organic-associated forms after 20 min. However, the chemical forms of Pb differed widely among soils after 3,600 h incubation (Fig. 20.5). Compared to semiarid soils, in the temperate soils with higher Fe–Mn oxides and OM contents and with lower pH and equivalent calcium carbonate (ECC), there was a noticeable conversion rate of Pb from more available forms to less available forms. In temperate soils after 3,600 h incubation, greater content of Pb was observed in Res (68 %), AOM (14 %), and AMnOx (5 %) fractions. In semiarid soils, however, greater content of Pb was observed in ACar (16 %) and Sol-Exch (8 %) fractions. The sum of AMnOx and AFeOx chemical forms for Pb in semiarid soils (7 %) was low compared to temperate soils (9 %). Thus, complex arrays of physical, chemical, and biological processes govern metal availability in soil and interactions with microorganisms. Soil microorganisms in unsterile soils compared to sterile soils had significant effect on AOM, AFeOx, and Res fractions of Pb. They not only increased AOM and AFeOx fractions of Pb in soils but also decreased Res fraction of Pb significantly (Safari Sinegani and Mirahamdi Araki 2010b). Hence, microbial activities have significant effects on

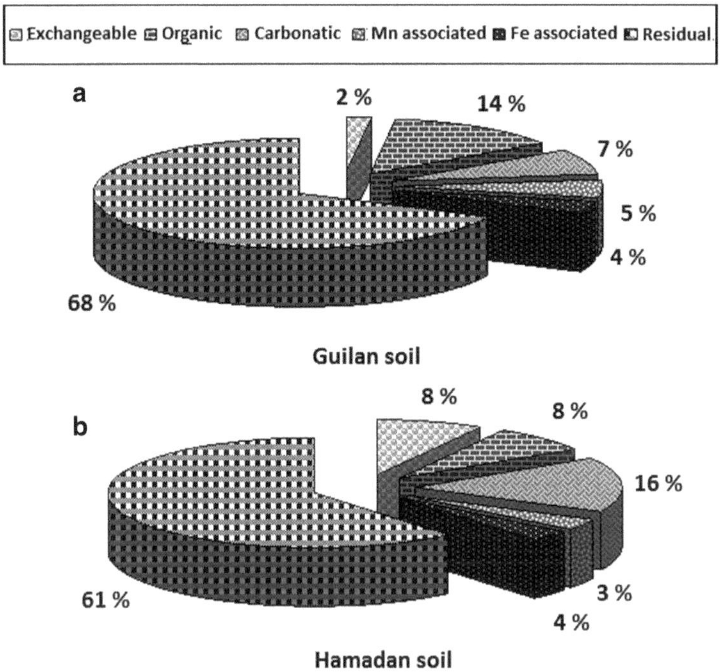

Fig. 20.5 Lead fractions in soils sampled from temperate (Guilan) and semiarid (Hamadan) regions and treated with $Pb(NO_3)_2$ after 3,600 h incubation at 27 °C in field capacity (Adopted from Safari Sinegani and Mirahamadi-Araki 2010a)

metal bioavailability and soil chemical changes affecting phytoextraction efficiency of heavy metals from soils.

20.4 Chelating Agents and Phytoextraction

The remediation of polluted soils and sediments by plants is a slow process which should be repeated for several years to achieve a considerable result. Soil salinity, low water holding capacity (WHC), toxic materials, and nutrition deficiency are some of the reasons why the efficiency of phytoextraction is low (Vangronsveld et al. 1996). Utilizing of chelating agents in these soils leads to increase in the bioavailability of elements, WHC, and therefore, the efficiency of soil remediation (Brown et al. 2003).

Continuous or *natural* phytoextraction and *induced*, *enhanced*, or *chemically assisted* phytoextraction are two basic strategies of phytoextraction suggested by Salt et al. (1998). The maximum efficiency which can be achieved in both strategies highly depends on the concentration of metals in aerial parts of the plant on the one hand and the biomass of harvestable parts on the other hand (Meers et al. 2008). To increase the phytoavailability of metals, the use of soil amendments (Huang et al. 1997; Kulli et al. 1999; Cooper et al. 1999; Blaylock and Huang 2000; Safari Sinegani and Ahmadi 2012), electric fields (Tahmasbian and Safari Sinegani 2013, 2014), and the control of soil pH have been suggested. Relative affinity of chelating agents to heavy metals, distribution and speciation of chelating agents and heavy metals in soil, mobilization of the new formed complexes, resistance, and half-life of complexes affect the actions of chelating agents in phytoremediation. An ideal chelating agent for improving metal phytoremediation is highly water soluble and resistant to biotransformation. It has the ability to reach the site of metal storage in soil, form nontoxic complex with heavy metal for plant, and remain bioavailable at different pHs of soil solution.

As the final purpose of using chelating agents is to increase the amount of pollutants taken up by plant (metals removed from the soil), it is necessary to know how to calculate the amount of metals taken up by plants. Eq. (20.1) reveals the different variables affecting the amount of metals removed from the soil:

$$\mathrm{Um} = \mathrm{Cmh} \times \mathrm{DWh} \tag{20.1}$$

where Um is representative of the metal taken up from the soil (metal uptake index or UI), Cmh shows the concentration of metals in shoots or harvestable parts of plant, and DWh is the dry weight of plants' shoots or harvestable parts. The equations reveal that the most important factors which should be increased in order to ameliorate the effectiveness of a phytoextraction system are increasing the concentration of metals in shoots and the biomass of shoots as well. Increasing the Cmh, in turn, depends on increasing the availability of metals in soil as well as increasing the translocation of metals from roots to shoots.

Translocation factor (TF) is used to quantify the translocation of metals from roots to shoots (Eq. (20.2)):

$$\mathrm{TF} = \mathrm{Cmh}/\mathrm{Cmr} \tag{20.2}$$

where Cmr shows the concentration of metals in plant roots, although the concentration of metals in shoots can be increased without increasing the TF when changes between the Cmh and Cmr are approximately equal (Tahmasbian and Safari Sinegani 2014). So, in phytoextraction, another factor named enrichment factor (EF) may be estimated according to Eq. (20.3) for comparison of phytoextraction efficiency in different conditions:

$$\mathrm{EF} = \mathrm{Cmh}/\mathrm{Cms} \tag{20.3}$$

where Cms shows the concentration of metals in soil. The enrichment factor is also called transfer factor. Enrichment factor may be calculated based on total concentration of metals in soil or available concentration of metals in soil. The estimated EF will be too low, if it estimates based on total concentration of metals in soil (Safari Sinegani and Khalilikhah 2011). In this state the differences between the data of treatments may be too small for analysis and comparison. When a soil is threatened with chelating agents, plants, microorganisms, etc., the total concentration of metal will not change markedly, but available forms of metal will change significantly. Since the total concentration of metals in a soil used in lab study is nearly constant, no more information will be gained and the result of EF analysis will be the same as the result of Cmh analysis. Thus it is suggested that it will be calculated based on available metal concentration in soil in different treatments.

20.4.1 Effects of Chelating Agents on Plant Dry Weight

In spite of reported successes in enhancing the available form of heavy metals using chelating agents mainly EDTA, there is a major concern about the excessive levels of heavy metals at plant's tissue and its toxicity. Many researchers have indicated that low biomass was produced when plants were exposed to both free metals (e.g., Cd and Pb) and free EDTA due to low seed germination, leaf wilt, chlorosis, shoot desiccation, and reduced transpiration (Vassil et al. 1998; Lombi et al. 2001; Romkens et al. 2002; Grc˘man et al. 2003; Duo et al. 2005; Nascimento et al. 2006). Reduction in harvestable parts of plants leads to a decrease in the total amount of removed metals from soil by plants (Quiroz et al. 2002; Safari Sinegani and Khalilikhah 2008).

There are several factors, however, affecting plant production, among which are the concentration of EDTA influencing EDTA/metal molar ratio, time of EDTA application, and plant specious. The results of the recent researches investigating the effects of chelants on phytoextraction confirmed that the application of high concentration of EDTA (higher than 0.5 g EDTA kg^{-1} soil) significantly reduced the plants' dry weight (Liphadzi and Kirkham 2006; Safari Sinegani and Khalilikhah 2008, 2010, 2011).

Safari Sinegani and Khalilikhah (2008) studied the Cd phytoextraction by *Helianthus annuus* grown in a mine calcareous soil treated with increasing concentrations of EDTA (0, 0.5, and 2 g EDTA kg^{-1} soil) or sheep manure extract (0, 0.5, and 2 g SME kg^{-1} soil) 30 or 10 days before sowing (T1, T2) and 10 or 30 days after sowing (T3, T4). They reported that increasing of EDTA concentration reduced sunflower seedling emergence and its dry weight significantly. Cadmium concentrations in harvested shoots increased with EDTA concentration, but the actual amount of phytoextracted Cd decreased at high EDTA concentrations, due to severe growth depression. It was found that after only 7 days of seedling growth, shoot biomass was decreased by all EDTA treatments compared to the control and SME treatments. The leaves showed chlorosis symptoms and then became necrotic (Fig. 20.1).

Lesage et al. (2005) reported that *Helianthus annuus* had suffered from heavy metal stress due to the significant increase in bioavailable metal fraction in the soil. The high EDTA dose of 2.0 g kg^{-1} soil reduced seedling emergence and height of the plants grown in the composted biosolids (Liphadzi and Kirkham 2006). Similar results were obtained in rapeseed study. Potential decreases in rapeseed dry weight were observed when the soil was treated with EDTA. EDTA significantly decreased root and shoot dry weight, whereas SME stimulated root and shoot dry weight compared to control.

Application time is also considered as an important factor affecting plant's biomass production. Treating the polluted soils with EDTA before germination or transplanting immediately after germination can severely inhibit plant growth as a result of free EDTA in soil or indirectly due to increased dissolution of metals (Chen and Cutright 2001; Meers et al. 2004; Safari Sinegani and Khalilikhah 2008, 2010, 2011). The addition of EDTA when plants have attained sufficient biomass may reduce the adverse impacts of this chelant on biomass productivity. Safari Sinegani and Khalilikhah (2008, 2010, 2011) reported that EDTA had lower effects on plant biomass when it was applied 30 days after sowing. Although the results showed that shoot dry weight was considerably high in the application of EDTA 30 days after sowing (T4), it was too late for an effective Cd absorption and a high Cd uptake. They suggested that 15 days after sowing (T3) is the best time for application of EDTA in calcareous contaminated soil for Cd phytoextraction by sunflower. But Saifullah et al. (2009) reported that EDTA can be added in low concentration in several increments during the entire growth period of plant for better phytoextraction. These effects, however, strongly depend on the plant, soil, chelating agent, heavy metal properties, and interactions. In study of rapeseed, the time of application of chelant and amendment on plant dry weight was also important. The negative effect of EDTA on rapeseed was more obvious when it was applied 30 and 10 days before sowing (T1 and T2). The plant biomass decreased with increasing the EDTA levels. However, in T4 (30 days after sowing), the negative effect of increasing concentration of EDTA was lower than in other times (Fig. 20.6).

On the other hand, natural chelants, e.g., cow, poultry, and sheep manures as well as their extracts, have been reported to increase the plant biomass (Safari Sinegani and Khalilikhah 2008, 2010, 2011; Tahmasbian Ghahfarokhi and Safari Sinegani 2013). In these studies, plant biomass was significantly higher in all manure and their extract treatments compared to control and plant dry weight increased with increasing of manure levels. Safari Sinegani and Khalilikhah (2008, 2010) reported that the effect of SME application time on root and shoot dry weight was not significant and as important as application time of EDTA treatments. However, the root was more sensitive to the time of SME application than shoot. The positive effect of SME on root dry weight was more obvious when it was applied 30 and 10 days before sowing (T1 and T2). Stimulating the plant biomass production in these treatments may be attributed to the improving soil physical, chemical, and biological characteristics.

Fig. 20.6 The effect of EDTA and SME application time on sunflower growth in a mine soil in pot culture at different stages of plant growth. (**a**) Application of 2 g EDTA kg^{-1} soil at 30 days before sowing, (**b**) application of 2 g EDTA kg^{-1} soil at 10 days before sowing, (**c**) application of 2 g EDTA kg^{-1} soil at 10 days after sowing, (**d**) application of 2 g SME kg^{-1} soil at 30 days before sowing (Adopted from Safari Sinegani and Khalilikhah 2010)

20.4.2 Effects of Chelating Agents on Metals' Availability

EDTA as a synthetic chelant has stimulating effects on increasing the soluble–exchangeable fraction ($NaNO_3$-extractable) of metals in soil. The history of the effect of EDTA on the solubility and phytoavailability of metals goes back to the time when Wallace et al. (1974) reported that metal–EDTA complexes could increase solubility and phytoavailability of metals in soils. The ability of EDTA to enhance the release of some metals such as Pb from insoluble or sparingly soluble compounds has been attributed to the higher binding capacity of EDTA for Pb as observed in numerous researches (e.g., Blaylock et al. 1997a, b; Huang et al. 1997; Wu et al. 1999). Tahmasbian and Safari Sinegani (2013) reported that application of 2 g kg^{-1} EDTA in a polluted soil significantly increased the amount of soluble–exchangeable Pb and Zn. Van Engelen et al. (2007) also reported that the utilization of EDTA in soil could increase the soluble form of heavy metals. A large fraction of the total metals is dissolved and becomes available for phytoextraction after the application of EDTA to soils (Elliott and Brown 1989; Haag-Kerwer et al. 1999).

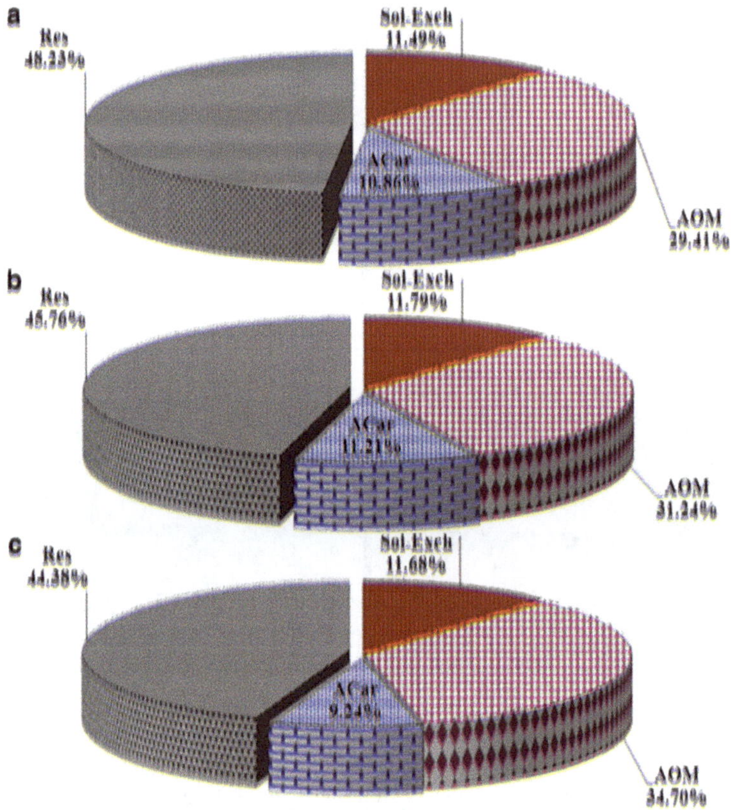

Fig. 20.7 Cd fractions in (**a**) unamended and uncultivated soil, (**b**) amended with manures and their extracts and uncultivated soil, and (**c**) amended with manures and their extract and cultivated *Cannabis sativa* soil after 8 weeks and plant harvest

Xian (1987) stated that AOM fraction was higher than the other fractions of Pb in slugged acid-amended soils. Lu et al. (2005) studied the speciation of Cu, Zn, Pb, and Cd in three soils and reported that high organic matter content resulted in increased AOM fraction. Safari Sinegani and Ahmadi (2012) found that available (DTPA-extractable) Pb and Cd and their soluble–extractable fraction rose after the utilization of cow and poultry manure and their extracts in soil. They studied the effects of four organic amendments on chemical forms of Pb and Cd in a contaminated calcareous soil. A pot experiment of control, unamended soil and soils amended with dry cow and poultry manures (20 g CM or PM kg^{-1} soil) and cow and poultry manure extracts (2 g CME or PME kg^{-1} soil) cultured with *Cannabis sativa* was done. After 8 weeks, sequential extraction scheme used in soil Pb and Cd fractionation showed that the addition of animal manures and their extracts increased Sol-Exch and AOM fractions and also DTPA-extractable Pb and Cd in soil significantly (Figs. 20.7 and 20.8). They found that the increase of AOM fraction was more obvious for Pb in the application of cow and poultry manure

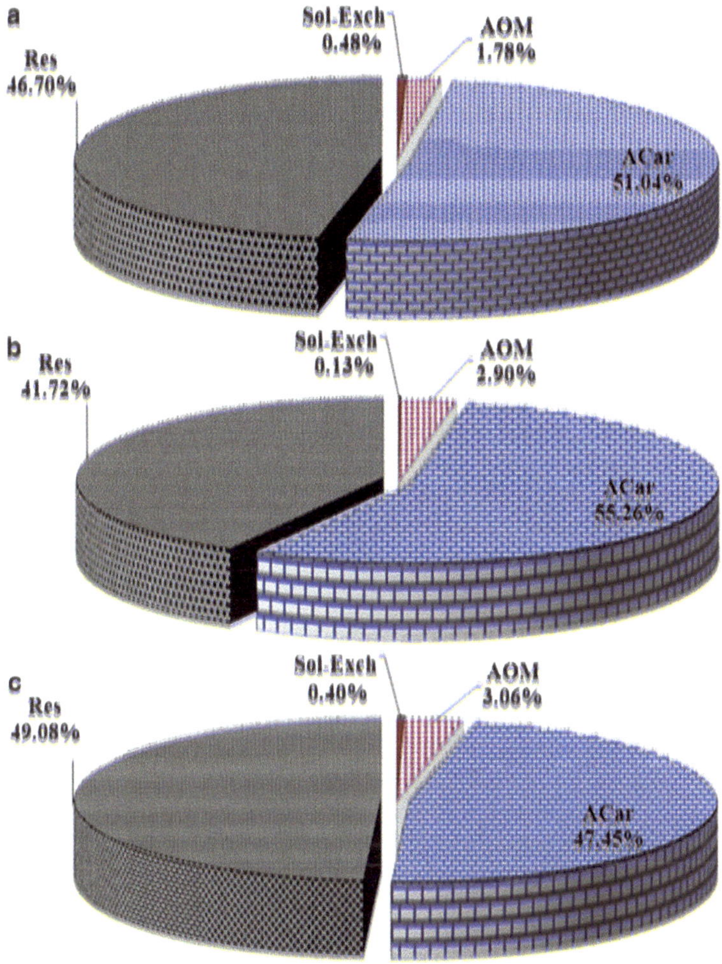

Fig. 20.8 Pb fractions in (**a**) unamended and uncultivated soil, (**b**) amended with manures and their extracts and uncultivated soil, and (**c**) amended with manures and their extract and cultivated *Cannabis sativa* soil after 8 weeks and plant harvest

extracts. The ACar chemical forms of Pb and Cd were also increased, but Res fractions decreased significantly by the application of manures and their extracts.

Against these findings, Harter (2006, 1983) found no correlation between soil organic matter content and the percentage of AOM fraction or metal retention in soil. He argued this lack of correlation to the fact that total organic matter of a wide variety of soils may not be sufficiently related to their reactive portions to generate a positive correlation. But organic matter in soils may vary in concentration, in the types of functional groups, and in solubility and activity. Such heterogeneity may be the reason for a different metal solubility and fractionation in soils with similar organic matter concentrations (Catlett et al. 2002). The research conducted by

Tahmasbian and Safari Sinegani (2013) also showed that poultry manure extract (PME) and cow manure extract (CME) applied to the soil significantly decreased the soluble–exchangeable fraction (Sol-Exch) of soil Pb, which can be explained by the increase in the organic-bound fraction (AOM) affected by these chelants.

20.4.3 Effects of Chelating Agents on Metal Concentration in Plant Shoots

It is worth mentioning different metals may act differently in soil and plant while facing chelating agents. A little Pb is taken up in the absence of chelants, while the uptake of Zn and Cu in low solution concentration is more efficient since in the absence of chelants, Zn and Cu accumulation in plants is governed by uptake of free metal ions in the symplastic pathway. When the metals were chelated in soil solution, the uptake would occur through the apoplastic pathways and Cu and Zn uptake would be reduced, whereas Pb uptake would be significantly increased (Nowack et al. 2006). Therefore, it becomes clear why there is an increase in Pb uptake in the presence of the chelants in most cases. In contrast, the few hydroponic data on Cu and Zn uptake in the presence of chelants mostly show a decrease in metal uptake. With a high dissolved metal concentration, translocation of all three metals to the shoots would increase in the presence of chelants because the nonselective uptake in the presence of chelants would exceed selective uptake along the symplastic pathway for both essential and nonessential metals. However, it depends on the plant species and may vary considerably.

In general EDTA as the most used synthetic chelating agents has positive effects on increasing the concentration of metals in plants' organs. EDTA can solubilize soil metals, and metal–EDTA complex can be easily absorbed by roots of plants (Lai and Chen 2004; Turgut et al. 2004; Lesage et al. 2005; Liphadzi and Kirkham 2006). Many researches showed that the concentration of metals in shoots had increased after the application of EDTA in soil (Lai and Chen 2005; Safari Sinegani and Khalilikhah 2010, 2011; Tahmasbian Ghahfarokhi and Safari Sinegani 2013).

It was revealed that applying 5 mmol EDTA kg^{-1} soil can significantly increase the Cd, Zn, or Pb concentrations in soil solution, thus increasing the accumulated metal concentrations in rainbow pink grass shoots (Lai and Chen 2004). Adding chelants to soil increases not only the total dissolved metal concentration but also changes the uptake mechanism and, depending on metal, plant species, and chelant concentration, significantly increases both plant Pb and plant Cu and Zn (Nowack et al. 2006).

The addition of 9 mmol EDTA kg^{-1} led to a 214-fold increase in soluble Pb content (as much as 60 % of mobilized Pb), and the addition of 9 mmol EDDS kg^{-1} led to a 33-fold increase in soluble Pb content compared to the control (Komárek et al. 2007). However, the competition of other metals (e.g., Fe and Mn from dissolving metal (hydr)oxides) in the soil for the ligand of EDTA is one of the most important efficiency-controlling parameters (Nowack et al. 2006; Komárek et al. 2007). It was reported that the significant lower Pb-extraction efficiency of EDDS is a great extent influenced by the competition of Cu to form the soluble metal–EDDS complex. The extraction efficiency of chelating agents is not only dependent on the stability constants (log K) but largely on the metals and ligand (chelant) concentrations. Therefore, the extraction efficiency of Pb in time is influenced by the partial dissolution of other metals like Fe oxides and hydroxides and consequent remobilization and complexation of the adsorbed metal (Nowack et al. 2006).

Natural organic chelating agents also showed stimulating effects on the concentration of metals in plants. Researchers working on natural organic chelating agents (Safari Sinegani and Khalilikhah 2008, 2010, 2011; Tahmasbian Ghahfarokhi and Safari Sinegani 2013) reported that manure extract (sheep, poultry, and cow manure) increases the concentration of metals in roots and shoots, although the plants affected by EDTA had a significantly higher concentration of metals in their shoots. A part of this difference between natural and synthetic chelating agents may be due to dilution effects or increase of plant growth in application of natural chelating agents and decrease of plant growth in application of synthetic chelating agents. This effect is so important that it may result in lower concentration of metal in aerial parts of plant in the application of natural chelating agents compared to control. All the discussed results indicated that the application of chelating agents, mainly EDTA, increases the concentration of metals in aerial parts of plant, which may have positive impact on phytoextraction efficiency.

Safari Sinegani and Khalilikhah (2008) reported that the stimulating effect of application of sheep manure extract (SME) on Pb concentration in root and shoot organs of the plant was significant ($P < 0.05$) compared to control. The time of application of this natural chelating agent was important. The increasing effect of SME on Pb concentration in root and shoot organs of the plant was more obvious when it applied in 30 days after sowing (T4). The rapid mineralization of SME and the high buffering capacity of the calcareous soil made SME less efficient in increasing the phytoextracted amounts of heavy metals when it applied 30 and 10 days before sowing (T1 and T2) and 15 days after sowing (T3) compared to T4.

Although heavy metal concentration increases in shoot and root of plants with increasing the levels of applied synthetic chelating agents in many studies, in some cases, it decreases with increasing the levels of applied natural chelating agents (Safari Sinegani and Khalilikhah, 2008). The concentration of Pb decreased in shoot and root of plants with increasing the levels of applied SME, especially at T1, T2 and T3. This decreasing trend by increasing SME level was not significant in application of it at 30 days after sowing (T4).

20.4.4 Effects of Chelating Agents on Translocation Factors of Metals (TF)

In addition to the concentration of metals in plant's organs which is reported in most of phytoextraction experiments, translocation of the metals to harvestable parts of plant (TF) as well as accumulation of them in aerial parts is important to assess the phytoextraction efficiency.

The application of chelating agents, e.g., EDTA, has been proved to enhance the translocation of metals from the root to shoot (Barber and Lee 1974; Hamon et al. 1995; Vassil et al. 1998; Gleba et al. 1999; Safari Sinegani and Khalilikhah 2011). Despite significant changes in phytoextraction efficiency, in some researches, TF had not changed after treatment (Tahmasbian and Safari Sinegani 2013, 2014). These results may associate with the equal changes in the concentration of metals in shoots and roots or the late time of the application of chelants. Organic chelants, on the other hand, are supposed to have different impacts on TF depending on the application time and concentration of chelant. Safari Sinegani and Khalilikhah (2010, 2011) found that sheep manure extract (SME) could not significantly alter TF of Cd and Pb when this chelant is applied 30 or 10 days before sowing and also 10 days after sowing. Application of SME 30 days after sowing in higher concentration, nevertheless, improved TF of the metals.

20.4.5 Effects of Chelating Agents on Metal Uptake

The removal of metals is analyzed by metal uptake index (UI) obtained by the concentration of metals in shoots (Cmh) multiplied by shoots' dry weight (DWh). Therefore, increasing the concentration of metals in harvestable parts of plants cannot solely maximize the phytoextraction efficiency. It is also possible that chelants decrease the UI when the toxic effects of this chelant reduce the biomass produced by plants. Tahmasbian Ghahfarokhi and Safari Sinegani (2013) found that despite the higher concentration of metals in shoots in EDTA-treated soils, not only could EDTA not improve the phytoextraction efficiency but UI decreased after using EDTA. They explained that the significant reduction in shoots biomass after the application of EDTA was the reason why this treatment had adverse impact on phytoextraction efficiency. Reversely, the plants treated with poultry manure extract (PME) in the referred study showed the highest UI since PME had significantly increased the biomass of aerial parts of the plants (Tahmasbian Ghahfarokhi and Safari Sinegani 2013). In a number of experiments, nevertheless, EDTA increased the removal of metals by plants, although it was used after sowing (Safari Sinegani and Khalilikhah 2008, 2010, 2011).

Safari Sinegani and Ahmadi (2012) indicated that soil cultivation with *Cannabis sativa* increased available, Sol-Exch, and AOM chemical forms of Pb in soil significantly compared to control soil. However, soil Pb and Cd in ACar fraction were decreased significantly by cannabis cultivation. Plant converts ACar fraction of heavy metals to more available forms in soil. These effects are the result of different processes occurring in rhizosphere.

20.5 Electrokinetic and Phytoextraction

Another approach to remove heavy metals, radionuclides, and organic contaminants from contaminated soil is electrodic and electrokinetic remediation. The principle of electrokinetic's operation is to use a direct electric field in soil to direct pollutants in the soil pores toward the electrodes (Wang et al. 2009). The major processes of electrokinetic remediation are electromigration and electroosmosis (Kim et al. 2002). Electromigration is defined as a directional movement in which ions run toward the electrode with opposite charge (Zhou et al. 2004, 2005). Therefore, metal ions and complexes with positive charges can move under electrokinetic treatments and be absorbed easily by plants cultured near cathode electrodes (Cang et al. 2011). The other process caused by applied electrical potential is known as electroosmosis, i.e., the movement of an ionic liquid under the action of an applied electric field through semipermeable membrane relative to a charged surface (Giannis et al. 2007). In recent years, the combination of electrokinetic remediation and phytoremediation to decontaminate metal-polluted soil has attracted many researches (Lim et al. 2004; Zhou et al. 2007; Cang et al. 2011; Bi et al. 2011).

In 2012, Tahmasbian Ghahfarokhi used a combined method to improve the phytoremediation efficiency of a contaminated calcareous soil. He employed DC electric fields in cooperation with chelating agents to move the cationic pollutant toward plant roots and subsequently into plant organs. In this study sunflower was planted in the center of plastic pots near the central cathodic electrode. Four graphite electrodes were placed around the pots as anodes. Thirty days after germination, the plants were treated by cow manure extract (CME), poultry manure extract (PME), and Na_2EDTA as the chelating agent (2 g kg^{-1}). Electrical treatment was applied in the last 2 weeks for an hour per day (0, 10, and 30 V). After 60 days from germination, at the flowering stage, the plants were carefully harvested for shoot and root analysis. The addition of the EDTA decreased the shoot and root dry weight, compared to the control, while PME increased them significantly. Root dry weight decreased significantly in the 30 V treatment. Metal concentration in shoot and root organs in the EDTA treatment was higher than that in the control, while metal uptake in the EDTA treatment is lower due to lower plant growth. Poultry extract increased the uptake index of metals significantly. In this experiment electric field had no significant effect on metal concentration and uptake index (Tahmasbian Ghahfarokhi 2012). This negative result may be due to the hydrolysis reaction of water during the electrokinetic process and production of hydroxyl ion at the cathode reducing metal bioavailability in calcareous soil. Although hydrogen ion may be produced at the anode (Kim et al. 2009), the buffering capacity of calcareous soil is so high that it prevents the suitable effects of that process. However, an acidic solution can be used to improve heavy metal bioavailability, movement, and the removal efficiency of metal from soils (Baek et al. 2009).

In another experiment the plants were charged as a cathode instead of the central cathodic electrode (Fig. 20.9). The other parts of the second experiment were

Fig. 20.9 (**a**) Diagram of the soil remediation system, (**b**) graphite electrodes placed around the pots as anodes, (**c**) stainless steel needle inserted in the lowest part of sunflower stem for charging the plant with negative charge (Adopted from Tahmasbian and Safari Sinegani 2014)

similar to those of the first experiment. In this experiment the application of electric field had also no significant impacts on the shoots and roots dry weight. But lead concentration and uptake index increased in the 10 V-EDTA treatment more than five times compared to that in the control. Cadmium uptake was enhanced by the electric field without application of chelating agent significantly. In contrast, electric field had no significant impact on Zn uptake by plants (Tahmasbian and Safari Sinegani 2014).

The obtained results can be explained by the fact that direct electric fields is proved to make biomembranes more permeable without damaging membrane structures (Rosenheck and Neumann 1972, 1973; Kinosita and Tsong 1997). The induced permeability results in a transient exchange of metals across the perturbed membrane structure (Rosenheck et al. 1975; Lindner et al. 1977) which in turn leads to enhanced uptake of metals. Furthermore, the arrangement of anodes around the pots makes the cationic ions move toward the plant roots which play the role of cathodes during the second experiment. These results might be strong reason to use this new technique to ameliorate the effects of chelating agents on phytoremediation, though the method needs improvements to be justifiable in the field.

References

Almas A, Singh BR, Salbu B (1999) Mobility of cadmium-109 and zinc-65 in soil influenced by equilibration time, temperature and organic matter. J Environ Qual 28:1742–1750

Aoyama M (1996) Fractionation of water-soluble organic substances formed during plant residue decomposition and high performance size exclusion chromatography of the fractions. Soil Sci Plant Nutr 42:31–40

Backes CA, McLaren RG, Rate AW, Swift RS (1995) Kinetics of cadmium and cobalt desorption from iron and manganese oxides. Soil Sci Soc Am J 59:778–785

Baek K, Kim DH, Park SW, Ryu BG, Batjargal T, Yang JS (2009) Electrolyte conditioning-enhanced electrokinetic remediation of arsenic-contaminated mine tailing. J Hazard Mater 161:457–462

Barber SA, Lee RB (1974) The effect of micro-organisms on the absorption of manganese by plants. New Phytol 73:97–106

Bar-Ness E, Chen Y, Hadar Y, Marschner H, Romheld V (1991) Siderophores of Pseudomonas putida as an iron source for dicot and monocot plants. Plant Soil 130:231–241

Bar-Ness E, Hadar Y, Chen Y, Romheld V, Marschner H (1992) Short-term effects of rhizosphere microorganisms on Fe uptake from microbial siderophores by maize and oat. Plant Physiol 100:451–456

Basta NT, Grand Whoil R (2000) Estimation of Cd, Zn, and Pb bioavailability in smelter-contaminated soils by a sequential extraction procedure. Soil Contam 9:149–194

Bergkvist B (1986) Leaching of metals from a spruce forest soil as influenced by experimental acidification. Water Air Soil Pollut 31:901–916

Bi R, Schlaak M, Siefert E, Lord R, Connolly H (2011) Influence of electrical fields (AC and DC) on phytoremediation of metal polluted soils with rapeseed (Brassica napus) and tobacco (Nicotianatabacum). Chemosphere 83:318–326

Blaylock MJ, Michael J, Salt DE, David ED, Duchencove S, Burt D, Raskin Y (1997a) Indian mustard as hyperaccumulated in soil. Environ Sci Technol 31:866–871

Blaylock MJ, Salt DE, Dushenkov S, Zakhrova O, Gussman C, Kapulnik Y, Ensley BD, Raskin I (1997b) Enhanced accumulation of Pb in Indian mustard by soil-applied chelating agents. Environ Sci Technol 13:860–865

Blaylock MJ (2000) Field demonstration of phytoextraction of lead contaminated soil and water. Lewis Publ, Boca Raton, FL, pp 1–12

Blaylock MJ, Huang JW (2000) Phytoextraction of metals. In: Raskin I, Ensley BD (eds) Phytoremediation of toxic metals: using plants to clean up the environment. Wiley, Toronto, p 303

Bolt GH, Bruggenvert MGM (1976) Soil chemistry: a basic elements. Elsevier, Amsterdam

Bril J, Salomons W (1990) Chemical composition of animal manure: a modeling approach. Neth J Agric Sci 38:333–351

Brown SL, Henry CL, Chaney R, Compton H, DeVolder PS (2003) Using municipal biosolids in combination with other residuals to restore metal-contaminated mining areas. Plant Soil 249:203–215

Cang L, Wang QY, Zhou DM, Xu H (2011) Effects of electrokinetic-assisted phytoremediation of a multiple-metal contaminated soil on soil metal bioavailability and uptake by Indian mustard. Sep Purif Technol 79:246–253

Catlett KM, Heil DM, Lindsay WL (2002) Soil chemical properties controlling Zn2+ activity in 18 Colorado soils. Soil Sci Soc Am J 66:1182–1189

Chefetz B, Hatcher PG, Hadar Y, Chen Y (1996) Chemical and biological characterization of organic matter during composting of municipal solid waste. J Environ Qual 25:776–785

Chen H, Cutright T (2001) EDTA and HEDTA effects on Cd, Cr, and Ni uptake by Helianthus annuus. Chemosphere 45:21–28

Cooper EM, Sims JT, Cunningham SD, Huang JW, Berti WR (1999) Chelate-assisted phytoextraction of lead from contaminated soils. J Environ Qual 28:1709–1719

Corie C, Panaviene Z, Loulergue C, Dellaporta SL, Briat JF, Walker EL (2001) Maize yellow stripe1 encodes a membrane protein directly involved in Fe(III) uptake. Nature 409:346–349

Cuerinot ML, Yi Y (1994) Iron: nutritious, noxious, and not readily available. Plant Physiol 104:815–820

Cunningham SC, Berti WR, Huang JW (1995) Phytoextraction of contaminated soils. TIBTECH 13:393–397

Del-Castilho P, Chardon WJ, Salomons W (1993) Influence of cattle-Manure slurry application on the solubility of cadmium, copper and zinc in a manure acidic, loamy-sand soil. J Eviron Qual 22:689–697

Duo LA, Gao YB, Zhao SL (2005) Heavy metal accumulation and ecological responses of turfgrass to rubbish compost with EDTA addition. J Integr Plant Biol 47:1047–1054
Elliott HA, Brown GA (1989) Comparative evaluation of NTA and EDTA for extractive decontamination of Pb-polluted soils. Water Air Soil Pollut 45:361–369
Giannis A, Gidarakos E, Skouta A (2007) Application of sodium dodecyl sulfate and humic acid as surfactants on electrokinetic remediation of cadmium-contaminated soil. Desalination 211:249–260
Gleba D, Borisjuk NV, Borisjuk LG, Kneer R, Poulev A, Skarzhinskaya M, Duskenkov S, Logendra S, Gleba YY, Raskin I (1999) Use of plant root for phytoextraction and molecular farming. Proc Natl Acad Sci 96:5973–5977
Glover LJ, Mathew H, Eick J, Brady PV (2002) Desorption kinetics of cadmium2+ and lead2+ from goethite: influence of time and organic acids. Soil Sci Soc Am J 66:797–804
Grcman H, Vodnik D, Velinkonja-Bolta Š, Leštan D (2003) Ethylenediaminedissuccinate as a new chelate for environmentally safe enhanced lead phytoextraction. J Environ Qual 32:500–506
Haag-Kerwer A, Schafer HJ, Heiss S, Walter C, Rausch T (1999) Cadmium exposure in Brassica juncea causes a decline in transpiration rate and leaf expansion without effect on photosynthesis. J Exp Bot 50:1827–1835
Hamon RE, Lorenz SE, Holm PE, Christensen TH, McGrath SP (1995) Changes in trace metal species and other components of the rhizosphere during growth of radish. Plant Cell Environ 18:749–756
Harter RD (1979) Adsorption of copper and lead by Ap and B2 horizons of several northeastern United States soils. Soil Sci Soc Am J 43:679–683
Harter RD (1983) Effect of soil pH on adsorption of lead, copper, zinc, and nickel. Soil Sci Soc Am J 47:47–51
Hashimoto Y, Matsufuru H, Sato T (2008) Attenuation of lead leachability in shooting range soils using poultry waste amendments in combination with indigenous plant species. Chemosphere 73:643–649
Hesterberg D, Bril J, Del Castilho P (1993) Thermodynamic modeling of zinc, cadmium and copper solubility in manure, acidic loamy-sand topsoil. J Environ Qual 22:1100–1106
Hider RC, Kong X (2010) Chemistry and biology of siderophores. Natur Product Rep 27:637–657
Hornick SB (1983) The interaction of soils with waste constituents. In: Parr JF, Marsh B, Kla JM (eds) Land treatment of hazardous waste. Noyes Data Corporation, Park Ridge, NJ, pp 4–19
Huang JW, Chen JJ, Berti WR, Cunningham SD (1997) Phytoextraction of lead-contaminated soils: role of synthetic chelates in lead phytoextraction. Environ Sci Technol 31:800–805
Jackson BP, Miller WP (2000) Soil solution chemistry of Soil Sci Soc. Fly ash poultry litter-and sewage sludge-amended soil. J Environ Qual 29:430–436
Japenga J, Harmsen K (1990) Determination of mass balances and ionic balances in animal manures. Neth J Agric Sci 38:353–367
Jurkevitch E, Hadar Y, Chen Y, Chino M, Mori S (1993) Indirect utilization of the phytosiderophore mugineic acid as an iron source to rhizosphere fluorescent Pseudomonas. BioMetals 6:119–123
Kim DH, Jeon CS, Baek K, Ko SW, Yang JS (2009) Electrokinetic remediation of fluorine-contaminated soil: conditioning of anolyte. J Hazard Mater 161:565–569
Kim SO, Kim KW, Stuben D (2002) Stueben, evaluation of electrokinetic removal of heavy metals from tailing soils'. J Environ Eng-ASCE 128:705–715
Kinosita KJR, Tsong TY (1997) Voltage-induced pore formation and hemolysis of human erythrocytes. Biochim Biophys Acta 471:227–242
Kirchner S (1957) Barium (ethylenediaminetetracetato) cobalt(III) 4-hydrate. Inorgan Synth 5:186–188
Knepper TP (2003) Synthetic chelating agents and compounds exhibiting complexing properties in the aquatic environment. Trend Anal Chem 22(10):725–733

Komárek M, Tlustoš P, Száková J, Chrastný V, Ettler V (2007) The use of maize and poplar in chelant-enhanced phytoextraction of lead from contaminated agricultural soils. Chemosphere 67:640–651

Kulli B, Balmer M, Krebs R, Lothenbach B, Geiger G, Schulin R (1999) The influence of nitrilotriacetate on heavy metal uptake of lettuce and ryegrass. J Environ Qual 28:1699–1705

Lai HY, Chen ZS (2004) Effects of EDTA on solubility of cadmium, zinc, and lead and their uptake by rainbow pink and vetiver grass. Chemosphere 55:421–430

Lesage E, Meers E, Vervaeke P, Lamsal S, Hopgood M, Tack F, Verlov M (2005) Enhanced phytoextraction: II. Effect of EDTA and citric acid on heavy metal uptake by Helianthus annuus from a calcareous soil. Int J Phytoremediation 7:143–152

Lim JM, Salido AL, Butcher DJ (2004) Phytoremediation of lead using Indian mustard (Brassica juncea) with EDTA and electrodics. Microchem J 76:3–9

Lindner P, Neumann E, Rosenheck K (1977) Kinetics of permeability changes induced by electric impulses in chromaffin granules. J Membr Biol 32:231–254

Linehan DJ, Shephard H (1979) A comparative study of the effects of natural and synthetic ligands on iron uptake by plants. Plant Soil 52:281–289

Liphadzi MS, Kirkham MB (2006) Availability and plant uptake of heavy metals in EDTA-assisted phytoremediation of soil and composted biosolids. S Afr J Bot 72:391–397

Lombi E, Zhao FJ, Dunham SJ, McGrath SP (2001) Phytoextraction of heavy metal-contaminated soils: natural hyperaccumulation versus chemically enhanced phytoextraction. J Environ Qual 30:1919–1926

Lu Y, Zhu F, Chen J, Gan H, Guo Y (2005) Chemical fraction of heavy metals in urban soils of Guangzhou, China. Environ Monit Assess 134:429–439

Martell AE, Calvin M (1958) The chemistry of metal-chelates (in German). VerlagChemie, Weinheim, Germany

Martin MJ, Camazano MS (1993) Adsorption and mobility of cadmium in natural, uncultivated soils. J Environ Qual 22:732–742

Masuda H, Suzuki M, Morikawa KC, Kobayashi T, Nakanishi H, Takahashi M, Saigusa M, Mori S, Nishizava NK (2008) Increase in iron and zinc in rice grains via the introduction of barley genes involved in phytosiderophore synthesis. Rice 1:100–108

McLean JE, Bledsoe BE (1992) Behavior of metals in soils. In: Walter W, Kovalick Jr, (ed.), Ground water issue. EPA/540/S-92/018. Technology Innovation Office. Office of Solid Waste and Emergency Response, US EPA, Washington, DC

Meers E, Hopgood M, Lesage E, Vervaeke P, Tack FMG, Verloo MG (2004) Enhanced phytoextraction: in search of EDTA alternatives. Int J Phytoremediat 6:95–109

Meers E, Tack FM, Van Slycken S, Ruttens A, Du Laing G, Vangronsveld J, Verloo MG (2008) Chemically assisted phytoextraction: a review of potential soil amendments for increasing plant uptake of heavy metals. Inter J Phytoex 10:390–414

Mori S (1998) Iron transport in graminaceous plants. In: Sigel A, Sigel H (eds) Metal ions in biological systems, vol 35. Marcel Dekker Inc., New York, pp 215–238

Mori S (1999) Iron acquisition by plants. Curr Opin Plant Biol 2:250–253

Nascimento CWA, Amarasiriwardena D, Xing B (2006) Comparison of natural organic acids and synthetic chelates at enhancing phytoextraction of metals from a multi-metal contaminated soil. Environ Pollut 140:114–123

Neumann E, Rosenheck K (1973) Potential difference across vesicular membranes. J Membr Biol 14:194–196

Rosenheck E, Neumann K (1972) Permeability changes induced by electric impulses in vesicular membranes. J Membr Biol 10:279–290

Nowack B, Schulin R, Robinson BH (2006) A critical assessment of chelant-enhanced metal phytoextraction. Environ Sci Technol 40:5225–5232

Prabhu V, Biolchini PF, Boyer GL (1996) Detection and identification of ferricrocin produced by ectendomycorrhizal fungi in the genus Wilcoxina. BioMetals 9:229–234

Quiroz A, Garcia FE, Ilangovan K (2002) Effects of natural hydrosoluble chelates of three plant species on the mobilization of heavy metals. Bull Environ Contam Toxicol 68:862–869

Romheld V (1991) The role of phytosiderophores in acquisition of iron and other micronutrients in graminaceous species: an ecological approach. In: Chen Y, Hadar Y (eds) Iron Nutrition and Interactions in plants. Kluwer Academic Publishers, Iloston, MA, pp 159–166

Romkens P, Bouwman L, Japenga J, Draaisma C (2002) Potentials and drawbacks of chelate-enhanced phytoextraction of soils. Environ Pollut 116:109–121

Rosenheck K, Lindner P, Pecht I (1975) Effect of electric fields on light-scattering and fluorescence of chromaffin granules. J Membr Biol 20(1–2):1–12

Sacher F, Lochow E, Brauch HJ (1998) Synthetic organic complexing agents-analysis and occurrence in surface waters. Vom Wasser 90:31–41

Safari Sinegani AA, Ahmadi P (2012) Manure application and cannabis cultivation influence on speciation of lead and cadmium by selective sequential extraction. Soil Sediment Contam 21 (3):305–321

Safari Sinegani AA, Khalilikhah F (2008) Phytoextraction of lead by *Helianthus annuus*: effect of mobilizing agent application time. Plant Soil Environ 10:434–440

Safari Sinegani AA, Khalilikhah F (2010) Effects of EDTA, sheep manure extract and their application time on Cd uptake by *Helianthus annuus* from a calcareous mine soil. Soil Sediment Contam 19(3):378–390

Safari Sinegani AA, Khalilikhah F (2011) The effect of application time of mobilising agents on growth and phytoextraction of lead by Brassica napus from a calcareous mine soil. Environ Chem Lett 9:259–265

Safari Sinegani AA, Mirahamdi Araki H (2009) Effects of sterilization and temperature on the decrease kinetic of lead bioavailability in two different soil groups. Soil Sediment Contam 18 (6):739–753

Safari Sinegani AA, Mirahamdi Araki H (2010a) Changes in chemical forms of lead in temperate and semiarid soils in sterile and unsterile conditions. Environ Chem Lett 8(4):323–330

Safari Sinegani AA, Mirahamdi Araki H (2010b) The effects of soil properties and temperature on the adsorption isotherms of lead on some temperate and semiarid surface soils of Iran. Environ Chem Lett 8(2):129–137

Saifullah E, Meers M, Qadir P, de Caritat FMG, Tack G, Du Laing G, Zia MH (2009) EDTA-assisted Pb phytoextraction: a review. Chemosphere 74:1279–1291

Salt DE, Smith RD, Raskin I (1998) Phytoextraction. Ann Rev Plant Physiol Plant Mol Biol 49:643–668

Sauvé S, Martnez CE, McBride M, Hendershot W (2000) Adsorption of free lead Pb^{2+} by pedogenic oxides, ferrihydrite, and leaf compost. Soil Sci Soc Am J 64:595–599

Schmidt U (2003) Enhancing phytoextraction: the effect of chemical soil manipulation on mobility, plant accumulation, and leaching of heavy metals. J Environ Qual 32:1939–1954

Schnitzer M, Desjardins JG (1969) Chemical characteristics of a natural soil leachate from a humic podzol. Can J Soil Sci 49:151–158

Shen ZG, Li XD, Wang CC, Chen HM, Chua H (2002) Lead phytoextraction from a contaminated soil with high-biomass plant species. J Environ Qual 31:1893–1900

Sposito G, Lund J, Change AC (1982) Trace metal chemistry in arid zone field soils amended with sewage sludge. I. Fractionation of Ni, Cu, and Pb in solid phases. Soil Sci Soc Am J 46:260–264

Tahmasbian Ghahfarokhi I (2012) The effect of application of some chelating agents and electrical field on phytoremediation of a heavy metals polluted soil. MSc thesis. Bu-Ali Sina University, Hamedan, Iran, MSc, p 150, In Persian

Tahmasbian I, Safari Sinegani AA (2013) Monitoring the effects of chelating agents and electrical fields on active forms of Pb and Zn in contaminated soil. Environ Monit Assess 185:8847–8860

Tahmasbian Ghahfarokhi I, Safari Sinegani AA (2013) The effects of some chelators and electrical fields on phytoremediation efficiency of a Pb polluted soil. Casp J Appl Sci Res 2(9):42–48

Tahmasbian I, Safari Sinegani AA (2014) Chelate-assisted phytoextraction of cadmium from a mine soil by negatively charged sunflower. Int J Environ Sci Technol 11:695–702

Tsang DCW, Lo IMC, Surampalli RY (2012) Chelating agents for land decontamination technology. Am Soc Civil Eng. ISBN 978-0-7844-7683-3. p 284

Turgut C, Pepe MK, Cutright TJ (2004) The effect of EDTA and citric acid on phytoremediation of Pb, Cr, and Ni from soil using Helianthus annuus. Environ Pollut 131:147–154

Van Engelen DL, Rachel E, Sharpe-Pedler C, Kevin KM (2007) Effect of chelating agents and solubility of cadmium complexes on uptake from soil by Brassica juncea. Chemosphere 68:401–408

Vangronsveld J, Colpaert JV, van Tichelen KK (1996) Reclamation of a bare industrial area contaminated by non-ferrous metals: physicochemical and biological evaluation of the durability of soil treatment and revegetation. Environ Pollut 94:131–140

Vassil AD, Kapulnik Y, Raskin I, Salt DE (1998) The role of EDTA in lead transport and accumulation in Indian mustard. Plant Physiol 117:447–453

Wallace A, Muller RJ, Cha JW, Alexander GV (1974) Soil pH excess lime and chelating agent on micronutrients in soybeans and bush beans. Agron J 66:698–700

Ward MK, Bitton G, Townsend T (2005) Heavy metal binding capacity (HMBC) of municipal solid waste landfill leachates. Chemosphere 60:206–215

Wang QY, Zhou DM, Chang L, Sun TR (2009) Application of bioassays to evaluate a copper contaminated soil before and after a pilot-scale electrokinetic remediation. Environ Pollut 157:410–416

Winkelmann G (2007) Ecology of siderophores with special reference to the fungi. BioMetals 20:379–392

Wu J, Hsu FC, Cunningham SD (1999) Chelate-assisted Pb phytoextraction: Pb availability, uptake, and translocation constraints. Environ Sci Technol 33:1898–1904

Xian X (1987) Chemical partitioning of cadmium, zinc, lead and copper in soils near smelter. J Environ Sci Health 6:527–541

Zhou LX, Wong JWC (2001) Effect of dissolved organic matter from sludge and sludge compost on soil copper sorption. J Environ Qual 30:878–883

Zhou DM, Deng CF, Cang L (2004) Electrokinetic remediation of a Cu contaminated red soil by conditioning catholyte pH with different enhancing chemical reagents. Chemosphere 56:265–273

Zhou DM, Deng CF, Akram LC, Alshawabkeh N (2005) Electrokinetic remediation of a Cu–Zn contaminated red soil by controlling the voltage and conditioning catholyte pH. Chemosphere 61:519–527

Zhou DM, Chen HF, Cang L, Wang YJ (2007) Ryegrass uptake of soil Cu/Zn induced by EDTA/EDDS together with a vertical direct-current electrical field. Chemosphere 67:1671–1676

Chapter 21
Remediation of Toxic Metal-Contaminated Soil Using EDTA Soil Washing

Domen Lestan

21.1 Introduction

When experts, politicians and stakeholders discuss global challenges such as achieving food security, they often fail to consider one of the most vital resources of all: soils. And no wonder the multiple functions of soil are not exactly obvious at first glance. However, an estimated 24 billion tons of soil are lost to degradation processes every year, and it takes hundreds of years for a few cm of lost soil to renew itself (the formation of a 2.5 cm layer of fertile soil takes approx. 500 years on agriculturally used land). While the area of arable land suitable for plant production is reducing, at the same time, the global demands for food, livestock feed and energy are increasing (Nonhebel and Kastner 2011). Global governance of soil resources (Montanarella and Vargas 2012) and revitalisations and remediation of degraded areas seem to be the necessary action for global sustainable development.

Land contamination exists in most industrialised countries and is a growing problem in the emerging economies. In China, for example, the land area degraded by mining activities had reached about 3.2 million ha by the end of 2004, and the trend is for this to increase at a rate of 46,700 ha year^{-1} (Li 2006). The European Environment Agency (EEA) estimates that there are some 3.5 million contaminated sites across Europe (EEA 2010). The largest and most affected areas are located in north-west Europe, from the north of France to Germany, across Belgium and the Netherlands, since these are regions with a high density of urban agglomerations and intensive industrial activities (ESBNEC 2005). Other regions include the north of Italy and, of the new Member States, Poland and the Czech and Slovak Republics. Most EU Member States have made the cleanup and restoration of the land a priority.

D. Lestan (✉)
Agronomy Department, Biotechnical Faculty, University of Ljubljana, Ljubljana, Slovenia
e-mail: domen.lestan@bf.uni-lj.si

I. Sherameti, A. Varma (eds.), *Heavy Metal Contamination of Soils*, Soil Biology 44, DOI 10.1007/978-3-319-14526-6_21

Many urbanised areas formed from an industrial nucleus and have inherited a long industrial history and an associated legacy of soil contamination (Wong and Li 2006). Even now the continuous growth of cities and suburbs integrates contaminated industrial areas, previously situated in the city limits, with usually low quality of public space and non-existing social cohesion. Living in such contaminated areas necessitates being in contact with contaminated soils in everyday life through soil dust inhalation, accidental mouth ingestion and consumption of foods grown in contaminated areas. The consequences of living with harmful substances in our living environment are not easy to comprehend, because they take an extended period to manifest themselves. Toxic metals (Pb, Zn, Cd, Ni, Cr, Cu and As as semimetal) are one of the major causes of concern as they are persistent in the soil and difficult to remove. Lead and other toxic metals accumulate continuously in the body with time. Even low levels of Pb are demonstrably toxic to the brain and mental development in children. There are an estimated 240,000 contaminated sites in western central and south-eastern Europe which are hazardous for humans and in need of immediate remedial treatment. In almost 40 % of these sites, toxic metals are the most important contaminants (EEA 2007). In the United States, toxic metals are present in 77 % of the Superfund sites (National Priorities List), in 72 % of the Department of Defence sites and in 55 % of the Department of Energy sites (US EPA 2004).

The proper remediation and management of soils contaminated with toxic metals is a difficult and costly issue. Historically, excavation to landfill (dig and dump) has been the solution, offering a quick removal mechanism of a pollution source in the soil. In Europe, the disposal of contaminated soil is now heavily restricted since the EU applies European Council directive 1999/31/EC (Council Directive 1999). Other treatments are needed to meet the need for remediation. The European Commission (EC) released a paper (EC 2011) where the lack of reports and evaluations of soil remediation technologies is addressed. This deficiency in soil remediation technologies is one of the reasons why the EC did not adopt the Soil Framework Directive (COM 2006). This might be attributed to scattered information regarding the application and performance of innovative technologies, lack of knowledge and lack of confidence.

Ex situ method of soil washing (extraction) with ethylenediaminetetraacetate (EDTA) chelating agent promises limited changes in soil properties compared to other potential extractants, such as strong mineral acids (Mulligan et al. 2001), and high metal removal efficiency especially from bioavailable and labile soil fractions (toxic metal bioavailability stripping). Even though EDTA-remediated soil effectively lowers the human health risk, there are several unresolved uncertainties associated with the method. For example, could soil remediated in such a way—soil deprived of micronutrients removed collaterally with toxic metals and most of the microbial life depressed in stringent conditions of soil extraction—be suitable habitat for soil fauna, plant cultivation or even agricultural use? What measures, if any, are needed for revitalisation of remediated soil to its inherent potentials and fertility? Can EDTA soil washing nevertheless qualify as sustainable method, focused on minimising the environmental impacts (emissions) of remediation activities and sustainable use of remediated soil? When returned to the site and exposed to environmental factors, is EDTA-remediated soil stable? Does ageing process shift residual toxic metals into more labile fractions, thus diminishing the effect of remediation? These questions and problems must be thoroughly addressed before remediation technology is applied in a wider scale.

This chapter appraises works of author's research group on development of efficient EDTA-based soil remediation technology and application of an integrated approach using pedological, chemical and physical analyses with biological tests and with plant experiments for assessing the quality, toxicity and functioning of remediated soil. In these works in most cases, Pb-, Zn- and Cd-contaminated soils from vegetable gardens in the Meža Valley, Slovenia, were used as model soils. Main findings of these studies are compared, explained and complemented with works of other researches.

21.2 EDTA-Based Soil Washing

Soil washing with an aqueous solution of EDTA is considered to be a remedial option with a potentially low impact on soil quality (Lestan et al. 2008). EDTA is a low-toxicity chemical produced commercially; it is cytotoxic and weakly genotoxic, but not carcinogenic. In animal studies, the lowest EDTA dose reported to cause toxicity was 150 mg kg^{-1} day^{-1} (Lanigan and Yamarik 2002). EDTA is today widely used in detergents, soil fertilisers (Fe- and Mn-EDTA) and personal-care products (typical concentration of use of EDTA in cosmetic formulations is less than 2 %) and even as a food additive as antioxidant to prevent metal ions that have entered from metallic processing equipment from causing rancidity.

In soil remediation, EDTA forms coordinate bonds with metals and facilitates their solubilisation from soil solid phases into the washing solution, which is then separated. An analogue to soil washing is human chelation therapy, in which EDTA is used for stripping Pb and other toxic metals from the human body (Kalia and Flora 2005). EDTA has capacity to chelate almost every positive ion in the periodic table. However, divalent and trivalent metals can be EDTA extracted from soil because the reported ordering of EDTA complex stability constants—$Na < Mg < Ca < Fe < Al < Zn < Cd < Pb < Ni < Cu < Hg$ (Martell and Smith 2003)—favours the removal of metal contaminants over natural hardness ions (Almaroai et al. 2013). Mobilisation of metals in soil by the formation of water-soluble toxic EDTA chelate (Karim and Khan 2012) poses a threat to the environment due to leaching into the subsurface (i.e. groundwater), and soil must be thoroughly rinsed to remove all mobilised toxic metal species and prevent emissions before remediated soil is returned to the site. Efficient soil rinsing requires use of large amounts of clean water.

Although EDTA is not particularly expensive (1.5 Euro kg^{-1} of industrial-grade chemical, Internet source http://www.alibaba.com, average price of five sellers), the cost of chelant use for soil remediation can be significant since low EDTA concentrations often do not extract toxic metals from soil effectively, especially from calcareous soils (Zhang et al. 2010b; Demir and Koleli 2013). Consequently, much research effort was directed towards developing effective EDTA recycling methods.

21.2.1 EDTA and Process Water Recycling

Demonstrated on a laboratory level, but not available commercially, there have been several proposals of how to recycle spent EDTA from the used washing solution. Toxic metals can be separated from EDTA with Na_2S under alkaline conditions, resulting in almost complete recovery of metals through precipitation in the form of insoluble metal sulphides (Zeng et al. 2005). Zero-valent bimetallic mixtures (Mg^0–Pd^0, Mg^0–Ag^0) can also be used to precipitate toxic metals from the solution while liberating EDTA in alkaline pH (Ager and Marshall 2001). Electrolytic recovery of toxic metals and EDTA from washing solution in a two-chamber electrolytic cell separated with a cation exchange membrane to prevent EDTA anodic oxidation has been reported (Juang and Wang 2000). In another electrochemical process, toxic metals and EDTA were separated in an electrolytic cell under alkaline conditions using a sacrificial Al anode. Al substitutes toxic metals in complex with EDTA, and the released metals are removed by electro-precipitation/coagulation (Pociecha and Lestan 2010). EDTA can also be recycled by substituting toxic metals with Fe^{3+} under acidic conditions, followed by precipitation of the released metals with phosphate at near-neutral pH. Fe^{3+} ions are then precipitated as hydroxides at high pH using NaOH, thus liberating the EDTA (Di Palma et al. 2003).

In addition to EDTA recycling, the generation of large amounts of waste water after soil washing, which needs treatment before safe disposal, has long remained an unsolved problem. We recently proposed and laboratory tested a method for both EDTA and clean process water recycling from waste soil washing solution. EDTA was recovered by acid precipitation coupled with initial alkaline trans-complexation and precipitation of toxic metals. The remaining EDTA and metals in the waste solution were removed using an electrochemical advanced oxidation process (EAOP) to yield clean process water (Pociecha and Lestan 2012a, b, c; Voglar and Lestan 2013). The recycling method relies on a combination of substitution/precipitation reactions in an imposed pH gradient for EDTA recovery and on electrochemical polishing of process waters for total water recycle in three steps and in a closed loop as shown in Fig. 21.1 and explained below. By using $Ca(OH)_2$ and H_2SO_4 as the base–acid pair, the accumulation of added reagents in the process water after multiple soil remediation batches/recycles was prevented by precipitation and removal of poorly soluble $CaSO_4$ during the process.

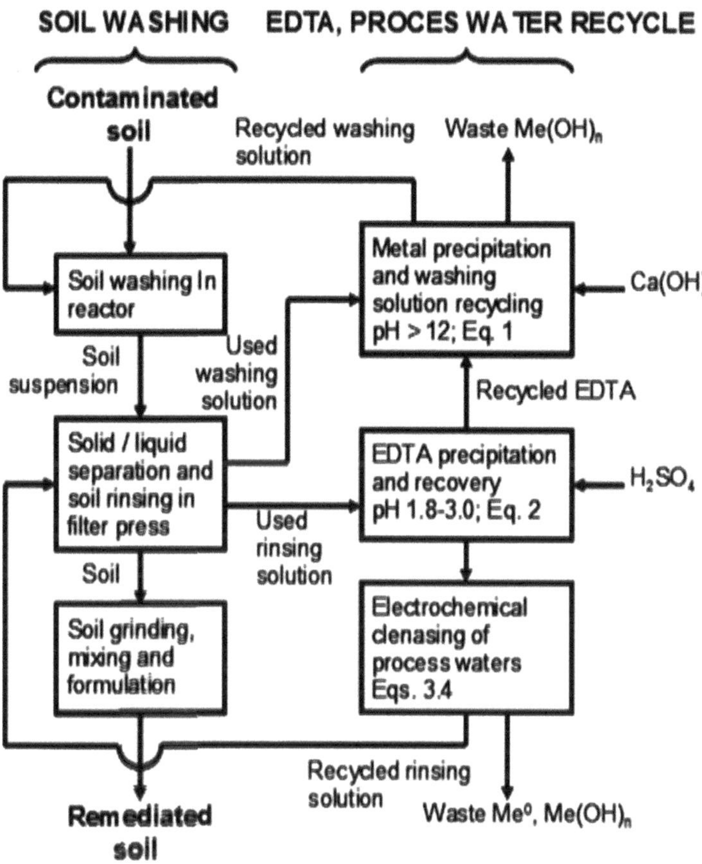

Fig. 21.1 Flowchart of the EDTA-based soil washing process and EDTA and process water recycling process. Equations (21.1)–(21.4) for process reactions are given within the text

21.2.1.1 Alkaline Metal Precipitation

At low pH, Fe interferes with acidic EDTA precipitation from the used washing solution and must be removed. Alkaline conditions make the formation of Fe-EDTA complexes less favourable (Kim et al. 2003), leading to Fe hydrolysis and $Fe(OH)_3$ precipitation (Pociecha and Lestan 2012a). Alkalisation with $Ca(OH)_2$ efficiently removed not only Fe but also Pb, Zn and Cd (Pociecha and Lestan 2012b), since Ca replaces toxic metals in the EDTA complex due to a higher stability constant of Ca-EDTA at high pH (Kim et al. 2003). However, the hydroxides of most toxic metals are amphoteric. They are increasingly soluble at both low and high pHs, and the optimum pH for precipitation depends on the metal: Pb (pH 9.5), Zn (pH 9.2) and Cd (pH 11.2) (Cushnie 2009). Since the pH must be raised to 12 and above before the stability constant for Ca-EDTA exceeds that of Pb-EDTA (Brown and Elliot 1991), part of the toxic metals remains in the solution after alkaline precipitation. Using NaOH did not remove any of the Pb, Zn or Cd from the solution due to the low affinity of Na to form complexes with EDTA.

$Ca(OH)_2$ has fewer free electrons than NaOH, making it less likely to change its structure when exposed to water, and is therefore less water soluble (0.173 g L^{-1} at 20 °C) and normally decreases with pH. However, Ca replacement of Fe and toxic metals in the EDTA complex at high pH pushes the chemical equilibriums towards $Ca(OH)_2$ dissolution. Released metals (Me) proteolytically react with waters, releasing H^+, which additionally enhances $Ca(OH)_2$ dissolution:

$$Me^{2+} + 2H_2O \rightarrow Me(OH)_2 + 2H^+ \tag{21.1}$$

The product of alkaline precipitation was process water with an active Ca-EDTA form of chelant. Fe and toxic metals substituted from the chelant precipitated as hydroxides.

21.2.1.2 Acidic Precipitation of EDTA

As mentioned above, removal of Fe in the alkaline phase prior to acidification significantly improved the yield of recovered EDTA, since the stability of Fe-EDTA complexes is relatively high in acidic conditions between pH 2 and 4 and higher than the stability of EDTA complexes with other metals. Protonated EDTA (H_4EDTA) is poorly soluble in acidic media below pH 3 (Di Palma et al. 2003) and, after acid (H_2SO_4) addition, precipitates from the solution:

$$\text{Me-EDTA} + 4H^+ \rightarrow H_4EDTA + Me^{2+} \tag{21.2}$$

Pb-EDTA complexes are increasingly less stable in more acidic conditions, leading to sharp $PbSO_4$ precipitation at $pH < 1.8$. To avoid contamination of recycled EDTA with precipitated Pb, the process water was not acidified below pH 2. In addition to EDTA, excess Ca in the process water was removed as poorly soluble $CaSO_4$. This prevented the accumulation of added reagents through consecutive soil remediation/EDTA recycling batches.

21.2.1.3 Electrochemical Cleansing of Process Waters

Several anode materials, such as graphite and noble metals, are known to successfully mediate the oxidation of organic compounds. According to the generally accepted mechanism, water is first discharged at anode active sites (M), producing adsorbed hydroxyl radicals, M(OH)ads, which are involved in the mineralisation of an organic compound (R) in aqueous solutions:

$$R_{aq} + (n/2)M(\cdot OH)_{ads} \rightarrow (n/2)M + \text{mineralisation products} + (n/2)H^+ + (n/2)e^- \quad (21.3)$$

where n is the number of electrons involved in the oxidation reaction of organics. Reaction (21.2) is in competition with the anodic discharge of these radicals to form dioxygen:

$$M(\cdot OH)_{ads} \rightarrow M + (1/2)O_2 + H^+ + e^- \quad (21.4)$$

The activity of these electrogenerated hydroxyl radicals is therefore strongly linked to their interaction with the anode surface; the weaker the interaction, the lower the electrochemical activity towards oxygen evolution (higher O_2 overpotential). Materials such as graphite, Pt and various noble metal oxides on a Ti core have sufficient oxygen overvoltage to produce reactive hydroxyl radicals. A boron-doped diamond anode has an extreme overvoltage of >3 V and has been proven to be particularly effective for oxidative degradation of EDTA (Finzgar and Lestan 2008). In relation to the EDTA degradation pathway in the electrolytic cell, EDTA is electro-oxidised through sequential removal of the acetate groups, until small-size hydrocarbon products are formed (Yamaguchi et al. 2003).

It was demonstrated in laboratory (Pociecha and Lestan 2012a, b, c) and also in a pilot plant scale (Voglar and Lestan 2014) that recycling processes (Fig. 21.1) enable efficient EDTA (up to 90 % of used chelant per batch) and complete process water recycle, supporting the feasibility of EDTA-based washing technology for soils contaminated with Pb, Zn and Cd. No waste water was generated, and solid wastes were efficiently bitumen stabilised before disposal. Stabilisation with bitumen efficiently prevented leaching of toxic metals and enables solid wastes disposal as non-hazardous material. The technology embodiment in a pilot-scale remediation plant (Voglar and Lestan 2014) comprised conventional process equipment: a mixer for soil extraction, screen for soil/gravel separation, filter chamber presses for soil/liquid and recycled EDTA separation and soil rinsing, continuous centrifuge separator for removal of precipitated metals and electrolytic cells for process water cleansing. The process removed up to 72, 25 and 66 % of Pb, Zn and Cd from garden soil from Meža Valley, Slovenia, contaminated with up to 6,960, 3,797 and 32.6 mg kg^{-1} of Pb, Zn and Cd, respectively, in nine 60 kg soil batches. Concentration of Pb and Zn remaining in the remediated soil and bioaccessible from the simulated human intestinal phase (unified bioaccessibility method, UBM, according to Wragg et al. 2011) was reduced by 97 and 96 % and was for Cd brought under the level of quantification of atomic absorption spectrophotometer. The novel embodiment of remediation technology is cost-effective; the material and energy costs of remediation amounted to 50.5 Euro ton^{-1} soil (Voglar and Lestan 2014).

21.3 Contamination and Remediation of Soils in Meža Valley, Slovenia

Soil contamination with Pb is a worldwide health problem. In their early study, Duggan and Inskip (1985) reported that children who resided in areas where soil contamination exceeded 500–1,000 mg Pb kg^{-1} often had increased blood Pb levels. One of the most affected locations in Slovenia is the Meža Valley, historically contaminated from Pb mining and smelting, which commenced in 1665 and continued until 1994. The Meža Valley encompasses approximately 50 km^2 in northern part of Slovenia and includes abandoned Pb mining and smelting industrial complex and numerous confined and unconfined deposits. During the long history of industrial activities in the Meža Valley, the soil has been the depository and integrator of released Pb and also of co-contaminating Zn (and to a lesser content Cd), and it still remains the most significant source of secondary contamination. Specific topology resulted in metal contamination of soils across all region. The site is home to more than 6,000 people in three main residential communities; cities of Mezica, Crna and Zerjav were developed around historically known locations of past Pb smelters.

The soil in the Meža Valley is pedologically diverse, composed of eutric fluvisol with patches of dystric leptisol and skeletal dystric cambisol. The predominate soil type is calcareous (developed on dolomite bedrock), rich in carbonate and with soil pH close to neutral. Soil capping has been applied on limited areas in the Meža Valley, but this is not a viable remedial option for small individually owned plots of land, which predominates. Several other remedial approaches to removing toxic metals from Meža soils have also failed: phytoextraction and chemically induced phytoextraction with various plants and soil extraction with mineral acid were not efficient. The most efficient plant *Cannabis sativa* phytoextracted only 0.6 % of the total soil Pb in a single growth cycle (Kos et al. 2003). Soil leaching with very high HCl concentration (4 M) did not remove more than 35 and 7 % of initial Pb and Zn and, as described below, significantly deteriorated soil quality (Udovic and Lestan 2012). Soils from the Meža Valley for the most part contain a high content of organic matter, clay and silt (Finzgar et al. 2007), so physical separation of contaminated fines from the soil bulk (Pearl et al. 2006) is also not feasible. However, soil washing using EDTA was found effective. In an early experiment with Meža's soil, using multi-step leaching with three 40 mmol EDTA doses per kg of dry soil removed 76.5 % of Pb and 17.5 % of Zn (Finzgar and Lestan 2007) despite the fact that the EDTA efficiency for extracting toxic metals from calcareous soils is generally lower than for non-calcareous samples. For example, Manouchehri et al. (2006) reported that for calcareous soil samples, the amount of Pb, Cu and Cd extracted by EDTA was reduced to 50 % of that extracted in non-calcareous soils.

Zn in soils from the Meža Valley is difficult to extract (Finzgar and Lestan 2007; Udovic and Lestan 2012). It is bound to non-labile soil fractions, mainly to the residual fraction according to Tessier's (Tessier et al. 1979) sequential extraction scheme (Lestan et al. 2005). Similar results have been reported by Dao et al. (2013). Zn in soil from a contaminated site in Galway, Ireland, was EDTA un-extractable and not bioavailable. However, Zn toxicosis is not a common medical problem, and there are no reports of human poisoning related to Zn soil contamination. Zn is in fact an essential element; its deficiency is causing dwarfism and hypogonadism in human males.

Based on successful development and scale-up of a feasible remediation technology (Pociecha and Lestan 2012a, b; Voglar and Lestan 2013, 2014) Finzgar et al. (2014) applied geostatistical technique to predicted spatial reduction of metal contamination after applying EDTA for extraction of soils in Meža Valley. The application of geostatistical techniques to describe the spatial distribution and variability of soil contaminants as a component of risk assessment has already been demonstrated in many papers (Dao et al. 2013; Wu and Zhang 2010). However, this was the first report on using geostatistics to provide an estimate of remediation outcome. Finzgar et al. (2014) collected and EDTA extracted a representative number of soil samples from 268 locations in the Meža Valley with Pb concentrations up to 8,955 mg kg^{-1}. On average, 63 % of Pb was removed with washing solution containing 60 mmol EDTA per kg of soil and 75 % with 120 mmol EDTA kg^{-1} soil. The range of soil extraction efficiencies of toxic metals was broad: some samples lost more than 90 % of Pb, which can be explained by the diversity of contamination sources. Some soils were presumably contaminated directly with ore and slag from the metallurgical process, with metals occluded into the mineral matrix and thus non-available for extraction (and thus less toxic). On the other hand, metals in soils contaminated indirectly by industrial gas and dust emissions were more likely to be removed by EDTA washing solution. Finzgar et al. (2014) used Empirical Bayesian kriging as a method of establishing a continuous prediction surface of contaminated topsoil and predicted spatial reduction of contamination after applying EDTA-based soil remediation and to map the probability of exceeding the regulatory threshold (action trigging and remediation goal) of contaminants. Spatial structure analysis revealed a good spatial structure and little spatial variation of data. The interpolation method distinguished three urban areas in the Meža Valley with increased Pb soil concentrations, well above critical value of 530 mg kg^{-1} stipulated by Slovenian environmental legislation (Fig. 21.2). These three areas correspond to historically known locations of past Pb smelters. As expected, the urban area around the central former Pb smelting plant in the town of Zerjav is the most contaminated, with Pb soil concentration up to 8,000 mg kg^{-1}. Geostatistical simulations also showed that the contaminated area covers 19.4 km^2 and that soil remediation (60 mmol EDTA kg^{-1}) has the potential to reduce the area with Pb above the critical regulatory threshold limit by 91 % to 1.8 km^2 (Fig. 21.2). Validated by pilot-scale remediation trials, soil extractions exhibit little scale dependency of extraction efficiency.

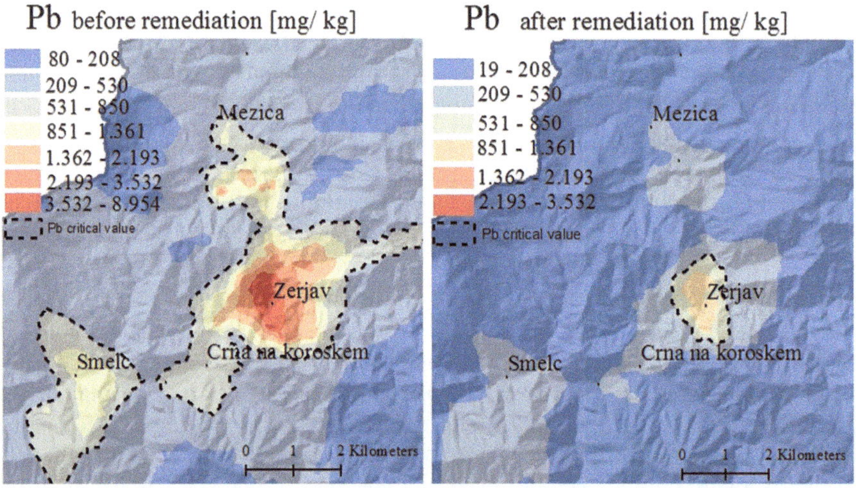

Fig. 21.2 Prediction map of Pb concentrations in soils of the Meža Valley, Slovenia, before and after remediation, with areas with critical Pb soil concentration indicated by a broken line

21.4 Bioaccessible Concentrations of Toxic Metals in Soil

Geostatistical simulations of Finzgar et al. (2014) revealed that soil extraction with EDTA to a large extent reduced the area with Pb above the critical regulatory threshold limit. However, how accessible for humans and soil fauna are toxic metals that remain in the soil after remediation? The risk posed by toxic metals depends more on their potential to be mobilised and then enter the human organism than on their total soil concentration. Numerous studies have also shown the importance of the soil reservoir of toxic metals in the cycling of toxic metals in the environment (Laidlaw et al. 2012). For Pb, the most common exposure pathways are hands-to-mouth behaviour with children, accidental soil ingestion and soil dust inhalation. Fruits and vegetables grown on contaminated soil may become contaminated as a result of plant uptake of toxic metals or direct deposition of soil dust onto plant surfaces.

Davis and Mirick (2006) have estimated soil ingestion by children to range between 37 and 207 mg day^{-1} and from 23 to 625 mg day^{-1} in adults. Metal oral bioaccessibility of ingested/inhaled soil and soil dust through the human gastrointestinal tract is therefore highly relevant for health risk assessment and management and decision making, i.e., about the need and efficiency of soil remediation (Glorennec and Declercq 2007). The unified bioaccessibility method (UBM), which is an in vitro method simulating the human digestive procedure, measures the fraction of a metal contaminant that is released from the soil matrix into solution by synthetic digestive fluids and thus represents the maximum amount of contaminant that is bioaccessible, hence available for intestinal absorption (Wragg et al. 2011). The physiologically based extraction test (PBET) is an earlier but still widely used method which also mimics condition in the human gastrointestinal tract Ruby et al. (1996). It was validated using animal models to measure bioaccessible fraction of Pb and As in soil. The absorption of nutrients (and toxic metals) takes place mainly in the small intestine (Ruby et al. 1996), so metal concentrations measured in the intestinal phase give a better evaluation of potential risks of ingestion/inhalation than the concentrations in the stomach phase.

Several simple chemical extraction tests are available to assess the availability of soil metals for plants. The diethylenetriaminepentaacetic acid (DTPA) extraction test is most often applied. DTPA test was originally designed to assess plant-accessible Zn, Fe, Mn and Cu in near-neutral and calcareous soil (Lindsay and Norvell 1978) and later adopted to assess phytoaccessibility and ecotoxicity of toxic metals (Conder et al. 2001). The toxicity characteristic leaching procedure (TCLP) is a regulatory testing procedure developed by the US EPA (1995) to determine whether solid wastes contain leachable (mobile) hazardous components. The TCLP often has been the de facto basis for evaluation of waste treatment efficacy, including soil remediation. The extractability of metals from soil with different extractants (deionised water, UBM, PBET, DTPA,TCLP and other solutions) depends on the distribution pattern of metals present in or bound to the different soil fractions (dissolved in the soil solution, exchangeable from the sesquioxides, bound to the organic matter, precipitated as sulphides, captured in the mineral aggregates). Fractionation of (toxic) metals is usually assessed by sequential extraction procedures such as Tessier's (Tessier et al. 1979) extraction scheme and different modifications of the original Tessier's sequential extractions (Lestan et al. 2003).

Finzgar et al. (2014) applied the UBM test to characterise the oral accessibility of Pb and Zn in five original (non-remediated) and remediated Meža Valley soils from the pilot-scale soil washing facility. EDTA extraction (120 mmol kg^{-1}) removed up to 85 and 65 % of total soil Pb and Zn concentration. As explained above, the absorption of toxic metals into the human body mainly takes place in the small intestine; in original soils, up to 11 and 1 % of Pb and Zn, respectively, were accessible from intestinal phase. As expected, Zn was a good deal (10 times) less bioaccessible than Pb, due to Zn association with more stable soil fractions (Lestan et al. 2005; Dao et al. 2013). After remediation, the pool of soil Pb bioaccessible from the intestinal phase was reduced by up to 99 % and the pool of bioaccessible Zn by up to 96 % indicating that EDTA first extracted toxic metals from more labile (and EDTA available) soil fractions (bioaccessibility stripping). The decrease in concentration of bioaccessible toxic metals after remediation was thus significantly higher than the reduction of the total soil Pb and Zn concentration. This presumably indicates that the EDTA soil washing reduced hazard from toxic metals to a greater extent than expected based on concentration of toxic metals which remained in soil after remediation.

Not only for humans, of vital interest are also the bioaccessibility and bioavailability of toxic metals from contaminated/remediated soil for terrestrial organisms and their potential trophic transfer. The selective chemical extractions such as UBM, PBET and DTPA cannot fully comprise the whole bioaccessible fraction, due to the dynamic and complex nature of metal–soil and metal–organism interactions. Some plants and animals, especially invertebrates, accumulate metals in proportion to the environmental concentrations and are therefore used as reliable bioindicators of metal availability in soil (Heikens et al. 2001). An example of widely used invertebrate bioindicator is terrestrial isopod *Porcellio scaber* which inhabits a wide range of soil habitats. Isopods accumulate the highest concentrations of toxic metals so far recorded in any soft tissue of terrestrial animals; the uptake from soil depends mainly on the bioavailability of metals in the ingested material (Odendaal and Reinecke 2004).

Udovic et al. (2009) extracted soil with increasing EDTA concentrations to significantly reduce Pb, Zn and Cd concentrations and the pool of metals in labile soil fractions assessed by PBET and other simple chemical extraction tests (TCLP and DTPA). When exposed to non-remediated and remediated soil, *P. scaber* accumulated Pb and Cd in a concentration-dependent manner from different soil fractions (modified Tessier's sequential extractions). The amounts of accumulated metals were, however, higher than expected on the basis of extraction (in vitro) tests. Contrary to expectations and indications given by chemical extractions, they observed nearly constant accumulation of metals even in the most EDTA-processed soils. This indicates to isopods a constant ratio between non-bioavailable and bioavailable forms of soil toxic metals. No effect of bioaccessibility (availability) stripping of toxic metals from remediated soil was therefore observed with *P. scaber* test. Results indicated that favourable results of extensively reduced bioaccessibility of toxic metals after remediation obtained by simple extraction tests need to be taken with caution.

Earthworms are also often used in soil ecotoxicological tests as bioindicators of accumulation and the effect of toxic substances. While acute (accumulation) tests do not provide an insight into the effects of metals on population dynamics, the avoidance behaviour of earthworms is relevant measurable endpoint for assessing the effect of metals in soil on earthworms at ecosystem level (Sousa et al. 2008). The presence of chemoreceptors makes earthworms highly sensitive to chemicals in their environment, allowing them to avoid unfavourable environments (Curry and Schmidt 2007). Udovic and Lestan (2010a) evaluated the feasibility of the standardised earthworm (*Eisenia fetida*) avoidance test for assessing the efficiency of soil remediation of Pb-, Zn- and Cd-polluted soil using EDTA extraction. The total soil concentration of metals and their mobility (TCLP), oral bioaccessibility (PBET) and phytoavailability (DTPA) were consistently reduced. The avoidance test showed no significant avoidance of polluted soil in favour of that which had been remediated. Pb, Zn and Cd accumulation in *E. fetida* mirrored the decreasing pattern of metal potential bioavailability gained by extraction of soil with increasing EDTA concentrations. Results indicated little sensitivity of *E. fetida* towards remediated soil.

21.5 The Long-Term Stability of Forms of Toxic Metals in EDTA-Remediated Soils

Toxic metals in soil are present in various species and forms (bound to different soil fractions) and as such are not entirely bioaccessible and bioavailable to organisms and chemically accessible to chelating agents. It is recognised that soil washing cannot remove toxic metals completely from the soil, even when very high EDTA concentrations are used. The reports on forms and availability of toxic metals left in soil after soil washing are not univocal. Barona et al. (2001) and Lei et al. (2008) found that the metals remaining in soil after chelant-based washing become more mobile and more weakly associated with soil components. Enhanced mobility of residual metals presumably takes place due to the metal detachment, due to chelant attack, soil dissolution or the cation exchange between the chelant complexes and the soil particles. On the other hand, experimental results with soils from Meža Valley (Udovic et al. 2007; Udovic and Lestan 2007; Jelusic and Lestan 2014) indicate that the residual metals left in the soil after remediation remain present in chemically stable species bound to non-labile soil fractions (modified Tessier's sequential extractions) and could be considered non-mobile and non-bioavailable and thus not toxic. However, with the reintroduction of remediated soil into the environment, we expose the soil to various environmental factors. Is therefore the reduced mobility and bioavailability of soil residual metals a permanent or only temporal achievement of soil remediation? Soil is a dynamic natural body and, after remediation, various abiotic (i.e. climatic, hydrological) and biotic soil (microorganisms and fauna) factors could initiate the transition of residual metals back to more labile, mobile and accessible forms to re-establish the disturbed equilibrium, although the availability of toxic metals in contaminated (and non-remediated) soil is considered to decrease with time (Han et al. 2003). Such a dynamic shift would increase the toxicity of the residual metals and consequently decrease and hamper the final efficiency of soil remediation.

Laboratory-scale studies on availability of toxic metals left in soil after EDTA soil washing were reviewed by Lestan and Udovic (2011). Studies on the effect of different earthworm species as key soil biotic environmental factors on residual toxic metals indicate that their activity can lead to significant changes of the fractionation, the mobility and the bioavailability of toxic metals. The increase/decrease in metal availability in soil depends on the conditions in the bulk soil relative to the conditions in the earthworm's gut (Sizmur and Hodson 2009). The effect, therefore, varied not only according to the type of (remediated) soil but also among earthworm ecological categories and species. For example, analysis of earthworm casts and the remainder of the soil indicated that endogeic *Lumbricus rubellus* and epigeic species *Eisenia fetida* did not significantly change Pb and Zn fractionation in non-remediated and remediated soil. However, *E. fetida* increased Pb bioaccessibility (PBET) up to 5.1 times in remediated soil; the effect of *L. rubellus* was not observed (Udovic and Lestan 2007). Udovic et al. (2007) reported that both *E. fetida* and endogeic species *Octolasium tyrtaeum* enriched the carbonate soil fraction in their casts with Pb and increased the Pb bioaccessibility (PBET) in the intestinal phase, but only *E. fetida* increased Pb leachability determined by TCLP test. Udovic and Lestan (2010b) reported small but statistically significant increase in Cu intestinal bioaccessibility (PBET) in EDTA-remediated vineyard soil processed by autochthonous endogeic *Lumbricus terrestris* species. A 50-year-old vineyard was regularly managed and treated with copper sulphate ($CuSO_4{\cdot}5H_2O$, Bordeaux mixture) as fungicide up to the Cu soil level 400 mg kg^{-1}.

Literature on the effect of abiotic soil ageing factors on availability of toxic metals in contaminated and remediated soil is scarce. Nevertheless, it has been shown that repetitive temperature changes lead to a decrease in soil pH and change the adsorption–desorption behaviour of toxic metals in soil, which affects their fractionation, availability and mobility (Rieuwerts et al. 1998). Temperature changes also affected the rates of toxic metal desorption from Fe and Mn oxides and the behaviour of the organic matter soil constituents (e.g. humic acids), enhancing metal mobility and solubility by complexation (Weng et al. 2002). Udovic and Lestan (2009) repetitively applied cycled high and low temperatures to the contaminated and EDTA-washed soil. They observed a significant increase of metals in exchangeable fractions of modified Tessier's sequential extraction scheme. The ageing treatment consistently lowered toxic metal phytoavailability (DTPA) in both soils. However, leachability test (TCLP) indicated that Pb, Zn and Cd behaved differently from each other; Pb mobility increased, Cd mobility decreased, while Zn mobility did not change.

21.5.1 Leachability and Accessibility of Toxic Metals from Remediated Soil in Agricultural Use

The possibility that earthworm activity may raise toxic metal bioaccessibility (Lestan and Udovic 2011) is of considerable relevance for the success of soil washing which removes only part of the metals (presumably accessible from labile soil fractions). It has been shown that earthworms can rapidly invade remediated soil (Langdon et al. 2001). With estimated annual rates of earthworm cast production ranging from 5 to more than 250 tons ha^{-1} (Bohlen 2002), earthworm activities might therefore, even in the short term, affect metal fractionation and bioavailability. When returned to the site and exposed to these and other environmental factors and intensive agricultural use, is the EDTA-remediated soil stable in terms of newly obtained properties?

Using soil from Meža Valley kindred to soil used in laboratory-scale studies on toxic metal bioaccessibility (Udovic et al. 2007; Udovic and Lestan 2007, 2010a) and applying similar remediation process, Jelusic and Lestan (2014) exposed the soil to environmental conditions and agricultural practice in larger experimental beds. Remediation reduced the metal burden by 80, 28 and 72 % for Pb, Zn and Cd, respectively. Initial metal leaching measured through lysimeters eventually ceased. Toxic metals associated with labile soil fractions (modified Tessier's sequential extractions) were largely removed: water-soluble and exchangeable fractions and fractions extracted in reduced conditions even fell below the level of quantification for Pb and Cd and decreased significantly for Zn. The share of almost all metal fractions decreased after remediation and remained practically unaltered throughout all 7 months of the field experiment. No shifts between labile and residual fractions were observed. A TCLP and DTPA procedure was employed to determine whether the remediated soil is potentially hazardous in terms of metal leaching and potential plant availability. After remediation, the mobility and phytoaccessibility of Pb and Zn dropped substantially and remained low (Fig. 21.3). No significant changes in the properties (standard pedological analysis) of remediated soil were observed during the course of the experiment. In order to characterise the oral bioaccessibility of Pb, Zn and Cd, the UBM test was applied. Values for the original soil stomach phase were very high, approximately 88, 64, and 100 % of the total metal concentration, for Pb, Zn, and Cd, respectively. Bioaccessibility in the stomach phase was reduced with remediation by more than 80 % for Pb and Cd and by 56 % for Zn. Accessibility was much lower in the intestinal phase for the original soil and was brought under the level of determination in the remediated soil. These results are quite opposite to the results obtained from a laboratory-scale soil ageing studies although extensive colonisation of remediated soil in experimental beds with earthworms was observed. Absolutely no trends were observed in a field experiment indicating metal shifts between fractions, leachability, phytoaccessibility and oral bioaccessibility between sample times of 0, 4 and 7 months, indicating no further soil matrix disturbance due to remediated soil exposure to natural environmental factors. The discrepancy between data obtained from the laboratory and field experiment could perhaps be explained by the fact that availability of toxic metals was in laboratory studies mostly assessed in earthworm casts with specific properties quite different from the bulk of the soil, where the toxic metal availability was measured in field experiments.

21.5.2 EDTA Emissions from Remediated Soil

EDTA is merely biodegradable and highly persistent in the soil, leading to the possible post-use emissions. Besides, leaching EDTA and metal complexes (chelates) that are formed during soil extraction can be absorbed by soil minerals, especially crystalline iron oxides (Nowack and Sigg 1996). Noren et al. (2009) suggested that EDTA is mainly absorbed by iron oxide through outer-sphere complexation. Outer-sphere surface complexes are relatively weak and more susceptible to leaching on changes of environmental conditions (Elzinga and Sparks 2002).

Jelusic et al. (2013) reported that 75 % of Pb and Cd and 30 % of Zn were removed by EDTA soil washing; however, up to 0.4, 0.8 and 0.3 % of the initial Pb,

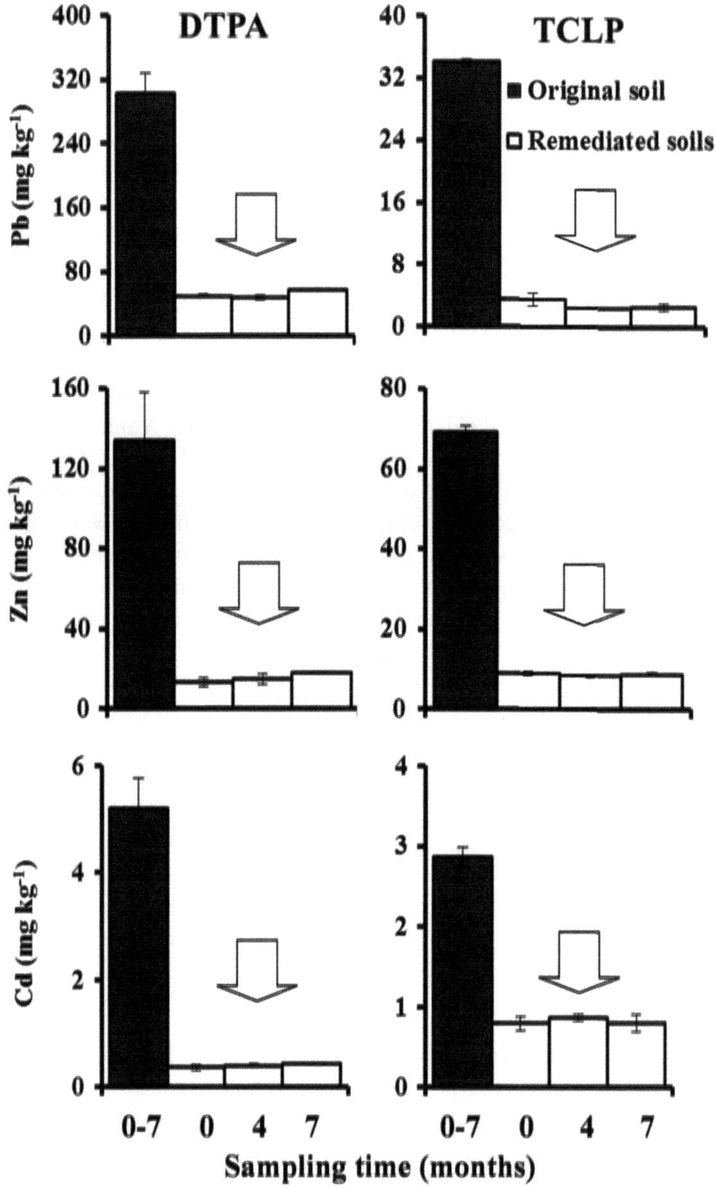

Fig. 21.3 Pb, Zn and Cd phytoaccessibility and leachability of original and remediated soils assessed by DTPA and TCLP extraction. For the original soil, the average is calculated from 0-, 4- to 7-month samples. Error bars represent standard deviation ($n = 9$ for original soil, $n = 3$ for remediated soils)

Cd and Zn soil concentration, respectively, were lost with leachate from soil columns, presumably as EDTA chelates. On the other hand, lysimeter study indicated that initial metal–EDTA chelate leaching ceased shortly after first rainfall (Jelusic and Lestan 2014). Pociecha and Lestan (2012a, b) reported that 15–25 % of EDTA remains in the soil after remediation in a laboratory scale. Voglar and Lestan (2014) calculated almost 30 % lost of EDTA from the chelant mass balance, mostly due to the chelant soil absorption during soil extraction in a pilot-scale remediation plant. The fate of retained EDTA, whether permanent absorption and gradual loss of chemical identity by becoming an integral part of soil organo-mineral complex (EDTA is an aminopolycarboxylic acid and as such structural analogue to the soil humic substances) or slow release of EDTA from the soil matrix, remains to be investigated.

21.6 Functioning of Remediated Soil as a Natural Substrate

The final aim of successful reclamation of contaminated soil is to demonstrate removal and permanent reduction of toxicity from soils without major intervention on the soil properties and quality. Remediation technologies, however, often concentrate only on metal removal efficiency and overlook the treated soil's overall health, functioning and potential use after remediation. It is desirable that, after reintroduction into the environment, remediated soil preserves the function of plant and microbial substrate and, consequently, can become quickly repopulated by soil organisms.

Among the available remediation techniques, soil washing with various solutions (chelants, acids, surfactants, solutions with different salts, organic solvents) is one of the most effective, but data about the impact on soil chemical and biological properties are still scarce. Udovic and Lestan (2012) studied the effect of two common washing agents, mineral hydrochloric acid (HCl) and a chelating agent (EDTA) on Pb, Zn and Cd removal and accessibility and on physico-chemical and biological properties in one calcareous, pH neutral soil from Meža Valley, Slovenia, and one non-calcareous acidic soil from Arnoldstein in Austria. Both contaminated sites were polluted from centuries of Pb mining and smelting. EDTA was a more efficient leachant compared to HCl: up to 133 times lower chelant concentration was needed for the same percentage of Pb removal. EDTA and HCl concentrations with similar Pb, Zn and Cd removal efficiency had different impacts on soil properties. As expected, HCl significantly affected soil chemical properties by dissolving carbonates from calcareous soil, while EDTA leaching increased the pH of the acidic soil. Biological activity assays showed that soil washing with HCl had a distinctly negative impact on soil microbial (glucose-induced respiration) and enzyme activity (dehydrogenase, acid phosphatase and β-glucosidase). EDTA was not only more efficient but also more soil-friendly extractant. Nevertheless, the results showed that chelant application still leads to negative consequences on indigenous soil biota in acidic soils from Arnoldstein.

21.6.1 *Effect of EDTA Soil Washing on Soil Chemical and Physical Properties*

Soil washing with EDTA only slightly changes soil chemical properties (Udovic and Lestan 2012). Besides expected reduction in chelate-forming metals, Zupanc et al. (2014) reported small increase in pH and base saturation (EDTA was added as Na_2-EDTA salt).

On the other hand, it is expected that washing process, especially intensive mixing of the soil slurry and soil compression after dewatering, significantly deteriorates physical properties of soil (Vulava and Seaman 2000). Tsang et al. (2007) reported that EDTA washing solution itself appreciably alters the soil physical properties by dissolving indigenous oxides, carbonates and organic matter. Zupanc et al. (2014) investigated the aggregate fractionation and stability, saturated hydraulic conductivity and water retention of the remediated soil. Soil washing with 60 mmol kg^{-1} EDTA removed 59, 16, and 56 % of Pb, Zn and Cd from original soil with 3,730, 2,140 and 26 mg kg^{-1} of Pb, Zn and Cd, respectively. In the final stage of the soil preparation, amorphous soil mass was rubbed through a 5 mm mesh sieve, thusly predetermining aggregate size. During (column) experiment, there was no shrinkage or swelling of remediated soil and the average bulk density (1.1 g cm^{-3}) only slightly decreased. There was also no significant difference between original and remediated soil in aggregate stability: 93 and 98 % stability for 2–4 mm fraction in original and remediated soil was reported using wet sieving method (Kemper and Rosenau 1986). Soil used in the experiment was garden soil with high organic matter content (7.8 and 7.7 % for original and remediated soils) and high calcium content, both in $CaCO_3$ form and as exchangeable cation. Calcium and organic matter content are important factors for aggregate formation (Bronick and Lal 2005), which could explain high aggregate stability measured. Saturated hydraulic conductivity in remediated soil was slightly higher (assessed as *K*-value according to the method by Klute and Dirksen 1986) than in original soils and in range of 102 m day^{-1}, which are values expected for sand and fine-sand soil texture (Bear 1972). Possible explanation for high *K*-value in remediated soil is the macropores formed by highly stable soil aggregates formed in turn by rubbing soil through mesh sieve (Fig. 21.4) as a part of remediation process. Concerning water retention characteristics, remediated soil had less average water content at the lower end of the over pressure (−0.033 MPa, field capacity) comparing to the original soils; however, the differences were not substantial. The amount of retained water in remediated soil at the −0.1 to −1.5 MPa over pressure was higher comparing to original soil. Average plant available water interval (PAW) of the original soil was significantly higher comparing to the remediated soils. Difference in PAW between original and remediated soils was explained by difference in soil texture as well as aggregate fractionation and aggregate stability. In the lower end of the over pressure (−0.033 to −0.1 MPa), the coarse structure influences water retention (Dexter 2004). Soil structure of the remediated soils was stable and coarse which resulted in less retained water. In the higher end of the over

Fig. 21.4 Artificial aggregates of remediated soil formed by rubbing soil through 5 mm mesh sieve

pressure (wilting point, −1.5 MPa), soil texture plays a decisive role (Dexter 2004). Remediation process broke soil particles, which resulted in decreased silt and sand content, and increased clay content which in turn retained higher amount of water. Overall, lower water content at field capacity and higher water content at wilting point resulted in narrower interval of the PAW in remediated soil. As rehabilitation measure, addition of hydrogel and peat was shown to efficiently increase PAW of the remediated soil (Zupanc et al. 2014).

21.6.2 Effect of Remediated Soil on Soil Biological Properties and Functioning

Udovic and Lestan (2010a) showed that earthworms generally did not avoid the EDTA extracted soil in favour of the control soil and concluded that the remediation regime applied to the polluted soil did not disturb its habitat functionality. Earthworms live in direct contact with the solid and pore-water soil phase and are thus exposed in a manner representative of soil bacteria and fungi. Jelusic et al. (2013) however reported that soil microbial biomass (indicated by total soil DNA) was significantly lower in the EDTA (30 and 60 mmol kg^{-1})-washed soil compared to the original one. This may be related to the stringent physical conditions during soil extraction. The study also showed a recovery of soil microbes with time: statistical evaluation of bacterial 16S rRNA and fungal ITS rRNA gene fragments showed a clear difference between original soil and EDTA-treated soil at the beginning of the experiment, whereas bacterial and fungal community structures in all soils were clustering together after 7 weeks. These results indicate that the effect of inherent soil properties and the same conditions of soil ageing overruled the initial effect of soil treatment.

Soil rhizosphere functioning is significantly affected by a plant's ability to form symbiotic mycorrhizal associations with ubiquitous arbuscular mycorrhizal (AM) fungi. AM fungi (phylum Glomeromycota) colonise two-thirds of plant species (Fitter and Moyersoen 1996). They are important in biogeochemical cycling of essential nutrients and minerals and maintenance of soil structure and facilitate plant mineral nutrient uptake (Houben et al. 2013). EDTA extraction (60 mmol kg^{-1}) was, however, destructive for AM fungi (Jelusic et al. 2014). The low values of infective mycorrhizal propagules such as fungal mycelium and viable spores and the reported intensity of mycorrhizae and the arbuscule abundance in the root cortex indicated that functional mycorrhizae did not develop in remediated soil over the course of the experiment. Authors concluded that additional inoculation of the soil either by natural (roots and soil collected in the field) or a commercial inoculum would be necessary for remediated soil in order to establish functional mycorrhizal symbiosis. It is well known that AM fungi and other plant-growth-promoting microorganisms are used as inoculants for biofertilisation and phytostimulation (Rana et al. 2012). In addition to using indigenous inocula (Turnau et al. 2008), a growing number of inocula are being marketed which may help to restore functionality of remediated soil. Indeed, Macek et al. (2014) characterised the biodiversity of AM fungi in a pot experiment by examination of 16S rRNA clone libraries, both in metal-polluted soil from Meža Valley and EDTA-remediated soil, before and after inoculation with indigenous and commercial inoculum. They found a significant increase in root colonisation with AM fungi in remediated soil after inoculation.

To estimate the effect of EDTA, soil washing on the functioning of remediated soil measurements of specific soil activities can be employed. Enzymatic activities in soil play an important role because all biochemical transformations and metabolic processes involved in the biogeochemical cycles of nutrients in soil are dependent on, or related to, the presence of enzymes. Enzyme activities are thus related to soil fertility and influence soil productivity. The advantage of those tests is not only the possibility of estimation of the direct effect of various quantities, substances and amendments on the environment but also the determination of the potential risk related with their mutual interactions. In principle, it is assumed that high values of enzymatic activity are evidence of good soil functioning and quality (Gianfreda et al. 2005). The origin of enzyme activity has not been precisely identified, although it is generally agreed that the microbial component is the main source of enzymes in soils (Mijangos et al. 2006).

Jelusic and Lestan (2014) applied soil respiration and specific soil enzyme activities as indicators of soil functioning to monitor impacts of soil remediation with EDTA. Soil washing (60 mmol EDTA kg^{-1}) successfully removed available forms of Pb, Zn and Cd and thus lowered the human and environmental hazards of the remediated soil. Substrate-induced respiration was used and is an index of active microbial biomass. A variety of enzyme tests were deployed to provide a thorough overview of soil functions: β-glucosidases catalyse the hydrolysis of cellulose, being the rate-limiting enzymes in the microbial degradation of cellulose to glucose, with a critical role in C cycling (Baker et al. 2011). Phosphatases catalyse the hydrolysis of phosphate esters to release orthophosphate and play an important role in plant P nutrition. Ureases catalyse the hydrolysis of urea to CO_2 and NH_4^+ and play an important role in the N cycle. Hydrolysis of fluorescein diacetate includes the activity of several enzymes: lipases, esterases and proteases. Dehydrogenases are cell-bound enzymes and are used as a measure of overall microbial activity (Alef and Nannipieri 1995).

Many studies of metals and soil enzymes have been performed, all indicating the toxic effect of metals on soil biology and biochemical processes and, consequently, inhibiting potential enzyme activities (Zhang et al. 2010a). However, Jelusic and Lestan (2014) reported that remediated soil (60 mmol EDTA kg^{-1})—soil with a reduced burden of toxic metals—demonstrated lower potential enzyme activities. The remediation process had obviously deprived the soil of certain metabolic activities to the extent that even after 7 months of plant cultivation and added fertilisers, the remediated soil could not restore itself. One of the reasons may lie in the chemical nature of the non-specific chelating agent EDTA. EDTA is known to act as a non-specific chelating agent and is therefore capable of extracting a variety of metal ions from the soil matrix, some of them being macro- and micronutrients essential for microbial nutrition. For example, micronutrient Zn plays a structural role in many enzymes and, as indicated by mobility (TCLP) and fractionation measurements (modified Tessier's sequential extractions), EDTA washing reduced bioavailable forms of Zn, rendering it unavailable for some biochemical soil processes. Jelusic et al. (2014) observed also approximately four times lower concentrations of Mn in remediated soil. Mn(II)-oxidising microorganisms are ubiquitous in soil and play an essential part in the biogeochemical cycling of manganese, iron, nitrogen, carbon, sulphur and several nutrients and trace metals. Mn is used as a terminal electron acceptor for bacterial respiration; it is required as a trace nutrient for a large number of cellular functions; it can serve as an antioxidant, and some bacteria even use it for protection from UV radiation, viral attack, predation or metal toxicity (Tebo et al. 2005). Responses of the same enzymes varied over time, due to different plant cover and changing seasonal and other environmental factors. However, measurements revealed that the remediation process significantly diminished potential soil enzyme activity and no trends were observed of the remediated soil developing the same enzymatic activity than the original soil.

Conflicting results were obtained by Udovic and Lestan (2012), since they observed a significant increase in some enzyme activities and substrate-induced respiration after relieving the soil of potentially toxic metals with two times lower EDTA dose (30 mmol kg^{-1}). Epelde et al. (2008) observed that the application of EDTA on non-polluted soil significantly diminished dehydrogenase activity but had no effect on β-glucosidase or acid phosphatase potential activity. Furthermore, EDTA had no effect on any of the enzyme activities when chelate was added to Pb-polluted soil. However, the relatively low concentrations of EDTA in the latter two studies were probably too small to extract the cations of essential metals needed for microbial activity.

21.7 Plant Growth on Remediated Soil

Plant performance in EDTA-washed soils can be affected by several factors. After soil washing and rinsing, some toxic metal–EDTA complexes are expected to remain bound into the soil solid phases and possibly effecting plants. Experiments dealing with induced metal phytoextraction have shown that also the application of EDTA can severely inhibit plant growth, due to enhanced dissolution of metals or, directly, through free EDTA itself (Chen and Cutright 2001; Meers et al. 2004). Direct toxicity of EDTA is expected to be caused by EDTA interference with cellular structures and functions. Plant growth on remediated soil may be inhibited due to the narrower interval of the plant available water (Zupanc et al. 2014) and loss of AM fungi (Jelusic et al. 2014). Most of all EDTA is expected to deprive soil of micronutrients. EDTA is a non-selective agent (Zhang et al. 2010b) that binds with many soil cations, whether they be toxic metals (Cd, Pb), essential metals (Mn, Mg, Ca, Fe) or both (Zn, Cu).

Jelusic et al. (2013) assessed growth of Chinese cabbage (*Brassica rapa*) in remediated soil using soil column experiment. EDTA soil washing removed up to 77, 29 and 72 % of metals from Meža Valley soil contaminated with 1,378, 578 and 8.5 mg kg^{-1} of Pb, Zn and Cd, respectively. Visually no reduction of plant growth on remediated soil compared to original (non-remediated) was observed. Furthermore, the fitness of plants grown on EDTA-washed soils (gas exchange, net photosynthesis; stomatal conductance; transpiration and fluorescence, photochemical efficiency; electron transport) was not compromised compared to *B. rapa* grown in non-washed substrate. Measurements showed a stable photochemical efficiency over the entire period of the experiment. Similarly, Ruley et al. (2006) studied the physiological response of *Sesbania drummondii* exposed to EDTA. They found a significant decrease in plant photosynthetic efficiency when chelant was added to soil free of toxic metals, but not when Pb was also present in the soil.

In the consequent study, Jelusic et al. (2014) studied performance of selected vegetables in a field experiment in two experimental plots of $4 \times 1 \times 0.3$ m, filled, one with EDTA-remediated and the other with original soil (contaminated with 1,585 mg kg^{-1} Pb, 525 mg kg^{-1} Zn and 8.8 mg kg^{-1} Cd). Remediation reduced the toxic metal burden by 80, 28 and 72 % for Pb, Zn and Cd, respectively. The total soil Mn was reduced fourfold and phytoaccessibility (DTPA) of micronutrients Cu, Fe and Mn for 54, 26 and 79 %, respectively. For selected vegetables, the essential metal (Mn, Cu and Fe) concentrations in their biomass were measured. Manganese concentrations in the green parts and leaves were significantly greater in vegetables grown in original soil than in those cultivated in remediated soil, indicating that plants suffered from Mn deficiency. Consequently, vegetables grown in original soil exhibited greater biomass yields than vegetables grown in remediated soils (Fig. 21.5). A deficiency of Mn can severely affect growth, since Mn is involved in important metabolic pathways. In photosynthetic reactions, Mn is associated with the oxygen evolving complex of photosystem II (Williams and Pittman 2010). Direct effects of Mn deficiency on C-assimilation cannot therefore be excluded.

Fig. 21.5 Remediated and original garden bed. Shown on the *left side* of the figure are the plants grown on remediated soil fertilised by 15 kg ha^{-1} of Mn (added as $MnSO_4$). On the *right side*, plants grown on original soil

The least affected was lettuce (*Lactuca sativa*) with 4 % lower biomass. The most evident difference was observed with spinach (*Spinacia oleracea*) which in remediated soil demonstrated symptoms of chlorosis and diminished growth and, consequently, a tenfold lower yield than spinach plants in original soil. As expected from the severe growth inhibition of spinach, its photosynthetic activity and fluorescence parameters were also drastically reduced. Photosynthetic parameters of other vegetables grown in original and remediated soil were similar, confirming results obtained with *B. rapa* (Jelusic et al. 20013).

Evidently, for some plants, effective means of revitalisation are needed to restore and reclaim EDTA-washed soil as a fertile substrate. Jelusic et al. (in press) attempted to regenerate remediated soil by fertilisation and different soil amendments. Since spinach (*S. oleracea*) was found to be particularly sensitive for growth on EDTA-washed soil (Jelusic et al. 2014), it was used as a bioindicator to evaluate the effects of revitalisation measures by assessing plant biomass and photosynthetic functions. Garden soil highly contaminated with 4,037, 2,527, and 26 mg kg^{-1} of Pb, Zn and Cd, respectively, was washed with a high EDTA dose (120 mmol kg^{-1} of soil) to final values of 1,246, 2,138 and 11 mg kg^{-1} of Pb, Zn and Cd, respectively. After remediation, the concentration of Mn decreased by 53 %. Concentrations of other essential metals, Cu, Fe, Mg and Ca, remained unchanged. Furthermore, remediation stripped the soil of the Mn-oxide fraction (assessed by modified Tessier's sequential extraction), leaving the remediated soil without the readily reducible Mn (Mn oxides) necessary for plant growth. To restore the diminished capacity of remediated soil to function as plant substrate, the soil was amended with fertiliser (N, K, Mn and Mg), manure, hydrogel, vermiculite, apatite and commercial metal absorbent and subjected to *S. oleracea* cultivation in a column experiment. Remediation reduced plant uptake of toxic metals but also diminished plant growth (9 times lower biomass compared to the original soil) and photosynthesis as observed before (Jelusic et al. 2014). Net photosynthetic rates of *S. oleracea* grown in the original soil exceeded 25 μmol $CO_2\,m^{-2}\,s^{-1}$; in remediated soil, however, it only reached half of this value, because of reduced photochemical efficiency and conductivity of stomata. Fertilisation partly restored net photosynthesis (to ca. 75 % of original), stomatal conductance, transpiration and yield of the tested plants. Fertilisation did not influence the photosynthetic rate of plants growing in the original substrate, clearly showing that Mn is the main essential nutrient in which availability is severely reduced after remediation process. Fertilisation of the original soil increased the plant biomass 1.7 times and in remediated soil 4.9 times, indicating again that the depletion of nutrients is a major contributing factor to poor plant growth. Growth depression was found to be mitigated further by hydrogel (up to 53 %). Hydrogel was selected to improve the soil's water characteristics (reduced PAW, Zupanc et al. 2014) and in turn nutrient availability, and consequently, improve conditions for plant cultivation (El-Hady and El-Dewiny 2006). The other tested amendments did not significantly affect the growth and photosynthetic functions of the test plant.

Zupanc et al. (2014) observed that white clover's (*Trifolium repens*) dry mass yield on original soil from the same contaminated site in Meža Valley was significantly higher (up to 40 %) than on EDTA-remediated soil. Poorer soil water-retention characteristics and the destruction of the natural structure of EDTA-washed soil could be potential growth-inhibiting factors. Hydrogel amendment however, with known positive effect on plant available water (Narjary et al. 2012), did not change plant yield. On the other hand, addition of manure as source of nutrients had positive influence on growth of *T. repens*, indicating the significance of nutritional imbalance in remediated soil. Again, chelating agent did not bind only to toxic metals, but soil washing also reduces the pool of soil micronutrients. Zupanc et al. (2014) experimented with soil which was formed on limestone and was rich in Ca and Mg. Soil washing with EDTA only slightly decreased the amount of available K. However, the elemental analysis however showed the decrease of essential elements Mo (original soil 40.6, remediated 9.8 mg kg^{-1}), Mn (original soil 767, remediated 381 mg kg^{-1}) and Cu (original soil 51.8, remediated 42.2 mg kg^{-1}). Decrease of Fe was not confirmed by analysis. The concentration of Mo was reduced for more than four times, the most of all measured essential elements. Molybdenum occurs in more than 60 enzymes catalysing diverse oxidation–reduction reactions. Because of its involvement in the process of N_2 fixation, nitrate reduction and the transport of nitrogen compounds in plants, Mo plays a crucial role in metabolism of plants (Hamlin 2006).

Results of these studies indicate that fertilisation helped to partly restore the soil nutrient pool. However, further effective means of revitalisation are needed to re-establish EDTA-remediated soil as a fertile plant substrate. For example, amending the soil with biochar has increasingly attracted widespread attention because of its chemical stability (biochar is ideally suited for sequestrating C,

Lehmann 2007), and as a soil conditioner, it increases the soil fertility and plant growth by retaining nutrients and improving the physical (structure, water retention) and biological properties of the soil and ameliorates acidity (Laird et al. 2010). Biochar was also reported to reduce bioavailable concentrations of toxic metals in soil, i.e. of Pb, Zn and Cd for 71, 87 and 92 %, respectively (Houben et al. 2013).

21.7.1 Phytoaccumulation of Toxic Metals from EDTA-Washed Soils

Toxic metal concentrations absorbed by plants vary according to soil metal concentrations, soil properties, plant species, plant organs and growing stages and various environmental factors. Pb and Cd are believed to be non-essential elements for flora (and fauna), elements not needed for the assembly of plant tissues or their functioning. Nevertheless, plants often do absorb these contaminants, for example, instead of their biologically antagonistic micronutrients (Pb and Ca, Cd and Zn) with a similar ionic radius.

The study of Jelusic et al. (2013) revealed the accumulation of toxic elements in green parts of the test plant from EDTA-washed soil is a potentially problematic aspect of the remediation technology applied. They reported that remediation reduced toxic metal–soil–root transfer by up to 61 % but did not prevent Pb and Cd accumulation in leaves of Chinese cabbage (*B. rapa*). The accumulation of toxic metals in the green parts of the plant's leaves was explained by a higher internal translocation of Pb and Cd from roots to leaves for plants grown in remediated soils compared to control soil. Similarly, Grcman et al. (2001) reported 1.7 and 3.5 and 3 times lower concentrations of Pb, Zn and Cd in the roots of *B. rapa* after the addition of 10 mmol EDTA kg^{-1} to soil, compared to plants grown in control soil. It has been suggested that enhanced plant uptake of metal–EDTA complexes can take place in points at which suberisation of the root cell walls has not yet occurred and at breaks in the root endodermis and the Casparian strip (Nowack et al. 2006; Vassil et al. 1998). In practice, this could mean that, after EDTA treatment, *B. rapa* will accumulate some amounts of toxic metals even when their total concentration in the soil is low. The mechanism might, however, be species specific and different for plants other than from the *Brassicae* family, which are known to be good toxic metal accumulators.

For instance, Jelusic et al. (in press) reported that remediation of highly contaminated soil (originally 4,037, 2,527, and 26 mg kg^{-1} of Pb, Zn and Cd, respectively) reduced spinach (*S. oleracea*) uptake of Pb, Zn and Cd in the green parts for 2.4, 1.8 and 6.4 times, respectively (Table 21.1). Interestingly, fertilisation of both the original and remediated soils further considerably reduced the plant Pb uptake (Table 21.1), presumably by decreasing the relative share of Pb in the total cation concentration in the soil solution. Rather high concentrations of Cd were observed in *S. oleracea* plants grown on the original soil, more than 40 mg kg^{-1}. *S. oleracea* is Cd tolerant and can accumulate high concentrations of this element (Salaskar

Table 21.1 Concentrations of Pb, Zn and Cd in edible parts of vegetables and green parts of horticultural plants cultivated on original and EDTA-remediated soil

	Pb (mg kg^{-1})		Zn (mg kg^{-1})		Cd (mg kg^{-1})	
Plant	Original	Remediated	Original	Remediated	Original	Remediated
Chinese cabbage[a] (*Brassica rapa*)	16 ± 8	17 ± 8	120 ± 26	126 ± 36	2.5 ± 1.3	3.5 ± 1.9
Chinese cabbage[b] (*Brassica rapa*)	6.6 ± 1.3	LOQ	66 ± 6	32 ± 4	1.4 ± 0.2	0.2 ± 0.1
Spinach[c] (*Spinacia oleracea*)	29 ± 5	11 ± 4	480 ± 80	260 ± 10	41 ± 16	7.0 ± 0.1
Spinach[b] (*Spinacia oleracea*)	13.5 ± 1.4	LOQ	110 ± 6	74 ± 0.6	9.3 ± 2.8	2.8 ± 0.4
Peas[b] (*Pisum sativum*)	LOQ	LOQ	56 ± 3	61 ± 3	0.7 ± 0.0	LOQ
Onion[b] (*Allium cepa*)	LOQ	LOQ	43 ± 2	17 ± 2	LOQ	1.3 ± 0.0
Cauliflower[b] (*Brassica oleracea*)	LOQ	LOQ	41 ± 6	29 ± 7	0.6 ± 0.0	LOQ
Lettuce[b] (*Lactuca sativa*)	7.8 ± 2.5	LOQ	84 ± 3	62 ± 3	3.4 ± 0.4	1.2 ± 0.1
Carrot[b] (*Daucus carota*)	5.5 ± 2.1	LOQ	37 ± 4	33 ± 0	1.6 ± 0.4	0.6 ± 0.4
Basil[b] (*Ocimum basilicum*)	1.5 ± 0.2	LOQ	122 ± 22	100 ± 15	1.4 ± 0.2	0.3 ± 0.0
Bell pepper[b] (*Capsicum annuum*)	LOQ	LOQ	35 ± 2	32 ± 1	1.9 ± 0.3	0.7 ± 0.1
White clover[d] (*Trifolium repens*)	LOQ	LOQ	56–66	43–49	0.5–1.8	LOQ
Garden verbena[e] (*Verbena hybrida*)	65 ± 9	20 ± 3	124 ± 28	87 ± 9	1.8 ± 0.5	1.2 ± 0.3

(continued)

Table 21.1 (continued)

Plant	Pb (mg kg^{-1})		Zn (mg kg^{-1})		Cd (mg kg^{-1})	
	Original	Remediated	Original	Remediated	Original	Remediated
Orchard grass[e] (*Dactylis glomerata*)	28 ± 8	4 ± 1	165 ± 30	85 ± 7	5.0 ± 0.2	1.5 ± 0.2

Data are given as means and standard deviations of at least three replicates
LOQ limit of quantification of instrument (atomic absorption spectrophotometer)
[a]Jelusic et al. (2013); soil with 1,378, 578 and 8.5 mg kg^{-1} Pb, Zn and Cd, respectively, treated with 60 mmol kg^{-1} EDTA; 77, 29 and 72 % of Pb, Zn and Cd removed; pot experiment
[b]Jelusic et al. (2014); soil with 1,585, 525 and 8.8 mg kg^{-1} Pb, Zn and Cd, respectively, treated with 60 mmol kg^{-1} EDTA; 80, 28 and 72 % of Pb, Zn and Cd removed; field experiment
[c]Jelusic et al. (in press); soil with 4,037, 2,527 and 26 mg kg^{-1} Pb, Zn and Cd, respectively, treated with 120 mmol kg^{-1} EDTA; 70, 15 and 58 % of Pb, Zn and Cd removed; pot experiment
[d]Zupanc et al. (2014); soil with 1,585, 525 and 8.8 mg kg^{-1} Pb, Zn and Cd, respectively, treated with 60 mmol kg^{-1} EDTA; 80, 28 and 72 % of Pb, Zn and Cd removed; pot experiment
[e]Unpublished results; soil with 4,037, 2,527 and 26 mg kg^{-1} Pb, Zn and Cd, respectively, treated with 120 mmol kg^{-1} EDTA; 70, 15 and 58 % of Pb, Zn and Cd removed; field experiment

et al. 2011). After remediation, concentrations of Cd in the plant dropped substantially (Table 21.1); however, the level in all treatments was still above the European Union commission regulation—setting maximum levels for certain contaminants in foodstuffs (EC 1881/2006). Cadmium is reported to be more plant accessible and mobile than Pb and less than Zn, an essential element with which Cd shares a transport path into the plant (Tsonev et al. 2012). Due to lowered Zn availability (Lestan et al. 2005), Cd could also have been more easily transferred into the plants grown in remediated soils (Ozturk et al. 2003). Broadley et al. (2001) reported that shoot Cd concentrations are generally higher in species from the Caryophyllales order, which explains the relatively high concentrations of Cd in spinach. The Pb concentrations in *S. oleracea* were below the EU legislation limits (Table 21.1).

Jelusic et al. (2014) rotated selected cultivars (four-crop rotation) over the course of 16 months in a field experiment and investigated prospects of potential re-use of remediated soil as an agricultural or garden substrate. Pb, Zn and Cd uptake of plants grown on original (1,585, 525, 8.8 mg kg^{-1} of Pb, Zn and Cd, respectively) and remediated soil (313, 378, 2.5 mg kg^{-1} of Pb, Zn and Cd, respectively) from Meža Valley was measured. Concentrations of toxic metals in edible parts of plants grown in original polluted soil were in all cases significantly higher than in plants grown in remediated soil (Table 21.1). Concentrations of Pb and Cd in edible plant tissues were compared with the EU regulation (EC 1881/2006). Zn, being an essential nutrient, is not restricted by European legislation. Maximum levels of Pb in EU vary according to plant species: onion, basil, pepper and carrot, 0.1; cauliflower, spinach, lettuce and Chinese cabbage, 0.3; and peas, 0.2 mg Pb per kg^{-1} fresh vegetable weight. The same applies for Cd, where the maximum values are onion and carrot, 0.1; pepper, cauliflower and Chinese cabbage, 0.05; and basil, spinach and lettuce 0.2 mg Cd kg^{-1} wet vegetable weight. In this study in the original soil, the levels of Pb in edible parts of spinach, cabbage, lettuce and carrot all exceeded the European regulation (Table 21.1). Vegetables grown on remediated soil, on the other hand, did not exceed Pb limits. Pb was not transported to edible parts of any vegetable grown on remediated soil. Cadmium was found in the edible parts of all vegetables grown in the original soil, except in onion. The element exceeded the levels set by the EU (EC 1881/2006) for peas, cauliflower, spinach, carrot and pepper. After remediation, cadmium concentrations in plants were reduced but were still higher than the maximum allowed EU levels for onion, pepper and spinach.

Different grass species and horticultural plants are commonly used for greening of degraded areas (Vangronsveld et al. 1995). In a field experiment, the highly contaminated and remediated soils were laid into two garden beds and subjected to cultivation of ornamental plant *Verbena hybrida* and grass *Dactylis glomerata*. Plants grown on remediated soil demonstrated similar biomass yield. Uptake of Pb, Zn and Cd decreased up to 7, 2 and 3 times, respectively, compared to plants cultivated on the original soil (Table 21.1). Zupanc et al. (2014) reported that concentration of Pb in the aboveground parts (shoots) of white clover (*T. repens*) was under detection in plants grown in original and EDTA-washed soil (Table 21.1), indicating that the Pb absorption from the contaminated soils in some plants is limited. Concentrations of Cd in shoots were under detection limit only in plants from remediated soils; Zn is a micronutrient; therefore, higher amount in plant tissue was expected. Nevertheless, the difference in Zn concentrations in shoots between original and remediated soil was significant (Table 21.1). The obtained results suggest that EDTA remediation produced soil suitable for use as a substrate in horticulture.

21.7.2 Potential to Use Remediated Soil for Food Production

Soils are the basis for more than 90 % of worldwide food production. As a result of growing demand for food and accelerating soil degradation processes, driven by anthropogenic factors and climate change, a further increase in pressure on limited soil resources will take place, increasing the need to use abandoned contaminated land, such as former landmine sites and urban areas. Already urban allotment gardens are being offered by local governments to encourage low-income and senior citizens to produce their own food (Tei et al. 2010). In the United Kingdom, 87 % of households tend a domestic garden (Gibbons et al. 2011). Metal contamination, especially with Pb, is however ubiquitous and generally higher in urban soils. Urban areas display a (older) housing stock with Pb-containing paint, high concentration of industry and heavy traffic. Since Pb is highly immobile in soils, concern about soil contamination persists despite the fact that most Pb was removed from residential paints and gasoline approximately 25 years ago.

Across all socio-economic groups and around the world, urban horticulture even on contaminated soil is booming, fulfilling diverse functions including food production, community building and reducing socio-economic tensions. Traditionally, the agricultural community has given little attention to the potential health effect of contaminated (urban) soil. Thus, crop samples from inner city and other urban vegetable gardens are frequently reported to exceed EU standards (EC 1881/2006) for toxic metal concentration in food crops. Exceedingly high levels of Pb were measured in more than half of crop samples from inner city vegetable gardens Berlin (Germany) (Säumel et al. 2012). Clark et al. (2008) surveyed 141 backyard gardens in Roxbury and Dorchester, MA, USA. Cultural practice of backyard gardening is characteristic for these two areas. More than 80 % of gardens had Pb level above the US EPA action limit of 400 mg kg^{-1}. Finster et al. (2004) surveyed the Pb levels of edibles grown in contaminated residential soils and reported that the majority of Pb was concentrated in roots. Edible portions of leafy vegetables and herbs also had Pb levels that, if consumed, could contribute to the total body burden of Pb. The Pb concentration in fruiting vegetables was usually below the limit of quantification. Jelusic et al. (2014b) reported the same pattern of Pb transference from soil to plants (Table 21.1). It is therefore of the vital importance to develop remedial strategies for urban gardens soils and ensure safety of edibles produced.

As shown in this chapter, soil washing with EDTA deteriorates some physical soil properties and deprives the soil of important micronutrients. It was also shown, however, that revitalisation measures have potential to restore quality of remediated soil as a plant substrate. EDTA soil washing efficiently removes toxic metals from the soil and decreases their availability. Further field experiments indicated no trends of shifts of toxic metals remaining in the soil after remediation towards more orally and phytoavailable and leachable forms. In soils from Meža Valley, EDTA soil washing also significantly reduced the transfer of Pb from soil to the edible parts of tested vegetables and herbs. Cd phytoaccumulation was however still substantial and in many tested plants above the level stipulated by the European Union legislation. Additional measures may need to be developed and introduced into the remediation scheme, for example, further immobilisation of residual Cd and other toxic metals into the solid matrix of remediated soil with phosphates, zeolites and other metal immobilising minerals. Zn is a micronutrient and its concentration in edibles is therefore not regulated. Nevertheless, Pb is the major and most prevalent metallic contaminant in many urban and other contaminated areas. Does therefore EDTA soil washing in the current stage of technology development promise safe use of Pb-contaminated and remediated soil as a substrate for food production? Presented results are indeed promising but solely based on remediation trials with Pb-, Zn- and Cd-contaminated soils from Meža Valley in Slovenia. They must therefore not be generalised to all soils contaminated with toxic metals. There are many different soil types with different properties, origins and history of contamination, and the remediation/revitalisation outcomes could also prove to be different. The decision on use of remediated soil for food production will probably depend on the case by case evaluation of the remedial efficiency, selections of cultivars, soil and site specifics, tradition and social acceptability of such venture.

References

Ager P, Marshall WD (2001) The removal of metals and release of EDTA from pulp wash water. J Wood Sci Technol 21:413–425
Alef K, Nannipieri P (1995) Methods in applied soil microbiology and biochemistry, 1st edn. Academic, London
Almaroai YA, Usman AR, Ahmad M, Kim KR, Vithanage M, Ok YS (2013) Role of chelating agents on release kinetics of metals and their uptake by maize from chromated copper arsenate-contaminated soil. Environ Technol 34:747–755
Baker LR, White PM, Pierzynski GM (2011) Changes in microbial properties after manure, lime, and bentonite application to a heavy metal-contaminated mine waste. Appl Soil Ecol 48:1–10
Barona A, Aranguiz I, Elias A (2001) Metal associations in soils before and after EDTA extractive decontamination: implications for the effectiveness of further clean up procedures. Environ Pollut 113:79–85
Bear J (1972) Dynamics of fluids in porous media. American Elsevier, New York
Bohlen PJ (2002) Earthworms. In: Lal R (ed) Encyclopedia of soil science. Marcel Dekker, New York, pp 370–373
Broadley MR, Willey NJ, Wilkins JC, Baker AJM, Mead A, White PJ (2001) Phylogenetic variation in heavy metal accumulation in angiosperms. New Phytol 152:9–27
Bronick CJ, Lal R (2005) Soil structure and management: a review. Geoderma 124:3–22
Brown GA, Elliot HA (1991) Influence of electrolytes on EDTA extraction of Pb from contaminated soil. Water Air Soil Pollut 62:157–165
Chen H, Cutright T (2001) EDTA and HEDTA effects on Cd, Cr, and Ni uptake by *Helianthus annuus*. Chemosphere 45:21–28
Clark H, Hausladen DM, Brabander DJ (2008) Urban gardens: lead exposure, recontamination mechanisms, and implications for remediation design. Environ Res 107:312–319
COM (2006) 232 Proposal for a Directive of the European Parliament and of the Council establishing a framework for the protection of soil
Conder JM, Lanno RP, Basta NT (2001) Assessment of metal availability in smelter soil using earthworms and chemical extractions. J Environ Qual 30:1231–1237
Council Directive (1999) 1999/31/EC relating to the landfill of waste
Curry JP, Schmidt O (2007) The feeding ecology of earthworms – a review. Pedobiol 50:463–477
Cushnie GC (2009) Pollution prevention and control technologies for plating operation, 2nd edn. National Center for Manufacturing Sciences, Ann Arbor
Dao L, Morrison L, Kiely G, Zhang C (2013) Spatial distribution of potentially bioavailable metals in surface soils of a contaminated sports ground in Galway, Ireland. Environ Geochem Health 35:227–238
Davis S, Mirick DK (2006) Soil ingestion in children and adults in the same family. J Expo Sci Environ Epidemiol 16:63–75
Demir A, Koleli N (2013) The sequential use of washing and an electrochemical reduction process for the remediation of lead-contaminated soils. Environ Technol 34:799–805
Dexter AR (2004) Soil physical quality: part I. Theory, effects of soil texture, density, and organic matter, and effects on root growth. Geoderma 120:201–214
Di Palma L, Ferrantelli P, Merli C, Biancifiori F (2003) Recovery of EDTA and metal precipitation from soil washing solutions. J Hazard Mater 103:153–168
Duggan MJ, Inskip MJ (1985) Childhood exposure to lead in surface dust and soil: a community health problem. Public Health Rev 13:1–54
EC (2011) More research needed into contaminated soil and water treatment. Science for Environment Policy. European Commission DG Environment News Alert Service, Brussels
EEA (2007) State of the environment No 1 chapter 2. Office for official publications of the European communities, Brussels
EEA (2010) The European environment. State and outlook 2010, soil thematic assessment. European Environment Agency, Copenhagen

El-Hady OA, El-Dewiny CY (2006) The conditioning effect of composts (natural) or/and acrylamide hydrogels (synthesized) on a sandy calcareous soil 1. Growth response, nutrients uptake and water and fertilizers use efficiency by tomato plants. J Appl Sci Res 2:890–898
Elzinga EJ, Sparks DL (2002) X-ray absorption spectroscopy study of the effects of pH and ionic strength on Pb(II) sorption to amorphous silica. Environ Sci Technol 36:4352–4357
Epelde L, Hernández-Allica J, Becerril JM, Blanco F, Garbisu C (2008) Effects of chelates on plants and soil microbial community: comparison of EDTA and EDDS for lead phytoextraction. Sci Total Environ 401:21–28
ESBNEC (2005) Soil Atlas of Europe. Office for Official Publications of the European Communities, Luxembourg
Finster ME, Gray KA, Binns H (2004) Lead levels of edibles grown in contaminated residential soils: a field survey. Sci Toltal Environ 320:245–257
Finzgar N, Lestan D (2007) Multi-step leaching of Pb and Zn contaminated soils with EDTA. Chemosphere 66:824–832
Finzgar N, Lestan D (2008) The two-phase leaching of Pb, Zn and Cd contaminated soil using EDTA and electrochemical treatment of the washing solution. Chemosphere 73:1484–1491
Finzgar N, Tlustos P, Lestan D (2007) Relationship of soil properties to fractionation, bioavailability and mobility of lead and zinc in soil. Plant Soil Environ 53:225–238
Finzgar N, Jez E, Voglar D, Lestan D (2014) Spatial distribution of metal contamination before and after remediation in the Meža Valley, Slovenia. Geoderma 217–218:135–143
Fitter AH, Moyersoen B (1996) Evolutionary trends in root-microbe symbioses. Philos Trans R Soc 351:1367–1375
Gianfreda L, Rao MA, Piotrowska A, Palumbo G, Colombo C (2005) Soil enzyme activities as affected by anthropogenic alterations: intensive agricultural practices and organic pollution. Sci Total Environ 341:265–279
Gibbons S, Mourato S, Resende G (2011) The amenity value of English nature: a hedonic price approach. London School of Economics and Political Sciences, London
Glorennec P, Declercq C (2007) Performance of several decision support tools for determining the need for systematic screening of childhood lead poisoning around industrial sites. Eur J Public Health 17:47–52
Grcman H, Velikonja-Bolta S, Vodnik D, Kos B, Lestan D (2001) EDTA enhanced heavy metal phytoextraction: metal accumulation, leaching and toxicity. Plant Soil 235:105–114
Hamlin R (2006) Molybdenum. In: Barker AV, Pilbeam DJ (eds) Handbook of plant nutrition. Taylor and Francis, Boca Raton, pp 375–394
Han FX, Banin A, Kingery WL, Triplett GB, Zhou LX, Zheng SJ, Ding WX (2003) New approach to studies of heavy metal redistribution in soil. Adv Environ Res 8:113–120
Heikens A, Pejinenburg WJGM, Hendriks AJ (2001) Bioaccumulation of heavy metals in terrestrial invertebrates. Environ Pollut 113:385–393
Houben D, Evrard L, Sonnet P (2013) Beneficial effects of biochar application to contaminated soils on the bioavailability of Cd, Pb and Zn and the biomass production of rapeseed (*Brassica napus* L.). Biomass Bioen 57:196–204
Jelusic M, Lestan D (2014) Effect of EDTA washing of metal polluted garden soils. Part I: toxicity hazards and impact on soil properties. Sci Total Environ 475:132–141
Jelusic M, Grcman H, Vodnik D, Suhadolc M, Lestan D (2013) Functioning of metal contaminated garden soil after remediation. Environ Pollut 174:63–70
Jelusic M, Vodnik D, Macek I, Lestan D (2014) Effect of EDTA washing of metal polluted garden soils. Part II: can remediated soil be used as a plant substrate? Sci Total Environ 475:142–152
Jelusic M, Vodnik D, Lestan D (2014) Revitalization of EDTA-remediated soil by fertilization and soil amendments. Ecol Eng 73:429–438
Juang RS, Wang SW (2000) Electrolytic recovery of binary metals and EDTA from strong complexed solutions. Water Res 34:3179–3185
Kalia K, Flora SJS (2005) Strategies for safe and effective therapeutic measures for chronic arsenic and lead poisoning. J Occup Health 47:1–21

Karim MA, Khan LI (2012) Enhancement of electrokinetic decontamination with EDTA. Environ Technol 33:2291–2298

Kemper WD, Rosenau RC (1986) Aggregate stability and size distribution. In: Klute A (ed) Methods of soil analysis. Part 1, physical and mineralogical methods. American Society for Agronomy Inc, Madison, pp 425–442

Kim C, Lee Y, Ong SK (2003) Factors effecting EDTA extraction of lead from lead-contaminated soils. Chemosphere 51:845–853

Klute A, Dirksen C (1986) Hydraulic conductivity and diffusivity: laboratory methods. In: Klute A (ed) Methods of soil analysis. Part 1, physical and mineralogical methods. American Society for Agronomy Inc, Madison, pp 687–734

Kos B, Grcman H, Lestan D (2003) Phytoextraction of lead, zinc and cadmium from soil by selected plants. Plant Soil 253:403–411

Laidlaw MAS, Zahran S, Mielke HW, Taylor MP, Filippelli GM (2012) Re-suspension of lead contaminated urban soil as a dominant source of atmospheric lead in Birmingham, Chicago, Detroit and Pittsburgh, USA. Atmos Environ 49:302–310

Laird D, Fleming P, Wang B, Horton R, Karlen D (2010) Biochar impact on nutrient leaching from a Midwestern agricultural soil. Geoderma 158:436–442

Langdon CJ, Piearce TG, Meharg AA, Semple KT (2001) Survival and behavior of the earthworms *Lumbricus rubellus* and *Dendrodrilus rubidus* from arsenate-contaminated and non-contaminated sites. Soil Biol Biochem 33:1239–1244

Lanigan RS, Yamarik TA (2002) Final report on the safety assessment of EDTA, calcium disodium EDTA, diammonium EDTA, dipotassium EDTA, disodium EDTA, TEA-EDTA, tetrasodium EDTA, tripotassium EDTA, trisodium EDTA, HEDTA, and trisodium HEDTA. Int J Toxicol 21:95–142

Lehmann J (2007) A handful of carbon. Nature 447:143–144

Lei M, Liao BH, Zeng QR, Khan S (2008) Fraction distributions of lead, cadmium, copper, and zinc in metal-contaminated soil before and after extraction with disodium ethylenediaminetetraacetic acid. Comm Soil Sci Plant Anal 39:1963–1978

Lestan D, Udovic M (2011) Mobility and availability of toxic metals after soil washing with chelating agents. In: Khan MS (ed) Biomanagement of metal-contaminated soils. Springer, Dordrecht, pp 343–364

Lestan D, Grcman H, Zupan M, Bacac N (2003) Relationship of soil properties to fractionation of Pb and Zn in soil and their uptake into *Plantago lanceolata*. Soil Sed Contam 12:507–522

Lestan D, Hanc A, Finzgar N (2005) Influence of ozonation on extractability of Pb and Zn from contaminated soils. Chemosphere 61:1012–1019

Lestan D, Luo C, Li X (2008) The use of chelating agents in the remediation of metal-contaminated soils: a review. Environ Pollut 15:13–33

Li MS (2006) Ecological restoration of mineland with particular reference to the metalliferous mine wasteland in China: a review of research and practice. Sci Total Environ 357:38–53

Lindsay WL, Norvell WA (1978) Development of a DTPA test for zinc, iron, manganese and copper. Soil Sci Soc Am J 42:421–428

Macek I, Šibanc N, Kavšček M, Lestan D (2014) Biodiversity of arbuscular mycorrhizal fungi in metal polluted and remediated garden soils before and after soil revitalisation with commercial and indigenous fungal inoculum. 33rd new phytologist symposium, Zurich, p 49

Manouchehri N, Besancon S, Bermond A (2006) Major and trace metal extraction from soil by EDTA: equilibrium and kinetic studies. Anal Chim Acta 559:105–112

Martell AE, Smith RM (2003) NIST critically selected stability constants of metal complexes Version 7.0. U.S. Department of Commerce, Gaithersburg

Meers E, Hopgood M, Lesage E, Vervaeke P, Tack FMG, Verloo MG (2004) Enhanced phytoextraction: in search of EDTA alternatives. Int J Phytorem 6:95–109

Mijangos I, Perez R, Albizu I, Garbisu C (2006) Effects of fertilization and tillage on soil biological parameters. Enzyme Microb Technol 40:100–106

Montanarella L, Vargas R (2012) Global governance of soil resources as a necessary condition for sustainable development. Curr Opin Terr Syst Environ Sustain 4:559–564

Mulligan CN, Yong RN, Gibbs BF (2001) Remediation technologies for metal-contaminated soils and groundwater: an evaluation. Eng Geol 60:193–207

Narjary B, Aggarwal P, Singh A, Chakraborty D, Singh R (2012) Water availability in different soils in relation to hydrogel application. Geoderma 187–188:94–101

EC No 1881 (2006) Setting maximum levels for certain contaminants in foodstuffs. Official Journal of the European Union L 364/5

Nonhebel S, Kastner T (2011) Changing demand for food, livestock feed and biofuels in the past and in the near future. Livest Sci 139:3–10

Noren K, Loring JS, Bargar JR, Persson P (2009) Adsorption mechanisms of EDTA at the water–iron oxide interface, implications for dissolution. J Phys Chem C 113:7762–7771

Nowack B, Sigg L (1996) Adsorption of EDTA and metal-EDTA complexes onto goethite. J Colloid Interface Sci 177:106–121

Nowack B, Schulin R, Robinsos BH (2006) Critical assessment of chelant-enhanced metal phytoextraction. Environ Sci Technol 40:5225–5232

Odendaal JP, Reinecke AJ (2004) Evidence of metal interaction in the bioaccumulation of cadmium and zinc in *Porcellio laevis* (Isopoda) after exposure to individual and mixed metals. Water Air Soil Pollut 156:1–4

Ozturk L, Karanlik S, Ozkutlu F, Cakmak I, Kochian LV (2003) Shoot biomass and zinc/cadmium uptake for hyperaccumulator and non-accumulator *Thlaspi* species in response to growth on a zinc-deficient calcareous soil. Plant Sci 164:1095–1101

Pearl M, Pruijn M, Bovendeur J (2006) The application of soil washing to the remediation of contaminated soils. Land Contam Reclamat 14:713–726

Pociecha M, Lestan D (2010) Electrochemical EDTA recycling with sacrificial Al anode for remediation of Pb contaminated soil. Environ Pollut 158:2710–2715

Pociecha M, Lestan D (2012a) Novel EDTA and process water recycling method after soil washing of multi-metal contaminated soil. J Hazard Mater 201–202:273–279

Pociecha M, Lestan D (2012b) Washing of metal contaminated soil with EDTA and process water recycling. J Hazard Mater 235–236:384–387

Pociecha M, Lestan D (2012c) Recycling of EDTA solution after soil washing of Pb, Zn, Cd and As contaminated soil. Chemosphere 86:843–846

Rana A, Joshi M, Prasanna R, Shivay YS, Nain L (2012) Biofortification of wheat through inoculation of plant growth promoting rhizobacteria and cyanobacteria. Eur J Soil Biol 50:118–126

Rieuwerts JS, Thornton ME, Farago ME, Ashmore MR (1998) Factors influencing metal bioavailability in soils: preliminary investigations for the development of a critical loads approach for metals. Chem Spec Bioavail 10:61–75

Ruby MV, Davis A, Schoof R, Eberle S, Sellstone CM (1996) Estimation of lead and arsenic bioavailability using a physiologically based extraction test. Environ Sci Technol 30:422–430

Ruley AT, Sharma NC, Sahi SV, Singh SR, Sajwan KS (2006) Effects of lead and chelators on growth, photosynthetic activity and Pb uptake in *Sesbania drummondii* grown in soil. Environ Pollut 144:11–18

Salaskar D, Shrivastava M, Kale SP (2011) Bioremediation potential of spinach (*Spinacia oleracea L.*) for decontamination of cadmium in soil. Curr Sci 101:1359–1363

Säumel I, Kotsyuk I, Hölscher M, Lenkereit C, Weber F, Kowarik I (2012) How healthy is urban horticulture in high traffic areas? Trace metal concentrations in vegetable crops from plantings within inner city neighbourhoods in Berlin, Germany. Environ Pollut 165:124–132

Sizmur T, Hodson ME (2009) Do earthworms impact metal mobility and bioavailability in soil? – a review. Environ Pollut 157:1981–1989

Sousa A, Pereira R, Antunes SC, Cachada A, Pereira E, Duarte AC, Gonçalves F (2008) Validation of avoidance assays for the screening assessment of soils under different anthropogenic disturbances. Ecotoxicol Environ Saf 71:661–670

Tebo BM, Johnson HA, McCarthy JK, Templeton AS (2005) Geomicrobiology of manganese (II) oxidation. Trends Microbiol 13:421–428
Tei F, Benincasa P, Farneselli M, Caprai M (2010) Allotment gardens for senior citizens in Italy: current status and technical proposals. Acta Hortic 881:91–96
Tessier A, Campbell PGC, Bisson M (1979) Sequential extraction procedure for the speciation of particulate trace metals. Anal Chem 51:844–851
Tsang DCW, Zhang WH, Lo IMC (2007) Copper extraction effectiveness and soil dissolution issues of EDTA-flushing of contaminated soils. Chemosphere 68:234–243
Tsonev T, Jose F, Lidon C (2012) Zinc in plants – an overview. Plant Sci 24:322–333
Turnau K, Anielska T, Ryszka P, Gawroński S, Ostachowicz B, Jurkiewicz A (2008) Establishment of arbuscular mycorrhizal plants originating from xerothermic grasslands on heavy metal rich industrial wastes – new solution for waste revegetation. Plant Soil 305:267–280
Udovic M, Lestan D (2007) The effect of earthworms on the fractionation and bioavailability of heavy metals before and after soil remediation. Environ Pollut 148:663–668
Udovic M, Lestan D (2009) Pb, Zn and Cd mobility, availability and fractionation in aged soil remediated by EDTA leaching. Chemosphere 74:1367–1373
Udovic M, Lestan D (2010a) *Eisenia fetida* avoidance behaviour as a tool for assessing the efficiency of remediation of Pb, Zn and Cd polluted soil. Environ Pollut 158:2766–2772
Udovic M, Lestan D (2010b) Fractionation and bioavailability of Cu in soil remediated by EDTA leaching and processed by earthworms (*Lumbricus terrestris* L.). Environ Sci Pollut Res 17:561–570
Udovic M, Lestan D (2012) EDTA and HCl leaching of calcareous and acidic soils polluted with potentially toxic metals: remediation efficiency and soil impact. Chemosphere 88:718–724
Udovic M, Plavc Z, Lestan D (2007) The effect of earthworms on the fractionation, mobility and bioavailability of Pb, Zn and Cd before and after soil leaching with EDTA. Chemosphere 70:126–134
Udovic M, Drobne D, Lestan D (2009) Bioaccumulation in *Porcellio scaber* (Crustacea, Isopoda) as a measure of the EDTA remediation efficiency of metal-polluted soil. Environ Pollut 157:2822–2829
US EPA (1995) Test methods for evaluation of solid waste, vol IA. Laboratory Manual Physical/ Chemical Methods, SW 86, 40 CFR Parts 403 and 503. US Government Printing Office, Washington, DC
US EPA (2004) Cleaning up the nation's waste sites: markets and technology trends, 46th edn. US Government Printing Office, Washington, DC
Vangronsveld J, Assche FV, Clijsters H (1995) Reclamation of a bare industrial area contaminated by non-ferrous metals: in situ metal immobilization and revegetation. Environ Pollut 87:51–59
Vassil AD, Kapulnik Y, Raskin I, Salt DE (1998) The role of EDTA in lead transport and accumulation by Indian mustard. Plant Physiol 117:447–453
Voglar D, Lestan D (2013) Pilot-scale washing of Pb, Zn and Cd contaminated soil using EDTA and process water recycling. Chemosphere 91:76–82
Voglar D, Lestan D (2014) Chelant soil-washing technology for metal-contaminated soil. Environ Technol 35:1389–1400
Vulava VM, Seaman JC (2000) Mobilization of lead from highly weathered porous material by extracting agents. Environ Sci Technol 34:4828–4834
Weng L, Temminghoff EJM, Lofts S, Tipping E, Van Riemsdijk WH (2002) Complexation with dissolved organic matter and solubility control of heavy metals in a sandy soil. Environ Sci Technol 36:4804–4810
Williams LE, Pittman JK (2010) Dissecting pathways involved in manganese homeostasis and stress in higher plant cells. In: Hell R, Mendel RR (eds) Cell biology of metals and nutrients. Springer, Berlin, Heidelberg, pp 95–118
Wong CSC, Li X (2006) Thornton I. Urban environmental geochemistry of trace metals, review. Environ Pollut 142:1–16

Wragg J, Cave M, Basta N, Brandon E, Casteel S, Denys S, Gron C, Oomen A, Reimer K, Tack K, Van de Wiele T (2011) An inter-laboratory trial of the unified BARGE bioaccessibility method for arsenic, cadmium and lead in soil. Sci Total Environ 409:4016–4030
Wu C, Zhang L (2010) Heavy metal concentrations and their possible sources in paddy soils of a modern agricultural zone, southeastern China. Environ Earth Sci 60:45–56
Yamaguchi Y, Yamanaka Y, Miyamoto M, Fujishima A, Honda K (2003) Hybrid electrochemical treatment for persistent metal complexes at conductive diamond electrodes and clarification of its reaction route. J Electrochem Soc 153:1123–1132
Zeng GR, Sauve S, Allen HE, Hendershot WH (2005) Recycling EDTA solutions used to remediate metal-polluted soils. Environ Pollut 133:225–231
Zhang FP, Li CF, Tong LG, Yue LX, Li P, Ciren YJ, Cao CG (2010a) Response of microbial characteristics to heavy metal pollution of mining soils in central Tibet, China. Appl Soil Ecol 45:144–151
Zhang W, Huang H, Tan F, Wang H, Qui R (2010b) Influence of EDTA washing on the species and mobility of heavy metals residual in soil. J Hazard Mater 173:369–376
Zupanc V, Kastelec D, Lestan D, Grcman H (2014) Soil physical characteristics after EDTA washing and amendment with inorganic and organic additives. Environ Pollut 186:56–62

Part V
Genetic Engineering: A Tool to Cleaning Up Contaminated Soils

Chapter 22
Genetic Engineering of Plants for Heavy Metal Removal from Soil

Umesh B. Jagtap and Vishwas A. Bapat

22.1 Introduction

Heavy metals are elements with metallic properties and have atomic mass >20 and specific gravity above 5 g cm^{-3} (Rascio and Navari-Izzo 2011). The most common heavy metal contaminants are arsenic (As), cadmium (Cd), chromium (Cr), copper (Cu), mercury (Hg), lead (Pb), and zinc (Zn) (Ha et al. 2014). Metals are natural components present in the soil (Tangahu et al. 2011). A large amount of hazardous materials including heavy metals were released into the environment from natural (e.g., geological erosion and saline seeps) and extensive anthropogenic (e.g., mining, agriculture, industry, wastewater treatment, construction) activities, which cause soil, air, and water pollution (Arthur et al. 2005). Some metals are micronutrients (requires at low concentrations) necessary for plant growth, such as Zn, Cu, manganese (Mn), nickel (Ni), and cobalt (Co), while others have unknown biological functions, such as Cd, Pb, and Hg (Appenroth 2010). At higher concentration these metals exert toxic effects on plant and animal health including human (Table 22.1). Unlike organic contaminants heavy metal does not undergo biodegradation and persist in the soil for a long time; therefore, its removal from the soil is receiving great attention.

The conventional (e.g., soil excavation, land filling, soil washing) and physico-chemical (e.g., thermal treatment, chemical extraction, encapsulation) methods were employed to remediate contaminated soil. These methods are expensive, inefficient especially for large-scale cleanup, and non-eco-friendly as they destroy natural habitat and leave unsightly scars on the landscape. Consequently, alternative biological methods such as bioremediation (use of living organisms), phytoremediation (use of plants), and zoo remediation (use of animals) (Gifford et al. 2006) are developed for environmental remediation. Of these,

U.B. Jagtap • V.A. Bapat (✉)
Department of Biotechnology, Shivaji University Vidyanagar, Kolhapur, MS 416004, India
e-mail: vabapat@gmail.com

I. Sherameti, A. Varma (eds.), *Heavy Metal Contamination of Soils*, Soil Biology 44, DOI 10.1007/978-3-319-14526-6_22

Table 22.1 General properties, sources, and toxic effects of some heavy metals on plant and human health

Heavy metal	General properties	Sources	Effects on plants	Effects on human
Arsenic (As)	Z: 33 at. wt.: 74.92 SD: 5.73 MP: 817 °C BP: 613 °C	Mining, ore dressing, and smelting of nonferrous metals, production of As and As compounds, petroleum and chemical industry, pesticides, beer, table salt, tap water, paints, pigments, cosmetics, glass and mirror manufacture, fungicides, insecticides, treated wood and contaminated food, dyestuff, and tanning industry	Inhibits photosynthesis, inhibits growth, biomass, and yield; death	As intake: leads to gastrointestinal symptoms; severe disturbances of the cardiovascular and central nervous systems; bone marrow depression; hemolysis; hepatomegaly; melanosis; polyneuropathy; encephalopathy; lung, bladder, kidney, and skin cancer; and other skin lesions
Cadmium (Cd)	Z: 48 at. wt.: 112.4 D: 8.65 g cm^{-3} MP: 321 °C BP: 765 °C A soft, silvery white, ductile metal with a faint bluish tinge	Industrial processes, farming practices, volcanic eruption, mining, ore dressing, smelting of nonferrous metals, battery manufacturing, cigarettes, processed and refined foods, large fish, shellfish, tap water, auto exhaust, plated containers, galvanized pipes, air pollution from incineration, occupational exposure	Chlorosis, growth inhibition, reduction in photosynthesis, water and nutrient uptake, browning of root tips, death	Cd inhalation can cause acute pulmonary and sporadic effects. Cd exposure may cause kidney damage. Carcinogenic
Chromium (Cr)	Z: 24 at. wt.: 51.996 D: 7.14 g cm^{-3} MP: 1,900 °C BP: 2,642 °C A lustrous, brittle, hard metal	Cr compound production, leather-working industry, metal and plastic electroplate, dyestuff and dying by acidic medium,	Alteration in germination, inhibition of plant growth, chlorosis, nutrient imbalance, wilting of tops, root injury,	Cr breathing: irritation on the lining of the nose, nose ulcers, runny nose, asthma, cough, etc. Skin contact:

(continued)

Table 22.1 (continued)

Heavy metal	General properties	Sources	Effects on plants	Effects on human
		production and application of dyestuff, metal Cr smelting	inhibition of chlorophyll biosynthesis, photosynthesis, yield	skin ulcers, skin redness, skin swelling. Cr exposure: liver and kidney damage, skin irritation
Copper (Cu)	Z: 29 at. wt.: 63.546 D: 8.94 g cm^{-3} MP: 1,356 °C BP: 2,868 °C A rosy-pink transition metal	Mining; milling; smelting; agriculture; waste disposal; local sources, such as foundries and smelters; application of fungicides and sewage sludge. Cu added to tap water; pesticides; intrauterine devices; dental amalgams; nutritional supplements, especially prenatal vitamins; birth control pills; weak adrenal glands; and occupational exposure	Chlorosis, necrosis, stunting, and inhibition of root and shoot growth. Inhibition of enzyme activity or protein function, impaired cell transport processes, and oxidative damage and metabolic disturbances	Cu toxicity: excessive oxidative stress and tissue damage, abdominal pain, nausea, vomiting, headache, lethargy, diarrhea, respiratory difficulties, hemolytic anemia, gastrointestinal bleeding, liver and kidney failure, and death
Lead (Pb)	Z: 82 at. wt.: 207 SD: 11.35 MP: 327.5 °C BP: 1,740 °C A bluish-white metal of bright luster and is soft, very malleable, ductile, and a poor conductor of electricity	Construction and application of pipes, used for batteries, cable coverings, plumbing, ammunition, fuel, additives, paint pigments, PVC plastics, X-ray shielding, crystal glass production, pesticides, tap water, cigarette smoke, hair dyes, paints, inks, glazes, etc.	Morphology, growth, photosynthesis, inhibition of enzyme activities, water imbalance, alteration in membrane permeability, oxidative stress	Pb exposure: headaches, irritability, abdominal pain, affects the nervous system, Pb encephalopathy, carcinogenic
Mercury (Hg)	Z: 80 at. wt.: 200.59 SD: 13.5 MP: −39 °C	Production and application of Hg catalyst in chemical industry, Hg	Obstruction of water flow, interference of mitochondrial activity,	Hg exposure: lung and kidney damage; other chronic

(continued)

Table 22.1 (continued)

Heavy metal	General properties	Sources	Effects on plants	Effects on human
	BP: −357 °C Metallic form, it volatilizes readily at room temperature	battery manufacturing, smelting and restoring of Hg, Hg compound production, pesticide and medicine making, production and application of fluorescent light and Hg lamps, Hg slime from caustic soda production, dental amalgams, large fish, shellfish, medications, manufacture of paper, chlorine, adhesives, fabric softeners, and waxes	oxidative stress, disruption of biomembrane lipids and cellular metabolism, affects photosynthesis	poisoning-like neurological and psychological symptoms, such as tremor; changes in personality, restlessness, anxiety, sleep disturbance, and depression
Nickel (Ni)	Z: 28 at. wt.: 58.71 D: 8.9 g cm^{-3} MP: 1,455 °C BP: 2,732 °C	Residue from the production of nickeliferous compounds; abandoned nickeliferous catalysts; nickeliferous residue and waste from electroplate technology; nickeliferous waste from analysis, assay, and testing activity; hydrogenated oils (margarine, commercial peanut butter, and shortening); shellfish; air pollution; cigarette smoke; plating; occupational exposure	Inhibits chlorophyll biosynthesis, chlorosis, necrosis, water and nutrient imbalance, disorder of cell membrane functions, wilting, browning of root tips	Ni exposure: skin allergies, lung fibrosis, kidney and cardiovascular system poisoning, stimulation of neoplastic transformation, carcinogenic
Zinc (Zn)	Z: 30 at. wt.: 65.39 D: 7.133 g cm^{-3} MP: 419.6 °C BP: 907 °C A bluish-white,	Mining, ore dressing, smelting of nonferrous metals, metal and plastic electroplate, pigment, beaded	Inhibition of root growth, senescence, chlorosis, oxidative stress	High Zn intake relative to Cu and induce Cu deficiency. In humans multiple adverse effects

(continued)

Table 22.1 (continued)

Heavy metal	General properties	Sources	Effects on plants	Effects on human
	relatively soft metal	paint and rubber working, Zn compound production, Zinoky battery product industry		include decrease in Cu-dependent enzymes such as superoxide dismutase, ceruloplasmin, and cytochrome C oxidase and changes in immunological parameters, cholesterol, and its lipoprotein distribution

Z atomic number, *at. wt.* atomic weight, *SD* specific density, *D* density, *MP* melting point, *BP* boiling point

References: Denkhaus and Salnikow (2002), Hasanuzzaman and Fujita (2013), Jarup (2003), Martin and Griswold (2009), Sunitha et al. (2013), Tangahu et al. (2011), Uriu-Adams and Keen (2005)

phytoremediation has emerged as inexpensive, eco-friendly, and publicly acceptable remediation technology that utilizes plants and associated microorganisms not only to remove, transform, or stabilize contaminants in the water and soil but also help in CO_2 sequestration, soil stabilization, watershed management, biodiversity improvement, providing diverse sources of energy, and aesthetics (Dickinson et al. 2009).

However, the slow rate of metal removal and incomplete metabolism, as an autotroph plant, lack catabolic enzyme machinery necessary to achieve degradation/full mineralization of xenobiotic substances which stymied the progress of phytoremediation. This results into the accumulation of toxic metabolites as it is into the plant tissues that could be released into the environment (Aken 2008). The possibility of the release of toxic substances into the food chain limits the widespread utilization of phytoremediation. Genetic engineering of plants has the potentiality to overcome these challenges. This chapter summarizes our current knowledge of transgenic plants for heavy metal removal from contaminated area.

22.2 Basic Strategies of Phytoremediation for Metals

The use of plants and their associated microbes to bioremediate contaminated soil, sediments, and water is known as phytoremediation (Arthur et al. 2005; Cherian and Oliveira 2005). The plants used several different phytoremediation strategies such as rhizofiltration, phytostabilization, phytodegradation, phytoextraction, and phytovolatilization to decontaminate soil and water contaminated with several organic pollutants (Arthur et al. 2005). Out of these, rhizofiltration-, stabilization-, extraction-/accumulation-, and volatilization-based phytoremediation strategies were utilized by the plants to remove metals from contaminant soils (EPA 1998).

In rhizofiltration, plant root system is employed for the elimination of metals from contaminated wastewater, where they absorb and accumulate metals in the roots (Dushenkov et al. 1995). Similarly, the hairy roots induced in some plants by the *Agrobacterium* infection were also utilized for rhizofiltration of radionuclides and heavy metals (Eapen et al. 2003; Straczek et al. 2009).

Phytostabilization simply prevents/reduces the mobility and bioavailability of metals in the environment through immobilization of soil by plant roots (Salt et al. 1995). Therefore, in phytostabilization, even if metal concentration is not reduced, the migration of metals in the surrounding environment is prevented (Li et al. 2000).

In phytoextraction, plants absorb metals from contaminated soils and concentrate/accumulate them into harvestable plant parts (shoot, leaves, etc.). After harvesting, the plant parts may be ashed or utilized for metal recovery followed by disposal of the ashes in a landfill (Kumar et al. 1995). It is the most effective among several phytoremediation methods, although technical difficulties exist in their applications (Krämer 2005).

Phytovolatilization involves the use of plants to take up the metals from the soil, transforming them into volatile form, and release them through transpiration into the atmosphere (Bizily et al. 2003; Rugh et al. 1998). The details of all of these technologies are summarized by Arthur et al. (2005).

22.3 Improving Metal Phytoremediation with Genetic Engineering of Plants

Widespread utilization of phytoremediation can be limited by the small habitat range or size of plants expressing remediation potential and insufficient abilities of native plants to tolerate and accumulate contaminants (Arthur et al. 2005). Several approaches such as agronomic practices (planting density, fertilization) (Chaney et al. 2000), use of soil amendments (organic acids, synthetic chelators) (Joner 2013), and conventional breeding are utilized to increase biomass and metal uptake capacity of the suitable plants used for metal phytoremediation (Pilon-Smits and Pilon 2002).

However, in order to achieve a better phytoremediation efficiency, plants should have the ability to grow outside their area of collection, extensive root system, and fast growth rate; accumulate high amounts of heavy metals in their easily harvestable parts; tolerate soil pollution; and also produce a great quantity of biomass in contamination condition (Pilon-Smits 2005). The development of plants having all these traits is not possible through conventional breeding methods as these are time-consuming and laborious and have several other ecological, physiological, and biological constraints. On the other hand, the precision of biotechnological approaches, mainly genetic engineering, contributed rapid and significant changes in the crop improvement by offering a wide array of novel genes and traits which can be effectively inserted into candidate plants to improve their phytoremediation potential for metal removal.

22.4 Transgenic Plants for Heavy Metal Removal

Transgenic plants expressing desirable genes from different organisms are developed to increase the heavy metal remediation efficiency of plants. Two principle strategies have been pursued in the phytoremediation techniques, i.e., improved *in planta* and *ex planta* metabolism that may lead to enhanced removal of xenobiotics/heavy metal (Fig. 22.1). *In planta* process includes uptake and diffusion through the roots, trunk, or leaves, sorption and transformation, and/or sequestration via tree metabolic activity manipulation. Alternatively, *ex planta* process includes genetic engineering of plants for extracellular synthesis of reactive enzymes, metal-selective ligands (phytosiderophores or chelating agents), or plant-associated microorganisms (bacteria and entophytes) in the rhizosphere (James and Strand 2009; Ma and Nomoto 1996; Raskin 1996).

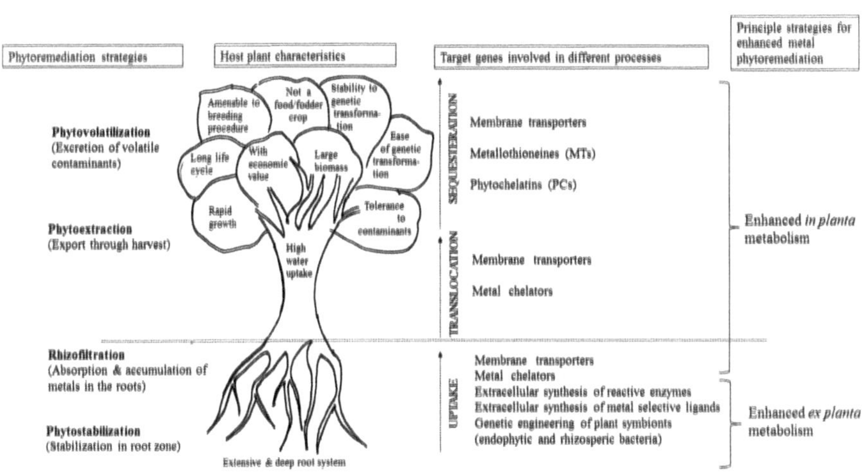

Fig. 22.1 Schematic representation of phytoremediation strategies and important characteristics of host plants along with targeted genes involved in various processes and suitable for genetic modification of plants to enhance phytoremediation of heavy metals

22.4.1 Host Systems

The plant system which is transformed for enhanced phytoremediation ability should possess certain characteristics such as extensive deep rooting system, fast growth rate with high biomass, tolerance to contaminants, ease of genetic transformation, stability of gene expression, and amenable to breeding procedures, and it should not be a food or fodder crop, and it has an advantage if it has an economic value (Kotrba et al. 2009). No single plant species can have all these characteristics. However, depending on phytoremediation strategies, types of heavy metal, and its intended final use, the plant species have to be selected.

22.4.2 Model Systems for Genetic Engineering Study

There is a great diversity in biological species present on the earth, and it is practically impossible to analyze each of them in detail. Since, phylogenetically related species display strong similarities in their genetic makeup, physiology, and behavior, at the molecular level, principle cellular signaling pathways are highly conserved. Biologists recognized the merits of employing model organisms as representatives of their species/subspecies to investigate the phenomena and mechanisms of development in great depth. Moreover, for most model organisms, protocols for transgenesis, full genome sequence information, and numerous bioinformatic resources are available, which offer significant informative data and tools that helps not only to improve plant phytoremediation properties in a highly targeted manner but also to disentangle biological complexity, to unravel networks of molecular interactions, and finally to predict mechanisms in distantly related species (Pitzschke 2013).

22.4.2.1 Whole Plant Systems

Arabidopsis and tobacco are considered as two of the best model species for phytoremediation-related transformation studies on metal tolerance and accumulation and for testing *in planta* expression of potential target gene constructs prior to transformation of other candidate species. The development of optimized tissue culture protocols and well-established genetic transformation methods makes *Arabidopsis* and tobacco ideal candidates for transformation studies. Additionally, this small nonfood plant with short generation time and prolific seed production provides an advantage to test and clarify the roles/functions of transgenes in a very short time.

22.4.2.2 In Vitro Cultures

In vitro culture involves growing plant cells and tissues aseptically on defined medium in environmentally controlled conditions (temperature, photoperiod, and darkness) (George et al. 2008). The in vitro dedifferentiated cells (callus or cell suspension), differentiated organs (roots and shoots), and genetically transformed (hairy roots and shooty teratomas) culture are convenient laboratory tools for phytoremediation studies related to metal removal (Doran 2009).

The in vitro culture systems offer a number of advantages over a whole plant system as follows. (1) Once established, these in vitro culture systems can be propagated indefinitely and are available on demand. (2) The in vitro (callus, cell suspensions) culture systems used as a suitable material to carry out nuclear transformation studies and also play an important role in the selection and confirmation of successful transformants having desired trait prior to time-consuming process of plant regeneration are carried out. (3) Use of in vitro cultures allows experiments to be carried out using material derived from the same parent plant avoiding the effects of variability between individual specimens. (4) An in vitro screening reduces not only the growth period and the treatment time length of the plants but also the space required for the experiments. (5) The in vitro systems allow the independent study of the complex interaction among plant/soil/microbiota to evaluate the participation of specific enzymes, organic compounds, transporters, or peptides involved in the plant response to the pollutants (Boominathan and Doran 2002; Doran 2009; Flocco and Giulietti 2007). (6) Cell cultures are also a useful system for metabolic engineering and for obtaining rapid evidence of the ecotoxicological behavior of chemicals and heavy metals in plants with less analytical expense (Golan-Goldhirsh et al. 2004). (7) Moreover, the environmental factor variability is also reduced, physiological activities can be increased by modifying the culture conditions (e.g., employing biotic and abiotic stress), and it is easier to isolate and analyze metabolites (Hu and Du 2006; Shanks and Morgan 1999). (8) When non-differentiated tissues are employed, genetic and epigenetic changes can be observed due to somaclonal variations (Lee and Phillips 1988). However, this variation and in vitro selection seem to be an appropriate technology for the development of new plant variants with enhanced metal accumulation and extraction properties (Herzig et al. 2003; Jan et al. 1997; Nehnevajova et al. 2007). (9) This approach also allows the analysis of metal accumulation properties of each organ (Kartosentono et al. 2001; Nedelkoska and Doran 2000a) and the possibility to develop industrial bioreactor models (Giri and Narasu 2000; Kim et al. 2002).

In spite of the number of advantages offered by in vitro culture systems in phytoremediation research, it is neither practical nor a feasible technology for direct large-scale phytoremediation applications. However, in vitro cell suspensions and hairy root culture are frequently used as model systems for understanding the metabolic and tolerance mechanisms that function in whole plants (Doran 2009).

22.4.3 Methods of Gene Transfer to the Plants

Significant achievements in vector-mediated/indirect and vectorless/direct gene transfer strategies have tremendously been assisting scientists to incorporate genes from any organisms to plants in a specific and targeted approach. Among the various vector-mediated transformation strategies, *Agrobacterium tumefaciens* plant-pathogenic soil bacterium has been widely used as a vector to create transgenic plants. This bacterium is known as "natural genetic engineer" of plants because these bacteria have natural ability to transfer a portion of its Ti (tumor inducing) plasmid and T-DNA (transfer DNA) to the genome of the host plant upon infection of cells at the wound site and cause an unorganized growth of a cell mass known as crown gall. Ti plasmids are used as gene vectors for delivering useful foreign genes into target plant cells and tissues. The foreign gene is cloned in the T-DNA region of Ti plasmid in place of unwanted sequences (Block 1993; Gelvin 2003). The T-DNA transfer and its integration into the plant genome are governed by various *Agrobacterium* and plant tissue-specific factors which include genotype of the plant, type of explant, plasmid vector, bacterial strain, composition and pH of culture medium, temperature and time of cocultivation, tissue damage, and suppression or elimination of *Agrobacterium* infection after cocultivation (Ziemienowicz 2014). *A. tumefaciens*-mediated transformation method has been extensively utilized for engineering dicotyledonous plants including tobacco (Gisbert et al. 2003; Pomponi et al. 2006), *A. thaliana* (Kim et al. 2005; Shukla et al. 2013), and Indian mustard (Gasic and Korban 2007; LeDuc et al. 2004) for heavy metal phytoremediation. However, monocotyledon plants and forest trees are considered as recalcitrant to *Agrobacterium*-mediated transformation. Therefore, considering the limitations of vector-mediated (*Agrobacterium*) gene transfer methods, the vectorless/direct gene transfer methods have been developed. The microprojectiles or biolistics or particle gun for gene transfer is widely employed in plant genetic engineering studies (Chen et al. 1998; Rugh et al. 1998). In this method, tungsten or gold microparticles coated with DNA of interest are accelerated to high velocity by a particle gun apparatus into living plant tissue. These particles with high kinetic energy penetrate the cells and membranes and carry foreign DNA inside of the bombarded cells (Block 1993).

Over the years, gene transfer techniques have been refined, and many underlying mechanisms have been decoded. However, there are several experimental prerequisites to be followed for effective gene delivery. Each plant has its own requirements, and it is necessary to establish a separate protocol of transformation individually. The choice of gene to be incorporated, type of vector, genotype, promoter, and reproducible regeneration system are some of the parameters that determine success of plant genetic transformation.

22.4.4 *Target Genes*

The plants were able to grow and colonize in heavy metal-contaminated sites by the development of several metal exclusion and tolerance mechanisms (Viehweger 2014). An integrated physiological, anatomical, biochemical, and molecular studies reveal the metal tolerance mechanism of plants as well as lead to the identification of target genes involved in various processes including metal uptake, translocation, and sequestration in plant. Therefore, the classical approach to engineer plants for enhanced heavy metal remediation consists in the strengthening of endogenous systems by upregulation of the expression or activity of target genes. Some of the target genes useful for genetic engineering of plants for enhanced heavy metal phytoremediation are enlisted below.

22.4.4.1 Membrane Transporters

Membrane transporters are likely to play a central role in translocation of heavy metals from source to root and to shoot and its further sequestration into specialized compartments (viz., vacuoles, chloroplast, mitochondria) to ensure sufficient levels to the necessary compartments while safely storing metals under times of excess (Cherian and Oliveira 2005; Palmer and Guerinot 2009). Recent developments in molecular biology, bioinformatics, omics, and sequencing technologies lead not only to the identification of novel vacuolar/plasma membrane-localized transporters but also understanding their role in heavy metal uptake, translocation, and sequestration. Many membrane transporters belonging to heavy metal ATPases, natural resistant-associated macrophage proteins (NRAMPs), cation diffusion facilitators (CDF), and ZIP (zinc-regulated transporter, iron-regulated transporter-related protein) gene families were identified in plant genome (Hall and Williams 2003; Krämer 2010). The manipulation of membrane transporter genes and its transfer to the non-accumulator plant potentially offer an enhanced phytoremediation process with increased heavy metal uptake and tolerance as shown in Table 22.2. Our knowledge of the molecular nature and regulation of transporters has expanded vastly over the past years. Fundamental research into transport mechanisms in plants is leading to the development of genetic-engineered plants provided with specialized membrane transporters can be geared up for cleanup of heavy metal-contaminated soils (phytoremediation), mining of rare metals (phytomining), and enhanced human nutrition through accumulation of nutritionally important metals in plant tissues (biofortification) which in turn could expand available arable land (Schroeder et al. 2013).

Table 22.2 Summary of transgenic plants developed for heavy metal phytoremediation

Gene family	Gene	Product	Origin	Target plant	Effect	Reference
Membrane transporters	AtMHX1	Vacuolar transporter	*A. thaliana*	Tobacco	Reduced tolerance to Mg and Zn	Shaul et al. (1999)
	AtCAX2	Vacuolar transporter	*A. thaliana*	Tobacco	15–20 % more metal ions in the shoots and higher root tonoplast transport in transgenic plants than in controls	Hirschi et al. (2000)
	ZAT1	Zn transporter	*A. thaliana*	*Arabidopsis*	Enhanced Zn tolerance and 2× higher Zn accumulation in roots	van der Zaal et al. (1999)
	ZntA	Heavy metal transporter	*E. coli*	*Arabidopsis*	Transgenic plant grew better than WT in medium with 0.7 mM Pb (II) or 70 μM Cd (II) and showed higher fresh weight	Lee et al. (2003)
	YCF1	Transport protein	Yeast	*Arabidopsis*, poplar	Tolerance to 1 mM Pb (II) and increased biomass and Cd tolerance on agar media	Song et al. (2003)
	AtNramp1	Fe transporter		*Arabidopsis*	At 600 μM iron concentrations, only the transgenic plants survived for longer periods	Curie et al. (2000)
	AtNramp3	Fe transporter		*Arabidopsis*	Increased accumulation of Fe, on Cd^{2+} treatment, and Cd hypersensitivity	Thomine et al. (2000)
	NtCBP4	Tobacco calmodulin-binding channel protein	*N. tabacum*	*N. tabacum*	Confers Ni^{2+} tolerance and Pb^{2+} hypersensitivity in transgenic plants	Arazi et al. (1999)

	CAX2	Low affinity Ca^{2+}, heavy metal cation/H^+ antiporter 2	*A. thaliana*	*N. tabacum*	Confers high Cd^{2+} transport and selectivity in tonoplast vesicles and increased Cd accumulation in roots of transgenic plants	Korenkov et al. (2007a)
	CAX4	Low affinity Ca^{2+}, heavy metal cation/H^+ antiporter 4	*A. thaliana*	*N. tabacum*	Confers tolerance to high toxic levels of Cd, Zn, and Mn in transgenic tobacco plants	Korenkov et al. (2007b)
	merC	Bacterial Hg^{2+} importer of unknown mode of transport and energy-coupling mechanism	*A. ferrooxidans*	*A. thaliana*	Hg^{2+} hypersensitivity biomass reduced by 6.4 times when grown on medium with 3 μM Hg^{2+}. Leaves submersed into test solution with 100 μM Hg^{2+} accumulated over a 3-h period 3.2 more Hg	Sasaki et al. (2006)
	PgIREG1	Vacuolar transporter	*Psychotria gabriellae*	Yeast and *Arabidopsis*	Confers Ni tolerance when expressed in yeast and in transgenic plants	Merlot et al. (2014)
	MerE	Mercury transporter	*E. coli*	*A. thaliana*	The transgenic *Arabidopsis* expressing MerE accumulated significantly more methylmercury and mercuric ions into plants than the WT *Arabidopsis*	Sone et al. (2013)
Phytochelatins (PCs)	TaPCS1	Phytochelatin synthase	Wheat	Tobacco	High Pb (1 mM) and Cd (50 mM) tolerance, longer roots, higher and greener leaves at seedling stage than WT	Gisbert et al. (2003)
	AtPCS1	Phytochelatin synthase	*A. thaliana*	*Arabidopsis*	Hypersensitivity to $CdCl_2$ and $ZnCl_2$	Lee et al. (2003)

(continued)

Table 22.2 (continued)

Gene family	Gene	Product	Origin	Target plant	Effect	Reference
	AtPCS1	Phytochelatin synthase	*A. thaliana*	*Arabidopsis*	20–100× more biomass on 250–300 mM arsenate and hypersensitivity to Cd	Li et al. (2004)
	OASTL	Cysteine synthase	*A. thaliana*	*Arabidopsis*	9× increase in Cd tolerance on medium with 200 mM $CdCl_2$	Domínguez-Solís et al. (2004)
	OASTL	Cysteine synthase	*A. thaliana*	*Arabidopsis*	Tolerance up to 400 mM $CdCl_2$ with exogenous cysteine supply and 72 % more Cd accumulation (mature plants grew on 250 mM $CdCl_2$ for 14 days)	Domínguez-Solís et al. (2004)
	OASTL	Cysteine synthase		Tobacco	Tolerance up to 300 mM Cd, 250 mM Se, and 500 mM Ni and produced higher biomass when grown on agar medium	Kawashima et al. (2004)
	AtPCS1	Phytochelatin synthase	*A. thaliana*	*N. tabacum*	2.2 times longer roots and 1.6 times higher Cd^{2+} accumulation in roots and shoots from hydroponic solution with 30 μM Cd^{2+} and 250 μM glutathione	Pomponi et al. (2006)
	AtPCS1	Phytochelatin synthase	*A. thaliana*	*B. juncea*	1.9 and 1.4 times longer roots on media with 100 μM Cd^{2+} and 500 μM AsO_4^{3-}, respectively	Gasic and Korban (2007)
	AtPCS1 and MTL4	Phytochelatin synthase and genetic fusion of four human metallothioneins	*A. thaliana* and *H. sapiens*	*M. huakuii*/*A. sinicus*	Roots of *A. sinicus* colonized with rhizobia *M. huakuii* producing AtPCS1 and	Ike et al. (2007)

					AtPCS1 + MT4 accumulated, respectively, 2.5 and 3 times more Cd from soil containing 1 ppm Cd. Colonized nodules increased Cd concentration [only] by 30 %	
	TaPCS1	Phytochelatin synthase	*Triticum aestivum*	*N. glauca*	1.6 times longer roots on media with 800 μM Pb^{2+} or 50 μM Cd^{2+}. Shoots of transformed line NgTP1 accumulated from polluted soil, respectively, 6.0, 3.3, 4.8, 18.2, and 2.6 times more Pb, Cd, Zn, Cu, and Ni	Martínez et al. (2006)
	CdPCS1	Phytochelatin synthase	*Ceratophyllum demersum*	*Arabidopsis*	*Arabidopsis* showed a significant enhanced accumulation of heavy metal(loid)s in aerial parts without significant difference in growth parameters in comparison to WT *Arabidopsis* plants	Shukla et al. (2013)
Metallothioneins	MT2	Metallothionein	Human	Tobacco, rapeseed	Enhanced Cd tolerance at the seedling stage	Misra and Gedamu (1989)
	MT1	Metallothionein	Mouse	Tobacco	Tolerated 200 mM $CdCl_2$ at the seedling level	Pan et al. (1994)
	CUP1	Metallothionein	Yeast	Cauliflower	Tolerated 400 mM $CdCl_2$ in hydroponic medium	Hasegawa et al. (1997)
	CUP1	Metallothionein	Yeast	Tobacco	2–3× higher Cu content than the control but no Cd tolerance	Thomas et al. (2003)

(continued)

Table 22.2 (continued)

Gene family	Gene	Product	Origin	Target plant	Effect	Reference
	PsMTA	Metallothionein	Pea	*Arabidopsis*	8× higher Cu accumulation	Evans et al. (1992)
	HisCUP1	CUP1 with hexahistidine extension	Recombinant fusion	*N. tabacum*	By 75–90 % higher Cd^{2+} accumulation from sandy soil with 0.2 ppm of Cd and humus soil with 0.4 ppm Cd (2.5 ppb exchangeable with Ca^{2+})	Macek et al. (2002)
					Transgenic plant tolerated up to 16.2 ppm Cd in sandy soil	Pavlíková et al. (2004)
	merP	Hg^{2+}-binding protein encoded within TnMERI1	*Bacillus megaterium*	*A. thaliana*	Capable of germination and growth on media with 12.5 μM Hg^{2+} accumulating 5.35 μg Hg^{2+} g^{-1} of fresh seedling weight	Hsieh et al. (2009)
	ScMTII	Metallothionein	*Saccharomyces cerevisiae*	*N. tabacum*	The transgenic tobacco plant accumulated 3.5–4.5-fold more Cd above the threshold level of 100 mg Cd kg^{-1}	Daghan et al. (2013)
Metal chelators	HvNAS1	Nicotianamine synthase gene	*Hordeum vulgare*	*Arabidopsis*	Confers enhanced tolerance of high levels of metals, in particular Ni	Kim et al. (2005)
				Tobacco	Transgenic tobacco plants with a high level of nicotianamine grew well in a Ni-enriched serpentine soil without developing any symptoms of Ni toxicity	Kim et al. (2005)

	Citrate synthase	Citrate synthase	–	*N. tabacum*	Overproduction of citrate was shown to result in aluminum (Al) tolerance in transgenic plants	de la Fuente et al. (1997)
				Carica papaya	Overproduction of citrate was shown to result in Al tolerance in transgenic plants	de la Fuente et al. (1997)
	NAAT	Nicotianamine aminotransferase	Barley	Rice	The transgenic plants released more phytosiderophores and grew better on iron-deficient soils	Takahashi et al. (2001)
Manipulation of enzymes involved in heavy metal phytoremediation	gshI	Bacterial γ-glutamylcysteine synthetase (γ-ECS)	*E. coli*	*B. juncea*	2.1 times longer roots in media with 200 μM Cd^{2+}. By 90 % higher shoot Cd levels when grown in media with 50 μM Cd^{2+}	Zhu et al. (1999a)
					When grown on polluted soil, shoots showed 1.5, 2.0, 2.0, and 3.1 times higher Cd, Zn, Cu, and Pb levels, respectively	Bennett et al. (2003)
				Populus	Cd accumulation from soil containing 225 ppm Cd higher by 2.5 and 3 times with plants producing γ-ECS to cytosol and plastids, respectively	Koprivova et al. (2002)

(continued)

Table 22.2 (continued)

Gene family	Gene	Product	Origin	Target plant	Effect	Reference
	gshI and arsC	Bacterial γ-glutamylcysteine synthetase and bacterial arsenate reductase	*E. coli*	*A. thaliana*	6 times higher biomass yield from medium with 200 μM AsO_4^{3-}. 3 times higher As accumulation from medium with 125 μM AsO_4^{3-}	Dhankher et al. (2002)
	gshII	Bacterial and yeast glutathione synthetase	*E. coli*	*B. juncea*	1.5 times longer roots on medium with 200 μM Cd^{2+}. By 20 % enhanced Cd^{2+} accumulation from media with 50 μM Cd^{2+}	Zhu et al. (1999b)
	GSH1	Bacterial and yeast glutathione synthetase	*S. cerevisiae*	*A. thaliana*	No effect on Cd^{2+}, AsO_4^{3-}, and AsO_2^- tolerance. Increased accumulation of Cd (4 times from media with 30 ppm Cd^{2+}) and As (2.5 and 4.4 times from media with 28 ppm AsO_4^{3-} and AsO_2^-, respectively)	Guo et al. (2008)
	GSH1 and AsPCS1	Bacterial and yeast glutathione synthetase and phytochelatin synthase of garlic	*S. cerevisiae* and *A. sativum*	*A. thaliana*	2 times longer roots on media with 50 μM Cd^{2+}, 150 μM AsO_4^{3-}, or 50 μM AsO_2^-. Increased accumulation of Cd (10 times from media with 30 ppm Cd^{2+}) and As (3 and 10 times from media with 28 ppm AsO_4^{3-} and AsO_2^-, respectively)	Guo et al. (2008)
	APS1	ATP sulfhydrylase	*A. thaliana*	*B. juncea*	1.5 times longer roots and 1.4 times higher biomass with plantlets grown on medium with 400 μM SeO_4^{2-}. Improved accumulation of Se and S: 3 times higher Se	Pilon-Smits et al. (1999)

					levels in shoots when plantlets grown on medium with 40 μM SeO_4^{2-}. Doubled levels of glutathione, both in shoots and roots	
	SMT	Selenocysteine methyltransferase	*A. bisulcatus*	*B. juncea*	No phytotoxicity of 25 μM SeO_3^{2-} in medium (97 % growth inhibition with WT). By 40 % reduced growth on medium with 25 μM SeO_4^{2-} (60 % inhibition with WT). Se accumulation from media with 200 μM SeO_4^{2-} and 100 μM SeO_3^{2-} increased 4 and 2 times, respectively	LeDuc et al. (2004)
	SMT and APS1	Selenocysteine methyltransferase and ATP sulfhydrylase (*Arabidopsis*)	*A. bisulcatus* and *A. thaliana*	*B. juncea*	Se accumulation from media with 200 μM SeO_4^{2-} increased 9 times (6 times compared to single-transformed APS1 plant)	LeDuc et al. (2006)
	merA and merB	Mercuric reductase and organomercurial reductase	*E. coli* Tn21	*N. tabacum* chloroplasts	Doubled biomass yield with seedlings grown on medium with 400 μM phenyl–Hg^{+}	Ruiz et al. (2003)
	merApe9	Encoding the same polypeptides as bacterial merA but with codons optimized for plants		*A. thaliana*	Germinating on media with 50–100 μM Hg^{2+} (25 μM Hg^{2+} is lethal to germination of WT seeds). 2.5 times higher rate of Hg volatilization from hydroponic medium with 5 μM Hg^{2+} (rate of 5 μg Hg^{0} $[g\ FW]^{-1}$ min^{-1})	Rugh et al. (1996)

(continued)

Table 22.2 (continued)

Gene family	Gene	Product	Origin	Target plant	Effect	Reference
				N. tabacum	Germinating on media with 100–350 μM Hg^{2+} (50 μM Hg^{2+} is lethal to germination of WT seeds). Increased rate of Hg volatilization from solution with 25 μM Hg^{2+}: 9 times by roots (23.8 μg Hg^0 $[g\ FW]^{-1}\ min^{-1}$), 8 times by leaves (6.9 μg Hg^0 $g^{-1}\ min^{-1}$), and 5 times by stem (4.1 μg Hg^0 $g^{-1}\ min^{-1}$)	He et al. (2001)
	merApe9 and merB	Encoding the same polypeptides as bacterial merA and merB, respectively, but with codons optimized for plants		*A. thaliana*	Seedlings with MerB localized in ER volatized Hg from solution with 25 μM phenyl–Hg^+ at rate of 760 ng Hg^0 $[g\ FW]^{-1}\ min^{-1}$.	Bizily et al. (2003)
	merBpe			*A. thaliana*	Germinating on media with 5 μM CH_3–Hg^+ or phenyl–Hg^+	Bizily et al. (1999)
	merApe9 and merBpe			*A. thaliana*	Germinating on media with 10 μM CH_3–Hg^+ (1 μM CH_3–Hg^+ is lethal to germination of WT seeds). Seedlings volatize Hg from solution with 25 μM phenyl–Hg^+ at a rate of 60 ng Hg^0 $[g\ FW]^{-1}\ min^{-1}$	Bizily et al. (2000)

				S. alterniflora	Callus culture can tolerate 500 μM Hg^{2+} and 100 μM phenyl–Hg^{+} (225 μM Hg^{2+} or 50 μM phenyl–Hg^{+} is lethal to WT callus)	Czakó et al. (2006)
	merA18			*L. tulipifera*	Germinating on media with 50 μM Hg^{2+} (25 μM Hg^{2+} is lethal to germination of WT seeds). Hg volatilization from hydroponic media with 10 μM Hg^{2+} improved 10 times (rate of 1.2 μg Hg^{0} $[g\ FW]^{-1}\ day^{-1}$)	Rugh et al. (1998)
	CGS1	Cystathionine-γ-synthase	*A. thaliana*	*B. juncea*	Doubled root length on medium with 200 μM SeO_3^{2-} but no effect on tolerance to SeO_4^{2-}. Doubled rate of Se volatilization from media with 40 μM SeO_3^{2-} (up to 0.23 g Se $[g\ FW]^{-1}\ day^{-1}$) or 40 μM SeO_4^{2-} (up to 0.3 g Se $[g\ FW]^{-1}\ day^{-1}$)	Van Huysen et al. (2003)
	Human CYP2E1 and glutathione *S*-transferase (GST)	Human CYP2E1 and glutathione *S*-transferase (GST)		Alfalfa	Alfalfa plants exhibited strong resistance toward the mixtures of Cd and trichloroethylene that were metabolized by the introduced GST and CYP2E1 in combination	Zhang and Liu (2011)

WT wild-type plants

22.4.4.2 Phytochelatins (PCs)

Phytochelatins (PCs) are low molecular weight cysteine-rich metal-binding peptides whose synthesis is induced by heavy metals. PCs are synthesized nontranslationally from glutathione by the enzyme phytochelatin synthase to form molecules of (γ-EC) nG (where $n \sim 2$–11) (Cobbett 2000; Krämer 2010). They chelate heavy metals with their thiol (-SH) group of cysteine, and the resulting metal–phytochelatin complexes are sequestered into vacuoles. This results into development of heavy metal tolerance in plants by decreasing free heavy metal ion concentration in plant fluids. A number of structural variants are identified in a wide variety of plant species, and different metals, including Cd, Hg, Ag, Cu, Ni, Au, Pb, and Zn, are found to induce PC production; however, Cd is by far the strongest inducer (Mejáre and Bülow 2001; Pal and Rai 2010). Transgenic plants with increased phytochelatin level through overexpression of phytochelatin synthase resulted in enhanced heavy metal tolerance (Table 22.2).

22.4.4.3 Metallothioneins (MTs)

Metallothioneins (MTs) are ubiquitous, low molecular weight (5–10 kDa), cysteine-rich proteins present in plants, animals, fungi, and cyanobacteria. In plants, MTs are suggested to be involved in metal tolerance or homeostasis, detoxification, and distribution as they are able to bind metal ions by the formation of mercaptide bonds with the numerous cysteine residues. Recent reports show that MTs are also involved in the scavenging of reactive oxygen species (Hassinen et al. 2011; Freisinger 2011; Leszczyszyn et al. 2013). Each MT exhibits preferences for a special metal ion due to coordination residues other than cysteine and differences in folding and stability in dependence on the bound metal (Leszczyszyn et al. 2007). However, various MT genes are cloned and transferred to several plants resulting in constitutive enhancement of heavy metal tolerance in transgenic plants (Eapen and D'Souza 2005) (Table 22.2).

22.4.4.4 Metal Chelators (MCs)

Metal chelators are mostly low molecular weight organic compounds, viz., organic acids (malate, citrate) and amino acids (nicotianamine, histidine), which can act as metal-binding ligands. MCs are involved in the uptake, transport, and sequestration of possibly toxic-free metal ions present in cytosol or plant fluids to the vacuolar compartment, providing metal tolerance to the plants (Haydon and Cobbett 2007). Overexpression of metal chelator genes in plants showed enhanced heavy metal tolerance (Table 22.2).

22.4.4.5 Manipulation of Enzymes Involved in Heavy Metal Phytoremediation

Genetic-engineered plants to express the enzymes responsible for detoxification, tolerance, and chelation of heavy metals, oxidative stress response/mechanisms, and plant architecture are developed and showed enhanced phytoremediation capacity (Table 22.2). Transgenic plants overexpressing the genes encoding the enzymes for histidine biosynthesis, ACC deaminase, Hg^2 reductase, glutathione synthetase, arsenate reductase, and aldolase/aldehyde reductase were shown to become more tolerant to the toxic levels of metals and carried out phytoextraction with increasing potential (Chatterjee et al. 2013).

22.4.5 Optimization of Transgene Expressions in Transgenic Plants

The expression levels of transgene in transgenic plants are generally low in the case of heterologous gene expression system. Since the additional transcriptional and translational enhancers and constitutive and tissue-specific promoters are needed to optimize expression of transgenes that results into increased phytoremediation capacity of transgenic plants.

22.4.5.1 Use of Constitutive Promoters

The promoter is used to manage gene expressions. The selection of a promoter in a particular expression system depends on the entry and integration of align gene into the host genome. Phytoremediation-related plant transformation studies have largely utilized the cauliflower mosaic virus 35S (CaMV 35S) (Thomine et al. 2000), and actin promoters (Dhankher et al. 2002) have been employed to drive constitutive high level of expression of integrated genes in most of the plant tissues.

22.4.5.2 Inducible Promoters

The performance of inducible promoters is not conditioned to endogenous factors but to environmental conditions and external stimuli that can artificially be controlled. The expression pattern of the gene may also be programmed to be only under certain environmental conditions (e.g., stress, light) by using different promoters. The double transgenic plant with strong tolerance to the As and enhanced As accumulation was developed by co-expressing two bacterial genes. The *E. coli* arsenate reductase, *arsC*, gene was expressed in leaves as driven by a light-induced soybean RuBisCO small subunit 1 (*SRS1*) promoter and results into arsenate reduction in leaves. In addition, the *E. coli* γ-glutamylcysteine synthetase, γ-*ECS*, was expressed in both roots and shoots, driven by strong constitutive *Actin2* promoter, and leads to enhanced biosynthesis of thiol-rich peptides for AsIII complexation. Thus, they provide increased AsV tolerance (Dhankher et al. 2002).

22.4.5.3 Tissue-Specific Expression of Transgenes

Tissue-specific promoters direct the expression of a gene in specific tissues (e.g., roots, leaves, tubers, fruits and seeds, etc.). The expression of metal hyper accumulator genes in easily harvestable plant tissues like leaves has an advantage in phytoextraction and phytomining. However, fruit-/seed-specific expression of transgene for heavy metal phytoremediation face objections due to fruits/seeds is eaten by birds and other organisms. This leads to the biomagnification of heavy metal in the food chain.

Addressing the problem of high Cd accumulation in rice plants including seeds, Li et al. (2007) focused their efforts on reducing Cd accumulation in rice seeds by silencing the expression of phytochelatin synthase (PCS) gene, *OsPCS1*. If *OsPCS1 gene* was constitutively silenced, the transgenic rice plant may become sensitive to Cd exposure that would affect the growth of plants on Cd-contaminated soil. Hence, the knockdown of *OsPCS1* gene by RNAi under the control of the seed-specific promoter ZMM1 (from maize) in rice showed reduced Cd accumulation in rice seeds as compared to the wild plants. Furthermore, this transgene-induced RNA interference approach can be used to control heavy metal accumulation in the edible part of the food crops that would be grown on metal-contaminated soils or irrigated with metal-contaminated water. Thus, it is possible to enhance tolerance to heavy metal and restrict its entry into the food chain.

22.4.5.4 Organelle Targeting

Transgene integration in the various targeted organelle compartments like chloroplast, mitochondria, and vacuoles facilitates sequestration/detoxification of toxic heavy metals in the organelle. This prevents adverse interaction of heavy metals with cytoplasmic environment. Plant chloroplast genetic engineering offers several advantages over nuclear transformation, i.e., very high levels of transgene expression to 46 % of total leaf protein, transgene containment by the maternal inheritance/cytoplasmic inheritance of the plastid genome, the absence of gene silencing and positioning effect, the ability to express multiple genes in a single transformation event, and the ability to express bacterial genes without codon optimization. However, chloroplast transformation technology success was limited by several challenges, viz., it requires a species-specific vectors and achieving homoplasmy (integration of transgene into each chloroplast genome and elimination of untransformed genome).

Phytoremediation of organomercurial compounds via chloroplast genetic engineering was achieved by integrating the native *mer* operon containing the *merA* (mercuric ion reductase) and *merB* (organomercurial lyase) genes into the tobacco chloroplast genome. The transgenic plants showed better growth and resistance to very high concentrations of PMA (phenylmercuric acetate) up to 400 μM as compared to control untreated plants (Ruiz et al. 2003). Therefore, chloroplast engineering could be a beneficial approach for Hg phytoremediation as well.

22.4.6 Genetic Engineering of Plant Symbionts

The plant-associated bacteria including endophytic, phyllospheric, and rhizospheric bacteria can affect plant growth and development by fixing atmospheric nitrogen, increasing bioavailability of essential mineral nutrients, and synthesizing phytohormones and enzymes involved in plant growth hormone metabolism (Weyens et al. 2009a, b). The potential of plants and their associated microorganisms in degradation and removal of environmental pollutants/heavy metals has been recently recognized. However, not every plant-associated bacterium possesses the ability to degrade every toxic compound, and not every bacterium that has a degrading capacity toward a contaminant limits their widespread application in phytoremediation (Newman and Reynolds 2005). However, the success rate of phytoextraction of heavy metals using endophytic bacteria remains slow because of the lack of proper strains with heavy metal resistance and detoxification capacities (Luo et al. 2011). To overcome these constraints, plant-associated bacteria are engineered with the appropriate characteristics to achieve enhanced phytoremediation (Menn et al. 2000).

Several plant-associated bacteria are reported to accelerate the phytoremediation in metal-contaminated soils by promoting plant growth and health (Miransari 2011; Rajkumar et al. 2012). On the other hand, the rate of phytoremediation is limited by metal availability, uptake, translocation, and phytotoxicity (Weyens et al. 2009a, b). Therefore, to improve the efficiency of phytoremediation of toxic metal-contaminated soils, plant-associated bacteria can be engineered with pathways for the synthesis of natural metal chelators such as citric acid to increase metal availability for plant uptake or alternatively with metal sequestration systems to reduce phytotoxicity and increase metal translocation to aerial plant parts (Sessitsch and Puschenreiter 2008). *Lupinus luteus* grown on a nickel-enriched substrate and inoculated with an engineered nickel-resistant bacterium *Burkholderia cepacia* L. S.2.4 *ncc-nre* showed significantly increased nickel concentration in roots (Lodewyckx et al. 2001).

A. thaliana PCS1 gene along with a genetic fusion of four mammalian MT-coding sequence was expressed in *Mesorhizobium huakuii* subsp. rengei (strain B3) and resulted in approximately 25-fold increase in Cd accumulation as compared to its natural capability (Rajkumar et al. 2012). The colonization of Chinese milk vetch (*Astragalus sinicus*) with the B3 strain in rice paddy soil containing

1 mg kg^{-1} Cd promoted uptake of the metal in roots but not in nodules, by three times. This strategy would be useful in the rhizofiltration or transient phytostabilization of heavy metals in soil.

The possible advantages of using genetic-engineered endophytic microorganisms to improve xenobiotic remediation were summarized by Newman and Reynolds (2005), a major advantage being where genetic engineering of a xenobiotic degradation pathway is required, bacteria are easier to manipulate than plants. In addition, quantitative gene expression of pollutant catabolic genes within the endophytic populations could be a useful monitoring tool for assessing the efficiency of the remediation process. The unique niche of the interior plant environment provides the xenobiotic degrader strain with an ability to reach larger population sizes due to the reduced competition. Another important advantage of using endophytic pollutant degraders is that any toxic xenobiotics taken up by the plant may be degraded *in planta*, thereby reducing phytotoxic effects and eliminating any toxic effects on herbivorous fauna residing on or near contaminated sites (Ryan et al. 2008).

22.5 Hairy Root Cultures for Heavy Metal Removal

Agrobacterium rhizogenes is a plant-pathogenic gram-negative soil bacterium, initiating hairy root disease in most of dicotyledonous plants upon infection. These hairy roots no longer require the continuous presence of the inciting bacterium for proliferation, demonstrating that the plant cells have been transformed. The molecular basis of this transformation revolves around the activities of a large (~200 kb) root-inducing (Ri) plasmid resident in *A. rhizogenes* virulent strains. Specifically, a portion of the Ri plasmid, the transferred DNA (T-DNA), delineated by 25 base pair border repeats, is transferred into the plant cells, integrated into the nuclear DNA, and expressed. This transformation process results into induction of hairy roots from infected plant tissue. A number of excellent review articles have been published describing *Agrobacterium* and plant genes involved in T-DNA transfer and integration (Gelvin 2000, 2003) and *Agrobacterium* and plant cell interaction (McCullen and Binns 2006); hence, these aspects are not described in this chapter.

The hairy root disease is characterized by plagiotropic root growth, a high degree of lateral branching, and the profusion of root hairs, although the tissue maintains a highly differentiated and functional root organ (Santos-Díaz 2013). "Hairy root" cultures have several properties—fast growth; greater genetic, phenotypic, and biochemical stability; growth in hormone-free media; and easily established—and maintenance and propagation in the laboratory offers a more reliable and reproducible in vitro model experimental system (cf. Sect. 22. 4.2.2), which have promoted their use in phytoremediation (Doran 2009; Georgiev et al. 2012; Shanks and Morgan 1999; Zhou et al. 2013). Table 22.3 highlights examples of hairy root cultures used for metal removal.

Table 22.3 Plant hairy root cultures used to remove metal

Plant species	*A. rhizogenes* strain used	Explant	Heavy metal	Remediation strategy/focus of study	Reference
Alyssum bertolonii	–	–	Nickel	Phytoextraction	Nedelkoska and Doran (2001)
Alyssum tenium	15834	–	Nickel	Phytoextraction	Boominathan and Doran (2002, 2003b)
Armoracia rusticana	A4	Callus	Uranium	Rhizofiltration	Soudek et al. (2011)
Brassica juncea	–	–	Uranium	Rhizofiltration	Eapen et al. (2003)
Brassica napus	LBA 9402	Leaf	Chromium	Rhizoremediation	Ontañon et al. (2014)
Calystegia sepium	–	Various plant organs	Cadmium	Phytoextraction	Metzger et al. (1992)
Chenopodium amaranticolor	–	–	Uranium	Rhizofiltration	Eapen et al. (2003)
Daucus carota	–	–	Uranium	Rhizofiltration	Straczek et al. (2009)
Euphorbia hirta	15834	–	Copper	Phytoextraction	Nedelkoska and Doran (2000b)
Hyptis capitata	A4	–	Copper	Phytoextraction	Nedelkoska and Doran (2000b)
Nicotiana tabacum	15834	Seedling	Cadmium	Phytoextraction	Boominathan and Doran (2003a, b)
Nicotiana tabacum	15834	–	Copper	Phytoextraction	Nedelkoska and Doran (2000b)
Nicotiana tabacum	15834	Seedling	Nickel	Phytoextraction	Boominathan and Doran (2002, 2003b)
Nicotiana tabacum cv. Wisconsin	LBA 9402	Leaf	Arsenic	Phytoextraction	Talano et al. 2014
Polycarpaea longiflora	15834	–	Copper	Phytoextraction	Nedelkoska and Doran (2000b)
Solanum nigrum	–	–	Zinc	Phytoextraction	Subroto et al. (2007)
Solanum nigrum	Bearing (Ri plasmid C58ci)	–	Cadmium	Rhizofiltration	Macek et al. (1994)

(continued)

Table 22.3 (continued)

Plant species	*A. rhizogenes* strain used	Explant	Heavy metal	Remediation strategy/focus of study	Reference
Thlaspi caerulescens	15834	Seedling	Cadmium	Phytoextraction	Nedelkoska and Doran (2000a)
Thlaspi caerulescens	15834	Seedling	Cadmium	Organic acid complexation, heavy metal distribution, and antioxidative defenses in hairy root culture	Boominathan and Doran (2003a, b)

22.6 Transgenic Plants: Lab to Land Transfer/Testing

The myriad of reports on transgenic plants with enhanced heavy metal accumulation, tolerance, and volatilization published in recent years could be used for phytoremediation of heavy metal-contaminated sites. However, the majority of phytoremediation studies are restricted to the model plants and have been carried out under controlled laboratory conditions using metal-spiked hydroponic system or agar medium. Up to now, these transgenic plants never have been tested/used in real metal-contaminated sites for remediation. On the other hand, phytoextraction capacity of transgenic phytochelatin-overproducing mustard plant was tested in a greenhouse study using metal-contaminated soil from Leadville, Colorado (Bennett et al. 2003). Both types of transgenics accumulated significantly higher levels of Cd and Zn in their shoots than untransformed plants.

In another greenhouse pot experiment using naturally seleniferous soil, the ATP sulfurylase (APS) transgenic *Brassica* accumulated threefold higher Se levels than wild-type (WT) *Brassica juncea*, while cystathionine-γ-synthase (CgS) transgenics contained 40 % lower Se levels than WT plant (Huysen et al. 2004). In a field experiment on Se-contaminated sediment in the San Joaquin Valley (CA, USA), the APS transgenics accumulated fourfold higher Se levels than wild-type *B. juncea* (Bañuelos et al. 2005), and on the same site, the selenocysteine lyase (cpSL) and selenocysteine methyltransferase (SMT) transgenic showed twofold higher Se accumulation compared to WT *B. juncea*, in agreement with earlier laboratory experiments (Bañuelos et al. 2007). The results obtained from the different transgenics under controlled laboratory conditions were essentially the same as those obtained with greenhouse and field experiment (Pilon-Smits and LeDuc 2009).

Additionally, the number of plants with enhanced heavy metal phytoremediation capability was developed through the genetic engineering tools and tested under the laboratory conditions; however, this is often where the process of evaluating phytoremediation ends. Ideally, these transgenic plants need to be tested in in situ field trials to assess its phytoremediation capability. Such established field trials are, therefore, immediately required to make it a commercially viable and acceptable technology (Cherian and Oliveira 2005).

22.7 Transgenic Plants for Phytoremediation: Key Considerations for Risk Assessment and Mitigation Strategies

Genetic-engineered plants are rapidly being developed which deal with heavy metal accumulation, tolerance, and resistance and have been used not only for phytoextraction of various metals (Pilon-Smits and Pilon 2002) but also to enhance crop productivity in areas with suboptimal soil metal levels or heavy metal-contaminated soil which could expand available arable land (Guerinot and Salt 2001). Furthermore, transgenic plants are also used for biofortification of crop plants (Guerinot and Salt 2001; Schroeder et al. 2013). However, each GE plants/ product undergoes an environmental risk assessment (ERA) prior to commercialization to assess potential harmful effects on the environment. The ERA directs the assessment based on the product concept, crop, trait(s), intended use (e.g., import versus cultivation), receiving environment, and potential interaction among these factors (Prado et al. 2014). The safety assessment strategy ensures that the safety of transgenic crops is reviewed by multiple regulatory agencies in accordance with different risk assessment strategies and with national and international safety assessment guidelines (Paoletti et al. 2008). This is essential to educate the community about the risks and benefits of transgenic plants for phytoremediation so that this technology can gain full regulatory and public acceptance and realize its full commercial potential (Linacre et al. 2003). The risk assessment scenario for transgenic plants is given in detail by Häggman et al. (2013) and Prado et al. (2014).

Some of the possible risks associated with transgenic plants for metal phytoremediation are accumulation of toxic metals in edible parts leading to metal entry into the food chain, which in turn affects animal and human health, and uncontrolled spread of transgene to the wild relatives leading to super weeds due to higher fitness offered by the transgene. The risk of biomagnification can be minimized by using nonedible plants and growing transgenic plants in restricted area. The various physical (placement of barriers to pollen spread, restriction of location or timing of crop planting, containment of engineered crops) and physiological (harvest engineered plants before flowering, engineer traits into plants that self-pollinate, engineer traits into plants that are sterile, perform genetic modifications on plastids) barriers are used to limit gene flow (through pollen) between related plant species (Daniell 2002; Häggman et al. 2013; Pilon-Smits and Pilon 2002; Ruiz and Daniell 2009). Additionally, transgenic for phytoremediation is likely to involve genetic use restriction technologies designed to impede transgene movement in the environment (Hills et al. 2007).

Further, considerations for the use of transgenics for the phytoremediation are the same as those involved with growing transgenics for other purposes and should also be evaluated and weighed against the risk of alternative remediation methods (Pilon-Smits and LeDuc 2009).

22.8 Future Perspectives and Conclusion

Phytoremediation is considered as an effective, low cost, environmental friendly, preferred remediation technology, and potentially applicable cleanup option to remove contaminants from soil- or water-contaminated areas (Dickinson et al. 2009). Transgenic plants expressing various genes from different sources have been developed as a means to increase heavy metal tolerance, accumulation, and volatilization that facilitates more effective heavy metal phytoremediation (Table 22.2). Most of the transgenic research was carried out on the model plant species, viz., *Arabidopsis*, tobacco, and *Brassica*. However, *B. juncea* (L.), an edible oil-producing crop, is being consumed by humans or animals in one form or another. Ecologically, use of edible crops for phytoremediation is not viable because the heavy metals enter into the food chain by consumption of either humans or animals. Therefore, further research is needed to explore more efficient nonedible candidate plant which is suitable for metal phytoremediation and genetic transformation. In this scenario, Gupta et al. (2013) suggested that aromatic plants (*Vetiveria zizanioides*, *Cymbopogon martinii*, *C. flexuosus*, *C. winterianus*, *Mentha sp.*, *Ocimum basilicum*) producing essential oils could be a better choice for the heavy metal phytoremediation as these plants are nonedible and are not being consumed directly by humans or animals. In addition, the essential oil obtained from aromatic plants is free from the risk of heavy metal accumulation from plant biomass which offers economic benefits. Furthermore, the use of halophytes is suggested as the optimal candidate for phytostabilization or phytoextraction of heavy metal-contaminated saline soils as many halophytes can accumulate or excrete heavy metals (Wang et al. 2014).

The advent of the genomic era, next-generation sequencing technologies, and rapid progress in molecular biology research have led to the fast screening and identification of novel genes and proteins responsible for metal tolerance, accumulation, and volatilization. This will open up further avenues for the creation of new transgenic plants having one or several genes (gene stacking) with desirable properties for heavy metal phytoremediation. Additionally, it might be possible to control transgene expressions in specific tissues/compartments under specific conditions (Pilon-Smits and Pilon 2002; Ruiz and Daniell 2009).

The recent novel omics approaches (genomics, transcriptomics, proteomics, metabolomics, secretomics, high-throughput and next-generation technologies) combined with new bioinformatics techniques will allow us to understand how integrated biological communities (plants and microbes) interact to adapt to contaminant stress and enhance soil remediation. Furthermore, the plant–microbe metaorganism approach can be modified to maximize growth, appropriate assembly of microbial communities, and, ultimately, phytoremediation activity (Bell et al. 2014; Schenk et al. 2012). To achieve further improvements in phytoremediation, it will need coordinated efforts from plant physiologists, agronomists, soil scientists, molecular biologists, microbiologists, chemists, environmental engineers, and government regulators (Lee 2013; Pilon-Smits 2005).

The enormous amount of knowledge regarding transgenic plants for phytoremediation is generated through emerging scientific tools and methodologies combined with established practices and techniques. However, translation of this knowledge into usable technologies is the need of the hour to accelerate phytoremediation as an eco-friendly and cost-effective technology.

Acknowledgment UBJ thanks University Grants Commission, New Delhi, India, for providing Dr D. S. Kothari postdoctoral fellowship, and VAB thanks Indian National Science Academy, New Delhi, India, for providing INSA Senior Scientist Fellowship.

References

Aken BV (2008) Transgenic plants for phytoremediation helping nature to clean up environmental pollution. Trends Biotechnol 26:225–227

Appenroth KJ (2010) Definition of heavy metals and their role in biological systems. In: Sherameti I, Varma A (eds) Soil heavy metals, vol 19, Soil biology. Springer, Berlin, pp 19–29

Arazi T, Sunkar R, Kaplan B, Fromm H (1999) A tobacco plasma membrane calmodulin-binding transporter confers Ni^{2+} tolerance and Pb^{2+} hypersensitivity in transgenic plants. Plant J 20:171–182

Arthur EL, Rice PJ, Rice PJ, Anderson TA, Baladi SM, Henderson KLD, Coats JR (2005) Phytoremediation - an overview. Crit Rev Plant Sci 24:109–122

Bañuelos G, LeDuc DL, Pilon-Smits EAH, Tagmount A, Terry N (2007) Transgenic Indian mustard overexpressing selenocysteine lyase or selenocysteine methyltransferase exhibit enhanced potential for selenium phytoremediation under field conditions. Environ Sci Technol 41:599–605

Bañuelos G, Terry N, LeDuc DL, Pilon-Smits EAH, Mackey B (2005) Field trial of transgenic Indian mustard plants shows enhanced phytoremediation of selenium- contaminated sediment. Environ Sci Technol 39:1771–1777

Bell TH, Joly S, Pitre FE, Yergeau E (2014) Increasing phytoremediation efficiency and reliability using novel omics approaches. Trends Biotechnol 32:271–280

Bennett LE, Burkhead JL, Hale KL, Terry N, Pilon M, Pilon-Smits EAH (2003) Analysis of transgenic Indian mustard plants for phytoremediation of metal-contaminated mine tailings. J Environ Qual 32:432–440

Bizily SP, Kim T, Kandasamy MK, Meagher RB (2003) Subcellular targeting of methylmercury lyase enhances its specific activity for organic mercury detoxification in plants. Plant Physiol 131:463–471

Bizily SP, Rugh CL, Meagher RB (2000) Phytodetoxification of hazardous organomercurials by genetically engineered plants. Nat Biotechnol 18:213–217

Bizily SP, Rugh CL, Summers AO, Meagher RB (1999) Phytoremediation of methylmercury pollution: merB expression in *Arabidopsis thaliana* confers resistance to organomercurials. Proc Natl Acad Sci U S A 96:6808–6813

Block M (1993) The cell biology of plant transformation: current state, problems, prospects and the implications for the plant breeding. Euphytica 71:1–14

Boominathan R, Doran PM (2002) Ni-induced oxidative stress in roots of the Ni hyperaccumulator *Alyssum bertolonii*. New Phytol 156:205–215

Boominathan R, Doran PM (2003a) Organic acid complexation, heavy metal distribution and the effect of ATPase inhibition in hairy roots of hyperaccumulator plant species. J Biotechnol 101:131–146

Boominathan R, Doran PM (2003b) Cadmium tolerance and antioxidative defenses in hairy roots of the cadmium hyperaccumulator *Thlaspi caerulescens*. Biotechnol Bioeng 83:158–167

Chaney RL, Li YM, Brown SL, Homer FA, Malik M, Angle JS, Baker AJM, Reeves RD, Chin M (2000) Improving metal hyperaccumulator wild plants to develop commercial phytoextraction systems: approaches and progress. In: Terry N, Bañuelos G (eds) Phytoremediation of contaminated soil and water. Lewis, Boca Raton, FL, pp 129–158

Chatterjee S, Mitra A, Datta S, Veer V (2013) Phytoremediation protocols: an overview. In: Gupta DK (ed) Plant-based remediation processes. Springer, Berlin, pp 1–18

Cherian S, Oliveira M (2005) Transgenic plants in phytoremediation: recent advances and new possibilities. Environ Sci Technol 39:9377–9390

Cobbett CS (2000) Phytochelatins and their roles in heavy metal detoxification. Plant Physiol 123:825–832

Curie C, Alonso JM, Le Jean M, Ecker JR, Briat JF (2000) Involvement of Nramp1 from *Arabidopsis thaliana* in iron transport. Biochem J 347:749–755

Czakó M, Feng X, He Y, Liang D, Márton L (2006) Transgenic *Spartina alterniflora* for phytoremediation. Environ Geochem Health 28:103–110

Daghan H, Arslan M, Uygur V, Koleli N (2013) Transformation of tobacco with ScMTII gene-enhanced cadmium and zinc accumulation. Clean Soil Air Water 41:503–509

Daniell H (2002) Molecular strategies for gene containment in transgenic crops. Nat Biotechnol 20:581–586

de la Fuente JM, Ramírez-Rodríguez V, Cabrera-Ponce JL, Herrera-Estrella L (1997) Aluminum tolerance in transgenic plants by alteration of citrate synthesis. Science 276:1566–1568

Denkhaus E, Salnikow K (2002) Nickel essentiality, toxicity and carcinogenicity. Crit Rev Oncol Hemat 42:35–56

Dhankher OP, Li Y, Rosen BP, Shi J, Salt D, Senecoff JF, Sashti NA, Meagher RB (2002) Engineering tolerance and hyperaccumulation of arsenic in plants by combining arsenate reductase and γ- glutamylcysteine synthetase expression. Nat Biotechnol 20:1140–1145

Dickinson NM, Baker AJM, Doronila A, Laidlaw S, Reeves RD (2009) Phytoremediation of inorganics: realism and synergies. Int J Phytorem 11:97–114

Domínguez-Solís JR, López-Martín MC, Ager FJ, Ynsa MD, Romero LC, Gotor C (2004) Increased cysteine availability is essential for cadmium tolerance and accumulation in *Arabidopsis thaliana*. Plant Biotechnol J 2:469–476

Doran PM (2009) Application of plant tissue cultures in phytoremediation research: incentives and limitations. Biotechnol Bioeng 103:60–76

Dushenkov V, Nanda Kumar PBA, Motto H, Raskin I (1995) Rhizofiltration: the use of plants to remove heavy metals from aqueous streams. Environ Sci Technol 29:1239–1245

Eapen S, D'souza SF (2005) Prospects of genetic engineering of plants for phytoremediation of toxic metals. Biotechnol Adv 23:97–114

Eapen S, Suseelan K, Tivarekar S, Kotwal S, Mitra R (2003) Potential for rhizofiltration of uranium using hairy root cultures of *Brassica juncea* and *Chenopodium amaranticolor*. Environ Res 91:127–133

EPA (1998) A citizen's guide to phytoremediation. EPA, Washington, DC, EPA publication 542-F-98-011

Evans K, Gatehouse J, Lindsay W, Shi J, Tommey A, Robinson N (1992) Expression of the pea metallothionein-like gene PsMTa in *Escherichia coli* and *Arabidopsis thaliana* and analysis of trace metal ion accumulation: implications for PsMTa function. Plant Mol Biol 20:1019–1028

Flocco CG, Giulietti AM (2007) *In vitro* hairy root cultures as a tool for phytoremediation research. In: Willey N (ed) Phytoremediation methods in biotechnology. Humana, Totowa, NJ
Freisinger E (2011) Structural features specific to plant metallothioneins. J Biol Inorg Chem 16:1035–1045
Gasic K, Korban SS (2007) Transgenic Indian mustard (*Brassica juncea*) plants expressing an *Arabidopsis* phytochelatin synthase (AtPCS1) exhibit enhanced As and Cd tolerance. Plant Mol Biol 64:361–369
Gelvin SB (2000) *Agrobacterium* and plant genes involved in T-DNA transfer and integration. Annu Rev Plant Physiol Plant Mol Biol 51:223–256
Gelvin SB (2003) Agrobacterium-mediated plant transformation: the biology behind the "Gene-Jockeying" tool. Microbiol Mol Biol Rev 67:16–37
George EF, Hall MA, De Klerk GJ (2008) Plant propagation by tissue culture, 3rd edn. Springer, Berlin
Georgiev MI, Agostini E, Ludwig-Müller J, Xu J (2012) Genetically transformed roots: from plant disease to biotechnological resource. Trends Biotechnol 30:528–537
Gifford S, Dunstan RH, O'Connor W, Koller CE, MacFarlane GR (2006) Aquatic zooremediation: deploying animals to remediate contaminated aquatic environments. Trends Biotechnol 25:60–65
Giri A, Narasu L (2000) Transgenic hairy roots: recent trends and applications. Biotechnol Adv 18:1–22
Gisbert C, Ros R, De Haro A, Walker DJ, Pilar Bernal M, Serrano R, Navarro-Aviñó J (2003) A plant genetically modified that accumulates Pb is especially promising for phytoremediation. Biochem Biophys Res Commun 303:440–445
Golan-Goldhirsh A, Barazani O, Nepovim A, Soudek P, Smrcek S, Dufkova L, Krenkova S, Yrjala K, Schröders P, Vanek T (2004) Plant response to heavy metals and organic pollutants in cell culture and at whole plant level. J Soils Sedim 4:133–140
Guerinot ML, Salt DE (2001) Fortified foods and phytoremediation. Two sides of the same coin. Plant Physiol 125:164–167
Guo J, Dai X, Xu W, Ma M (2008) Overexpressing gsh1 and AsPCS1 simultaneously increases the tolerance and accumulation of cadmium and arsenic in *Arabidopsis thaliana*. Chemosphere 72:1020–1026
Gupta AK, Verma SK, Khan K, Verma RK (2013) Phytoremediation using aromatic plants: a sustainable approach for remediation of heavy metals polluted sites. Environ Sci Technol 47:10115–10116
Ha H, Olson J, Bian L, Rogerson PA (2014) Analysis of heavy metal sources in soil using Kriging interpretation on principle components. Environ Sci Technol 48:4999. doi:10.1021/es405083f
Häggman H, Raybould A, Borem A, Fox T, Handley L, Hertzberg M, Lu M-Z, Macdonald P, Oguchi T, Pasquali G, Pearson L, Peter G, Quemada H, Séguin A, Tattersall K, Ulian E, Walter C, McLean M (2013) Genetically engineered trees for plantation forests: key considerations for environmental risk assessment. Plant Biotechnol J 11:785–798
Hall JL, Williams LE (2003) Transition metal transporters in plants. J Exp Bot 54:2601–2613
Hasanuzzaman M, Fujita M (2013) Heavy metals in the environment current status, toxic effects on plants and phytoremediation. In: Anjum NA, Pereira ME, Ahmad I, Duarte AC, Umar S, Khan NA (eds) Phytotechnologies: remediation of environmental contaminants. CRC Press Taylor & Francis Group, Boca Raton, FL, pp 7–74
Hasegawa I, Terada E, Sunairi M, Wakita H, Shinmachi F, Noguchi A, Nakajima M, Yazaki J (1997) Genetic improvement of heavy metal tolerance in plants by transfer of the yeast metallothionein gene (CUP1). Plant Soil 196:277–281
Hassinen VH, Tervahauta AI, Schat H, Kärenlampi SO (2011) Plant metallothioneins – metal chelators with ROS scavenging activity. Plant Biol 13:225–232
Haydon MJ, Cobbett CS (2007) Transporters of ligands for essential metal ions in plants. New Phytol 174:499–506

He YK, Sun JG, Feng XZ, Czakó M, Márton L (2001) Differential mercury volatilization by tobacco organs expressing a modified bacterial merA gene. Cell Res 11:231–236

Herzig R, Guadagnini M, Rehnert A, Erismann KH (2003) Phytoextraction efficiency of *in vitro* bred tobacco variants using a non-GMO approach. In: Vanek T, Schwitzguebel JP (eds) Phytoremediation inventory – COST Action 837 view. UOCHB AVČR, Prague

Hills MJ, Hall L, Arnison PJ, Good AG (2007) Genetic use restriction technologies (GURTs): strategies to impede transgene movement. Trends Plant Sci 12:177–183

Hirschi KD, Korenkov VD, Wilganowski NL, Wagner GJ (2000) Expression of *Arabidopsis* CAX2 in tobacco. altered metal accumulation and increased manganese tolerance. Plant Physiol 124:125–133

Hsieh J, Chen C, Chiu M, Chein M, Chang J, Endo G, Huang CC (2009) Expressing a bacterial mercuric ion binding protein in plant for phytoremediation of heavy metals. J Hazard Mater 161:920–925

Hu ZR, Du M (2006) Hairy root and its application in plant genetic engineering. J Integr Plant Biol 48:121–127

Huysen T, Terry N, Pilon-Smits EAH (2004) Exploring the selenium phytoremediation potential of transgenic *Brassica juncea* overexpressing ATP sulfurylase or cystathionine-γ-synthase. Int J Phytorem 6:111–118

Ike A, Sriprang R, Ono H, Murooka Y, Yamashita M (2007) Bioremediation of cadmium contaminated soil using symbiosis between leguminous plant and recombinant rhizobia with the MTL4 and the PCS genes. Chemosphere 66:1670–1676

James CA, Strand SE (2009) Phytoremediation of small organic contaminants using transgenic plants. Curr Opin Biotech 20:237–241

Jan VV, Demacedo CC, Kinet JM, Bouharmont J (1997) Selection of Al-resistant plants from a sensitive rice cultivar using somaclonal variation, *in vitro* and hydroponic cultures. Euphytica 97:303–310

Jarup L (2003) Hazards of heavy metal contamination. Brit Med Bull 68:167–182

Joner EJ (2013) Effects of biotic and abiotic amendments on phytoremediation efficiency applied to metal-polluted soils. In: Anjum NA, Pereira ME, Ahmad I, Duarte AC, Umar S, Khan NA (eds) Phytotechnologies: remediation of environmental contaminants. CRC Press Taylor & Francis Group, Boca Raton, FL, pp 283–288

Kartosentono S, Nuraida A, Indrayanto G, Noor CZ (2001) Phytoremediation of $Sr^{2+}$and its influence on the growth, $^{Ca2+}$ and solasodine content of shoot culture of *Solanum laciniatum*. Biotechnol Lett 23:153–155

Kawashima CG, Noji M, Nakamura M, Ogra Y, Suzuki KT, Saito K (2004) Heavy metal tolerance of transgenic tobacco plants over-expressing cysteine synthase. Biotechnol Lett 26:153–157

Kim J, Wyskiyzuk BE, Weathers PJ (2002) Secondary metabolism of hairy root cultures in bioreactors. In Vitro Cell Dev Biol Plant 38:1–10

Kim S, Takahashi M, Higuchi K, Tsunoda K, Nakanishi H, Yoshimura E, Mori S, Nishizawa N (2005) Increased nicotianamine biosynthesis confers enhanced tolerance of high levels of metals, in particular nickel, to plants. Plant Cell Physiol 46:1809–1818

Koprivova A, Kopriva S, Jäger D, Will B, Jouanin L, Rennenberg H (2002) Evaluation of transgenic poplars over-expressing enzymes of glutathione synthesis for phytoremediation of cadmium. Plant Biol 4:664–670

Korenkov V, Hirschi K, Crutchfield JD, Wagner GJ (2007a) Enhancing tonoplast Cd/H antiport activity increases Cd, Zn, and Mn tolerance, and impacts root/shoot Cd partitioning in *Nicotiana tabacum* L. Planta 226:1379–1387

Korenkov V, Park S, Cheng N, Sreevidya C, Lachmansingh J, Morris J, Hirschi K, Wagner GJ (2007b) Enhanced Cd^{2+} -selective root-tonoplast-transport in tobaccos expressing *Arabidopsis* cation exchangers. Planta 225:403–411

Kotrba P, Najmanova J, Macek T, Ruml T, Mackova M (2009) Genetically modified plants in phytoremediation of heavy metal and metalloid soil and sediment pollution. Biotech Adv 27:799–810

Krämer U (2005) Phytoremediation: novel approaches to cleaning up polluted soils. Curr Opin Biotech 16:133–141
Krämer U (2010) Metal hyperaccumulation in plants. Annu Rev Plant Biol 61:517–534
Kumar PBAN, Dushenkov V, Motto H, Raskin I (1995) Phytoextraction: the use of plants to remove heavy metals from soils. Environ Sci Technol 29:1239–1245
Chen L, Marmey P, Taylor NJ, Brizard J, Espinoza C, D'Cruz P, Huet H, Zhang S, Kochko A, Beachy RN, Fauquet CM (1998) Expression and inheritance of multiple transgenes in rice plants. Nat Biotechnol 16:1060–1064
LeDuc DL, AbdelSamie M, Móntes-Bayon M, Wu CP, Reisinger SJ, Terry N (2006) Overexpressing both ATP sulfurylase and selenocysteine methyltransferase enhances selenium phytoremediation traits in Indian mustard. Environ Pollut 144:70–76
LeDuc DL, Tarun AS, Montes-Bayon M, Meija J, Malit MF, Wu CP, AbdelSamie M, Chiang C, Tagmount A, deSouza M, Neuhierl B, Böck A, Caruso J, Terry N (2004) Overexpression of selenocysteine methyltransferase in *Arabidopsis* and Indian mustard increases selenium tolerance and accumulation. Plant Physiol 135:377–383
Lee JH (2013) An overview of phytoremediation as a potentially promising technology for environmental pollution control. Biotechnol Bioprocess Eng 18:431–439
Lee M, Phillips RL (1988) The chromosomal basis of somaclonal variation. Annu Rev Plant Physiol Plant Mol Biol 39:413–437
Lee S, Moon JS, Ko T, Petros D, Goldsbrough PB, Korban SS (2003) Overexpression of *Arabidopsis* phytochelatin synthase paradoxically leads to hypersensitivity to cadmium stress. Plant Physiol 131:656–663
Leszczyszyn OI, Imam HT, Blindauer CA (2013) Diversity and distribution of plant metallothioneins: a review of structure, properties and functions. Metallomics 5:1146–1169
Leszczyszyn OI, Schmid R, Blindauer CA (2007) Toward a property/function relationship for metallothioneins: histidine coordination and unusual cluster composition in a zinc-metallothionein from plants. Prot Struct Funct Bioinform 68:922–935
Li JC, Guo JB, Xu WZ, Ma M (2007) RNA Interference-mediated silencing of phytochelatin synthase gene reduce cadmium accumulation in rice seeds. J Integr Plant Biol 49:1032–1037
Li Y, Dhankher OP, Carreira L, Lee D, Chen A, Schroeder JI, Balish RS, Meagher RB (2004) Overexpression of phytochelatin synthase in *Arabidopsis* leads to enhanced arsenic tolerance and cadmium hypersensitivity. Plant Cell Physiol 45:1787–1797
Li YM, Chaney RL, Angle JS, Baker AJM (2000) Phytoremediation of heavy metal contaminated soils. Environ Sci Poll Con Ser 22:837–857
Linacre NA, Whiting SN, Baker AJM, Scott Angle J, Ades PK (2003) Transgenics and phytoremediation: the need for an integrated risk assessment, management, and communication strategy. Int J Phytorem 5:181–185
Lodewyckx C, Taghavi S, Mergeay M, Vangronsveld J, Clijsters H, van der Lelie D (2001) The effect of recombinant heavy metal resistant endophytic bacteria in heavy metal uptake by their host plant. Int J Phytorem 3:181–185
Luo S, Wan Y, Xiao X, Guo H, Chen L, Xi Q, Chen J (2011) Isolation and characterization of endophytic bacterium LRE07 from cadmium hyperaccumulator Solanum nigrum L. and its potential for remediation. Appl Microbiol Biotechnol 89:1637–1644
Ma JF, Nomoto K (1996) Effective regulation of iron acquisition in graminaceous plants. The role of mugineic acids as phytosiderophores. Physiol Planta 97:609–617
Macek T, Kotrba P, Suchova M, Skacel F, Demmerova K, Rumi T (1994) Accumulation of cadmium by hairy root cultures of *Solanum niger*. Biotechnol Lett 16:621–624
Macek T, Macková M, Pavlíková D, Száková J, Truksa M, Cundy A, Kotrba P, Yancey N, Scouten WH (2002) Accumulation of cadmium by transgenic tobacco. Acta Biotechnol 22:101–106
Martin S, Griswold W (2009) Human health effects of heavy metals. Environ Sci Technol Brief Cit 15:1–6

Martínez M, Bernal P, Almela C, Vélez D, García-Agustín P, Serrano R et al (2006) An engineered plant that accumulates higher levels of heavy metals than *Thlaspi caerulescens*, with yields of 100 times more biomass in mine soils. Chemosphere 64:478–485

McCullen CA, Binns AN (2006) *Agrobacterium tumefaciens* and plant cell interactions and activities required for interkingdom macromolecular transfer. Annu Rev Cell Dev Biol 22:101–127

Mejáre M, Bülow L (2001) Metal binding proteins and peptides in bioremediation and phytoremediation of heavy metals. Trends Biotechnol 19:67–73

Menn F-M, Easter JP, Sayler GS (2000) Genetically engineered microorganisms and bioremediation. In: Klein J (ed) Biotechnology, vol 11b. Wiley, New York, NY, pp 441–463

Merlot S, Hannibal L, Martins S, Martinelli L, Amir H, Lebrun M, Thomine S (2014) The metal transporter PgIREG1 from the hyperaccumulator *Psychotria gabriellae* is a candidate gene for nickel tolerance and accumulation. J Exp Bot 65:1551–1564

Metzger L, Fouchault I, Glad C, Prost R, Tepfer D (1992) Estimation of cadmium availability using transformed roots. Plant Soil 143:249–257

Miransari M (2011) Hyperaccumulators, arbuscular mycorrhizal fungi and stress of heavy metals. Biotechnol Adv 29:645–653

Misra S, Gedamu L (1989) Heavy metal tolerant transgenic *Brassica napus* L. and *Nicotiana tabaccum* L. plants. Theor Appl Genet 78:161–168

Nedelkoska T, Doran PM (2000a) Hyperaccumulation of cadmium by hairy roots of *Thlaspi caerulescens*. Biotechnol Bioeng 67:607–615

Nedelkoska T, Doran PM (2000b) Characteristics of heavy metal uptake by plant species with potential for phytoremediation and phytomining. Miner Eng 13:549–561

Nedelkoska TV, Doran PM (2001) Hyperaccumulation of nickel by hairy roots of *Alyssum* species: comparison with whole regenerated plants. Biotechnol Prog 17:752–759

Nehnevajova E, Herzig R, Erismann KH, Schwitzguébel JP (2007) *In vitro* breeding of *Brassica* juncea L. to enhance metal accumulation and extraction properties. Plant Cell Rep 26:429–437

Newman LA, Reynolds CM (2005) Bacteria and phytoremediation: new uses for endophytic bacteria in plants. Trends Biotechnol 23:6–8

Ontañon OM, González PS, Ambrosio LF, Paisio CE, Agostini E (2014) Rhizoremediation of phenol and chromium by the synergistic combination of a native bacterial strain and *Brassica napus* hairy roots. Int Biodeter Biodegrad 88:192–198

Pal R, Rai JPN (2010) Phytochelatins: peptides involved in heavy metal detoxification. Appl Biochem Biotechnol 160:945–963

Palmer CM, Guerinot ML (2009) Facing the challenges of Cu, Fe and Zn homeostasis in plants. Nat Chem Biol 5:333–340

Pan A, Yang M, Tie F, Li L, Chen Z, Ru B (1994) Expression of mouse metallothionein-I gene confers cadmium resistance in transgenic tobacco plants. Plant Mol Biol 24:341–351

Paoletti C, Flamm E, Yan W, Meek S, Renckens S, Fellouse M, Kuiperf H (2008) GMO risk assessment around the world: some examples. Trends Food Sci Technol 19:S70–S78

Pavlíková D, Macek T, MacKová M, Száková J, Balík J (2004) Cadmium tolerance and accumulation in transgenic tobacco plants with a yeast metallothionein combined with a polyhistidine tail. Int Biodeter Biodegrad 54:233–237

Pilon-Smits E (2005) Phytoremediation. Annu Rev Plant Biol 56:15–39

Pilon-Smits EAH, Hwang S, Lytle CM, Zhu Y, Tai JC, Bravo RC, Chen Y, Leustek T, Terry N (1999) Overexpression of ATP sulfurylase in Indian mustard leads to increased selenate uptake, reduction, and tolerance. Plant Physiol 119:123–132

Pilon-Smits EA, Pilon M (2002) Phytoremediation of metals using transgenic plants. Crit Rev Plant Sci 21:439–456

Pilon-Smits EAH, LeDuc DL (2009) Phytoremediation of selenium using transgenic plants. Curr Opin Biotechnol 20:207–212

Pitzschke A (2013) From bench to barn: plant model research and its applications in agriculture. Adv Genet Eng 2:110. doi:10.4172/2169-0111.1000110

Pomponi M, Censi V, DiGirolamo V, DePaolis A, di Toppi LS, Aromolo R, Costantino P, Cardarelli M (2006) Overexpression of *Arabidopsis* phytochelatin synthase in tobacco plants enhances Cd^{2+} tolerance and accumulation but not translocation to the shoot. Planta 223:180–190

Prado JR, Segers G, Voelker T, Carson D, Dobert R, Phillips J, Cook K, Cornejo C, Monken J, Grapes L, Reynolds T, Martino-Catt S (2014) Genetically engineered crops: from idea to product. Annu Rev Plant Biol 65:769–790

Rajkumar M, Sandhya S, Prasad MN, Freitas H (2012) Perspectives of plant-associated microbes in heavy metal phytoremediation. Biotechnol Adv 30:1562–1574

Rascio N, Navari-Izzo F (2011) Heavy metal hyperaccumulating plants: how and why do they do it? and what makes them so interesting? Plant Sci 180:169–181

Raskin I (1996) Plant genetic engineering may help with environmental cleanup [commentary]. Proc Natl Acad Sci U S A 93:3164–3166

Rugh CL, Senecoff JF, Meagher RB, Merkle SA (1998) Development of Transgenic yellow poplar for mercury phytoremediation. Nat Biotechnol 16:925–928

Rugh CL, Wilde HD, Stack NM, Thompson DM, Summers AO, Meagher RB (1996) Mercuric ion reduction and resistance in transgenic *Arabidopsis thaliana* plants expressing a modified bacterial merA gene. Proc Natl Acad Sci U S A 93:3182–3187

Ruiz ON, Daniell H (2009) Genetic engineering to enhance mercury phytoremediation. Curr Opin Biotech 20:213–219

Ruiz ON, Hussein HS, Terry N, Daniell H (2003) Phytoremediation of organomercurial compounds via chloroplast genetic engineering. Plant Physiol 132:1344–1352

Ryan RP, Germaine K, Franks A, Ryan DJ, Dowling DN (2008) Bacterial endophytes: recent developments and applications. FEMS Microbiol Lett 278:1–9

Salt DE, Prince RE, Pickering IJ, Raskin I (1995) Mechanism of cadmium mobility and accumulation ion Indian mustard. Plant Physiol 109:1427–1433

Santos-Díaz MS (2013) Metal remediation via *in vitro* root cultures. In: Gupta DK (ed) Plant-based remediation processes, vol 35, Soil biology. Springer, Berlin, pp 101–116

Sasaki Y, Hayakawa T, Inoue C, Miyazaki A, Silver S, Kusano T (2006) Generation of mercury hyperaccumulating plants through transgenic expression of the bacterial mercury membrane transport protein MerC. Transgenic Res 15:615–625

Schenk PM, Carvalhais LC, Kazan K (2012) Unraveling plant–microbe interactions: can multi-species transcriptomics help. Trends Biotechnol 30:177–184

Schroeder JI, Delhaize E, Frommer WB, Guerinot ML, Harrison MJ, Herrera-Estrella L, Horie T, Kochian LV, Munns R, Nishizawa NK, Tsay YF, Sanders D (2013) Using membrane transporters to improve crops for sustainable food production. Nature 497:60–66

Sessitsch A, Puschenreiter M (2008) Endophytes and rhizosphere bacteria of plants growing in heavy metal-containing soils. In: Dion P, Nautiyal CS (eds) Microbiology of extreme soils. Springer, New York, NY, pp 317–332

Shanks JV, Morgan J (1999) Plant 'hairy root' culture. Curr Opin Biotech 10:151–155

Shaul O, Hilgemann DW, de-Almeida-Engler J, Van Montagu N, Inze D, Galili G (1999) Cloning and characterization of a novel Mg2+/H+ exchanger. EMBO J 18:3973–3980

Shukla D, Kesari R, Tiwari M, Dwivedi S, Tripathi RD, Nath P, Trivedi PK (2013) Expression of *Ceratophyllum demersum* phytochelatin synthase CdPCS1 in *Escherichia coli* and *Arabidopsis* enhances heavy metal(loid)s accumulation. Protoplasma 250:1263–1272

Sone Y, Nakamura R, Pan-Hou H, Sato MH, Itoh T, Kiyono M (2013) Increase methylmercury accumulation in *Arabidopsis thaliana* expressing bacterial broad-spectrum mercury transporter MerE. AMB Exp 3:52

Song W, Sohn EJ, Martinoia E, Lee YJ, Yang Y, Jasinski M, Forestier C, Hwang I, Lee Y (2003) Engineering tolerance and accumulation of lead and cadmium in transgenic plants. Nat Biotechnol 21:914–919

Soudek P, Petrová S, Benešová D, Vaněk T (2011) Uranium uptake and stress responses of *in vitro* cultivated hairy root culture of *Armoracia rusticana*. Agrochimica 60:15–28

Straczek A, Wannijn J, Van Hees M, Thijs H, Tiry Y (2009) Tolerance of hairy roots of carrots to U chronic exposure in standardized *in vitro* device. Environ Exp Bot 65:82–89

Subroto MA, Priambodo S, Indrasti NS (2007) Accumulation of zinc by hairy root cultures of *Solanum nigrum*. Biotechnology 6:344–348

Sunitha MSL, Prashant S, Anilkumar S, Rao S, Narasu ML, Kavi Kishore PB (2013) Cellular and molecular mechanisms of heavy metal tolerance in plants: a brief overview of transgenic plants overexpressing phytochelatin synthase and metallothionein genes. Plant Cell Biotech Mol Biol 14:33–48

Takahashi M, Nakanishi H, Kawasaki S, Nishizawa NK, Mori S (2001) Enhanced tolerance of rice to low iron availability in alkaline soils using barley nicotianamine aminotransferase genes. Nat Biotechnol 19:466–469

Talano MA, Oller ALW, González P, González SO, Agostini E (2014) Effects of arsenate on tobacco hairy root and seedling growth, and its removal. In Vitro Cell Dev Biol Plant 50:217–225

Tangahu BV, Abdullah SRS, Basri H, Idris M, Anuar N, Mukhlisin M (2011) A Review on heavy metals (As, Pb, and Hg) uptake by plants through phytoremediation. Int J Chem Eng 2011, Article ID: 939161, pp 1–31

Thomas JC, Davies EC, Malick FK, Endreszl C, Williams CR, Abbas M, Petrella S, Swisher K, Perron M, Edwards R, Osenkowski P, Urbanczyk N, Wiesend WN, Murray KS (2003) Yeast metallothionein in transgenic tobacco promotes copper uptake from contaminated soils. Biotechnol Prog 19:273–280

Thomine S, Wang R, Ward JM, Crawford NM, Schroeder JI (2000) Cadmium and iron transport by members of a plant metal transporter family in *Arabidopsis* with homology to Nramp genes. Proc Natl Acad Sci U S A 97:4991–4996

Uriu-Adams JY, Keen CL (2005) Copper oxidative stress and human health. Mol Aspects Med 26:268–298

van der Zaal BJ, Neuteboom LW, Pinas JE, Chardonnens AN, Schat H, Verkleij JA, Hooykaas PJ (1999) Overexpression of a novel *Arabidopsis* gene related to putative zinc-transporter genes from animals can lead to enhanced zinc resistance and accumulation. Plant Physiol 119:1047–1055

Van Huysen T, Abdel-Ghany S, Hale KL, LeDuc D, Terry N, Pilon-Smits EAH (2003) Overexpression of cystathionine-γ-synthase enhances selenium volatilization in *Brassica juncea*. Planta 218:71–78

Viehweger K (2014) How plants cope with heavy metals. Bot Stud 55:35–46

Wang H-L, Tian C-Y, Jiang L, Wang L (2014) Remediation of heavy metals contaminated saline soils: a halophyte choice? Environ Sci Technol 48:21–22

Weyens N, Lelie D, Taghavi S, Newman L, Vangronsveld J (2009a) Exploiting plant–microbe partnerships to improve biomass production and remediation. Trends Biotechnol 27:591–598

Weyens N, Lelie D, Taghavi S, Vangronsveld J (2009b) Phytoremediation: plant–endophyte partnerships take the challenge. Curr Opin Biotech 20:248–254

Zhang Y, Liu J (2011) Transgenic alfalfa plants co-expressing glutathione S-transferase (GST) and human CYP2E1 show enhanced resistance to mixed contaminates of heavy metals and organic pollutants. J Haza Mat 189:357–362

Zhou ML, Tang YX, Wu YM (2013) Plant hairy roots for remediation of aqueous pollutants. Plant Mol Biol Rep 31:1–8

Zhu YL, Pilon-Smits EA, Jouanin L, Terry N (1999a) Overexpression of glutathione synthetase in Indian mustard enhances cadmium accumulation and tolerance. Plant Physiol 119:73–80

Zhu YL, Pilon-Smits EA, Tarun AS, Weber SU, Jouanin L, Terry N (1999b) Cadmium tolerance and accumulation in Indian mustard is enhanced by overexpressing γ-glutamylcysteine synthetase. Plant Physiol 121:1169–1178

Ziemienowicz A (2014) Agrobacterium-mediated plant transformation: factors, applications and recent advances. Biocatal Agric Biotechnol 3:95. doi:10.1016/j.bcab.2013.10.004

Chapter 23
Characterization of Plant Growth-Promoting Rhizobacteria (PGPR): A Perspective of Conventional Versus Recent Techniques

Ruchi Agrawal, Alok Satlewal, and Ajit Varma

23.1 Introduction

Plant growth-promoting rhizobacteria (PGPRs) are root-colonizing bacteria that form symbiotic relationships with many plants. PGPRs colonize the rhizosphere, rhizoplane (root surface), or the root itself (within radicular tissues). Bacteria of diverse genera have been identified as PGPR, of which *Bacillus* and *Pseudomonas* spp. are predominant. They act in both direct (by synthesizing phytohormones or by facilitating the uptake of nutrients from the environment) and indirect ways (by inducing resistance to phytopathogens) to promote the plant growth. They have been exploited for improving the growth of several plants/crops like maize, wheat, oat, barley, peas, canola, soy, potatoes, tomatoes, lentils, radicchio, and

R. Agrawal (✉)
Department of Biochemistry, College of Basic Sciences and Humanities, Govind Ballabh Pant University of Agriculture and Technology, Pantnagar, U.S. Nagar, Uttarakhand 263145, India

Present affiliation: Department of Bioenergy, DBT-IOC Centre for Advanced Bioenergy Research, Sector 13, Faridabad, Haryana 121007, India
e-mail: dr.ruchiagrawal010@gmail.com

A. Satlewal
Present affiliation: Department of Bioenergy, DBT-IOC Centre for Advanced Bioenergy Research, Sector 13, Faridabad, Haryana 121007, India

Department of Microbiology, College of Basic Sciences and Humanities, Govind Ballabh Pant University of Agriculture and Technology, Pantnagar, U.S. Nagar, Uttarakhand 263145, India

A. Varma
Amity Institute of Microbial Technology, Amity Science, Technology and Innovation Foundation (ASTIF), Amity University, Noida, Uttar Pradesh, India

I. Sherameti, A. Varma (eds.), *Heavy Metal Contamination of Soils*, Soil Biology 44, DOI 10.1007/978-3-319-14526-6_23

cucumber (Gray and Smith 2005). PGPRs have gained worldwide importance and acceptance for agricultural benefits (do Vale Barreto Figueiredo et al. 2010).

Different PGPR strains use different strategies to improve the plant growth: some produce phytohormones (auxin, cytokinin, indole acetic acid), some act as biocontrol agents against specific phytopathogens, some control the plant ethylene levels, and some help in nitrogen fixation. Therefore, critical screening and selection of PGPR is essential, according to the crop, soils, and environmental properties, so that optimal bacterial strains could be selected and tailor-made solutions could be provided. PGPR strains interact in synergistic manner, and precise consortiums of selected PGPR strains are currently being developed for field application. A variety of phenotypic and genotypic approaches are now available for characterization of different rhizobacteria.

In this study, methods for PGPR detection and characterization are discussed with their pros and cons. More emphasis has been levied on recently developed methods for quick and sensitive PGPR detection.

23.2 Soil Toxicity and Importance of Bioremediation

Soil is the basic environmental element constituting the ecosystem, and it is important for human survival. With industrial revolution and urbanization, contamination of soil has also increased rapidly. The soil contamination by heavy metals has become one of the environmental problems that polluted widely and harm severely. It is a threat to environment, food safety, and agriculture, and its remediation is essential (Yao et al. 2012).

23.3 Different Techniques for Soil Remediation

A variety of in situ and ex situ techniques are available for the treatment of contaminated soil. The in situ techniques involve applying chemical, biological, or physical processes to the subsurface to degrade, remove, or immobilize contaminants without removing the bulk soil. In comparison to ex situ treatment (excavation), in situ techniques are cheaper in costs and address deep contamination (EPA 542/F-06/013 2006).

23.4 Traditional Techniques

Traditional techniques for metal contamination in soils are expensive and cost prohibitive (Department of Agriculture Natural Resources Conservation Service, Soil Quality Institute, United States 2000). The ex situ or extractive technologies are rarely adopted because of high risks and costs related to the use of hazardous reactants and the consequent need of treating secondary effluents. The principal feature of many in situ treatment technologies is delivery and recovery of fluids or other reactants to the subsurface. The ability to control and monitor the delivery and recovery of these fluids or reactants is central to the effectiveness of in situ technologies in treating the contamination (EPA 542/F-06/013 2006). The in situ techniques are categorized into three major groups based on the primary mechanism by which remediation is achieved: physical, chemical, and biological.

23.5 Physical Remediation

The physical remediation mainly involves soil replacement method. The soil replacement means using clean soil to replace or partly replace the contaminated soil with the aim to dilute the pollutant concentration, increase the soil environmental capacity, and thus remediate the soil. The soil replacement is also divided into three types, including soil replacement, soil spading, and new soil importing. The soil replacement can effectively isolate the soil and ecosystem and thus decrease its effect on environment. However, this technology is large in working volume, costs a lot, and is suitable for severely polluted soil with small area (Yao et al. 2012).

23.6 Chemical Remediation

Chemical treatment includes soil vapor extraction, solidification/stabilization, soil flushing, chemical oxidation, and electrokinetic separation (EPA 542/F-06/013 2006).

23.7 Biological Remediation

As remediation of contaminated soils by physical and chemical methods is still considered as one of the most challenging and cost-intensive problems to be solved (Xu and Lu 2012), therefore, the development of sustainable, environmentally friendly, and economically feasible remediation techniques is essential (Montinaro et al. 2012). Biological remediation or bioremediation may provide an alternative solution for this challenge. It uses two different approaches: phytoremediation is a strategy that uses plants to degrade, stabilize, and/or remove soil contaminants, while microbial remediation by PGPR involves microbes (Tang et al. 2010). Bioremediation of contaminated soil involves low cost, causes less interference with the soil structure, and has a higher public acceptance than other approaches.

23.8 PGPR Detection and Characterization

An understanding of the microenvironment of rhizosphere is vital for the effective utilization of PGPR in biotechnology. In rhizospheric soil system, bacteria live in close association of the soil and plant roots. Many studies (Enebak et al. 1997) have tested the efficiency of PGPR in various conditions, observing that PGPRs are efficient under determined conditions only (Bashan and Holguin 1998). Therefore, a reliable dispersion method, either physical or chemical or combination of both, is required to disintegrate the links among the bacteria and the substrate for the isolation of bacteria. It is also crucial to study the effect of selected PGPR inoculation in given soil conditions, as they have significant impact on the composition and structure of microbial communities, and these changes must be evaluated carefully, since they have been related to the inefficiency of biofertilizers when applied to plant roots (Ramos et al. 2003).

After isolation of PGPR, its subsequent testing for in vitro abilities is done. However, there are some limitations because some of the biochemical traits shown in vitro are inducible; that is, they are expressed in certain conditions but not in others. Therefore, a bacterial PGPR trait could be expressed in the laboratory in a culture media but not in the rhizosphere. This is true for PGPR traits related to plant nutrition, such as phosphate solubilization and siderophore production, which are not expressed in phosphorous-rich and iron-rich soils, respectively. There are also problems with a bacterial property called phase variation, which produces strong genetic variations in bacteria by an enzyme called site-specific invertase. Hence, when these genetic variations occur, a strong phenotypical change occurs. It may be the case that a bacterial culture exhibits a PGPR trait, but after a time does not because of these phase variations (van der Woude 2006). Hence, their characterization is of utmost importance before their application in the field (do Vale Barreto Figueiredo et al. 2010).

Identification of different bacterial strains could be done based on the following three different approaches: (1) morphological and biochemical characterization (phenotypic), (2) chemotaxonomic characterization (chemotypic), and (3) molecular or genomic characterization (genotypic) (Table 23.1). It is difficult to classify all bacterial species based on only phenotypic characteristics; therefore, polyphasic approach employing all phenotypic, chemotypic, and genotypic characteristics is suggested. A complete integrated information would allow a confident classification and a reliable grouping of the organism.

Table 23.1 Merits and demerits of different approaches for PGPR characterization

Approaches	Techniques/tools	Merits	Demerits
Phenotypic	Gram staining, microbial respiration (i.e., substrate-induced respiration, SIR), ATP level assay biochemical tests (e.g., API kits, VITEK cards, and Biolog plates), phospholipid fatty acid analysis (PLFA) etc.	Minimal input of time	Reproducibility is difficult, change in phenotypic characters with environmental conditions
Chemotaxonomic	FAME profiling, PAGE analysis, polar lipid analysis, quinone content, cell wall diamino acid content, pyrolysis mass spectrometry, FTIR, Raman spectroscopy, and MALDI-TOF	Confirmation of putative taxonomic classification obtained from morphological study	Chemical variability, misperception about phylogeny as biosynthetic pathways are unstable, lack of fossil evidence
Molecular	Ribosomal RNA sequencing, DNA and/or RNA hybridization, DNA reassociation, PCR (PCR-RAPD, ERIC-PCR, BOX-PCR, and REP-PCR), GþC percentages, TGGE, and DGGE	Rapid and specific, very sensitive, accurate homology indexing	Complex analysis, absence of regulatory factors, and temperature changes under in vitro conditions, live material is often a prerequisite
	DNA microarrays, BioMEMS, AFM, biosensors/immunosensors	Highly sensitive and high throughput	Expensive, since most techniques are of relatively recent origin and so the data accumulated thus far are limited

23.9 Phenotypic Characterization

The phenotypic characteristics of microorganisms are defined by their morphological, physiological, and biochemical properties. Traditional phenotypic tests comprise colony morphology (color, dimensions, form) and microscopic appearance of the cells (shape, endospore, flagella, inclusion bodies); characteristics of the organism on different growth substrates; growth range of microorganisms on different conditions of salt, pH, and temperature; and susceptibility toward different kinds of antimicrobial agents. Even if cell wall composition is analyzed, the Gram reaction is still a valuable diagnostic character. Biochemical tests in bacterial identification include the relationship with oxygen, fermentation reactions, and nitrogen metabolism. Other tests may be performed, depending on the bacterial strains studied (Rodríguez-Díaz et al. 2008).

Miniaturized versions of traditional biochemical tests (e.g., API kits, VITEK cards, and Biolog plates) are available for taxonomical studies and mostly contain a battery of dehydrated reagents. Addition of a standardized inoculum initiates the reaction (growth, production of enzymatic activity, etc.). The results are interpreted as recommended by the manufacturer and are readily accessible with a minimal input of time. The phenotypic fingerprinting systems API 50CH—composed of 49 different carbohydrates and one negative control—have been used to identify *Bacillus* (Logan and Berkeley 1984) and *Paenibacillus* strains (Seldin and Penido 1986), while the API 20NE system has yielded the highest rate of correct identification of *Pseudomonas* species (Barr et al. 1989). In the same way, Biolog assay is considered a much less laborious system for bacterial identification (Miller and Rhoden 1991). This technique is based on the differential utilization of 95 carbon sources, and a redox dye, tetrazolium violet, permits colorimetric determination of the increased respiration that occurs when cells are oxidizing a carbon source. The Biolog system was very useful for the identification of PGPR strains belonging to the species *P. azotofixans* (Pires and Seldin 1997).

There are several ways to approach these parameters including direct counting under a microscope (e.g., by using acridine orange dye) (Bååth and Arnebrant 1994), microbial respiration (i.e., substrate-induced respiration, SIR (Anderson and Domsch 1978), and ATP level assay (Eiland 1983), counting viable cells with the most probable number (MPN) (Reichardt et al. 2001), using biomarkers such as lipids (Hopkins et al. 1991), and fumigating soil with chloroform (Vance et al. 1987). Bacteria are grown on plates in these studies and are used to calculate the soil bacterial diversity, by observing the number and abundance of each species. Diversity indexes, such as the Shannon index (H), the Simpson index, and the equitability index (J), have all been used to describe the structure of communities (Atlas and Bartha 1993).

As the percentage of culturable microorganisms in soil is very low, some researchers suggest that only 10 % (Campbell and Greaves 1990) or 1 % (Atlas and Bartha 1993) (or even lower (between 0.2 and 0.8 %)) (Bakken and Olsen 1989) of bacteria could be cultured. Thus non-culturable techniques are used where it is not necessary to culture microorganisms on plates. One such technique is the phospholipid fatty acid analysis (PLFA) (Lucas García et al. 2004). Phospholipids are integrated in the bacterial cell membranes (Tunlid and White 1992). Different groups of microorganisms possess different fatty acid patterns. It is not usually possible to detect specific strains or species, but changes in the concentration of specific fatty acids can be correlated to changes in specific groups of microorganisms. Another approach to non-culturable diversity is through techniques of molecular genetics, which, in the past 20 years, has revealed new information about soil microbial communities (Head et al. 1998).

However, there are certain limitations of this method as discussed below:

1. Reproducibility of results from phenotypic examination of different microorganisms in different laboratories is a major challenge. Hence, only standardized procedure should be used during execution of experiment.
2. Other major disadvantage with phenotypic methods is the conditional nature of gene expression wherein the same organism might show different phenotypic characters in different environmental conditions. Therefore, phenotypic data must be compared with similar set of data from type strain of closely related organism(s).

23.10 Chemotaxonomic Characterization

Chemotaxonomy is the method of biological classification based on similarities in the structure of certain compounds (proteins, amino acids, nucleic acids, peptides, etc.) among the organisms being classified. Chemotaxonomy allows the confirmation or support of putative taxonomic classification obtained from morphological study.

Different chemotaxonomic fingerprinting techniques used for PGPR detection include FAME profiling, PAGE analysis of whole-cell proteins, polar lipid analysis, quinone content, cell wall diamino acid content, pyrolysis mass spectrometry, Fourier transform infrared spectroscopy, Raman spectroscopy, and matrix-assisted laser desorption/ionization time-of-flight (MALDI-TOF) mass spectrometry.

Fatty acids are the major constituents of lipids and lipopolysaccharides and have been used extensively for taxonomic purposes. FAME analysis is presently the only chemotaxonomic technique that is linked to a commercial database for identification purposes. Fatty acid profiles showing variability in chain length, double-bond position, and substituent groups are perfectly suitable for taxon description and also for comparative analyses of profiles that have been obtained under identical growth conditions (Suzuki et al. 1993).

Sodium dodecyl sulfate-PAGE of whole-cell proteins requires standardized conditions of growth, combined with a rigorously standardized procedure for analysis, and normalization of the data for computer-assisted comparison of the results. Nevertheless, it has made important contributions to polyphasic taxonomic studies among the aerobic endospore formers (Logan et al. 2009).

Determination of the cell wall composition has traditionally been important in Gram-positive bacteria which contain various peptidoglycan types. The peptidoglycan type of Gram-negative bacteria is rather uniform and provides little information. Preparation of cell wall samples and determination of peptidoglycan structure are usually carried out using the methods described by Schleifer and Kandler (1972).

Isoprenoid quinones occur in the cytoplasmic membranes of most prokaryotes and play important roles in electron transport, oxidative phosphorylation, and, possibly, active transport. There are two major structural groups, the naphthoquinones (subdivided into two types: the phylloquinones and the menaquinones) and the benzoquinones. The large variability of the side chains (differences in length, saturation, and hydrogenation) can be used to characterize bacteria at different taxonomic levels (Collins and Jones 1981).

The taxonomic importance of polar lipids has now been demonstrated for some novel genera among *Bacillaceae*, although many polar lipids detected have not yet been structurally characterized. Likewise, quinones (MK-7, MK-8, and MK-9) have so far been reported for representatives of *Bacillaceae* (Logan et al. 2009).

Finally, pyrolysis mass spectrometry, Fourier transform infrared spectroscopy, and UV resonance Raman spectroscopy are sophisticated analytical techniques which examine the total chemical composition of bacterial cells. These methods have been used for taxonomic studies of particular groups of bacteria, including the members of the family *Bacillaceae* (Logan et al. 2009).

However, chemotaxonomy has limitations as biosynthetic knowledge about chemical characters used in taxonomy is inadequate and chemical compounds and/or biosynthetic pathways are unstable which leads to misperception about phylogeny. Also, sometimes it is possible that chemical and morphological features have evolved at different speeds due to differing selection pressures, and thus one might find "primitive" chemical characters together with advanced morphology in a taxon (Mannheimer 1999). Thus, the information obtained from chemotaxonomy must be used in conjunction with other sources of taxonomic evidence, and thus a multidisciplinary approach is required in order to establish a system of classification which reflects natural relationships as accurately as possible.

23.11 Molecular Characterization

Molecular methods are based on the analysis of DNA, RNA, or protein molecules. These methods have transformed the bacterial identification and taxonomial system. Different techniques are being used for the identification of bacteria up to strain level, such as restriction fragment length polymorphism (RFLP), plasmid profiling, ribotyping, amplified ribosomal DNA restriction analysis (ARDRA), pulsed field gel electrophoresis (PFGE), and randomly amplified polymorphic DNA (RAPD). These tools are helpful for the quick and reliable characterization of the PGPR (Depret and Laguerre 2008; Monteiro et al. 2009; Rodríguez-Díaz et al. 2008; Logan et al. 2009). The determination of the mole percent guanine plus cytosine (GþC percentage) is one of the classical genotypic methods. Generally, the range observed is not more than 3 % within a well-defined species and not more than 10 % within a well-defined genus (Stackebrandt and Goebel 1994).

DNA–DNA hybridization or DNA–DNA reassociation technique is based on the fact that at high temperatures DNA can be denatured, but the molecule can be brought back to its native state by lowering down the temperature (reassociation). This technique considers the comparison between whole genome of two bacterial species (Stackebrandt and Liesack 1993). A bacterial species, generally, would include the strain with 70 % or greater DNA–DNA hybridization values. There are many different methods for DNA–DNA hybridization, but it is important to state that this technique gives the relative % of similarity but not the actual sequence identity.

The gene sequences of ribosomal RNA (rRNA) molecules are used to compare the evolutionary similarities among strains (phylogenetic comparisons). All the three kinds of rRNA molecules, i.e., 5S, 16S, and 23S, and spacers between these can be used for phylogenetic analyses, but 16S rRNA gene (1,650 bp) is the most commonly used marker. It has a universal distribution, highly conserved nature, fundamental role of ribosome in protein synthesis, no horizontal transfer, and its rate of evolution which represents an appropriate level of variation between organisms. The 16S rRNA molecule comprises of variable and conserved regions, and universal primers for the amplification of full 16S rRNA gene are usually chosen from conserved region, while the variable region is used for comparative taxonomy. The 16S rRNA gene sequence is deposited in databases such as Ribosomal Database Project II (http://rdp.cme.msu.edu/) and GenBank (http://www.ncbi.nlm.nih.gov/). Sequences of related species for comparative phylogenetic analysis can also be retrieved from these databases. Thereafter, sequences comparing software packages such as BLAST and CLUSTAL X are used for the alignment of 16S rRNA gene sequence. The extent of relatedness between bacterial species can be scrutinized by the construction of phylogenetic trees or dendrograms. The phylogenetic tree ascertains the genus to which the strain belongs, and its closest neighbors, i.e., those sharing the clade or showing >97 % 16S rRNA gene sequence similarity, are obtained from various culture collections to perform further genotypic, chemotaxonomic, and phenotypic analysis. At present, by correlation with experimental data obtained in the comparison of total genomic DNA (DNA–DNA hybridization), it is proposed that a similarity below 98.7–99 % on the 16S rRNA gene sequences of two bacterial strains is sufficient to consider them as belonging to different species. On the other hand, two strains showing similarities above the 98.7 % threshold may represent two different species. In these cases, total genome DNA–DNA hybridization must be performed, and those strains for which similarities are below 70 % are considered to belong to different species (Stackebrandt and Goebel 1994).

The sequences of other highly conserved housekeeping or other protein encoding genes, such asrpoB, gyrB, and recA, have also great potential for taxonomic analysis at the species level. For example, Mota et al. (2005) obtained clustering patterns for *Paenibacillus* based upon rpoB sequence comparisons that were similar to those obtained with 16S rRNA gene sequences. Moreover, Wang et al. (2007) included gyrB sequence comparisons in the studies of the *B. subtilis* group, and Cerritos et al. (2008) included recA sequence comparisons in the work that led to the proposal of a new *Bacillus* species (do Vale Barreto Figueiredo et al. 2010).

The most common genetic techniques are PCR-RAPD (randomly amplified polymorphic DNA), ERIC-PCR, BOX-PCR, and REP-PCR. They all compare the bacterial genomes and establish a homology index among them. These techniques allow the formation of groups of bacteria with very similar genomes and thus with supposed similar PGPR abilities (Ramos Solano et al. 2007).

At present, temperature gradient gel electrophoresis (TGGE) and denaturing gradient gel electrophoresis (DGGE) are predominantly being used. These are based on the direct extraction of DNA or RNA from soil; the amplification of this DNA is done by PCR, followed by electrophoretic separation in a temperature gradient for the former, or by using chemical denaturing substances for the latter. These techniques allow the separation of DNA fragments of exactly the same length but with different sequences, based on their melting properties.

Each of the methods described above possesses its own distinctive advantages and disadvantages. Generally, the more selective the method, the less able it is to detect global changes in communities and vice versa. These tools can provide an estimate of the rhizosphere diversity in the soil.

23.12 DNA Microarrays

DNA microarrays offer a rapid and sensitive detection of PGPR microorganisms from agricultural and environmental samples in a high-throughput manner. Numerous DNA fragments can be hybridized on a single microarray chip, and it can be used for analyzing multiple samples at a time in high-throughput manner (Lee et al. 2008). Microarrays are developed by immobilizing the nucleic acids (genomic DNA, cDNA, or oligonucleotides) in an ordered two-dimensional matrix using glass microslides or nylon membranes (Epstein and Butow 2000). They can be prepared by synthesizing DNA in situ on a glass surface using combinational chemistry or by robotic microdeposition of cDNAs (0.5–2 kb) amplified by PCR (Lemarchand et al. 2004). A fluorescent or enzymatically labeled test DNA is allowed to hybridize with the microarray library, and the detection is carried out by fluorescence scanning or enzyme-mediated detection (de Boer and Beumer 1999).

Although it is an expensive methodology, it can give resolution up to strain level. DNA microarrays provide a high sensitivity, an ability to simultaneously detect diverse individual sequences in complex DNA samples, and a capacity to do comparative analysis of a large number of samples (Lemarchand et al. 2004). However, there are few limitations like sample size, matrix-associated inhibitors, nonspecific binding, and cross-hybridization which need to be addressed for the detection and differentiation of pathogens in environmental samples (Lee et al. 2008). Moreover, microarrays cannot distinguish between nonviable, culturable, and viable but non-culturable (VBNC) microbes (Masters et al. 1994).

23.13 Bio-microelectromechanical Systems (BioMEMS)

The BioMEMS are designed based on recent techniques by micro-/nanoscale fabrication. These are being exploited for the identification, immobilization, growth, separation, purification, and manipulation of single or multiple cells, biomolecules, toxins, and other chemical/biological species (Kua et al. 2005). They can be used to monitor the shelf life of an agricultural product, to realize smart drug delivery systems in plants and animals, to detect traces of biological contaminants between samples, and to monitor real-time air and water characteristics to reduce pollution (Karunakaran and Jayas 2005; Bhattacharya et al. 2007).

23.14 Atomic Force Microscopy (AFM)

AFM is utilized for the identification and characterization of PGPR in both quantitative and qualitative manner. The quantitative enumeration of PGPR is carried out by estimating them in AFM images. The qualitative characterization of microbes is based on their morphological properties evaluated by different statistical and mathematical tools available to measure the cell size and cell width after staining.

Although AFM can be applied in characterizing PGPR, still there are some challenges which need to be addressed. Sample preparation is one of the critical problems for imaging using AFM. To protect the bacteria from force generated by probe, they need to be fixed on solid surface or substrate. But sometimes this kind of sample fixation is not possible (as in case of live bacteria); therefore, porous membranes are used to immobilize the microorganisms for imaging live microorganisms (Dufrêne 2002).

The imaging force is another challenge for imaging the live and delicate bacteria. Sometimes the interaction force between the tip and the sample is quite large, and it could break the sample molecule. It could also contaminate and affect the next sample scanning (Yang et al. 2005). Thus controlling the imaging force and the size, shape, and composition of the AFM probe is very important for obtaining the high-quality images (Dufrêne 2002).

23.15 Immunosensors (Biosensors Based on Immune Response)

A large number of lipopolysaccharides present on the cell surface of PGPRs (e.g., pseudomonads) have a property to induce immune responses. Agrawal et al. (2012) utilized this inherent property of bacteria for monitoring PGPRs by ELISA (the serological markers) and developed a piezoelectric biosensor (immunosensor) based on specific antigen–antibody interaction.

Quartz crystal microbalance (QCM)-based immunosensors are important analytical tools for monitoring antibody–antigen reactions, allowing rapid and continuous analysis of the binding event without the requirement for added reagents or separation/washing steps. QCM-based piezoelectric immunosensors are based on the principle that if the sensor surface is coated with specific antibodies and the sensor is allowed to come in contact with the bacterial solution, the bacteria will bind to it with high specificity. Due to this, the mass of the crystal increases, while the resonance frequency of oscillation decreases proportionally. This fall in frequency indicates the presence of particular strain, and the shift in frequency is inversely proportional to the concentration of the culture.

As the piezoelectric crystals are little expensive therefore, nowadays most of the biosensors are being designed for multiple reuses without any significant loss in their activity. Immunosensors are very attractive systems that, in principle, may be used for a wide range of applications in the food industry, environmental monitoring, clinical diagnostics, and biotechnology. Earlier, many piezoelectric immunosensors have been developed for monitoring *Vibrio cholera* (Carter et al. 1995), *Candida albicans* (Muramatsu et al. 1986), *L. monocytogenes* (Jacobs et al. 1995), *Salmonella typhimurium* (Ye et al. 1997), porcine reproductive and respiratory syndrome virus (PRRSV) (Su et al. 2000), and mycobacteria (Kumar 2000).

Choudhary et al. (2010) made use of silver nanoparticles to enhance the efficiency of QCM biosensor by employing nanomaterials. Graham and Sabelnikov (2004) have reported that sensors based on antibody (AB)–antigen binding (plus a transducer) are much more general and versatile than sensors based on complementary binding of nucleotides, such as PCR-based and DNA chips. They also discussed that recently developed technology to discover and design AB with extremely high affinities and specificity for antigens (e.g., combinatorial AB library, or phage display AB library techniques) should significantly improve the selectivity of antigen selection. Sensitive, real-time detection and identification of microorganisms by AB requires a reporting system, which can evaluate discrete microbe-AB binding events, amplifying these reactions and transducing them into signals (Helfinstine et al. 2006).

23.16 Conclusions

The growing interest in exploring the microbial diversity of PGPR illustrates their importance for the ecosystem. PGPR strains have been conventionally used for improving the growth and yield of different crops and plant varieties. Deeper understanding of the PGPR diversity will allow us to apply specialized PGPR strains for solving the specific environmental or agricultural issues. The development of new methods for the PGPR identification, characterization, and application is the need of the hour. These techniques are based on molecular approaches and could reduce the PGPR detection time to hours instead of number of days. They are highly selective, reliable, and specific for the PGPR detection. Using these methods will decipher the interlinkage between PGPR structure and function.

References

Agrawal R, Satlewal A, Chaudhary M, Verma A, Singh R, Verma AK, Kumar R, Singh KP (2012) Rapid detection of cadmium-resistant plant growth promotory c: a perspective of ELISA and QCM-based immunosensor. J Microbiol Biotechnol 22(6):849–855

Anderson JPE, Domsch KH (1978) Soil Biol Biochem 10:215–221

Atlas RM, Bartha R (1993) Microbial ecology: historical development in microbial ecology, fundamentals and applications. The Benjamin/Cummings Publishing Company, Menlo Park, Redwood City, CA, pp 3–20

Bååth E, Arnebrant K (1994) Soil Biol Biochem 26(8):995–1001

Bakken LR, Olsen RA (1989) Soil Biol Biochem 21:789

Barr JG, Emmerson AM, Hogg GM, Smyth E (1989) API-20NE and sensititre autoidentification systems for identifying Pseudomonas spp. J Clin Pathol 42:1113–1114

Bashan J, Holguin G (1998) Soil Biol Biochem 30(8/9):1225–1228

Bhattacharya S, Jang J, Yang L, Akin D, Bashir R (2007) Biomems and nanotechnology-based approaches for rapid detection of biological entities. J Rapid Meth Autom Microbiol 15:1–32

Campbell R, Greaves MP (1990) Anatomy and community structure of the rhizosphere. In: Lynch JM (ed) The rhizosphere. Wiley, Essex, pp 11–34

Carter RM, Mekalanos JJ, Jacobs MB, Lubrano GJ, Guilbault GG (1995) Quartz crystal microbalance detection of Vibrio cholerae O139 serotype. J Immunol Methods 187:121–125

Cerritos R, Vinuesa P, Eguiarte LE, Herrera-Estrella L, Alcaraz-Peraza LD, Arvizu-Go´mez JL, Olmedo G, Ramirez E, Siefert JL, Souza V (2008) Bacillus coahuilensis sp. nov., a moderately halophilic species from a desiccation lagoon in the Cuatro Cie´negas Valley in Coahuila, Mexico. Int J Syst Evol Microbiol 58:919–923

Choudhary MK, Agrawal R, Kumar R, Singh P, Gupta BRK, Singh KP (2010) Detection of cadmium-resistant rhizobacteria using piezoelectric nanobiosensor. Int J Nanosci 9:461–469

Collins MD, Jones D (1981) Distribution of isoprenoid quinine structural types in bacteria and their taxonomic implications. Microbiol Rev 45:316–354

De Boer E, Beumer RR (1999) Methodology for detection and typing of foodborne microorganisms. Int J Food Microbiol 50:119–130

Depret G, Laguerre G (2008) Plant phenology and genetic variability in root and nodule development strongly influence genetic structuring of Rhizobium leguminosarum biovar viciae populations nodulating pea. New Phytol 179:224–235

do Vale Barreto Figueiredo M, Seldin L, Fernando de Araujo F, de Lima Ramos Mariano R (2010) Plant growth promoting rhizobacteria: fundamentals and applications. Plant growth and health promoting bacteria. Microbiol Monogr 18, DOI: 10.1007/978-3-642-13612-2_2, http://www.springer.com/978-3-642-13611-5

Dufrêne YF (2002) Atomic force microscopy, a powerful tool in microbiology. J Bacteriol 184:5205–5213

Eiland F (1983) Soil Biol Biochem 15:665–670

Enebak SA, Wei G, Kloepper JW (1997) Forest Sci 44:139–144

EPA 542/F-06/013 (2006) In situ treatment technologies for contaminated soil, United States Environmental Protection Agency, Solid Waste and Emergency Response 5203P, EPA 542/F-06/013, Nov 2006. www.epa.gov/tio/tsp
Epstein CB, Butow RA (2000) Microarray technology – enhanced versatility persistent challenge. Curr Opin Biotechnol 11:36–41
Graham TW, Sabelnikov AG (2004) How much is enough: real-time detection and identification of biological weapon agents. J Homeland Sec Emerg Manag 1:1–13
Gray EJ, Smith DL (2005) Intracellular and extracellular PGPR: commonalities and distinctions in the plant-bacterium signaling processes. Soil Biol Biochem 37:395–412
Head IM, Saunders JR, Pickup RW (1998) Microb Ecol 35:1–21
Helfinstine SL, Lavrentovich OD, Woolverton CJ (2006) Lyotropic liquid crystal as a real-time detector of microbial immune complexes. Lett Appl Microbiol 43:27–32
Hopkins DW, Macnaughton SJ, O'Donnell AG (1991) Soil Biol Biochem 23:217
Jacobs MBR, Cater M, Lubrano GJ, Guilbault GG (1995) A piezoelectric biosensor for Listeria monocytogenes. Am Lab 27:26–28
Karunakaran C, Jayas DS (2005) Nanotechnology – an emerging technology for use in agricultural and food research. Proceedings of the CSAE/SCGR, paper no. 05–001
Kua CH, Lam YC, Yang C, Toumi KY (2005) Review of bio-particle manipulation using dielectrophoresis. Report on innovation in manufacturing systems and technology. http://hdl.handle.net/1721.1/7464. Accessed 15 Jan 2005
Kumar A (2000) http://www.tms.org/pubs/journals/JOM/0010/Kumar/Kumar-0010.htmc
Lee DY, Lauder H, Cruwys H, Falletta P, Beaudette LA (2008) Development and application of an oligonucleotide microarray and real-time quantitative PCR for detection of wastewater bacterial pathogens. Sci Total Environ 398:203–211
Lemarchand K, Masson L, Brousseau R (2004) Molecular biology and DNA microarray technology for microbial quality monitoring of water. Crit Rev Microbiol 30:145–172
Logan NA, Berkeley RCW (1984) Identification of Bacillus strains using the API system. J Gen Microbiol 130:1871–1882
Logan NA, Berge O, Bishop AH, Busse HJ, De Vos P, Fritze D, Heyndrickx M, Kampfer P, Rabinovitch L, Salkinoja-Salonen MS, Seldin L, Ventosa A (2009) Proposed minimal standards for describing new taxa of aerobic, endospore-forming bacteria. Int J Syst Evol Microbiol 59(8):2114–2121
Lucas García JA, Domenech J, Santamaría C, Camacho M, Daza A, Gutiérrez Mañero FJ (2004) Environ Exp Bot 52(3):239–251
Mannheimer CA (1999) An overview of chemotaxonomy, and its role in creating a phylogenetic classification system. Agricola 87–90
Masters CI, Shallcross JA, Mackey BM (1994) Effect of stress treatments on the detection of Listeria monocytogenes and enterotoxigenic Escherichia coli by the polymerase chain reaction. J Appl Bacteriol 77:73–79
Miller JM, Rhoden DL (1991) Preliminary evaluation of Biolog, a carbon source utilization method for bacterial identification. J Clin Microbiol 29:1143–1147
Monteiro JM, Vollú RE, Coelho MRR, Alviano CS, Blank AF, Seldin L (2009) Culture-dependent and -independent approaches to analyze the bacterial community of different genotypes of Chrysopogon zizanioides (L.) Roberty (vetiver) rhizospheres. J Microbiol 47:363–370
Montinaro S, Concas A, Pisu M, Cao G (2012) Remediation of heavy metals contaminated soils by ball milling. Chem Eng Trans 28
Mota FF, Gomes EA, Paiva E, Seldin L (2005) Assessment of the diversity of Paenibacillus species in environmental samples by a novel rpoB-based PCR-DGGE method. FEMS Microbiol Ecol 53:317–328
Muramatsu H, Kajiwara K, Tamiya E, Karube I (1986) Piezoelectric immunosensor for the detection of Candida albicans microbes. Anal Chem Acta 188:257–261
Pires MN, Seldin L (1997) Evaluation of Biolog system for identification of strains of Paenibacillus azotofixans. Antonie Leeuwenhoek 71:195–200

Ramos Solano B, Pereyra de la Iglesia MT, Probanza A, Lucas García JA, Megías M, Gutiérrez Mañero FJ (2007) Plant Soil 287:59–68
Ramos B, Lucas García JA, Probanza A, Barrientos ML, Gutiérrez Mañero FJ (2003) Environ Exp Bot 49:61–68
Reichardt W, Briones A, de Jesus R, Padre B (2001) Appl Soil Ecol 17:151–163
Rodríguez-Díaz M, Rodelas B, Pozo C, Martínez-Toledo MV, González-López J (2008) A review on the taxonomy and possible screening traits of plant growth promoting rhizobacteria. In: Ahmad I, Pichtel J, Hayat S (eds) Plant-bacteria interactions: strategies and techniques to promote plant growth. Wiley, Weinheim
Schleifer KH, Kandler O (1972) Peptidoglycan types of bacterial cell walls and their taxonomic implications. Bacteriol Rev 36:407–477
Seldin L, Penido EGC (1986) Identification of Bacillus azotofixans using API tests. Antonie Leeuwenhoek 52:403–409
Department of Agriculture Natural Resources Conservation Service, Soil Quality Institute, United States (2000) Soil Quality – Urban Technical Note No. 3, heavy metal soil contamination. Department of Agriculture Natural Resources Conservation Service, Soil Quality Institute, United States
Stackebrandt E, Goebel BM (1994) Taxonomic note: a place for DNA-DNA reassociation and 16S rRNA sequence analysis in the present species definition in bacteriology. Int J Syst Bacteriol 44:846–849
Stackebrandt E, Liesack W (1993) Nucleic acids and classification. In: Goodfellow M, O'Donnell AG (eds) Handbook of new bacterial systematics. Academic, London, pp 151–194
Su XD, Li SFY, Kwang J, Low S (2000) Piezoelectric quartz crystal based screening test for porcine reproductive and respiratory syndrome virus infection in pigs. Analyst 125:725–730
Suzuki K, Goodfellow M, O'Donnell AG (1993) Cell envelopes and classification. In: Goodfellow M, O'Donnell AG (eds) Handbook of new bacterial systematics. Academic, London, pp 195–250
Tang JC, Wang RG, Niu XW, Wang M, Chu HR, Zhou QX (2010) Characterisation of the rhizoremediation of petroleum-contaminated soil: effect of different influencing factors. Biogeosciences 7:3961–3969
Tunlid A, White DC (1992) Biochemical analysis of biomass community structure, nutritional status, and metabolic activity of microbial communities in soil. In: Stotzky G, Bollag JM (eds) Soil biochemistry. Marcel Dekker, New York, pp 229–262
van der Woude MW (2006) FEMS Microbiol Lett 254(2):190–197
Vance ED, Brookes PC, Jenkinson DS (1987) Soil Biol Biochem 19(6):703–707
Wang L-T, Lee F-L, Tai C-J, Kasai H (2007) Comparison of gyrB gene sequences, 16S rRNA gene sequences and DNA-DNA hybridization in the Bacillus subtilis group. Int J Syst Evol Microbiol 57:1846–1850
Xu M, Lu N (2012) Research on removing heavy metals from mine tailings. Disaster Adv 5:116–120
Yang H, An H, Feng G, Li Y (2005) Visualization and quantitative roughness analysis of peach skin by atomic force microscopy under storage. LWT Food Sci Technol 38:571–577
Yao Z, Li J, Xie H, Yu C (2012) Review on remediation technologies of soil contaminated by heavy metals. Procedia Environ Sci 16:722–729
Ye J, Letcher SV, Rand AG (1997) Piezoelectric biosensor for detection of Salmonella Typhimurium. J Food Sci 62:1067–1071

Index

A
Abandoned mine sites
  chemical and data analysis, 161–162
  description, 159
  ecological restoration, 160
  HMs and trace elements, 162–163
  legume species, 159
  Mediterranean grassland species (*see* Mediterranean grassland species)
  metal accumulation capacity, 176
  phytoremediation (*see* Phytoremediation)
  plant communities, 159
  pollution, 175–176
  and sampling, 160–161
  soil-plant processes, 160
  trace element and Al contents, 160
ABC family. *See* ATP-binding cassette (ABC) family
Acid reaction, 146, 147
Actinorhizal plants
  in Africa, 363
  C. equisetifolia plantation, 357
  Casuarinaceae and Betulaceae families, 357
  environmental pollutants, 356
  HMs-contaminated soils, 355, 358–359
  inorganic substances, toxic, 355
  mycorrhizal fungi (*see* Mycorrhizal fungi)
  pedo-climatic parameters, 363
  root nodules, 357, 359
  soil fertility, 362–363
  stressed environments, 358
  toxic HMs, 363
Activation energies, copper geochemistry
  energy transfer (E_T), 72
  mass transfer and diffusion reaction control, 73
  reaction rate theory, 71
Adsorption equilibrium
  cations and anions, 110
  Cd, 107–108
  description, 105
  elution tests, 105
  equation, 106–107
  experiment, 107
  hazardous metals, 105
  humic substances, 108–109
  Pb(II), 109–110
Aerobic conditions microbes, 121, 122
Agricultural soils
  anthropogenic and natural activities, 23
  atmospheric deposition, 24
  bioremediation, 38
  chemometric and GIS approaches, 32–33
  earth ecosystem, 23
  ecological risks, 29–32
  fertilizer/pesticide application, 24
  metabolic functioning, plants, animals and humans, 24
  metal biogeochemical cycles, 24
  metal sources (*see* Heavy metals (HMs))
  mitigation measures
    bacterial bioremediation, 37–38
    mycoremediation, 33–35
    phytoremediation, 35–37

I. Sherameti, A. Varma (eds.), *Heavy Metal Contamination of Soils*, Soil Biology 44, DOI 10.1007/978-3-319-14526-6

Agricultural soils (*cont.*)
 nutrient and organic wastes recycling, 23
 reclamation strategies, 25
 sewage sludge application, 24
 vehicle emissions, 24
 wastewater irrigation, 24
Alkaline metal precipitation, 399–400
Allozymes, 204
AMF. *See* Arbuscular mycorrhizal fungi (AMF)
Amino polycarboxylate chelating agents, 370
Anoxic conditions, As(V), 121, 122
Aquatic ecosystems, 130, 132–133
Arbuscular mycorrhizal fungi (AMF)
 adaptation potential, 218, 220
 contamination, Zn, 220
 Glomus and *Acaulospora,* 218, 220
 occurrence and diversity, tropical soils, 218, 219
 and rhizobia, symbiotic relationships
  As accumulation, 237
  HMs, effects, 231–236
  legume trees, 230
  limestone and calcium silicate, 230, 236
  mechanisms, 236–237
  nodules, formation, 230
 sewage sludge, long-term application, 220
 tolerance to HMs
  Cd and Zn, 223
  constitutive, adaptive and induced, 223
  contamination effects, 223–229
  germination, spores, 229
  Pb and Zn, 223, 229
  recalcitrant glycoprotein, 229
ARMs. *See* Arsenic-resistant microbes (ARMs)
Arsenic (As)
 As(III), 114
 As(V), 114
 biological samples, 115
 bioremediation techniques, 122–123
 deposition, agricultural soils, 114
 Earth's crust, 114
 EEA, 113
 environment, 113–114
 fate and mobility, 115–117
 hydrothermal fluids, 114
 inorganic and organic forms, 114
 metals and metalloids, soil contamination, 113
 oxidating organisms, 122
 pH range, 114
 reducing organisms, 121
 soil enzymes, 119, 120
 soil microbial biomass, 117–119
 species and inter-transformation, 115
 tolerance and adaptation, 119, 121
Arsenic-resistant microbes (ARMs), 121, 122
Asteraceae family, 164, 167–168
Atomic force microscopy (AFM), 481
ATP-binding cassette (ABC) family, 189
Average well color development (AWCD), 248, 252
Avian tissues, 127, 137
AWCD. *See* Average well color development (AWCD)

B

Bacterial bioremediation , 37–38
Bioaccessible concentrations
 chemical extractions, 405–406
 DTPA, 405
 earthworms, 406
 geostatistical simulations, 404
 P. scaber test, 406
 Pb and Zn, 405–406
Bio-concentration factor (BCF), plants, 305
Bioindication, *P. sylvestris*
 adaptive potential, 196
 cytology and reproduction, 200–202
 in homogeneous age group, 196–197
 morphological alteration and HMs, 198–199
 Pb and Cd accumulation, 197
 properties, 197
 tolerance patterns, 196
 Upper Silesia, 196
Biolog assay
 bacterial identification, 476
 in C source, 246
 Ecoplate, 246
 GN plate, 246
 limitations, 247
 microbial catabolic activity, 248
 plate design, 246
Bio-microelectromechanical systems (BioMEMS), 481
Biomonitoring studies
 agricultural and industrial waste, 128
 aquatic ecosystems, 130, 132–133
 avian tissues, 127
 biosphere, 128
 birds, 128 (*see also* Birds)
 chemical use, 130
 eggs

accumulated mass, 139
bioindicator, 138
in birds, 138, 139
cadmium, 139
composition, 128
contaminant concentrations, 138
different areas and bird species, 130–132
egg-laying female birds, 127, 129
eggshell, 137
geographical difference, 136–137
interspecific variations, 133–134
Pb, 139
temporal variations, 135–136
trophic levels, 134
HMs, 128
interspecies differences, 129
municipal waste, 128
noninvasive technique, 127
Pb and Cd concentration, 129
spatiotemporal differences, 129
terrestrial ecosystems, 130, 132
Bioremediation
arsenic, 122–123
industrial revolution and urbanization, 472
metal abatement strategies, 38
Birds, soil characteristics and metal bioavailability, 129–130
Brassica species, 182
BR-4406 (recommended strain for *Enterolobium* spp.), 221

C
CAO. *See* Chemo-litho autotrophic arsenite oxidizers (CAO)
Carbon (C) sources
AWCD, 248
biochemical categories, 249
Biolog and MicroResp techniques, 246
functional approaches, 246
use rate measurement, 246
Casuarina equisetifolia plantation
magnesite-mined outlands, 357
mycorrhizal roots, 357
Cation-diffusion facilitator family, 190
Cation exchange capacity (CEC), 184, 315, 319
CEC. *See* Cation exchange capacity (CEC)
Chelating agents
description, 367, 369
EDTA, 367, 368
elements, 367
HMs, soil environment, 375–377
K_{ML}, 368
metal solubility, 367, 369
natural, 371–375
and phytoextraction (*see* Phytoextraction)
phytoremediation, 368
soil fertility, 368
synthetic, 369–371
Chemical fractionation methods, 116
Chemo-litho autotrophic arsenite oxidizers (CAO), 122
Chemo-remediation
categories, 300
chelates, 300
chemicals, 299–300
EDDS and NTA, 300
extractants (*see* Extractants)
organic/inorganic materials, 300
Chemotaxonomy, 477–478
Chromium (Cr), 152–153
CLPPs. *See* Community-level physiological profiles (CLPPs)
Community-level physiological profiles (CLPPs)
AWCD value, 248
Shannon-Weaver diversity index *(H′), 248–249*
Composting, 262–263
Conifers
MTE
antioxidative pathway, 339–340
distribution, 342–343
growth and development, 338–339
phytoextraction process, 337–338
scavenging, 340–341
phytoextraction
angiosperms *vs.* gymnosperms, 337
boreal area, 337
mycorrhization, 345–346
Pinus sylvestris, 337
Pinus taeda, 338
plant growth, 344
Pseudotsuga menziesii, 338
soil amendment, 344–345
Copper, geochemistry
activation energies, 71–73
adsorption and desorption and hysteresis approach
Freundlich coefficients, 62, 63
partition parameters, 62
soil/solution ratios, 57, 58
in contaminated and uncontaminated soils
metallurgical industries, 48
organo-mineral soil complex, 49

Copper (*cont.*)
simulative natural diffusion studies, 49
solubility and desorption, soil systems, 49
surface complexes' formation, 49, 50
"trace elements", 48
CO_2 sequestration mediated mineral solubilisation, 44–45
dynamic phase evaluation, soils
activation energies, 63
adsorption and desorption and hysteresis approach, 61–63
batch tests, 57
Elovich equation, 60–61
Freundlich kinetic adsorption and desorption, 70
geogenesis, 43–44
ion exchange reactions, soils, 70
kinetic reaction (adsorption/desorption) rate constants, 69
Langmuir and Freundlich models (isotherms), 58–59
physical and chemical soil tests, 51, 54–56
site characterisation and soil sampling procedure, 51–53
soil colloid surface *vs.* equilibrium concentration, 65
solid-soil solution interface, 45–48
sorption and surface charge densities, 60
sorptive sites distribution, soil colloids, 64
surface-based sorption density isotherms, 65Cytology and reproduction, *P. sylvestris*
abnormal mitosis, 201–202
chromosome aberrations, 202
DNA damage, 200, 201
industrial emissions, 200
MI, stress indicator, 201
model plants, 200
NA, 202
Pb, Cu and Cd, 200
Zn and Cu, 200–201

D
DAPS. *See* Dissimilatory arsenate reducing prokaryotes (DAPS)
Diethylene triamino pentaacetic acid (DTPA), 300, 301, 370, 371, 382, 405
Direct disposal, harvested biomass waste, 263
Dissimilatory arsenate reducing prokaryotes (DAPS), 121, 122
Dissolved organic carbon (DOC), 316
DNA damage and repair, Scots pine
abnormal mitosis types, 201–202
chromosome aberrations, 202
microsatellite DNA, 205–206
molecular marker, 205
DNA microarrays, 480
DOC. *See* Dissolved organic carbon (DOC)
DTPA. *See* Diethylene triamino pentaacetic acid (DTPA)
Dunnett's T3 test, 317

E
Earth's crust, 114
Earthworms
bioaugmentation, 292
as ecological engineers
casts, 283–284
environmental functioning, 282
gallery network, 283
geographic origins and taxonomy, 282
physical activities, 282–283
plant biomass effects, 280, 281
and HM content, 284–285
and phytoremediation (*see* Phytoremediation)
plants and microorganisms, interactions, 292–293
Earthworms in Cd-polluted (EwCd) soil, 253
Ecotoxicology, 169, 172, 174–175
EDDHA. *See* Ethylenediamine-*N,N'*-bis (*o*-hydroxyphenyl)acetic acid (EDDHA)
EDDS. *See* Ethylene diamine disuccinate (EDDS)
EDTA. *See* Ethylenediaminetetraacetic acid (EDTA)
EDTA-based soil washing
description, 395
industrial nucleus, 396
land contamination, 395
remediation and management, 397
soil solid phases, 397
water recycling, 398–401
EEA. *See* European Environment Agency (EEA)
Egg box, 341
Egg-laying female birds, 127
Eggshell, 137
EGTA. *See* Ethylene glycol tetraacetic acid (EGTA)
Electromigration, 387
Electroosmosis, 387

Environment Agency (EEA), 395
Ethylene diamine disuccinate (EDDS), 300, 384, 385
Ethylenediamine-*N,N'*-bis (*o*-hydroxyphenyl) acetic acid (EDDHA), 300, 371
Ethylenediaminetetraacetic acid (EDTA), 301, 345, 368, 371, 379–381, 384–387, 396. *See also* EDTA-based soil washing
Ethylene glycol tetraacetic acid (EGTA), 300, 371
European Environment Agency (EEA), 113
Extracellular enzymes, 145, 148, 149, 154
Extractants
  AB-DTPA, 301
  EDTA, 301

F

Fabaceae family, 164, 167–168
Fluctuating asymmetry (FA), *P. sylvestris* needles, 199
Fodder contamination and livestock safety, 32
FOM. *See* Free organic matter (FOM)
Food and Agriculture Organization of the United Nations (FAO), 162
Food chain contamination and human health concerns
  abnormal biochemical responses, 31
  arsenic excess dosage, 31
  wastewater irrigation, 30
Frankia strains, 359–360
Free organic matter (FOM), 319
Fulvic acid, 371, 372
Functional diversity, soils microorganisms
  biochemical substrate categories, 249–250
  Biolog and MicroResp assay, 246–247
  CLPP approach, 248–249
  C sources, 246
  cultivable bacteria, 247–248
  environmental factors
    root exudation, 252–253
    soil pH, 252
    vegetation, 252
  PICT, 250–252
  Shannon-Weaver diversity index (*H'*), 248

G

Genetic engineering, plants
  membrane transporters, 443–453
  MTs, 453
  PCs, 453
Glutamate dehydrogenase (GDH), 204
Glutamate oxaloacetate transaminase (GOT), 204
Glutathione-S-transferases (GST), 340
GST. *See* Glutathione-S-transferases (GST)

H

Hairy root cultures, 458–460
HAO. *See* Heterotrophic arsenite oxidizers (HAO)
Heavy metal ATPase (HMA), 181, 187, 190
Heavy metals (HMs)
  contamination
    in agricultural soils, 218
    AMF (*see* Arbuscular mycorrhizal fungi (AMF))
    degradation, 216–217
    excessive concentrations, 261
    incineration, 263
    inorganic and phosphate fertilisers, 217
    mining activities, 215
    partitioning, 264
    phytoremediation, 218, 262
    rhizobia (*see* Rhizobia, tolerance to HMs)
    seedling growth, effect on, 262
    sterile materials, 216
  vegetal growth, 261
    Zn processing, Brazilian, 215–216
  post-catastrophic soils
    accidents, metal mining (*see* Metal mining)
    floods, effect, 11–12
    hurricanes, 12–14
    volcano eruptions, 14–15
    wildfire effect, 15–17
  removal, soil
    hazardous materials, 433
    organic contaminants, 433
    properties, sources and toxic effects, 434–437
    transgenic plants (*see* Transgenic plants)
    xenobiotic substances, 436
  sources
    agricultural soils (*see* Agricultural soils)
    atmospheric deposition, 28–29
    pesticide/fertilizer application, 25–27
    wastewater irrigation, 27–28
HEDTA. *See* Hydroxy ethyl-ethylene diamine triacetic acid (HEDTA)
HEIDA. *See* N-(2-hydroxyethyl)iminodiacetic acid (HEIDA)
Heterotrophic arsenite oxidizers (HAO), 122

Hurricanes, HMs content
pollution, 14
sediment redistribution, 13
storm system, 12
wastewater contamination, 13
Hydroxy carboxylates chelating agents, 370
Hydroxy ethyl-ethylene diamine triacetic acid (HEDTA), 300, 370, 371

I

Immunosensors, 481–482
Incineration, HMs-contaminated biomass
chemical analysis, flue gas, 265
management, metal-enriched plant material, 264
operating temperature, combustion equipment, 264
residual ashes, fly ash and emitted gas, 266
solid combustion products, 264
volatility, metals, 264, 265
X-ray diffraction analysis, 266
Induced method remediation, 302
Inducible promoters, 455–456
Inductively coupled plasma mass spectrometry (ICP-MS), 107
Inter-simple sequence repeat (ISSR), 205
Intracellular enzymes, 145, 149, 155, 156

L

Lead (Pb), 105, 108, 117, 150–152, 182
Least significant difference (LSD) test, 316–317
Legume species, 159
Long-term stability, EDTA
emissions, 409–411
leachability and accessibility, 408–409
LSD test. *See* Least significant difference (LSD) test

M

Mann-Whitney test, 317
MATE family. *See* Multidrug and toxic compound extrusion (MATE) family
MDS. *See* Minimum data set (MDS)
Mean weight diameter (MWD), 315
Medicago truncatula model, 185
Mediterranean grassland species
element concentrations, plant shoots, 164, 172
HMs and As concentrations
Fabaceae and Asteraceae families, 164, 167–168
native herb species, 164, 169–171
Poaceae family, 164–166
woody species, 164, 169–171
J. montana, 169
Mercury (Hg), 128, 130, 137
Metal accumulation, eggs, 133, 135, 136
Metal chelators (MCs), 454
Metallothioneins (MTs)
description, 454
phytoremediation process, 447
root-to-shoot transport, 203
Metal mining
drainage system, 5
heavy rainfall, 9
land, flooding, 9
natural detoxfication, 7
physico-chemical properties, agricultural land, 11
sediment quality evolution, 8
sequential indicator simulation, 4
soil and vegetation, 5
soil contamination, 10
watering, 6
Metal trace elements (MTE)
advantages, 333
conifers (*see* Conifers)
human activities, 333
Juniperus sp., 347
Mediterranean basin, 347
natural origin, 333
phytoextraction (*see* Phytoextraction)
phytoremediation, 347
Pinus pinaster, 347
Pinus sylvestris, 347
ROS production, 347
Meza Valley, Slovenia
Bayesian kriging method, 403
contaminated fines, 402
eutric fluvisol, 402
feasible remediation technology, 403
Pb concentration, 403, 404
Tessier's extraction, 402–403
Microbial biomass, soil, 117–119
Microbial siderophores, 373
MicroResp assay
and Biolog, 246
design, 246–247
microbial catabolic activity, 248
Microsatellite DNA, 205–206
Mine tailing soils

chemical characteristics, 319, 321–322
metal content, 325, 326
physical characteristics, 319
samples, 314–315
sites, 328
Minimum data set (MDS), 317
Mitotic index (MI), in Scots pine, 201
MTE. *See* Metal trace elements (MTE)
MTs. *See* Metallothioneins (MTs)
Multidrug and toxic compound extrusion (MATE) family, 181, 187, 188
MWD. *See* Mean weight diameter (MWD)
Mycoremediation
cost-benefit method, 33–34
HMs contamination, 33
metal uptake efficiency, 35
soluble and insoluble metal species, 34
tolerance and biosorption analysis, 34, 35
Mycorrhizal fungi
AMF and EMF, 361–362
arbuscular (*see* Arbuscular mycorrhizal fungi (AMF))
ectomycorrhizal fungi, 360–361
Frankia strains, 359–360
HMs effects, 231–236

N
Natural method remediation, 302
Natural resistance-associated macrophage protein (NRAMP) family, 181, 185–186, 190
Natural substrate
EDTA soil washing, 412–413
remediated soil, biological properties, 413–415
Neutral/alkaline soil reaction, 149, 151
N-(2-hydroxyethyl)iminodiacetic acid (HEIDA), 300, 371
Nickel (Ni), 153, 325, 360, 373
Nitrilotriacetic acid (NTA), 300, 370, 371
Nitrogen-fixing bacteria, rhizobia. *see* Rhizobia, tolerance to HMs
Noninvasive technique leads, 127
NTA. *See* Nitrilotriacetic acid (NTA)
Nuclear microsatellites, 205
Nucleolar activity (NA), Scots pine, 202
Nutrient and organic wastes recycling, 23, 423

O
Oligopeptide transporter family, 188

P
PCA. *See* Principal component analysis (PCA)
Phenotypic characterization, 475–477
Phosphoglucomutase (PGI), 204
6-Phosphogluconate dehydrogenase (6PGD), 204
Physiologically based extraction test (PBET), 404–405
Phytochelatins (PCs)
genetic engineering, plants, 443
HMs tolerance, 454
root-to-shoot transport, 203
Phytochemical remediation
chemo-remediation (*see* Chemo-remediation)
contaminated soils, 299
efficiency and translocation factors, plants, 305–306
hemp, maize and elephant grass, 304–305
hyperaccumulator plants, 303–304
phyto-remediation, 301–303
sunflower and sorghum, 304–305
Phytoextraction
chelating agents
application, 381
Cd fractions, 382
continuous/natural, 377
and electrokinetics, 387–388
enrichment factor, 378–379
metal concentration, 384–385
metals' availability, 381–384
metal uptake, 378, 386
Pb fractions, 382, 383
phytoremediation, metal, 378
plant dry weight, 379–381
soil pH, 378
translocation factors, 378, 385–386
water holding capacity, 377
conifers (*see* Conifers)
intraspecific genetic variation, 338
MTE phytoaccumulation, 337
non-hyperaccumulating trees, 336–337
scarcity trees, 334–336
Phytoremediation
advantages, 279–280
bioaccumulation capacity, 172–173
composting, 262–263
description, 437–438
direct disposal, 263
and earthworms
AMF, 286
and HMs, 288–289

Phytoremediation (*cont.*)
hyperaccumulating plants, growth of, 286, 287
PGPB, 286
in soil enzyme activities, 289–290
and soil microorganisms, 290–291
ecotoxicological assessment, 169, 172
vs. ecotoxicology, 174–175
hyperaccumulators, 301–302
in situ and ex situ experiments, HMs reduction, 35
incineration, 263
leaching, 263
lead, bullet manufacture, 36
limitations, 280
macro and trace metal potential, 36
metal leaching, 37
phytoextraction, HMs-contaminated soil, 37
phytovolatilization, 438
plants, microorganisms and earthworms, 292–293
pyrolysis, 263
rhizofiltration, 438
rhizosphere role, 280
risk assessment and mitigation strategies, 461–462
tolerance, resistance and toxicity, 173–174
Phytovolatilization, 438
PICT. *See* Pollution-induced community tolerance (PICT)
Pinus pinaster, 347
Pinus sylvestris
antioxidative pathway, 339
bioindication (*see* Bioindication, *P. sylvestris*)
conifers physiology, MTE effect, 38
cytology and reproduction, 200–202
GSH levels, needles, 340
metal tolerance
adaptability, 203
description, 203
mechanisms and stability, DNA, 204–206
metabolic pathways, 204
population genetic effects, 206–208
protein build-up, 203–204
morphological characteristics
double-needled fascicles, 199
FA measurements, 199
in leaf growth, reduction, 198
needles, differences, 198, 199
plantlets, 342, 346
pollution maps, 337
Pinus taeda, 206, 338, 341
Plant growth, remediated soil
Chinese cabbage (*Brassica rapa*), 416
DTPA, 416
fertilisation, 418
food production, 422–423
hydrogel amendment, 418
original garden bed, 417
photosynthesis, 418
phytoaccumulation, 419–422
Plant growth-promoting bacteria (PGPB), 286, 292
Plant growth-promoting rhizobacteria (PGPR)
Bacillus and *Pseudomonas*, 471
biological remediation, 473–474
chemical remediation, 473
description, 471–472
detection and characterization, 474–475
molecular characterization, 478–480
physical remediation, 473
phytohormones, 472
soil remediation, 472
toxicity and bioremediation, soil, 472
Plants
Brassica species, 182
copper transporter family, 186
in cytosol, 189
efficiency and translocation factors, 305–306
enzymatic/nonenzymatic., 190
GSH, 190
hyperaccumulator, 303–304
MATE protein family, 188
metabolism, 181–182
metal ion binding
cell wall, 183–184
root exudates, 183
microbial communities, 182
NRAMP family, 185–186
oligopeptide transporter family, 188
plasma membrane transporters, 184
P-type ATPase transporter family, 187
at root-shoot level, 187
smelting and coal-fired power plants, 182
transporters
ABC family, 189
cation-diffusion facilitator family, 190
HMA and NRAMP family, 190
in vacuole, 189
ZIP family, 184–185
Plant symbionts (genetic engineering), 457–458
Plasma membrane transporters, 184
PME. *See* Poultry manure extract (PME)
Poaceae family, 164–166
Pollution-induced community tolerance (PICT)
in bacterial communities, 251

Biolog microplate, 251
Cr^{3+} and Cr^{6+}, toxic effects, 251
description, 250–251
metal stress, Zn-PICT, 252
Population genetics, *P. sylvestris*
adaptive strategy, 207–208
environmental contamination, 206
vs. P. nigra, 207, 208
population diversity, 206–207
Poultry manure extract (PME), 386, 387
Principal component analysis (PCA), 317
Promoters gene, 455
Pseudotsuga menziesii, 338
P-type ATPase transporter family, 187
Pyrolysis, 263, 477, 478

R

Random amplified polymorphic DNA (RAPD), 205
RDB. *See* Relative Deviation to Background (RDB)
Reactive oxygen species (ROS), 181
Relative Deviation to Background (RDB), 162
Rhizobia, tolerance to HMs
and AMF, symbiotic relationships
effective nodules, formation, 230
mitigating effect, calcium silicate, 236
Sesbania virgata and *Azorhizobium doebereinerae*, 230
biosorption capacity, 223
BR-4406 and UFLA-01-457, 221
Cd, Pb, Cu and Zn, 221, 222
Cupriavidus necator strains, 222
definition, 220
medium 79, 221
tolerance/resistance mechanisms, 222–223
Rhizofiltration, 191, 437–438, 458
ROS. *See* Reactive oxygen species (ROS)

S

SCOPUS database, 118
Scots pine. *See Pinus sylvestris*
Settling pond soils
chemical characteristics, 319–321
metal content, 324
physical characteristics, 318
sites, 327–328
Sheep manure extract (SME), 386
Shikimate dehydrogenase (ShDH), 204
Simple sequence repeats (SSR), 205, 206
SME. *See* Sheep manure extract (SME)
SOC. *See* Soil organic carbon (SOC)
Soil enzyme activity
alkaline soil reaction, 149
C, N and P, 149
characterization, 146–147
dehydrogenases, 146, 150–156
description, 157
ecosystems, 145
environmental pollution, 148
extracellular enzymes, 145, 149
HMs, 148–149
intracellular enzymes, 149
invertase, 147, 152, 155
long-term influence, HMs
alkaline phosphatase and urease, 154
dehydrogenase activity, 154–155
ecosystem restoration, 156
inactivation, dehydrogenases, 154
metal smelters, 154
microbial population, 155
mining/metallurgy areas, 154
nonferrous metals, 154
physiological adaptation, microorganisms, 155
urease and invertase activities, 155
natural environmental factors, 147–148
phosphatase, 146, 150–152, 154, 155
plant growth, 145
protease, 147
short-term influence, HMs
biochemical assessment, 154
chromium, 152–153
incubation, 149
inhibitory effect, metals, 153
nickel, 153
synergistic interaction, 153
toxic effect, 154
Zn, Pb and Cd contamination, 150–152
urease, 146–147, 150–152, 154–156
Soil microorganisms
catabolic activity, 248
functional diversity (*see* Functional diversity, soils microorganisms)
microbial communities, 245
phytoremediation and earthworms, 290–291
significance, 245
Soil organic carbon (SOC), 316, 319
Soil quality indexes (SQI)
cation exchange capacity, 319
chemical properties, 315–316
Cu concentration, 325
and *E. globulus* Labill, 312
eucalyptus/pines, 325
GRL, 324
indicators, 311

Soil quality indexes (SQI) (*cont.*)
  mine tailing soils (*see* Mine tailing soils)
  organic and inorganic carbon, 319
  and *P. pinaster* Aiton, 312
  physical properties, 315
  phytoavailable and Olsen P, 324
  planted trees, 318
  principal component analysis, 329
  pseudototal Ni, Pb and Zn concentrations, 325
  reclamation treatment, 328–329
  settling pond soils (*see* Settling pond soils)
  soil sampling, 313
  statistical analysis, 316–318
  in Touro mine, 311, 312
  waste amendments, 318
Solid combustion systems
  cadmium, 270
  chemical composition, residues, 269, 271
  cyclone dust collection system, 274
  flue gas composition, 268
  fluidized bed combustion technology, 264, 265
  ICP spectrometry, 269
  incinerated wood waste, 271–273
  incineration (*see* Incineration, HMs-contaminated biomass)
  leachability, 266
  logged *vs*. chipped oak and poplar samples, 268
  measured fractions, HMs, 269, 270
  measurement apparatus, 268–269
  metallic vapours in hot flue gas, 275
  mining area, 267
  operational parameters, 268
  partitioning, HM, 272, 273
  phytoremediation, 267
  residues, 264
  sample fuel and ash samples, 271, 272
Solid-soil solution interface
  adsorption/desorption processes, 46
  metals solid-phase/aqueous-phase interactions, 45
  metal transport prediction, 45
  nonlinear least square analysis, 46
  rate-determining steps, chemical reactions, 46
  reaction rate theory, 47
  soil chemistry, 45
  sorption and buffering parameters, 47, 48
Soluble organic carbon (SOC), 372
SQI. *See* Soil quality indexes (SQI)
Sterile materials, 216

T
Temporal variation, egg, 135–136
Terrestrial ecosystems, 130, 132
Tissue-specific expressions, 456
Toxic element contamination, urban agriculture. *See* Urban agriculture
Transgenic plants. *See also* Genetic engineering, plants; Phytoremediation
  ex planta process, 439
  expressions
    constitutive promoters, 455
    organelle targeting, 456–457
    tissue-specific, 456
  gene transfer, 442
  host systems, 440
  in vitro culture, 441
  lab to land transfer/testing, 460–461
  phytoremediation strategies, 439
  whole plant systems, 440
Translocation factors (TFs), 305–306, 378, 385–386

U
UFLA-01-457 (isolated from contaminated soil), 221
Unified bioaccessibility method (UBM), 404
Urban agriculture
  concentrations and spatial distribution
    anthropogenic sources, 91
    pollution prevention and reduction strategies, 91
    of PTEs, 87–90
    urbanization and industrialization, 91
  contamination sources
    atmospheric deposition, hazardous contaminants, 86
    household wastes, 85–86
    human activities, 83
    industrial emissions, 85
    traffic emissions, 84
  crop and livestock goods, 81
  ecosystem, 82
  food webs, 83
  geochemical fractions and mobility, PTE, 91–92
  human activities, 82
  intensive production methods, 81
  soil biological properties, PTE
    enzyme activities, 93–95
    microbial biomass and diversity, 92–93
  worldwide distribution, 82

V
Volcano eruptions, HMs content, 14–15

W
Water recycling
  acidic precipitation, EDTA, 400
  alkaline metal precipitation, 399–400
  electrochemical cleansing, 400–401
  electrolytic cell, 398
  electro-precipitation/coagulation, 398
  flowchart, 398, 399
Water-stable aggregates (WSA), 315
Whole plant systems, 440
Wildfire, HMs content
  classification, 15
  siero humic soils, 17
Woody species, 164, 169–171
WSA. *See* Water-stable aggregates (WSA)

X
X-ray diffraction analysis, 266

Y
Yeast mannitol agar (YMA) medium, 221

Z
Zinc (Zn), 151–153, 156, 185, 338, 373
ZRT-IRT-like protein (ZIP) family, 184–185

MIX
Papier aus verantwortungsvollen Quellen
Paper from responsible sources
FSC® C105338

If you have any concerns about our products,
you can contact us on
ProductSafety@springernature.com

In case Publisher is established outside the EU,
the EU authorized representative is:
Springer Nature Customer Service Center GmbH
Europaplatz 3, 69115 Heidelberg, Germany

Printed by Libri Plureos GmbH
in Hamburg, Germany